Earth Science:
Earth Materials and Resources

Earth Science: Earth Materials and Resources

Volume 1

Editor

Stephen I. Dutch, Ph.D.

University of Wisconsin-Green Bay

Salem Press

A Division of EBSCO Publishing

Ipswich, Massachusetts

Copyright © 2013, by Salem Press, A Division of EBSCO Publishing, Inc.
All rights reserved. No part of this work may be used or reproduced in any manner whatsoever or transmitted in any form or by any means, electronic or mechanical, including photocopy, recording, or any information storage and retrieval system, without written permission from the copyright owner. For permissions requests, contact proprietarypublishing@ebscohost.com.

The paper used in these volumes conforms to the American National Standard for Permanence of Paper for Printed Library Materials, Z39.48-1992 (R1997).

Library of Congress Cataloging-in-Publication Data
Earth science. Earth materials and resources / editor, Steven I. Dutch, Ph. D., University of Wisconsin-Green Bay.
    pages cm
Includes bibliographical references and index.
ISBN 978-1-58765-989-8 (set) – ISBN 978-1-58765-981-2 (set 3 of 4) – ISBN 978-1-58765-982-9 (volume 1) – ISBN 978-1-58765-983-6 (volume 2) 1. Geology, Economic. 2. Mines and mineral resources. I. Dutch, Steven I. II. Title: Earth materials and resources.
  TN260.E25 2013
  553–dc23

2012027580

PRINTED IN THE UNTED STATES OF AMERICA

CONTENTS

PUBLISHER'S NOTE

Salem Press's *Earth's Materials and Resources* provides a two-volume introduction to the major topics of study in the earth's physical processes and structures and offers a comprehensive revision and update to an earlier edition, published by Salem Press in 2001.

The essays in this collection cover a wide range of subject areas, including economic and energy resources; minerals, soil, and rock; and issues surrounding such fields as petroleum geology and engineering. Areas of special attention include developments in resource use and its environmental impact, which is a major field of study that continues to show rapid growth. The editor, Steven I. Dutch, Ph.D., has reviewed each article for scientific authority and ensured each essay's currency.

Designed for high school and college students and their teachers, these volumes provide hundreds of expertly written essays supplemented by illustrations, charts, and useful reference materials, resulting in a comprehensive overview of each topic. Librarians and general readers alike will turn to this reference work for both foundational information and current developments.

Each essay topic begins with helpful reference information, including a summary statement that explains its significance in the study of the earth and its processes. *Principal Terms* define key elements or concepts related to the subject. The background and history of each subject are provided and detail important contextual information on the topic. The text itself is organized with informative subheadings that guide readers to areas of particular interest. An annotated *Bibliography* closes each essay and refers the reader to external sources for further study that are useful to both students and nonspecialists. Finally, a list of cross-references directs the reader to other subject-related essays within the set. At the end of each volume, several appendices are designed to assist in the retrieval of information, including a *Glossary* that defines key terms contained in each set, a *Periodic Table of Elements*, the *Mohs Scale* of mineral hardness, and an *Energy Source Comparison*.

Salem Press's *Earth's Materials and Resources* is part of a series of Earth science books that includes *Physics and Chemistry of the Earth, Earth's Surface and History,* and *Earth's Weather, Water, and Atmosphere.*

Many hands went into the creation of these volumes. Special mention must be made of its editor, Steven I. Dutch, who played a principal role in shaping the reference work and its contents. Thanks are also due to the academicians and professionals who communicated their expert understanding of Earth science to the general reader; a list of these individuals and their affiliations appears at the beginning of the volume. The contributions of all are gratefully acknowledged.

INTRODUCTION

In the eleven years since this volume was first published, there have been many dramatic changes in our understanding of Earth materials and in the availability of Earth resources. This revised volume reflects those changes.

First is that many statistics on resource availability needed to be updated; articles mentioning oil at $35 per barrel seem almost quaint in 2012. Resources that were nearly unknown in 2001 have emerged as vital today. Foremost among them are lithium, used in rechargeable batteries, and other rare earth elements used in electronics. Known supplies of lithium are sufficient for consumer electronics, but if lithium batteries become widely used for powering automobiles, these lithium resources could become severely strained. Rare earth elements are produced in many places, but virtually all of the world's refining capacity is in China, a geopolitical fact with far-reaching strategic consequences.

Resource issues are tied to social issues as well. Conflict or "blood" diamonds were used to finance civil wars in parts of Africa and were frequently mined by slave labor. A system of international regulations on the import of diamonds has significantly curtailed the conflict diamond trade. Although many are aware of blood diamonds, notably from films like *Die Another Day* (2002) and *Blood Diamond* (2006), not many know of a similar problem with rare earth minerals used in electronics. This issue is most acute in the Congo region of central Africa where small, hand-dug mines extract columbite and tantalite, known as "col-tan" in mining circles. The col-tan is frequently stolen by military and paramilitary groups and sold either to buy weapons or simply to enrich the thieves.

Other important updates relate to the future availability of resources. Oil is just as finite in 2012 as it was in 2001, and despite new discoveries and improved methods of exploiting unconventional resources, we are still using oil several times faster than it is being discovered. For example, in June 2011, an oil field was discovered in the Gulf of Mexico that would provide 700 million barrels of oil, which was the largest discovery in the Gulf in twelve years. However, the United States uses 20 million barrels of oil *per day*, so to stay abreast of U.S. energy demand, we would have to find an oil field that size *every thirty-five days*, not every twelve years.

One development that was all but unknown in 2001 but that is very much in the news today is "fracking," or fracturing reservoir rocks to liberate oil and natural gas. One potential belt of rocks suitable for fracking extends along the west side of the Appalachians through highly populated states, where the notion has raised fears of contamination of groundwater and other hazards. "Peak oil"—the point where global demand will exceed the physical capability to extract oil from the ground—is widely expected to occur in the first half of the twenty-first century. But a new resource peak has begun to cause concern: peak phosphorus. Our ability to feed our population is critically dependent on phosphorus fertilizers, and five countries hold 90 percent of the world's known phosphate rock. Much of our phosphate fertilizer is lost to runoff and carried out to sea by rivers. There, it can cause the eutrophication of coastal waters and create dead zones, and much of it ends up in deep water where it remains indefinitely. Fortunately, phosphorus, unlike oil, can be recycled from human and animal wastes or from plant matter through no-till agriculture. Although the idea of recycling wastes for fertilizer may seem unappetizing, it is a millennia-old practice in many regions.

Resource use also entails risk, and several dramatic disasters in the past decade require inclusion in this revised volume. Foremost is probably the Deepwater Horizon drilling disaster of 2010, in which an oil rig in the Gulf of Mexico blew up, burned, and sank, releasing the largest oil spill in history. The great Japanese earthquake of 2011 caused immense loss of life from tsunamis, but it also severely damaged the nuclear power plant at Fukushima, releasing radiation comparable to that from the infamous Chernobyl disaster of 1986. The Japanese disaster, however, was better contained, and most of the radiation dispersed into the Pacific rather than over populated landmasses. Mining mishaps continue to claim lives, notably in China, which accounts for 80 percent of the world's mining fatalities and where over 2,400 miners died in 2010.

In editing the revised volume, my goals were first to ensure that the coverage was up-to-date. Several articles that were appropriate for the first edition were no longer relevant, or they required such extensive revision that they needed to be rewritten. A number of new essays have also been included to reflect recent developments, and statistics and data have been updated throughout. I have also strived to keep the target audience in mind at all times. This reference set is intended primarily for students at secondary and introductory college levels, and in reviewing the articles, the foremost question in my mind was "Would this material make sense to a high school or college student without an extensive science background?" For the most part, the authors of the original articles satisfied that objective admirably, but there were several instances where changes were made to better reflect the needs of the target audience, and concepts and material that may have been too advanced or potentially confusing for nonspecialists were clarified.

My personal philosophy when choosing material to include in courses, textbooks, or reference works is that the material satisfies at least one of four criteria:

The material is of broad general interest. Gemstones are a relatively small part of the world's mineral wealth, but they interest nonscientists in a manner that is far greater than their total economic value.

The material is of great practical value. Most of the articles in this volume satisfy this criterion easily. Civilization is based on oil, iron, copper, aluminum, and a host of other metals. Soils and groundwater are not glamorous topics, but they are absolutely essential to human survival.

The material is needed to understand subjects in either of the first two categories. In order to understand how iron is deposited, for example, it is essential to know that iron loses electrons, or ionizes, in two different ways, one of which is highly soluble in water and one of which is not.

The material illustrates how science works. All the articles include sections on how the particular subject of the article is researched, so there are many references to geologic mapping, the petrographic microscope, X-ray diffraction, and so on. Geologic mapping is how geologists synthesize information from a vast region into a concise and easily visualized form. The petrographic microscope and other analytical techniques are precise ways of identifying minerals.

It is our hope that this revised Earth science encyclopedia will give readers an accessible and current picture of the geological sciences.

Steven I. Dutch, Ph.D.
University of Wisconsin–Green Bay

CONTRIBUTORS

Richard W. Arnseth
Science Applications International

Michael P. Auerbach
Marblehead, MA

N. B. Aughenbaugh
University of Mississippi

John L. Berkley
State University of New York, Fredonia

David M. Best
Northern Arizona University

Rachel Leah Blumenthal
Somerville, MA

Alan Brown
Livingston University

Joseph I. Brownstein
Atlanta, GA

Michael Broyles
Collin County Community College

James A. Burbank, Jr.
Western Oklahoma State College

Scott F. Burns
Louisiana Tech University

Robert E. Carver
University of Georgia

Dennis Chamberland
Independent Scholar

Habte Giorgis Churnet
University of Tennessee at Chattanooga

Mark Cloos
University of Texas at Austin

James R. Craig
Virginia Polytechnic Institute and State University

Robert L. Cullers
Kansas State University

Larry E. Davis
Washington State University

Ronald W. Davis
Western Michigan University

Albert B. Dickas
University of Wisconsin

Steven I. Dutch
University of Wisconsin—Green Bay

Richard H. Fluegeman, Jr.
Ball State University

Annabelle M. Foos
University of Akron

Charles I. Frye
Northwest Missouri State University

Pamela J. W. Gore
De Kalb Community College

Martha M. Griffin
Columbia College

William R. Hackett
Idaho State University

Gina Hagler
Washington, DC

Edward C. Hansen
Hope College

Clay D. Harris
Middle Tennessee State University

David F. Hess
Western Illinois University

J. D. Ho
Independent scholar

René De Hon
University of Louisiana at Monroe

Robert A. Horton, Jr.
California State University at Bakersfield

Micah L. Issitt
Independent scholar

Kyle L. Kayler
Richard H. Gorr and Associates

Richard S. Knapp
Belhaven *College*

Ralph L. Langenheim, Jr.
University of Illinois at Urbana-Champaign

Donald W. Lovejoy
Palm Beach Atlantic College

Gary R. Lowell
Southeast Missouri State University

David N. Lumsden
Memphis State University

Sergei A. Markov
Austin Peay State University

Paul S. Maywood
Bridger Coal Company

Lance P. Meade
American Institute of Mining Engineers

Nathan H. Meleen
American Geophysical Union

Randall L. Milstein
Oregon State University

Otto H. Muller
Alfred University

Brian J. Nichelson
U.S. Air Force Academy

Bruce W. Nocita
University of South Florida

Edward B. Nuhfer
University of Wisconsin—Platteville

Michael R. Owen
St. Lawrence University

Susan D. Owen
Bonneville Power Administration

Robert J. Paradowski
Rochester Institute of Technology

Jeffrey C. Reid
North Carolina Geological Survey

J. Donald Rimstidt
Virginia Polytechnic Institute and State University

James L. Sadd
Occidental College

Dorothy Fay Simms
Northeastern State University

Paul P. Sipiera
Harper College

Joseph L. Spradley
Wheaton College, Illinois

Ronald D. Stieglitz
University of Wisconsin—Green Bay

Frederick M. Surowiec
Society of Professional Journalists

Eric R. Swanson
University of Texas at San Antonio

Stephen M. Testa
California State University, Fullerton

Donald J. Thompson
California University of Pennsylvania

Ronald D. Tyler
Independent Scholar

Dermot M. Winters
American Institute of Mining Engineers

James A. Woodhead
Occidental College

COMMON UNITS OF MEASURE

Notes: Common prefixes for metric units—which may apply in more cases than shown below—include giga- (1 billion times the unit), mega- (1 million times), kilo- (1,000 times), hecto- (100 times), deka- (10 times), deci- (0.1 times, or one tenth), centi- (0.01, or one hundredth), milli- (0.001, or one thousandth), and micro- (0.0001, or one millionth).

Unit	Quantity	Symbol	Equivalents
Acre	Area	ac	43,560 square feet 4,840 square yards 0.405 hectare
Ampere	Electric current	A *or* amp	1.00016502722949 international ampere 0.1 biot *or* abampere
Angstrom	Length	Å	0.1 nanometer 0.0000001 millimeter 0.000000004 inch
Astronomical unit	Length	AU	92,955,807 miles 149,597,871 kilometers (mean Earth-sun distance)
Barn	Area	b	10^{-28} meters squared (approximate cross-sectional area of 1 uranium nucleus)
Barrel (dry, for most produce)	Volume/capacity	bbl	7,056 cubic inches; 105 dry quarts; 3.281 bushels, struck measure
Barrel (liquid)	Volume/capacity	bbl	31 to 42 gallons
British thermal unit	Energy	Btu	1055.05585262 joule
Bushel (U.S., heaped)	Volume/capacity	bsh *or* bu	2,747.715 cubic inches 1.278 bushels, struck measure
Bushel (U.S., struck measure)	Volume/capacity	bsh *or* bu	2,150.42 cubic inches 35.238 liters
Candela	Luminous intensity	cd	1.09 hefner candle
Celsius	Temperature	C	1° centigrade
Centigram	Mass/weight	cg	0.15 grain
Centimeter	Length	cm	0.3937 inch
Centimeter, cubic	Volume/capacity	cm³	0.061 cubic inch
Centimeter, square	Area	cm²	0.155 square inch
Coulomb	Electric charge	C	1 ampere second

UNIT	QUANTITY	SYMBOL	EQUIVALENTS
Cup	Volume/capacity	C	250 milliliters 8 fluid ounces 0.5 liquid pint
Deciliter	Volume/capacity	dl	0.21 pint
Decimeter	Length	dm	3.937 inches
Decimeter, cubic	Volume/capacity	dm³	61.024 cubic inches
Decimeter, square	Area	dm²	15.5 square inches
Dekaliter	Volume/capacity	dal	2.642 gallons 1.135 pecks
Dekameter	Length	dam	32.808 feet
Dram	Mass/weight	dr *or* dr avdp	0.0625 ounce 27.344 grains 1.772 grams
Electron volt	Energy	eV	$1.5185847232839 \times 10^{-22}$ Btu $1.6021917 \times 10^{-19}$ joule
Fermi	Length	fm	1 femtometer 1.0×10^{-15} meter
Foot	Length	ft *or* '	12 inches 0.3048 meter 30.48 centimeters
Foot, cubic	Volume/capacity	ft³	0.028 cubic meter 0.0370 cubic yard 1,728 cubic inches
Foot, square	Area	ft²	929.030 square centimeters
Gallon (British Imperial)	Volume/capacity	gal	277.42 cubic inches 1.201 U.S. gallons 4.546 liters 160 British fluid ounces
Gallon (U.S.)	Volume/capacity	gal	231 cubic inches 3.785 liters 0.833 British gallon 128 U.S. fluid ounces
Giga-electron volt	Energy	GeV	$1.6021917 \times 10^{-10}$ joule
Gigahertz	Frequency	GHz	—
Gill	Volume/capacity	gi	7.219 cubic inches 4 fluid ounces 0.118 liter

UNIT	QUANTITY	SYMBOL	EQUIVALENTS
Grain	Mass/weight	gr	0.037 dram 0.002083 ounce 0.0648 gram
Gram	Mass/weight	g	15.432 grains 0.035 avoirdupois ounce
Hectare	Area	ha	2.471 acres
Hectoliter	Volume/capacity	hl	26.418 gallons 2.838 bushels
Hertz	Frequency	Hz	$1.08782775707767 \times 10^{-10}$ cesium atom frequency
Hour	Time	h	60 minutes 3,600 seconds
Inch	Length	in *or* "	2.54 centimeters
Inch, cubic	Volume/capacity	in^3	0.554 fluid ounce 4.433 fluid drams 16.387 cubic centimeters
Inch, square	Area	in^2	6.4516 square centimeters
Joule	Energy	J	$6.2414503832469 \times 10^{18}$ electron volt
Joule per kelvin	Heat capacity	J/K	$7.24311216248908 \times 10^{22}$ Boltzmann constant
Joule per second	Power	J/s	1 watt
Kelvin	Temperature	K	-272.15 degrees Celsius
Kilo-electron volt	Energy	keV	$1.5185847232839 \times 10^{-19}$ joule
Kilogram	Mass/weight	kg	2.205 pounds
Kilogram per cubic meter	Mass/weight density	kg/m^3	$5.78036672001339 \times 10^{-4}$ ounces per cubic inch
Kilohertz	Frequency	kHz	—
Kiloliter	Volume/capacity	kl	—
Kilometer	Length	km	0.621 mile
Kilometer, square	Area	km^2	0.386 square mile 247.105 acres
Light-year (distance traveled by light in one Earth year)	Length/distance	lt-yr	5,878,499,814,275.88 miles 9.46×1012 kilometers
Liter	Volume/capacity	L	1.057 liquid quarts 0.908 dry quart 61.024 cubic inches

UNIT	QUANTITY	SYMBOL	EQUIVALENTS
Mega-electron volt	Energy	MeV	—
Megahertz	Frequency	MHz	—
Meter	Length	m	39.37 inches
Meter, cubic	Volume/capacity	m^3	1.308 cubic yards
Meter per second	Velocity	m/s	2.24 miles per hour 3.60 kilometers per hour
Meter per second per second	Acceleration	m/s^2	12,960.00 kilometers per hour per hour 8,052.97 miles per hour per hour
Meter, square	Area	m^2	1.196 square yards 10.764 square feet
Metric. See unit name			
Microgram	Mass/weight	mcg or μg	0.000001 gram
Microliter	Volume/capacity	μL	0.00027 fluid ounce
Micrometer	Length	μm	0.001 millimeter 0.00003937 inch
Mile (nautical international)	Length	mi	1.852 kilometers 1.151 statute miles 0.999 U.S. nautical mile
Mile (statute or land)	Length	mi	5,280 feet 1.609 kilometers
Mile, square	Area	mi^2	258.999 hectares
Milligram	Mass/weight	mg	0.015 grain
Milliliter	Volume/capacity	mL	0.271 fluid dram 16.231 minims 0.061 cubic inch
Millimeter	Length	mm	0.03937 inch
Millimeter, square	Area	mm^2	0.002 square inch
Minute	Time	m	60 seconds
Mole	Amount of substance	mol	6.02×10^{23} atoms or molecules of a given substance
Nanometer	Length	nm	1,000,000 fermis 10 angstroms 0.001 micrometer 0.00000003937 inch

Unit	Quantity	Symbol	Equivalents
Newton	Force	N	0.224808943099711 pound force 0.101971621297793 kilogram force 100,000 dynes
Newton-meter	Torque	N·m	0.7375621 foot-pound
Ounce (avoirdupois)	Mass/weight	oz	28.350 grams 437.5 grains 0.911 troy or apothecaries' ounce
Ounce (troy)	Mass/weight	oz	31.103 grams 480 grains 1.097 avoirdupois ounces
Ounce (U.S., fluid or liquid)	Mass/weight	oz	1.805 cubic inch 29.574 milliliters 1.041 British fluid ounces
Parsec	Length	pc	30,856,775,876,793 kilometers 19,173,511,615,163 miles
Peck	Volume/capacity	pk	8.810 liters
Pint (dry)	Volume/capacity	pt	33.600 cubic inches 0.551 liter
Pint (liquid)	Volume/capacity	pt	28.875 cubic inches 0.473 liter
Pound (avoirdupois)	Mass/weight	lb	7,000 grains 1.215 troy or apothecaries' pounds 453.59237 grams
Pound (troy)	Mass/weight	lb	5,760 grains 0.823 avoirdupois pound 373.242 grams
Quart (British)	Volume/capacity	qt	69.354 cubic inches 1.032 U.S. dry quarts 1.201 U.S. liquid quarts
Quart (U.S., dry)	Volume/capacity	qt	67.201 cubic inches 1.101 liters 0.969 British quart
Quart (U.S., liquid)	Volume/capacity	qt	57.75 cubic inches 0.946 liter 0.833 British quart
Rod	Length	rd	5.029 meters 5.50 yards

UNIT	QUANTITY	SYMBOL	EQUIVALENTS
Rod, square	Area	rd²	25.293 square meters 30.25 square yards 0.00625 acre
Second	Time	s or sec	1/60 minute 1/3,600 hour
Tablespoon	Volume/capacity	T or tb	3 teaspoons 4 fluid drams
Teaspoon	Volume/capacity	t or tsp	0.33 tablespoon 1.33 fluid drams
Ton (gross or long)	Mass/weight	t	2,240 pounds 1.12 net tons 1.016 metric tons
Ton (metric)	Mass/weight	t	1,000 kilograms 2,204.62 pounds 0.984 gross ton 1.102 net tons
Ton (net or short)	Mass/weight	t	2,000 pounds 0.893 gross ton 0.907 metric ton
Volt	Electric potential	V	1 joule per coulomb
Watt	Power	W	1 joule per second 0.001 kilowatt $2.84345136093995 \times 10^{-4}$ ton of refrigeration
Yard	Length	yd	0.9144 meter
Yard, cubic	Volume/capacity	yd³	0.765 cubic meter
Yard, square	Area	yd²	0.836 square meter

COMPLETE LIST OF CONTENTS

Volume 1

Volume 2

CATEGORY LIST OF CONTENTS

Earth Science:
Earth Materials and Resources

ALUMINUM DEPOSITS

Aluminum is a ubiquitous metal that is critical to many industries. Because of the characteristics of aluminum ore, it was not practical to produce until the late nineteenth century. Aluminum production remains power-intensive, so most aluminum is recycled.

PRINCIPAL TERMS

- **carbonate rock:** a rock composed mainly of calcium carbonate
- **electrolysis:** process by which liquid or dissolved metals are separated by electromagnetic attraction
- **heterogeneous mixture:** a nonuniform mixture
- **ion:** an atom with a net charge due to the addition or loss of electrons
- **karst:** a region formed by the weathering of underlying rock, typically limestone or dolomite
- **mineral:** a naturally occurring solid with a specific chemical composition
- **ore:** chemical compounds of a desired material that are economically viable to exploit
- **pressure vessel:** a container designed to hold gases or liquids at a much higher pressure than the surrounding pressure
- **rock:** a mixture of minerals
- **strip mining:** a method of mining that occurs on the earth's surface

ALUMINUM ORE

Aluminum is an important metal because of its high strength, low density, and high corrosion resistance. It is used in items ranging from cans to airframes. However, extensive use of aluminum by humans is a new phenomenon.

Historically, aluminum production has required more energy than could be provided. Because of the chemical characteristics of aluminum, the metal could not be produced on a large scale until the development of the Hall-Heroult process in 1886, in conjunction with dependable and inexpensive electricity.

BAUXITE ORE

Aluminum is highly reactive and is seldom found in its elemental state. As such it is found as ore,

with bauxite (discovered in 1821) as the primary source of aluminum. Bauxite is a rock composed of several aluminum-rich minerals, such as gibbsite, boehmite, and diaspore, in a heterogeneous mixture with various iron oxides and the clay mineral kaolin. Kaolin itself is rich in aluminum, and it has been considered as a potential source. However, given the abundance of bauxite, kaolin processing is not of economic interest.

There are two primary forms of bauxite: lateritic and karst. Lateritic bauxite is silicate based, whereas karst bauxite tends to be carbonate. Karst bauxite forms above carbonate rocks, such as limestone and dolomite, in a karst region. A karst region is a large area of carbonate rock formed during periods in Earth's history called marine incursions, when a shallow ocean covers a region that was formerly dry land. For the karst region to have formed, the ocean must have been shallow.

Lateritic bauxites are formed atop silicates. Laterite weathering is intense tropical weathering that produces a soil called laterite. Weathering, the process by which soil is formed, requires the bedrock to be broken down by the conjunction of natural forces such as wind, rain, and seasonal temperature fluctuations.

Equally, chemistry plays a large role in the process known as chemical weathering. Organisms play a role as well, with many plants secreting chemicals to break down rocks. Weathering is distinct from erosion, which moves materials; weathering occurs in situ.

Soil is produced in areas of wet weather and vigorous plant life—namely, in the tropics. Tropical soil is rich in many important minerals and, particularly in the Amazon, is poor in nutrients. This type of soil occurs throughout the tropics or areas that once were tropics. Thus, many tropical countries are good sources for bauxite mining. Bauxite is often found in large deposits near the surface that covers large areas; it also can be found in deposits deeper underground.

In addition to bauxite, other potential sources of aluminum are kaolin clay, oil shale, anorthosite, and coal. However, these sources are not economically viable, and given the widespread availability of bauxite, there is no economic incentive to make use of them.

MINING

Because most bauxite deposits are on the surface, bauxite mining often takes the form of strip mining. First, the land above the deposit is cleared of vegetation. Explosives are then used to loosen the soil before heavy equipment digs up the bauxite-rich soils. This method often makes use of bucket-wheel excavators and chain-bucket excavators.

Bucket-wheel and chain-bucket excavators are some of the largest machines made by humans. While the specific mechanisms are different, each has a series of excavating buckets on a loop that pick up the dirt and place it on a belt that leads to a central hub. Once there, the soil is placed into trucks or onto a belt that takes the bauxite to a refinery. These machines are effective but can be expensive to repair. Often, bulldozers and backhoes are used to the same effect. Although efficient, such mining practices disrupt the environment, particularly in tropical regions. In the process, land must be cleared over the active mine area. The mined area must then be restored, or the land will face erosion. Even with soil restoration, the environment is damaged.

BAYER PROCESS

Bauxite is only 30 to 54 percent alumina. Once mined, bauxite ore is heated to about 150 to 200 degrees Celsius (302 to 392 degrees Fahrenheit) in a pressure vessel with a solution of sodium hydroxide. The sodium hydroxide solution converts the bauxite into aluminum hydroxide, which then dissolves in the hydroxide solution. The rest of the bauxite does not dissolve, forming a waste product called red mud. It is formed mostly of oxidized iron.

Little can be done with red mud, as it cannot be built upon and is highly basic. It must be dried in settlement ponds and stored in landfills. One of the main problems associated with bauxite refining is the management of this red mud. Additionally, should a

settlement pond fail, the resultant slide of red mud can destroy towns and pollute the environment.

Once the solution cools, the dissolved aluminum hydroxide precipitates out. When it is heated to 980 degrees Celsius (1,800 degrees Fahrenheit), the aluminum hydroxide decomposes to alumina, releasing water vapor in the process. This process, invented in 1887 by Carl Josef Bayer and named for him, turns bauxite to aluminum oxide. Bayer had been working on supplying alumina to the textile industry, where the alumina was used to fix colors in the cotton-dying process. (The role is called mordant.) Several processes had already been invented for turning bauxite to aluminum oxide, but the Bayer process was more efficient.

Alumina has been used in many industries. Because it is chemically inert, it is often used as filler in plastics and sunscreen. It also can be used as an abrasive as a substitute for diamonds. Alumina also is used in many kinds of sandpaper. Additionally, alumina can be used as a catalyst to help convert hydrogen sulfide into sulfur. This helps clean emissions, including the emissions of factories.

HALL-HEROULT PROCESSING

Alumina must be processed to make aluminum; this is done through the Hall-Heroult process, whereby aluminum is formed through electrolysis. While most such processes can use water, the high reactivity of aluminum makes the use of water impossible. Thus, alumina must be dissolved in cryolite (sodium hexafluoroaluminate), a light-colored rare mineral in which alumina is soluble.

An interesting property of cryolite is that its refractive index is close to that of water. This means it bends light the same way water does and thus would be near invisible if put in a bowl of water. Cryolite is used as flux. A flux is a substance used to preserve purity and to clean and improve flow in the metal being worked with. When smelting, fluxes remove impurities. This is especially important with aluminum because of its high reactivity.

Both alumina and cryolite have high melting points of more than 1,000 degrees Celsius (1,832 degrees Fahrenheit). The resultant mixture is then electrolyzed, with the liquid aluminum being precipitated on the cathode. A voltage of between 3 and 5 volts is required, with the rate of precipitation proportional to the current. The liquid aluminum is then siphoned off.

The process releases carbon dioxide and hydrogen fluoride, which are vented. Hydrogen fluoride is neutralized to sodium fluoride. The process requires much electricity, both for deposition and for heating purposes. As a result, most aluminum smelters are located near sources of relatively inexpensive electricity, such as hydroelectric stations. Because of the high energy requirements for producing aluminum, the metal was treated like gold before the development of the Hall-Heroult process.

The Hall-Heroult process was discovered nearly simultaneously by two chemists, leading to the process's name. One chemist, Charles Martin Hall, invented the process in 1886. Heroult came up with the idea a few months later. Hall, however, patented the process in 1889 and then founded the Pittsburgh Reduction Company, later to become the Aluminum Company of America, or Alcoa.

RECYCLING

Even after the development of the Hall-Heroult process, producing new aluminum remained energy intensive. As a result, it is far more efficient, even today, to reuse already processed aluminum. Recycling takes only 5 percent of the energy needed to make new aluminum. Because of widespread recycling, 75 percent of the aluminum produced since 1888 is still in use.

First, in recycling, cans and other aluminum components are removed from waste. This is often done by a form of electromagnetic sorting called an eddy current separator, which works by producing a powerful magnetic field and moving the metals through it. Electromagnetic induction causes the metal to fly into a collection bin.

Ferrous, or magnetic, metals can cause damage to such as system and are usually removed first. After being separated, the aluminum parts are cleaned to prevent oxidation during smelting; oxidation ruins the aluminum. Only then is the aluminum melted and recast; gases such as chlorine or nitrogen remove any hydrogen released in the process.

In general, a recycled can is made into a new can within sixty days. Because of the massive energy and financial savings, the recycling process is economically significant. An empty aluminum can is worth about 1 U.S. cent, and the energy savings from recycling one single can is enough to run the average television set for three hours.

PLACES

Geologists locate bauxite by prospecting for it. They take core samples of soils around the world in regions suitable for bauxite and examine them for evidence of the ore. Through study of the samples, they can learn the quality and quantity of the bauxite at the site.

Global resources are estimated to be between 55 and 75 billion tons, sufficient for current demand for several centuries. Of these resources, 33 percent is in Africa, 24 percent in Oceania, 22 percent in South America and the Caribbean, and 15 percent in Asia. Jamaica has a particularly high concentration, and reports suggest that there could be large deposits in Vietnam (of about 11,000 megatons), which would makes these sites the largest bauxite reserves in the world. The largest producer now is Australia (producing one-third of the world's bauxite), followed by China, Brazil, Guinea, and India. Bauxite production is a major industry in the regions where it is mined.

Gina Hagler

FURTHER READING

Batty, Lesley C., and Kevin B. Hallberg. *Ecology of Industrial Pollution.* New York: Cambridge University Press, 2010. A comprehensive guide to industrial pollution, providing an understanding of the effects of bauxite mining and processing, including red mud production and strip mining, in their ecological context. Scholarly, but useful for all readers.

Brandt, Daniel A., and Jairus C. Warner. *Metallurgy Fundamentals: Ferrous and Nonferrous.* Tinley Park, Ill.: Goodheart-Willcox, 2009. A metallurgy textbook that explains the details and concerns of making metal in general. Recommended for all readers. Clearly written. Includes diagrams.

Menzie, W. D. *The Global Flow of Aluminum from 2006 through 2025.* Reston, Va.: U.S. Geological Survey, 2010. Provides a look at aluminum production, recycling, and use. Recommended for all readers.

Reece, Erik. *Lost Mountain: A Year in the Vanishing Wilderness: Radical Strip Mining and the Devastation of Appalachia.* New York: Riverhead Books, 2007. This environmental work details the total destruction of a mountain during mining operations in Kentucky. Examines the problems and the pressures associated with strip mining.

Schlesinger, Mark E. *Aluminum Recycling*. Boca Raton, Fla.: CRCelsius Press, 2007. A comprehensive study of aluminum recycling, from the sources of aluminum to the cleaning processes, to the structure of the scrap and recycling industry.

Smith, George D. *From Monopoly to Competition: The Transformations of Alcoa, 1888-1986*. New York: Cambridge University Press, 2003. Details the history of Alcoa, which has been central to the development and use of aluminum. Also provides a general history of technology and an understanding of the economic and social factors involved in the making of the metal.

See also: Dolomite; Geologic Settings of Resources; Iron Deposits; Radioactive Minerals; Silicates; Uranium Deposits.

ANDESITIC ROCKS

Andesite is an intermediate extrusive igneous rock. Active volcanoes on the earth erupt andesite more than any other rock type. Andesites are primarily associated with subduction zones along convergent tectonic plate boundaries.

PRINCIPAL TERMS

- **Bowen's reaction principle:** a principle by which a series of minerals forming early in a melt react with the remaining melt to yield a new mineral in an established sequence
- **extrusive rock:** igneous rock that has been erupted onto the surface of the earth
- **groundmass:** the fine-grained material between phenocrysts of a porphyritic igneous rock
- **intermediate rock:** an igneous rock that is transitional between a basic and a silicic rock, having a silica content between 54 and 64 percent
- **phenocryst:** a large conspicuous crystal in a porphyritic rock
- **plutonic rock:** igneous rock formed at a great depth within the earth
- **porphyry:** an igneous rock in which phenocrysts are set in a finer-grained groundmass
- **stratovolcano:** a volcano composed of alternating layers of lava flow and ash; also called a composite volcano
- **subduction zone:** a convergent plate boundary
- **viscosity:** a substance's ability to flow; the lower the viscosity, the greater the ability to flow

INTEREST TO GEOLOGISTS

Andesite takes its name from lavas in the Andes mountains of South America. To most geologists, andesites are light gray porphyritic volcanic rocks containing phenocrysts of plagioclase, very little quartz, and no sanidine or feldspathoid. Despite their lackluster appearance, andesites are of great interest to geologists for several reasons. First, active volcanoes on the earth erupt andesite more than any other rock type; andesite is the main rock type at 61 percent of the world's active volcanoes. Second, andesites have a distinctive tectonic setting. They are primarily associated with convergent plate boundaries and occur elsewhere only in limited amounts. Of the active volcanoes that occur within 500 kilometers of a subduction zone, 78 percent include andesite; only three active volcanoes not near a convergent plate boundary erupt it. Third, andesites have bulk compositions similar to estimates of the composition of continental crust. This similarity, in association with the tectonic setting of andesites, suggests that they may play an important role in the development of terrestrial crust. Fourth, the development and movement of andesitic magma seem to be closely related to the formation of many ore deposits, including such economically important ores as molybdenum and porphyry copper deposits.

VISCOSITY

Rocks are classified chemically according to how much silica they contain; rocks rich in silica (more than 64 percent), such as rhyolite, are called silicic. They consist mostly of quartz and feldspars, with minor amounts of mica and amphibole. Rocks low in silica (less than 54 percent), with no free quartz but high in feldspar, pyroxene, olivine, and oxides, are called basic. Basic rocks, free of quartz, tend to be dark, while silicic rocks are lighter and contain only isolated flecks of dark minerals. Basalts are examples of basic volcanic rocks. Andesites, having a silica content of about 60 percent, are volcanic rocks termed intermediate. Andesite's plutonic equivalent is diorite. There is no cut-and-dried difference between basalt and andesite, or between andesite and rhyolite. Instead, there is a broad transitional group of rocks that carry names such as "basaltic andesite" or "andesitic rhyolite."

Nevertheless, some generalizations can be applied to the lavas and magmas that form these rocks. One generalization has to do with viscosity. Andesite lavas are more viscous than basalt lavas and less viscous than rhyolite lavas. This difference is primarily an effect of the lava's composition, and to a certain extent it is a result of the high portion of phenocrysts present in the more viscous lavas.

Different minerals crystallize at different temperatures. As a basalt magma cools, a sequence of minerals appears. The first mineral to crystallize is usually olivine, which continues to crystallize as the magma cools until a temperature is reached at which a second mineral, pyroxene, begins to crystallize. As the temperature continues to drop, these two continue to crystallize. Cooling continues until a temperature is reached when a third mineral, feldspar, crystallizes. This chain of cooling and mineral crystallization is known as Bowen's reaction principle. Often, olivine and pyroxene crystallize out early in the process, so they may be present in the final rock as large crystals up to a centimeter across. These crystals are called phenocrysts. The size of these phenocrysts is in direct contrast to the fine-grained crystals of the groundmass. Igneous rocks with phenocrysts in a fine-grained groundmass are known as porphyries. Most volcanic rocks contain some phenocrysts. The groundmass crystals form when the lava cools upon reaching the surface. If the lava has a low viscosity, reaches the surface, spreads out, and cools quickly, individual crystals do not have enough time to grow. The overall rock remains fine-grained. The phenocrysts crystallized out much earlier, while the magma was still underground. There, they had plenty of time to grow and were then carried to the surface with the magma during eruption.

ANDESITE FLOWS

Basalts, as a result of their low viscosity, tend to produce thin lava flows that readily spread over large areas. They rarely exceed 30 meters in thickness. Andesite flows, by contrast, are massive and may be as much as 55 meters thick. The largest single andesite flow described, which is in northern Chile, has an approximate volume of 24 cubic kilometers. Because of their low viscosity, basalt flows can advance at considerable rates; speeds up to 8 kilometers per hour have been measured. Andesite flows often move only a matter of tens of meters over several hours. As a result of their higher viscosity, andesite flows show none of the surface features of more "liquid" lavas. Flow features such as wave forms, swirls, or the ropy textures often associated with basalt flows never occur in andesite flows. Andesite flows tend to be blocky with large, angular, smooth-sided chunks of solid lava. The flow tends to behave as a plastic rather than a liquid. An outer, chilled surface develops on the slow-moving flow, with the interior still molten. Plasticity within the flow increases toward the still-molten inner portion. As the flow slowly shifts, moves, and cools, the hard, brittle outer layer breaks into the large angular blocks characteristic of andesite lava flows. The flow slowly moves, with the blocks colliding and overriding one another to form piles of angular andesite blocks.

The viscous nature of andesite lavas is responsible for many classic volcanic features. Most notable is the symmetrical cone shape of the stratovolcanoes of the circum-Pacific region. Short, viscous andesitic flows pouring down the flanks of these volcanoes are alternately covered by pyroclastic material and work to build the steep central cone characteristic of composite stratovolcanoes. When an andesite's silica content rises to a point that it approaches rhyolite composition, it is termed a dacite. Dacite is often so viscous that it cannot flow and blocks the vent of the volcano. This dacite plug is called a lava dome. Such a dome can be seen in pictures of Mount St. Helens. If the plugged volcano becomes active again, the lava dome does not allow for a release of accumulating pressure and explosive gases. Pressure builds until the volcano finally erupts with great force and violence.

ANDESITE LINE

The differences between basalts and andesites reflect their differences in composition, which is a function of the environment in which their source lavas occur. Basalts are typically formed at midocean ridges and build up oceanic crust. The generation of andesite magma is characteristic of destructive plate margins. Here, oceanic plates are being subducted below continental plates. Destructive plate boundaries tend to produce a greater variety of lavas than spreading zones (ocean ridges). The close association of andesite with convergent plate boundaries is the significance of the "Andesite Line" often drawn around the Pacific Ocean basin. This fairly well-defined line separates two major petrographic regions. Inside this line and inside the main ocean basin, no andesites occur. All active volcanoes inside the line erupt basaltic magma, and all volcanic rocks associated with dormant volcanoes within this region are basaltic. Outside the line, andesite is common. The Andesite Line parallels the western and northern boundary of the Pacific plate and the eastern boundary of the Juan de Fuca, Cocos, and Nazca plates. The Andesite Line parallels the major island arc systems, the subducting edges of the tectonic plates listed above, and a chain of prominent and infamous stratovolcanoes known as the Ring of Fire. These stratovolcanoes are exemplified by the following: Mount Rainier and other volcanoes of the western United States, El Chichón of Mexico, San

Pedro of Chile, Mounts Egmont and Taupo of New Zealand, Krakatau and Tambora of the Indonesian Arc, Fujiyama of Japan, Bezymianny of Kamchatka, and the Valley of Ten Thousand Smokes in Alaska.

BENIOFF ZONE

The Benioff zone is a plane of seismic activity dipping at an angle of about 45 degrees below a continent, marking the path of a subducting oceanic plate at the convergent plate margin. At the Benioff zone, the subducting plate heats up and gives off water, which lowers the melting point of the overlying mantle and causes it to melt. As magma is formed at the Benioff zone and begins its slow rise toward the surface, it passes through and comes in contact with regions of the mantle and continental crust. During its rise, the magma also comes in contact with circulating meteoric water (water originating on the surface), and new fluids derived from magma are formed. Because andesitic magma rises through a variety of host rocks, the variety of minerals that enrich the new fluids is increased. The rising magma also stresses the surrounding host rocks, causing them to fracture. These fractures and similar open spaces are filled with mineral-rich water and fluids derived from magma. As the fluids cool, mineral ores precipitate from the solutions and fill the open spaces. These filled spaces often become mineral-rich dikes and sills.

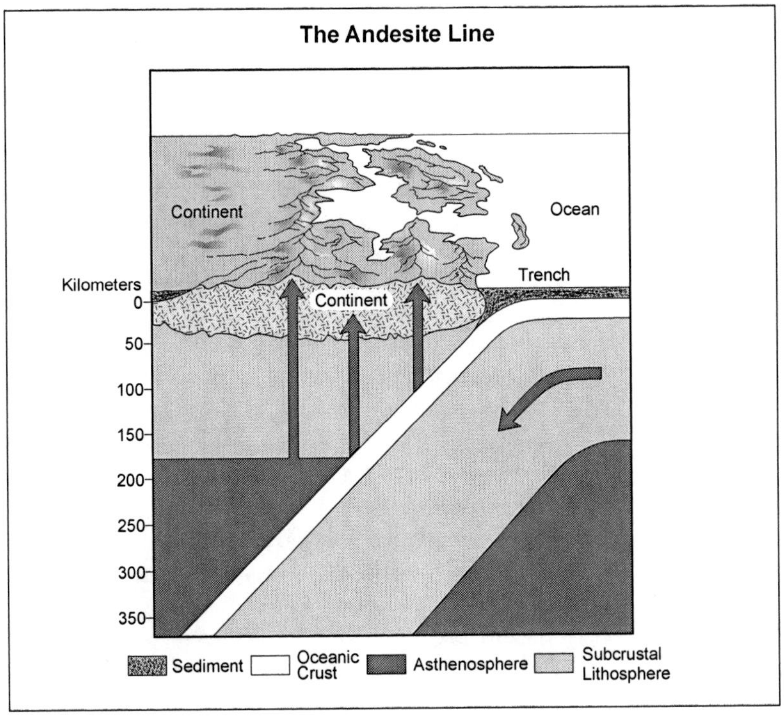

FIELD STUDY

The study and interpretation of andesitic rocks are accomplished in three basic ways: fieldwork, in which researchers travel to locations to assess and interpret a specific region of the earth that is known or suspected of being andesitic terrain; petrological studies, which use all available methods of study to ascertain the history, origin, conditions, alterations, and weathering of the rocks collected during fieldwork; and mineralogical studies, which through intensive laboratory investigations identify the specific mineral characteristic of a sample.

If a scientist suspects that a region of the earth is andesitic terrain, the first step is to read the information already available about the region. If travel to that region for fieldwork is considered useful, the researcher will most likely collect a large number of samples. The location of each sample is carefully recorded on a map and additional data describing the geologic setting of the sample are recorded. These data include thickness, areal extent, weathering, and strike and dip of the location. Photographs are also commonly used to record the surrounding environment of the sample location. Samples are usually given preliminary study at the researcher's field camp.

The Andesite Line

Kilometers

Continent

Ocean

Trench

Continent

0
50
100
150
200
250
300
350

Sediment | Oceanic Crust | Asthenosphere | Subcrustal Lithosphere

PETROLOGICAL AND MINERALOGICAL STUDIES

When the fieldwork is over, the researcher returns with the samples to a laboratory setting, where petrological and mineralogical investigations begin. A petrological study of an andesitic sample will include both petrographic and petrogenetic analyses. The petrographic analyses will describe the sample and attempt to place it within the standard systematic classification of igneous rocks. Identification of the sample is accomplished by means of examining a thinly sliced and polished portion of the sample under a petrographic microscope—a process known as thin-section analysis. The information obtained by microscopic examination gives a breakdown of the type and amount of mineral composition within the sample. More advanced analysis techniques, such as X-ray fluorescence and electron microprobe, are often used to obtain very precise chemical compositions of samples. Knowing the conditions under which these minerals form allows the researcher to make a petrogenetic assessment.

Petrogenesis deals with the origin and formation of rocks. If a mineral is known to form only at certain depths, temperatures, or pressures, its presence within a sample allows the researcher to draw some specific conclusions concerning the rock's formation and history. Mineralogical studies of field-gathered samples aid in the petrogenetic portion of the analysis. Mineralogy involves the study of how a mineral forms—its physical properties, chemical composition, and occurrence. Minerals can exist in a stable form only within a narrow range of pressure and temperature. Experimental confirmation of this range enables the researcher to make a correlation between the occurrence of a mineral in a rock and the conditions under which the rock was formed. In this sense, mineralogical and petrological analyses complement each other and work to formulate a concise history of a given sample.

When data gathered from the field and laboratory are combined, the geological history of a given field area can begin to be interpreted. When areas of similar igneous rock types, ages, and mineral compositions are plotted on a map, they form a petrographic province. A petrographic province indicates an area of similar rocks that formed during the same period of igneous activity. On a global scale, the Andesite Line marks the boundary between two great provinces: the basaltic oceanic crust and the andesitic continental crust. Both crustal forms have distinctly different geological histories and mineral compositions.

Randall L. Milstein

FURTHER READING

Bowen, N. L. *The Evolution of Igneous Rocks.* Mineola, N.Y.: Dover, 1956. An unmatched source and reference book on igneous rocks, written by the father of modern petrology. The basis of Bowen's reaction principle. Written for graduate students and professional scientists. A classic but difficult work.

Carmichael, I. S. E., F. J. Turner, and J. Verhoogen. *Igneous Petrology.* New York: McGraw-Hill, 1974. A college-level textbook on the formation and development of igneous rocks. Very detailed and complete.

Faure, Gunter. *Origin of Igneous Rocks: The Isotopic Evidence.* New York: Springer-Verlag, 2010. Discusses chemical properties of igneous rocks, and isotopes within these rocks formations. Specific locations of igneous rock formations and the origins of these rocks are provided. There are many diagrams and drawings along with the overview of isotope geochemistry in the first chapter to make this accessible to undergraduate students as well as professionals.

Gill, J. B. *Orogenic Andesite and Plate Tectonics.* New York: Springer-Verlag, 1981. A well-documented summary of the entire field of andesite genesis. Written for graduate students and professional earth scientists.

Klein, Cornelis, and Cornelius S. Hurlbut, Jr. *Manual of Mineralogy.* 23d ed. New York: John Wiley & Sons, 2008. An introductory college-level text on mineralogy. Discusses the physical and chemical properties of minerals and describes the most common minerals and their varieties, including gem mineral varieties. Contains more than 500 mineral name entries in the mineral index and describes in detail about 150 mineral species.

Perchuk, L. L., ed. *Progress in Metamorphic and Magmatic Petrology.* New York: Cambridge University Press, 1991. Although intended for the advanced reader, several of the essays in this multiauthored volume will serve to familiarize new students with the study of igneous rocks. In addition, the bibliography will lead the reader to other useful material.

Ross, Pierre-Simon, et al. "Basaltic to Andesitic Volcaniclastic Rocks in the Blake River Group, Abitibi Greenstone Belt: 2. Origin, Geochemistry, and Geochronology." *Canadian Journal of Earth Sciences.* 48 (2011): 757-777. Discusses the rock formations in the Archean Blake River Group. Physical characteristics, age relationships, and geochemistry data are measured and interpreted to determine the origins of these rock formations. The authors hypothesize the volcanic processes that resulted in these formations. Best suited for graduate students or researchers studying volcanism, petrology, or mineralogy.

Science Mate: Plate Tectonic Cycle. Fremont, Calif.: Math/Science Nucleus, 1990. Part of the Integrating Science, Math, and Technology series of manuals intended to help teachers explain basic scientific concepts to elementary students. Provides a basic understanding of plate tectonics, earthquakes, volcanoes, and general geology for the student with no background in the Earth sciences.

Sutherland, Lin. *The Volcanic Earth: Volcanoes and Plate Tectonics, Past, Present, and Future.* Sydney, Australia: University of New South Wales Press, 1995. Provides an easily understood overview of volcanic and tectonic processes, including the role of igneous rocks. Includes color maps and illustrations, as well as a bibliography.

Tamura, Y., et al. "Andesites and Dacites from Daisen Volcano, Japan: Partial-to-Total Remelting of an Andesite Magma Body." *Journal of Petrology* 44 (2003): 2243-2260. Discusses the genesis of andesitic and dacitic rocks from the Daisen Volcano. Lava characteristics and the chemical and physical characteristics of local rocks are compared to determine the stages of the rock forming process.

Williams, H., and A. R. McBirney. *Volcanology.* San Francisco: Freeman, Cooper, 1979. A classic textbook on volcanoes and volcanology. Well illustrated and very descriptive. Written for the undergraduate or graduate student.

Windley, B. F. *The Evolving Continents.* 3d ed. New York: John Wiley & Sons, 1995. An excellent reference book on plate tectonics and tectonic processes. Written for the college-level reader.

See also: Anorthosites; Basaltic Rocks; Batholiths; Carbonatites; Earth Resources; Extraterrestrial Resources; Granitic Rocks; Hazardous Wastes; Igneous Rock Bodies; Kimberlites; Komatiites; Metamictization; Nuclear Power; Nuclear Waste Disposal; Orbicular Rocks; Plutonic Rocks; Pyroclastic Rocks; Rocks: Physical Properties; Strategic Resources; Xenoliths.

ANORTHOSITES

Anorthosites are coarse-grained, intrusive igneous rocks composed principally of plagioclase feldspar. They are useful for what they reveal about the early crustal evolution of the earth, and they are the source of several economic commodities.

PRINCIPAL TERMS

- **crust:** the upper layer of the earth and other "rocky" planets; it is composed mostly of relatively low-density silicate rocks
- **gabbro:** coarse-grained, iron-magnesium-rich, plutonic igneous rock; anorthosite is an unusual variety of gabbro
- **hypersthene:** a low-calcium pyroxene mineral
- **magma:** molten silicate liquid plus any crystals, rock inclusions, or gases trapped in that liquid
- **mantle:** a layer in the earth extending from about 5 to 50 kilometers below the crust
- **massif:** a French term used in geology to describe very large, usually igneous intrusive bodies
- **norite:** gabbro in which hypersthene is the principal pyroxene; it is commonly associated with anorthosites
- **peridotite:** the most common rock type in the upper mantle, where basalt magma is produced
- **plagioclase:** a silicate mineral found in many rocks; it is a member of the feldspar group
- **plutonic:** formed by solidification of magma deep within the earth and crystalline throughout
- **Precambrian:** The span of geologic time extending from early planetary origins to about 540 million years ago

ANORTHOSITE COMPOSITION

Anorthosites are igneous rocks that are composed primarily of plagioclase feldspar (calcium-sodium-aluminum silicate). Minor minerals in these rocks may include pyroxene minerals (calcium-iron-magnesium silicates), iron-titanium oxides, and, in metamorphosed varieties, garnet. All anorthosites are coarse-grained, plutonic rocks (they crystallize at depth), and they may have plagioclase crystals 10 centimeters or more in length. Because the color of plagioclase changes with minor changes in chemical composition, anorthosites come in a variety of colors. Light gray is the most common, but dark gray, black, light blue, green, and brown varieties are known.

ANORTHOSITE TYPES

In the earth, there are two types of anorthosite: massif-type anorthosites and layered intrusive anorthosites. The former occur as large lens- or dome-shaped intrusions that may be exposed by erosion over an area of several square kilometers. The latter variety is associated with intrusions of gabbroic (iron-rich, low-silica) magma that has segregated into mineralogically distinct layers, with anorthosite occurring in the uppermost layers. The lighter areas on the moon, known as the "lunar highlands," are composed of anorthosite that originated in a similar manner to terrestrial, layered intrusive anorthosites.

Massif-type anorthosites are unusual among igneous rocks in their restricted distribution in time and space and their composition. They contain at least 85 percent plagioclase, with the other silicate minerals being either augite (high-calcium pyroxene), hypersthene (low-calcium pyroxene), or both. Ilmenite and apatite commonly occur as minor accessory minerals. Massif anorthosites are found almost exclusively in a wide belt from the southwestern United States, through Labrador and on the other side of the Atlantic through Sweden and Norway. Another belt extends from Brazil through Africa (Angola, Tanzania, Malagasy), Queen Maud Land in Antarctica, and across Bengal, India, to Australia (These regions were once adjacent before continental drift dispersed them). The best examples of massif anorthosites in the United States are in the Adirondack Mountains in northeastern New York State. Another large anorthosite body, the Laramie Anorthosite, is in Wyoming. In Canada, the Nain and Kiglapait intrusions in northern Labrador are notable for their excellent surface exposures of anorthosite and associated rocks.

ANORTHOSITE FORMATION

Ages determined with radiometric dating techniques demonstrate that nearly all anorthosites, including the layered-intrusion varieties, are Precambrian; most are between about 1,700 million and 1,100 million years old. No lava flows of anorthositic composition are known; they are exclusively plutonic. Additionally, experimental evidence shows that liquids of anorthositic composition cannot be produced by any known process of melting at depth in the earth. Although specific mechanisms are not fully understood, it is generally accepted that anorthosites form from concentrated plagioclase crystals that have previously crystallized from gabbroic or other magmas. Massif anorthosites are commonly associated with gabbroic rocks called "norites" (gabbro in which the pyroxene mineral is hypersthene), and it can be demonstrated in most cases that anorthosite is produced by the norite magma becoming progressively richer in plagioclase.

Anorthosites in layered intrusives clearly result from the same processes that produce the other layered rocks associated with the anorthosite layers. Layered intrusives, commonly termed "layered complexes," are generally tabular bodies that vary in thickness from a few hundred to thousands meters thick. They consist of extremely iron- and magnesium-rich rocks called peridotites in their lower reaches, but they grade upward into gabbros and, in some cases, anorthosites in their upper extents. This layering of different rock types is attributed to the gravitationally controlled settling of dense minerals to the bottom of a large body of originally gabbroic magma. Rock layers of different compositions are built up over time as different minerals crystallize and are deposited on the chamber floor. Plagioclase generally crystallizes relatively late in this process and, depending on its composition and that of the enclosing silicate liquid, may actually float to the top of the magma chamber instead of settling to the bottom. Anorthosite deposits are known to have been formed by both settled and floated plagioclase. In both cases, anorthosite does not form directly as magma, but instead results from the concentration of plagioclase by different densities as the magma evolves.

ANORTHOSITE DISTRIBUTION

Layered intrusive anorthosites occur in the early Precambrian Stillwater Complex of western Montana and the late Precambrian Duluth Complex that parallels the north shore of Lake Superior in northern Minnesota. The Stillwater contains three anorthosite units, each about 400 to 500 meters thick. In contrast, the Duluth Complex, about 40 kilometers thick, is mostly anorthosites and "gabbroic anorthosites" (somewhat richer in iron-magnesium silicates than anorthosites), with a mostly unexposed, relatively thin peridotite unit at its base. Norites and troctolites (olivine plus plagioclase) occur as minor associated rocks. Other layered anorthosites occur in the Bushveld complex in South Africa, the Fiskenaesset Complex in Greenland, and the Dore Lake and Bell River complexes in Ontario, Canada, among others. All of the layered intrusions in the United States are Precambrian.

Without question, the most obvious (but least accessible) exposures of anorthosite occur on the moon. Relatively rare on Earth, anorthosite is the dominant rock type on the moon, where it forms the bulk of the rocks in the lunar highlands. The lunar highlands are the moon's ancient, highly cratered, light-colored areas. The dark areas are basalt (iron-rich silicate rock) lava flows that fill huge craters blasted in the highlands by large meteorite collisions. Ever since the first moon rocks were returned to Earth by the Apollo 11 astronauts in 1969, scientists have studied lunar anorthosites and associated rocks for clues to how the moon's crust and interior originated and evolved. The generally accepted model postulates that early in its history, more than 4 billion years ago, the entire surface of the moon became molten to a depth of several kilometers. This "magma ocean" then cooled, and as it cooled, it crystallized various silicate and oxide minerals. The heavier, iron-rich minerals, being denser than the silicate magma, sank to form deep-seated layers, similar to the peridotite layers in the Stillwater and Bushveld complexes described earlier. Plagioclase, in contrast, was less dense than the iron-rich liquid, so it floated to the top, solidifying to form the early lunar highland anorthosites. Over time, the highlands became increasingly cratered by the meteorite impacts, until they acquired their present appearance about 3 billion years ago. Since that time, meteorite impacts have been sporadic. Interestingly, lunar highland anorthosite occurs with minor quantities of norite and troctolite, two rock types that also are commonly associated with terrestrial anorthosites. To lunar scientists, this peculiar group of associated rocks is the "ANT suite"; ANT is an acronym for anorthosite, norite, and troctolite.

ASSESSING ANORTHOSITE ORIGIN AND SIGNIFICANCE

Paradoxically, the origin of terrestrial anorthosites is not as clear-cut as the origin of lunar anorthosites, and many models and hypotheses have been offered to explain their unique composition and space-time relationships. No explanation is readily accepted as applying to all or most anorthosite occurrences, but enough is known about anorthosite bodies to allow some good, educated guesses. Scientific work on anorthosites has included numerous field-mapping projects to determine their spatial extent and structure. Laboratory work has included radiochemical dating studies to determine the absolute ages of anorthosites and associated rocks and experimental studies that have explored possible parent materials and melting environments of anorthosite magmas.

Several factors must be considered in assessing the origin and significance of anorthosite bodies—specifically the massif varieties. First, anorthosites are restricted, for the most part, to the very narrow time band between about 1.7 and 1.1 billion years before the present, as determined mostly by potassium-argon, uranium-lead, and rubidium-strontium dating. Second, these bodies occur in belts where orogeny (mountain building) may have occurred before or after their emplacement but where they intruded during anorogenic times or during rifting. They are unequivocally igneous as opposed to metamorphic; that is, they did not result from some other type of rock that changed in the solid state but rather from molten magmas. This fact is determined in the field by examining anorthosite contacts with older rocks to see whether these rocks show evidence of thermal heating caused by intrusion or hot magma. Anorthosites lack minerals that contain water in their structures, so water was not an important constituent of anorthosite parent magmas. Finally, many anorthosite complexes, such as those in New York's Adirondack Mountains, show an association not only with norites and troctolites but also with pyroxene-bearing granitic rocks called charnockites.

Although experimental evidence shows that a liquid of plagioclase composition cannot be generated under high pressures from any known earth materials, magmas can form from rocks in the mantle or lower crust that, under anhydrous (water-free) conditions, are capable of generating the plagioclase "mush" (crystals plus a small amount of liquid)

by plagioclase flotation. Laboratory melting experiments show that one type of magma, called "quartz diorite" (relatively siliceous magma of the type erupted in 1980 by Mount St. Helens), could produce a plagioclase mush at great pressures. Intrusion of this mush at higher levels in the crust could squeeze out some of the interstitial liquid, which would crystallize as charnockites. Norites and gabbroic anorthosites represent cases where the plagioclase mush has trapped greater or smaller amounts of more iron-rich silicate liquid.

Whatever the precise mechanism might have been, it produced large volumes of a very unusual suite of rocks over a geologically brief time interval. No doubt this extensive melting event required higher heat flow in the crust and mantle than is now present in areas of anorthosite's occurrence, or in most other areas of the earth. The cause of this brief episode of high heat flow is still uncertain.

DECORATIVE AND PRACTICAL APPLICATIONS

Anorthosites are not very common in the earth's crust, and most people are not as familiar with this stone as they are with the more common building stones, granite and marble. Ironically, anyone who has ever seen the moon has seen huge spans of anorthosite; it is the principal rock type composing the lunar highlands, the dominant light-colored areas of the moon.

Anorthosite is used as a building stone or for decorative building facings. A beautiful variety composed mostly of the high-calcium plagioclase known as labradorite is in particular demand. Labradorite exhibits blue, violet, or green iridescence that varies in color with the angle of incidence of light. This rock, a dark gray rock that exhibits bluish flashes, is called larvikite, after Larvik, Norway. Anorthosite composed of labradorite plagioclase is sometimes polished and used in tabletops and floor or wall panels. It has other decorative and practical applications as well. Like other feldspar-rich rocks, anorthosite can be used to make ceramic products. Porcelain bath fixtures, insulators, and dining ware are made from finely pulverized feldspar that is heated to very high temperatures. The resistance to heat and electricity and the general durability of porcelain come from the same properties inherent in feldspar.

Anorthosites and associated rocks are also the sites of economically exploitable iron-titanium deposits in some localities. The ore occurs mostly as ilmenite (iron-titanium oxide) and magnetite (iron oxide), mostly concentrated in associated rocks such as norite. Some notable deposits occur in the following places: Lofoten, in northern Norway; Allard Lake, Quebec; Iron Mountain, in the Laramie Range of Wyoming; Duluth, Minnesota; and Sanford Lake (near Tahawus) in the Adirondacks of New York State.

John L. Berkley

FURTHER READING

Bédard, Jean H. "Parental Magmas of the Nain Plutonic Suite Anorthosites and Mafic Cumulates: A Trace Element Modeling Approach." *Contributions to Mineralogy and Petrology* 141 (2001): 747-771. Discusses the trace elements found within mafic and anorthositic rock formations. Draws conclusions on the melt origins of these cumulates and the details of crystallization.

Carmichael, Ian S. E., F. J. Turner, and John Verhoogan. *Igneous Petrology.* New York: McGraw-Hill, 1974. Designed for upper-level undergraduates and graduate students; the section on anorthosites does an excellent job of summarizing their occurrences and the problems associated with their interpretation and geologic significance. Appropriate for anyone with a minimal background in geology. Critical references from the scientific literature are listed in full near the end of the book.

Duchesne, J. C., and A. Korneliussen, eds. *Ilmenite Deposits and Their Geological Environment. Norges Geologiski Undersokelse.* Special publication 9. Trondheim, Norway: Geological Survey of Norway, 2003. A compilation of articles discussing the chemical composition, physical properties, and origins of anorthosites in the Rogaland Anorthosite Province of Scandinavia. Provides information on the processes involved in emplacement and formation of these rocks. Over forty locations within the province are described. Highly technical and best suited for individuals with a strong background in geology.

Isachsen, Yngvar W., ed. *Origin of Anorthosite and Related Rocks.* Memoir 18. Albany, N.Y.: State Department of Education of New York, 1968. A massive compendium of articles presented at the Second Annual George H. Hudson Symposium, held at Plattsburg, New York, in 1966. One of the most extensive and comprehensive works on anorthosites ever prepared. Includes a detailed appendix, and each article has an extensive references list at the end. Best place to start for students who are interested in the details of anorthosites.

Klein, Cornelis, and Cornelius S. Hurlbut, Jr. *Manual of Mineralogy.* 23d ed. New York: John Wiley & Sons, 2008. An introductory college-level text on mineralogy. Discusses the physical and chemical properties of minerals and describes the most common minerals and their varieties, including gem mineral varieties. Contains more than 500 mineral name entries in the mineral index and describes in detail about 150 mineral species.

Norman, M. D., L. E. Borg, L. E. Nyquist, and D. D. Bogard. "Chronology, Geochemistry, and Petrology of a Ferroan Noritic Anorthosite Clast from Descartes Breccia 67215: Clues to the Age, Origin, Structure, and Impact History of the Lunar Crust." *Meteoritics and Planetary Science* 38 (2003): 645-661. Discusses anorthosites in context of the lunar crust. A highly technical paper best suited for readers with a strong understanding of geochemistry and of trace elements in particular.

Perchuk, L. L., ed. *Progress in Metamorphic and Magmatic Petrology.* New York: Cambridge University Press, 1991. Although intended for the advanced reader, several of the essays in this multiauthored volume will serve to familiarize new students with the study of igneous rocks. In addition, the bibliography will lead the reader to other useful material.

Science Mate: Plate Tectonic Cycle. Fremont, Calif.: Math/Science Nucleus, 1990. Part of the Integrating Science, Math, and Technology series of manuals intended to help teachers explain basic scientific concepts to elementary students. Provides an explanation of plate tectonics, earthquakes, volcanoes, and general geology.

Sutherland, Lin. *The Volcanic Earth: Volcanoes and Plate Tectonics, Past, Present, and Future.* Sydney, Australia: University of New South Wales Press, 1995. Provides an easily understood overview of volcanic and tectonic processes, including the role of igneous rocks. Includes color maps and illustrations, as well as a bibliography.

Taylor, G. Jeffrey. *A Close Look at the Moon.* New York: Dodd, Mead, 1980. Designed for high school readers. Offers a good section on lunar rocks by one of the world's authorities on lunar rocks and lunar crustal evolution. Excellent cross-sectional diagrams trace the history of the lunar highlands (anorthosite-rich areas) and the lunar crust in general. Contains photographs of moon rocks, including some taken through a microscope.

Windley, Brian F. *The Evolving Continents*. 3d ed. New York: John Wiley & Sons, 1995. An advanced text that focuses on the evolution of the earth's continental crust as revealed by its complex geology. The section on anorthosite bodies is very descriptive and should be comprehensible to most nonspecialists. Contains a brief description of economic commodities associated with anorthosites; references to various anorthosite occurrences may be found in the index. Includes a complete listing of technical references for those who wish to dig deeper.

Winter, John D. *An Introduction to Igneous and Metamorphic Petrology*. Upper Saddle River, N.J.: Prentice Hall, 2001. Provides a comprehensive overview of igneous and metamorphic rock formation. A basic understanding of algebra and spreadsheets is required. Discusses volcanism, metamorphism, thermodynamics, and trace elements; focuses on the theories and chemistry involved in petrology. Written for students taking a university level course on igneous petrology, metamorphic petrology, or a combined course.

See also: Andesitic Rocks; Basaltic Rocks; Batholiths; Carbonatites; Granitic Rocks; Igneous Rock Bodies; Kimberlites; Komatiites; Orbicular Rocks; Plutonic Rocks; Pyroclastic Rocks; Rocks: Physical Properties; Xenoliths.

B

BASALTIC ROCKS

"Basalt" is the term applied to dark, iron-rich volcanic rocks that occur everywhere on the ocean floors, as oceanic islands, and in certain areas on continents. It is the parent material from which nearly any other igneous rock can be generated by various natural processes.

PRINCIPAL TERMS

- **augite:** an essential mineral in most basalts, a member of the pyroxene group of silicates
- **crust:** the upper layer of the earth and the other "rocky" planets; it is composed mostly of relatively low-density silicate rocks
- **lithospheric plates:** giant slabs composed of crust and upper mantle; they move about laterally to produce volcanism, mountain building, and earthquakes
- **magma:** molten silicate liquid, including any crystals, rock inclusions, or gases trapped in that liquid
- **mantle:** a layer beginning at about 5 to 50 kilometers below the crust and extending to the earth's metallic core
- **oceanic ridges:** a system of mostly underwater rift mountains that bisect all the ocean basins; basalt is extruded along their central axes
- **olivine:** a silicate mineral found in mantle and some basalts, particularly the alkaline varieties
- **peridotite:** the most common rock type in the upper mantle, where basalt magma is produced
- **plagioclase:** one of the principal silicate minerals in basalt, a member of the feldspar group
- **subduction zones:** areas marginal to continents where lithospheric plates collide

BASALT COMPOSITION

Basalt is a dark, commonly black, volcanic rock. It is sometimes called "trap" or "traprock," from the Swedish term *trapp*, which means "steplike." On cooling, basalt tends to form hexagonal columns, which in turn form steplike structures after erosion. Excellent examples can be seen at Devils Postpile, in the Sierra Nevada in central California, and the Giant's Causeway, in County Antrim, Ireland. The term "trap," however, is used mostly by miners and nonspecialists; scientists prefer the word "basalt" for the fine-grained, volcanic rock that forms by the solidification of lava flows. Basaltic magma that crystallizes more slowly below the earth's surface, thus making larger mineral grains, is called "gabbro."

The importance of basalt to the evolution of planets like Earth cannot be overemphasized. Basalt is considered to be a "primary silicate liquid," in that the first liquids to form by the melting of the original minerals that made up all the so-called rocky planets (those composed mostly of silicates) were basaltic in composition. In turn, basalt contains all the necessary ingredients to make all the other rocks that may eventually form in a planet's crust. Furthermore, many meteorites (which are believed to represent fragments of planetoids or asteroids) are basaltic or contain basalt fragments, and the surfaces of the moon and the planets Mercury, Venus, and Mars are known to be covered to various degrees by basaltic lava flows.

Like most rocks in the earth's crust, basalt is composed of silicate minerals, substances whose principal component is the silica molecule. Compared with other silicate rocks, basalts contain large amounts of iron and magnesium and small amounts of silicon. This characteristic is reflected in the minerals in basalts, which are mostly pyroxene minerals (dark-colored, calcium-iron-magnesium silicates such as augite) and certain feldspar minerals called plagioclase (light-colored, calcium-sodium-aluminum silicates). Pyroxenes and plagioclase are essential minerals in basalts, but some types of basalt also contain the mineral olivine (a green, iron-magnesium silicate). The high abundance of dark green or black pyroxene and, in some cases, olivine gives basalts their characteristic dark color. This color can be mainly attributed to the high iron content of pyroxene and olivine.

15

OCEANIC BASALT DISTRIBUTION

Basalt is the most common type of igneous rock (that is, rock formed by the crystallization of magma) on or near the surface of the earth. It is the principal rock in the ocean floors and is common, though less abundant, on the continents as well. Drilling into the sea floor by specially designed oceanographic research ships reveals that basalt invariably lies just below a thin cover of fine, sedimentary mud. Ocean floor basalt flows out of midocean ridges, a system of underwater mountain ranges that spans the globe. These ridges commonly trend roughly down the middle of ocean basins, and they represent places where the earth's lithospheric plates are being literally split apart. In this process, basalt magma is generated below what is referred to as rift mountains; it flows onto the cold ocean floor and solidifies. Although oceanic ridges are normally hidden from view under the oceans, a segment of the Mid-Atlantic Ridge emerges above the waves as the island of Iceland.

Oceanic islands are also composed of basalt. The islands of Hawaii, Fiji, Mariana, Tonga, and Samoa, among others, are large volcanoes or groups of volcanoes that rise above water from the ocean floor. Unlike the basalts that cover the ocean floors, however, these volcanoes do not occur at oceanic ridges but instead rise directly from the sea floor. Basalt is also a fairly common rock type on island arcs, volcanic islands that occur near continental margins. These curvilinear island chains arise from melting along subduction zones, areas where the earth's lithospheric plates are colliding. This process generally involves material from the ocean basins diving under the more massive continents; andesite (a light-colored rock) volcanoes are the main result, but some basalt erupts there as well. The Japanese, Philippine, and Aleutian island chains are examples of island arcs, as are the island countries of New Zealand and Indonesia.

CONTINENTAL BASALT DISTRIBUTION

Basaltic lava flows are not nearly as common on continents as in oceanic areas. Andesites, rhyolites, and related igneous rocks are far more abundant than basalt in continental settings. In North America, basaltic lava flows and volcanoes are most exposed to view in the western United States and Canadian provinces, western Mexico, Central America, and western South America. The greatest accumulations of basaltic lava flows in the United States are in the Columbia Plateau of Washington, Oregon, and Idaho. In this large area, a series of basaltic lava flows have built up hundreds of feet of nearly flat-lying basaltic flows over a few million years. These "fissure flows" result from lava pouring out of long cracks in the crust, or fissures. They are similar in many respects to the basalt flows produced at oceanic ridges, because no actual volcanic cones are produced—only layer after layer of black basalt. The Snake River Plain in southern Idaho has a similar origin, and other extensive basalt plateaus occur in the Deccan area of southwestern India, the Karroo area of South Africa, and Paraná State in Brazil.

Not all basalt is erupted as lava flows. If the lava is particularly rich in volatiles such as water and carbon dioxide, it will be explosively ejected from the volcano as glowing fountains of incandescent particles that rain down on the surrounding area. Conical volcanoes composed almost exclusively of basalt ejecta particles are called cinder cones. Good examples of cinder cones are Sunset Crater in northern Arizona, and the numerous cinder cones in Hawaii and Iceland.

BASALT CLASSIFICATION

Although basaltic rocks may all look alike to the nonspecialist, there are actually many different kinds of basalt. They are arranged by scientists into a generally accepted classification scheme based on chemistry and, to some extent, mineralogy. To begin with, basalt can be distinguished from the other major silicate igneous rocks by its relatively low silica content (about 50 percent). Within the basalt clan itself, however, other means of classification are used. Basalts are divided into two major groups: the alkaline basalts and the subalkaline basalts. Alkaline basalts contain large amounts of the alkali metal ions potassium and sodium but relatively small amounts of silica. In contrast, subalkaline basalts contain less potassium and sodium and more silica. As might be expected, this chemical difference translates into differences in the mineral content of the basalt types as well. For example, all alkaline basalts contain one or more minerals called "feldspathoids" in addition to plagioclase feldspar. They also commonly contain significant olivine. Subalkaline basalts, in contrast, do not contain feldspathoids, although some contain olivine, and they may be capable of crystallizing very tiny amounts of the mineral quartz. The presence of this very silica-rich mineral reflects the relatively high silica content of subalkaline basalt magmas versus alkaline basalt magmas. Within these two major groups are many subtypes, too numerous to discuss here.

STUDY TECHNIQUES

Like other igneous rocks, basalts are analyzed and studied by way of many techniques. Individual studies may include extensive field mapping, in which the distribution of various types of basalt is plotted on maps. The history of magma generation and its relation to tectonic history (earth movements) can be reconstructed for a particular area by correlating. Geologic maps of basalt types with absolute ages are determined through radiometric dating techniques. Good examples of such studies are those conducted in recent years on the Hawaiian Islands. These studies indicate that the alkaline basalts on any given island are generally older than the subalkaline basalts, showing that magma production has moved upward, to lower-pressure areas, in the mantle with time. This finding supports the idea that oceanic island basalts such as those in Hawaii are generated within so-called mantle plumes—roughly balloon-shaped, slowly rising masses of mantle material made buoyant by localized "hot spots."

Samples of basalt are also analyzed in the laboratory. The age of crystallization of basalt is obtained by means of radiometric dating techniques that involve the use of mass spectrometers to determine the abundances of critical isotopes, such as potassium-40 and argon-40, or rubidium-87 and strontium-86 and -87. To obtain information on how basalt magma is generated and how it subsequently changes in composition before extrusion as a lava flow, scientists place finely powdered samples in metallic, graphite, or ceramic capsules and subject them to heating and cooling under various conditions of pressure. Such procedures are known as experimental petrology. Such studies prove that nearly any other igneous magma composition can be derived from basalt magma by the process of crystal fractionation. Widely believed to be the major factor influencing chemical variation among igneous rocks, this process results in ever-changing liquid compositions, as the various silicate minerals crystallize and are thus removed from the liquid over time. Basalt's parental role gives

it enormous importance in the discipline of igneous petrology.

TRACE ELEMENT ANALYSIS

Another useful avenue of research is the trace element analysis of basalts. Trace elements occur in such low abundances in rocks that their concentrations must usually be expressed in terms of parts per million or even parts per billion. Among the most useful substances for tracing the history of basalt are the rare-earth elements. The elements chromium, vanadium, nickel, phosphorus, strontium, zirconium, scandium, and hafnium are also used. There are many methods for measuring these elements, but the most common, and most accurate, is neutron activation. This method involves irradiating samples in a small nuclear reactor and then electronically counting the gamma-ray pulses generated by the samples. Since different elements tend to emit gamma rays at characteristic energies, these specific energies can be measured and the intensity of gamma pulses translated into elemental concentrations.

Once determined for a particular basalt sample, trace element abundances are sensitive indicators of events that have transpired during the evolution of the basalt. There are two reasons for this sensitivity. First, trace elements are present in such low concentrations (as compared with major elements—iron, aluminum, calcium, silicon, and the like) that any small change in abundance caused by changes in the environment of basalt production will be readily noticed. Second, different minerals, including those crystallizing in the magma and those in the source peridotite, incorporate a given trace element into their structures or reject it to the surrounding liquid to widely varying degrees. Therefore, trace element concentrations can be used to show the minerals that were involved in producing certain observed chemical signatures in basalts and those that were likely not involved.

For example, it is well known that the mineral garnet readily accepts the rare-earth element lutetium into its structure but tends to reject most lanthanum to any adjacent liquid, even though these elements are very similar chemically. Basalts with very little lutetium but much lanthanum were therefore probably derived by the melting of garnet-bearing mantle rocks. Since garnet-bearing mantle rocks can exist only at great depths, basalts with such trace element patterns must have originated by melting at these depths in the mantle. In fact, that is one of the most important lines of evidence to support that alkaline basalts originate at high-pressure regions in the mantle.

ECONOMIC APPLICATIONS

Basaltic islands, particularly in the Pacific basin, are some of the most popular tourist stops in the world. More importantly, however, basalt magma contains low concentrations of valuable metals that, when concentrated by various natural processes, provide the source for many important ores. Copper, nickel, lead, zinc, gold, silver, and other metals have been recovered from ore bodies centered in basaltic terrains. Some of the richest mines of metallic ores in the world are located in Canada, where ores are found associated with extremely old basaltic rocks, called "greenstones," from long-vanished oceans. The richest of these mines is Kidd Creek in northern Ontario. These ore-bearing basalts were first extruded more than 2 billion years ago, during what geologists call Precambrian times (the period from 4.6 billion to about 600 million years ago). Other notable ore deposits include the native, or metallic, copper in late Precambrian basalts that were mined for many years in the Keweenaw Peninsula of northern Michigan. The island of Crete in the Mediterranean Sea has copper mines that were mined thousands of years ago during the "copper" and "bronze" ages of human history. The basalt enclosing these ores is believed to have erupted from an ancient midocean ridge formerly located between Africa and Europe.

Basalt can also be used as a building stone or raw material for sculptures, but its high iron content makes it susceptible to rust stains. It is also ground up to make road gravel, especially in the western United States, and it is used as decorative stone in yards and gardens.

John L. Berkley

FURTHER READING

Ballard, Robert D. *Exploring Our Living Planet.* Rev. ed. Washington, D.C.: National Geographic Society, 1993. Covers every aspect of the earth's volcanic and tectonic features; lavishly illustrated with color photographs, illustrations, and diagrams. The sections on "spreading" and "hot spots" largely deal with basalt volcanism and its relationship to plate tectonic theory. Well written and indexed, the text will be easily understood and appreciated by specialists and laypersons alike.

Decker, Robert, and Barbara Decker. *Volcanoes.* 4th ed. New York: W. H. Freeman, 2005. Gives a comprehensive treatment of volcanic phenomena. Illustrated with numerous black-and-white photographs and diagrams. Chapters 1, 2, 3, and 6 deal almost exclusively with basalt volcanism. The last four chapters deal with human aspects of volcanic phenomena, such as the obtaining of energy and raw materials, and the effect of volcanic eruptions on weather. Includes an excellent chapter-by-chapter bibliography. Suitable for high school and college students.

Faure, Gunter. *Origin of Igneous Rocks: The Isotopic Evidence.* New York: Springer-Verlag, 2010. Discusses the chemical properties of igneous rocks and the isotopes within these rock formations. Specific locations of igneous rock formations and the origins of these rocks are provided. Includes many diagrams and drawings, along with an overview of isotope geochemistry in the first chapter, to make this accessible to professionals and undergraduate students.

Hofmann, W. "2.03: Sampling Mantle Heterogeneity Through Oceanic Basalts: Isotopes and Trace Elements." In *Treatise on Geochemistry.* Volume 2. Edited by K. K. Turekian and H. D. Holland. San Diego: Elsevier Inc., 2003. Part of a ten-volume reference set on geochemistry. Chemical composition and trace elements of midocean ridge and seamount basalts are discussed. Written in a highly technical manner, requiring prior knowledge and understanding of geochemistry fundamentals. Early sections of this chapter do provide some background information on isotopes and trace elements.

Lewis, Thomas A., ed. *Volcano.* Alexandria, Va.: Time-Life Books, 1982. Part of the Planet Earth series; written with the nonspecialist in mind. Wonderful color photographs, well-conceived color diagrams, and a readable narrative guide the reader through the world of volcanism. Describes past eruptions and their effects on humankind. Basalt is covered mainly in the chapter on Hawaii and the chapter on Heimaey, Iceland. Has a surprisingly extensive bibliography and index for a book of this kind.

Lutgens, Frederick K., Edward J. Tarbuck, and Dennis Tasa. *Earth: An Introduction to Physical Geology.* 10th ed. Upper Saddle River, N.J.: Prentice Hall, 2010. Provides a clear picture of the earth's systems and processes suitable for high school or college readers. Offers an accompanying computer disc that is compatible with either Macintosh or Windows. Bibliography and index.

Macdonald, Gordon A. *Volcanoes.* Englewood Cliffs, N.J.: Prentice-Hall, 1972. Written by one of the premier volcanologists in the world; ideal for those desiring a serious but not overly technical treatment. Covers every conceivable aspect of volcanic phenomena, but the sections on basalt (particularly as it occurs in Hawaii) are particularly good. Includes suggested readings, a comprehensive list of references, a very good index, and an appendix that lists the active volcanoes of the world.

Perchuk, L. L., ed. *Progress in Metamorphic and Magmatic Petrology.* New York: Cambridge University Press, 1991. Although intended for the advanced reader, several of the essays in this multiauthored volume will serve to familiarize the new student with the study of igneous rocks. In addition, the bibliography will lead the reader to other useful material.

Philpotts, Anthony, and Jay Aque. *Principles of Igneous and Metamorphic Petrology.* 2d ed. New York: Cambridge University Press, 2009. An easily accessible text for students of geology. Discusses igneous rock formations of flood basalts and calderas. Covers processes and characteristics of igneous and metamorphic rock.

Putnam, William C. *Geology.* 2d ed. New York: Oxford University Press, 1971. A comprehensive and accessible classic text. Covers fundamental topics still considered relevant. Chapter 4, "Igneous Rocks and Igneous Processes," uses a vivid description of the 1883 eruption of Krakatau as a way of introducing the formation processes of igneous rocks. Other famous, historic volcanic eruptions are also discussed. Describes rock classification and composition in detail. The chapter concludes with a list of references. Illustrated.

Ross, Pierre-Simon, et al. "Basaltic to Andesitic Volcaniclastic Rocks in the Blake River Group, Abitibi Greenstone Belt: 2. Origin, Geochemistry, and Geochronology." *Canadian Journal of Earth Sciences* 48 (2011): 757-777. Discusses the rock formations in the Archean Blake River Group. Physical characteristics, age relationships, and geochemistry data are measured and interpreted to determine the origins of these rock formations. The authors hypothesize the volcanic processes that resulted in these formations. Best suited for graduate students or researchers studying volcanism, petrology, or mineralogy.

Science Mate: Plate Tectonic Cycle. Fremont, Calif.: Math/Science Nucleus, 1990. Part of the Integrating Science, Math, and Technology series of manuals, intended to help teachers explain basic scientific concepts to elementary students. Provides a basic understanding of plate tectonics, earthquakes, volcanoes, and general geology for the student with no background in Earth sciences.

Sobolev, A. V., et al. "The Amount of Recycled Crust in Sources of Mantle-Derived Melts." *Science* 316 (2007): 412-417. Discusses basaltic crust mixing with peridotitic mantle due to plate tectonics processes. Examines the chemical composition of melts.

Sutherland, Lin. *The Volcanic Earth: Volcanoes and Plate Tectonics, Past, Present, and Future.* Sydney, Australia: University of New South Wales Press, 1995. Provides an easily understood overview of volcanic and tectonic processes, including the role of igneous rocks. Includes color maps and illustrations, as well as a bibliography.

Tarbuck, Edward J., Frederick K. Lutgens, and Dennis Tasa. *Earth: An Introduction to Physical Geology.* 10th ed. Upper Saddle River, N.J.: Prentice Hall, 2010. Aimed at the reader with little or no college-level science experience. Includes a chapter devoted to igneous rocks and their textures, mineral compositions, classification, and formation. Illustrated with photographs and diagrams. Includes review questions and list of key terms.

See also: Andesitic Rocks; Anorthosites; Batholiths; Building Stone; Carbonatites; Feldspars; Granitic Rocks; Igneous Rock Bodies; Industrial Metals; Iron Deposits; Kimberlites; Komatiites; Orbicular Rocks; Plutonic Rocks; Pyroclastic Rocks; Rocks: Physical Properties; Xenoliths.

BATHOLITHS

Batholiths are gigantic bodies of granitic rock located in mobile belts surrounding the ancient cores of the continents. The growth of continental crust during the past 2.5 billion years is intimately related to the origin and emplacement of major volumes of granitic magma that solidify as batholiths.

PRINCIPAL TERMS

- **crystallization:** the solidification of molten rock as a result of heat loss; slow heat loss results in the growth of crystals, but rapid heat loss can cause glass to form
- **granitic/granitoid:** descriptive terms for plutonic rock types having quartz and feldspar as major mineral phases
- **I-type granitoid:** granitic rock formed from magma generated by partial melting of igneous rocks in the upper mantle or lowermost crust
- **magma:** molten rock material that crystallizes to form igneous rocks
- **migmatite:** a rock exhibiting both metamorphic and plutonic textural traits
- **mobile belt:** a linear belt of igneous and deformed metamorphic rocks produced by plate collision at a continental margin; relatively young mobile belts form major mountain ranges; synonymous with orogenic belt
- **partial melting:** a process undergone by rocks as their temperature rises and metamorphism occurs; magmas are derived by the partial melting of preexisting rock; also known as ultrametamorphism or anatexis
- **pluton:** a generic term for an igneous body that solidifies well below the earth's surface; plutonic rocks are coarse-grained because they cool slowly
- **S-type granitoid:** granitic rock formed from magma generated by partial melting of sedimentary rocks within the crust

BATHOLITH FORMATION

Batholiths are large composite masses of granitoid rock formed by numerous individual bodies of magma that have risen from deep source areas in molten form and solidified near enough to the surface to be exposed by erosion. The resulting rocks are relatively coarse-grained in texture and markedly heterogeneous in chemical and mineralogical composition. A well-studied example is the coastal batholith of Peru, which forms an almost continuous outcrop 1,100 kilometers long and 50 kilometers wide along the western flank of the Andes. This enormous body has steep walls and a flat roof. It is composed of more than one thousand individual plutons emplaced along a narrow belt parallel to the present coastline during a volcanic-plutonic event that extended over a period of 70 million years. Many such batholiths are known in the mountainous areas of the world, but few are as large or as magnificently exposed to view as that in Peru. Geological glossaries often define a batholith as a "coarse-textured igneous mass with an exposed surface area in excess of 100 square kilometers." This description has the virtue of simplicity, but it is misleading; the description encompasses granitic plutons, and even nongranitic plutons, which form under conditions quite removed from those associated with the world's major batholiths.

In most instances, there is evidence to indicate that the individual plutons of a batholith were emplaced as hot, viscous melts containing suspended crystals. This molten material is called magma. Cooling and crystallization occur during the ascent of magma toward the surface and gradually transform it to solid rock, which prevents further upward movement. The depth at which total solidification occurs varies and is strongly dependent upon the initial temperature and water content of the magma. Extreme levels of ascent, within 3 to 5 kilometers of the surface, are possible only for very hot magmas with very low initial water contents. Most granitic plutons complete their crystallization at depths in the range of 8 to 20 kilometers. The characteristic coarse textures observed in most granitic plutons are the result of slow cooling, which, in turn, implies that the rate of magma ascent is also slow. These traits distinguish plutonic rocks from their volcanic counterparts. As would be expected, the formation of a batholith is a complex and lengthy event that is the sum of the processes responsible for the emplacement of each member pluton. Each member pluton has an individual history involving the generation of magma in the source region, ascent and partial crystallization, the physical displacement of overlying solid rock, chemical interaction with the solid rocks encountered during ascent, and the terminal crystallization phase. Consequently, each member pluton of a batholith can be expected to exhibit a unique combination of textural, mineralogical, and chemical variations.

MOBILE BELTS

It has long been recognized that major batholiths are confined to narrow zones elongated parallel to present or former continental margins. In such zones, granitic melts intrude either thick sequences of chemically related volcanic rocks or highly deformed and metamorphosed sedimentary rocks. These granite-dominated zones are called mobile belts. The ancient cores of continents are all more than 2.5 billion years old. They are surrounded by mobile belts that become successively younger away from the core. The most recent mobile belts form major mountain chains along continental margins. The resulting age pattern clearly shows that continents grow larger with time by the marginal accretion of mobile belts. In the late 1960's, the emergence of plate tectonic theory provided a basis for understanding how mobile belts form and are accreted to preexisting continent margins. The impetus provided by this theory sparked intensive study of the world's mobile belts. These studies amply show that logical time-space relationships exist between plate collisions, deformation styles, and rock types that occur in mobile belts. Two distinct types of mobile belts are now recognized, and each is dominated by granitic batholiths. These batholiths, however, are very different in terms of granitic rock types, modes of pluton emplacement, rock associations, metamorphic effects, and the metallic ores they host. The batholiths of the two mobile belt types are called I-type and S-type batholiths. I-type batholiths are those that seem to have derived mostly from igneous sources—that is, by melting of the mantle—and S-type batholiths seem to have derived from remelting of metamorphic rocks that were originally sedimentary.

Mobile belts along the eastern margin of the Pacific Ocean contain I-type batholiths exclusively. Their size and collective volume is staggering. The Peruvian batholith is an example already mentioned. Others of this type include the Sierra Nevada batholith, the Idaho batholith, and the tremendous Coast Range batholith, which extends from northern Washington to the Alaska-Yukon border. In contrast, the western margin of the Pacific Ocean is dominated by mobile belts with S-type batholiths, although some I-types are also present. The batholiths of Western Europe are also mainly of the S-type. In southeastern Australia, where the two types of batholiths were first recognized, S-type and I-type granitoids form a paired belt parallel to the coastline. Although their geographical distribution is uneven, both I-type and S-type batholiths occur worldwide.

I-TYPE BATHOLITHS

The most distinctive trait of I-type batholiths is the broad range of granitic rock types they contain. In these batholiths, the rock types gabbro-diorite, quartz diorite-granodiorite, and granite occur in the approximate proportions of 15:50:35. This means that quartz diorite (also called tonalite) and granodiorite jointly compose 50 percent of I-type batholiths, and true granite is a subordinate component in them. This wide compositional spectrum not only characterizes an entire I-type batholith but also is typical of the individual member plutons. Usually the major plutons are concentrically zoned with small central cores of true granite enveloped by extensive zones of granodiorite, which grade outward into margins of quartz diorite. Small plutons in the compositional range of gabbro-diorite are common but subordinate to the zoned granitic bodies. Most member plutons of I-type batholiths have domal or cylindrical shapes and very steep contacts with the surrounding rock. Others may have a steeply tilted sheetlike form, but regardless of shape most I-type plutons cut through the preexisting rock layers at a steep angle. The emplacement of these plutons appears to be controlled by near-vertical fractures that may extend downward to the base of the crust. In younger I-type batholiths such as that in Peru, the granitoids have intruded into a roof of chemically related volcanic rocks that show the same compositional spectrum as the granitic plutons. This volcanic pile, dominated by andesite, may be 3 to 5 kilometers thick at the time of pluton emplacement. Gradually, this volcanic roof is stripped away by erosion, so that volcanic rocks are generally absent in older, deeply eroded batholiths. The grade of regional metamorphism in the rocks enclosing I-type batholiths is relatively low, and there is little evidence of large-scale horizontal compression or crustal shortening. Structural displacements and the movements of rising plutons are dominantly vertical and typically occur over a time span of 50 to 100 million years.

S-TYPE BATHOLITHS

S-type batholiths contrast with I-types in almost every respect. To begin with, the ratio of gabbro-diorite to quartz diorite-granodiorite to granite is 2:18:80 in S-type batholiths. These plutonic complexes are very much dominated by true granite, and gabbro-diorite plutons are rare or absent. In many cases, S-type granites are the distinctive "two mica granites," which contain both biotite and muscovite, and are frequently associated with major tin and tungsten ore deposits. The batholiths of northern Portugal are typical examples of this association. S-type batholiths, as well as their member plutons, lack the concentric zoning that characterizes I-type plutons. Compositional homogeneity is their trademark. S-type plutons are intruded into thick sequences of regionally metamorphosed sedimentary rocks. The metamorphic grade ranges from moderate to very high, and, frequently, the granites are located within the zone of highest metamorphic grade. In such cases, migmatites are often present. The enclosing metamorphosed rocks are intensely folded in response to marked crustal compression, and volcanic rocks are conspicuously absent. S-type batholiths are smaller in volume and form over a shorter period of time (usually less than 20 million years) than their I-type counterparts.

DIFFERING ORIGINS OF BATHOLITH TYPES

The many contrasting traits of I-type and S-type batholiths are an indication that the conditions of magma generation and emplacement are very different in the mobile belts in which they are found. In the case of I-type batholiths, it appears that magmas are generated at relatively great depths and above the subduction zones that form at destructive plate boundaries. The magmas are derived by partial melting of upper-mantle basic igneous rocks and, perhaps, lower crustal igneous rocks. The melts rise along the steep fractures produced by crustal tension over the subduction zone. The igneous ancestry of these melts is the reason for calling the resulting plutons "I-type." The hottest and driest of these I-type magmas will reach the surface to produce extensive fields of volcanic rocks and large calderas. In some

cases, like the Peruvian batholith, the rise of magma was "passive" in the sense that room was provided for the rising plutons by gravitational subsidence of the overlying roof rock. This is the process of cauldron subsidence. In the case of the Sierra Nevada batholith, however, it appears that I-type magmas were emplaced by "forceful injection." In this process, rising magma makes room for itself by shouldering aside the surrounding solid rock.

Conversely, the evidence suggests that S-type magmas originate by partial melting of metamorphosed sedimentary rocks. This sedimentary parentage of the magmas is the reason for designating them as "S-type." Melting is made possible by dehydration of water-bearing minerals under conditions of intense metamorphism. The frequent presence of migmatites (the complex intermingling of igneous and metamorphic rocks) is evidence for this transition from metamorphic conditions to magmatic conditions. The essential requirements for relatively high-level crustal melting are high temperatures, intense horizontal compression to produce deep sedimentary basins, and a thick pile of sediments to fill these basins. Such conditions are best met in back-arc basins, which form at subduction margins but considerably inland from the volcanic-plutonic environment of I-type batholiths. This may explain why some I-type and S-type batholiths occur in paired belts parallel to a continental margin, as in southeastern Australia. The collision environment that arises when continent meets continent in the terminal stage of subduction may also provide suitable conditions for S-type magma production. The magmas that result will be relatively cool and wet and will not be able to rise far above their zone of melting. During this limited ascent, the S-type magmas tend to assume the shape of a light bulb, with a neck tapering down to the zone of melting. Because of their limited capacity for vertical movement, S-type plutons require no special mechanisms to provide additional space for them.

FIELD STUDY

The study of a batholith begins with the study of its individual member plutons. This always involves fieldwork, laboratory analysis of rock samples returned from the field, and comparison of the resulting data with those obtained from other batholiths. Because of their great size, batholiths present special problems for field study. The most informative studies are those in mountainous areas, such as the Peruvian Andes, where erosion has exposed the batholith roof contact and cut steep canyons between and through individual plutons. High topographic relief is desirable if the geologist is to learn anything about the variation in shape and composition within the plutons.

Study of even a small portion of a major batholith requires several well-trained geologists working intensively during short field seasons over a period of several years. Geologists traverse on foot across and around the individual plutons as topography permits, and they record the textural, mineralogical, and structural features observed on maps or aerial photographs of the area. These maps eventually reveal the overall shape of plutons and the patterns of concentric zoning within them. Special maps are prepared to show the distribution of fractures and flow structures within individual plutons. These features indicate how fluid the magma was at the time of emplacement. Contacts between the plutons and older enclosing rocks are closely examined for deformation effects and evidence of thermal and chemical interaction with the magma. Fragments of older rock engulfed by magma are often preserved in a recrystallized state, and these are scrutinized carefully, since they provide clues as to whether the emplacement process was passive or forceful. As the end of the field season draws near, the mapped plutons are sampled. Large, fresh samples must be collected from each recognized zone of each pluton for subsequent laboratory study. The number of samples collected from a single pluton depends upon its size and homogeneity but is frequently in the range of one hundred to five hundred samples. A smaller number of samples is collected from the host rocks at varying distances from the plutonic contact in order to study the thermal effects produced by the pluton. The field description, identifying number, and exact location of each sample site must be meticulously recorded. If, at the end of the field season, several plutons have been studied and sampled, there may be several thousand rock samples to label, pack securely, and ship to the laboratory, where they will receive further study.

LABORATORY STUDY

At the laboratory, the samples are usually cut in half and labeled in a permanent fashion. One half

of each sample is stored for future reference, and the other is prepared for the laboratory procedures. Paper-thin slices of each sample are glued to glass slides for examination under a petrographic microscope. The microscopist identifies the mineral phases present in each slide and determines the abundance of each. The texture of each rock, as revealed under the microscope, is carefully described and interpreted in terms of crystallization sequence and deformation history. When the microscopic study is complete, certain samples, perhaps fifty to one hundred, are chosen for chemical analysis. Most will be analyzed because they are judged to be representative of major zones of a pluton; a few may be analyzed because they exhibit unusual minerals or some peculiar trait not explained by the microscopic study. If the age of a pluton is not known, a few samples (one to ten) will be shipped to a laboratory that specializes in age determinations by radioisotope methods.

DETERMINING ORIGIN AND EMPLACEMENT

Finally, on the basis of the field observations, microscopic examinations, and chemical data, the investigators will assemble rival hypotheses or scenarios for the origin and emplacement of the plutons that have been studied. Any scenario that conflicts seriously with known facts is discarded. Those remaining are compared with well-known laboratory melting-crystallization experiments on synthetic and natural rock systems. The size and shape of the plutons, as determined by the field mapping, can be compared with those of "model plutons" derived through sophisticated, but idealized, centrifuge experiments in laboratory settings. The investigators will compare their data in detail with data reported in the geological literature by workers in other parts of the world. They will also compare their results with earlier studies of the same plutons, or studies in the same region, if they exist. As additional plutons are studied in detail, more constraints on the mode of origin and emplacement of the batholith are obtained.

ROLE IN CRUSTAL GROWTH

Mobile belts, dominated by immense granitic batholiths, have been systematically accreted to the ancient continental cores for the last 2.5 billion years of Earth history. Modern plate tectonic theory has provided the basis for understanding the periodic nature of mobile belt accretion and the ways in which crustal and mantle materials are recycled. It is evident that the emplacement of batholiths is at present—and has been for at least 2.5 billion years—the major cause for progressive crustal growth. It is also clear that the rate at which batholiths formed during this lengthy period has far exceeded the rate of continental reduction by erosion. The generation of large volumes of granitic magma and its subsequent rise to form batholiths a few kilometers below the crustal surface must be viewed as fundamental to crustal growth. Batholiths play a major role in the formation of mountain systems and are the most important element in the complex rock and metallic ore associations of mobile belts. The very existence of continents is, in fact, a result of the long-standing process of batholith emplacement.

Gary R. Lowell

FURTHER READING

Atherton, Michael P., and J. Tarney, eds. *Origin of Granite Batholiths: Geochemical Evidence.* Orpington, England: Shiva Publishing, 1979. A summary of the views of major authorities on the origin of batholiths. Suitable for advanced students of geology.

Bennison, George M., Paul A. Olver, and A. Keith Moseley. *An Introduction to Geological Structures and Maps.* 8th ed. London: Hodder Education, 2011. Provides fundamental information on geological structures with colorful photographs and maps. Contains a glossary and index.

Best, Myron G. *Igneous and Metamorphic Petrology.* 2d ed. Malden, Mass.: Blackwell Science Ltd., 2003. A popular university text for undergraduate majors in geology. A well-illustrated and fairly detailed treatment of the origin, distribution, and characteristics of igneous and metamorphic rocks. Chapter 4 covers granite plutons and batholiths.

Grotzinger, John, et al. *Understanding Earth.* 5th ed. New York: W. H. Freeman, 2006. Covers the formation and development of the earth. Appropriate for high school students, as well as general readers. Includes an index and glossary of terms.

Hamilton, Warren B., and W. Bradley Meyers. *The Nature of Batholiths.* Professional Paper 554-C. Denver, Colo.: U.S. Geological Survey, 1967. Classic paper proposes the controversial "shallow batholith model." Short but very descriptive account; can be followed by college-level readers. Influenced many geologists to abandon the traditional view of batholiths.

_____. "Nature of the Boulder Batholith of Montana." *Geological Society of America Bulletin* 85 (1974): 365-378. The authors apply their 1967 model to the Boulder batholith and reply to the heated criticism of colleagues who did the fieldwork on this batholith. Aimed at professionals but can be understood by those with moderate knowledge of plutonic processes.

Hill, Mary. *Geology of the Sierra Nevada.* Rev. ed. Berkeley: University of California Press, 2006. Covers fundamental topics in geology with a focus on the Sierra Nevada region. Contains a rock identification key, a table of geological features, and maps. Very clear writing accessible to anyone with some science background and an interest in geology.

Jerram, Dougal, and Nick Petford. *The Field Description of Igneous Rocks.* 2d ed. Hoboken, N.J.: Wiley-Blackwell, 2011. Begins with a description of field skills and methodology. Contains chapters on lava flow and pyroclastic rocks. Designed for student and scientist use in the field.

Judson, S., and M. E. Kauffman. *Physical Geology.* 8th ed. Englewood Cliffs, N.J.: Prentice-Hall, 1990. A traditional text for beginning geology courses. Simplified but suitable for high school readers. Contains a good index, illustrations, and an extensive glossary. Chapter 3 treats igneous processes and rocks. Chapters 4, 7, 9, and 11 examine fundamental processes related to mountain building, metamorphism, volcanism, and plate tectonics.

Klein, Cornelis, and Cornelius S. Hurlbut, Jr. *Manual of Mineralogy.* 23d ed. New York: John Wiley & Sons, 2008. An introductory college-level text on mineralogy. Discusses the physical and chemical properties of minerals and describes the most common minerals and their varieties, including gem mineral varieties. Contains more than 500 mineral name entries in the mineral index and describes in detail about 150 mineral species.

Meyers, J. S. "Cauldron Subsidence and Fluidization: Mechanisms of Intrusion of the Coastal Batholith of Peru into Its Own Volcanic Ejecta." *Geological Society of America Bulletin* 86 (1975): 1209-1220. Possibly the best available account of a major batholith. The excellent cross-section diagrams clearly indicate that the author was influenced by the model of Hamilton and Myers. Aimed at professionals but can be understood by college-level readers with some background in geology.

Perchuk, L. L., ed. *Progress in Metamorphic and Magmatic Petrology*. New York: Cambridge University Press, 1991. Although intended for the advanced reader, several of the essays in this multiauthored volume will serve to familiarize the new student with the study of igneous rocks. In addition, the bibliography will lead the reader to other useful material.

Press, F., and R. Siever. *Earth*. 4th ed. New York: W. H. Freeman, 1986. Chapter 5, "Plutonism," is a more thorough treatment of the subject than that in Judson and Kauffman's text, but requires a slightly higher level of comprehension. This is one of the best university-level texts for a first course in geology.

Smith, David G., ed. *The Cambridge Encyclopedia of Earth Sciences*. New York: Crown, 1981. More of a supertext than an encyclopedia. The authors skillfully place their fields of expertise in the plate tectonic context and provide a modern overview of the entire field of Earth science. Includes comprehensive index and glossary as well as high-quality maps, tables, and photographs. For both general and college-level readers.

Sutherland, Lin. *The Volcanic Earth: Volcanoes and Plate Tectonics, Past, Present, and Future*. Sydney, Australia: University of New South Wales Press, 1995. Provides an easily understood overview of volcanic and tectonic processes, including the role of igneous rocks. Includes color maps and illustrations, as well as a bibliography.

Winter, J. D. *Principles of Igneous and Metamorphic Petrology*. 2d ed. New York: Pearson Education, Inc, 2010. Covers techniques of modern petrology. Recommended for readers with a geological dictionary on hand, as some parts are very technical.

See also: Andesitic Rocks; Anorthosites; Basaltic Rocks; Carbonatites; Granitic Rocks; Igneous Rock Bodies; Kimberlites; Komatiites; Orbicular Rocks; Plutonic Rocks; Pyroclastic Rocks; Rocks: Physical Properties; Ultrapotassic Rock; Xenoliths.

BIOGENIC SEDIMENTARY ROCKS

Biogenic sedimentary rocks represent the accumulation of skeletal material, produced by the biochemical action of organisms, to form limestone, chert, and phosphorites, or the accumulation of plant material to form coal. Biogenic sedimentary rocks are economically important as sources of energy, fertilizers, and certain chemicals, and they contain evidence of former life on Earth.

PRINCIPAL TERMS

- **carbonate rocks:** the general term for rocks containing calcite, aragonite, or dolomite
- **chert:** multicolored, fine-grained sedimentary rock composed of silica and formed in the ocean
- **coal:** dark brown to black sedimentary rock formed from the accumulation of plant material in swampy environments
- **diagenesis:** the chemical, physical, and biological changes undergone by a sediment after its initial deposition
- **limestone:** light gray to black sedimentary rock composed of calcite and formed primarily in the ocean
- **lithification:** the process whereby loose material is transformed into solid rock by compaction or cementation
- **opal:** a form of silica containing a varying proportion of water within the crystal structure
- **phosphorite:** sedimentary rock composed principally of phosphate minerals
- **polymorph:** minerals having the same chemical composition but a different crystal structure

LITHIFICATION AND COALIFICATION

Numerous organisms, particularly those living in the oceans, produce a shell or some type of skeletal material composed of the minerals calcite or aragonite (polymorphs of calcium carbonate), silica (silicon dioxide), or phosphate minerals, principally apatite. After the organism dies and the soft tissue decays, the skeletal material accumulates as sediment. This sediment then undergoes lithification, a process involving compaction and chemical cementation. The time necessary for lithification is highly variable and ranges from a few decades to millions of years. Biogenic sedimentary rocks composed of calcite or aragonite minerals are commonly referred to as limestones; those composed of silica are cherts, and those composed of phosphate minerals are known as phosphorites or sedimentary phosphate deposits. Limestones and cherts may form as the result of direct chemical precipitation of calcium carbonate or silica from a saturated solution, but the primary mode of formation is biochemical. Limestones are the most abundant of the biogenic sedimentary rocks and, economically, the most important. Iron formations, our principal source of iron ore, were also precipitated biogenically early in the earth's history. They are discussed in a separate article.

Coal also is a biogenic sedimentary rock. Most coal forms from the accumulation of woody plant material in anoxic (oxygen-lacking) environments, such as stagnant lakes, swamps, and bogs. The formation of coal is referred to as coalification and represents microbiological, physical, and chemical processes whereby the percentage of carbon is increased and the volatile content decreased. Limestones, cherts, phosphorites, and coal are often closely associated and may occur as alternating beds in a sequence of sedimentary rocks.

PHOSPHORITES

Because of their biological affinities, biogenic sedimentary rocks contain important evidence of former life on Earth, a record of organic evolution (changes in organisms through geologic time), and clues to the past environments in which the rocks were formed. Limestones may serve as reservoir rocks for oil and gas, which form from the chemical breakdown of microscopic organisms that accumulate with the sediment at the bottom of the ocean and in some lakes. Phosphates, which are contained in phosphorites, are one of the chief constituents of fertilizers and are widely used in the chemical industry. Important evidence of past swamp communities, including both plants and animals, has been preserved in coal beds.

Phosphorus is one of the essential elements of life, and when chemically linked with other elements to form phosphate minerals (especially apatite) it becomes one of the major constituents of all vertebrate skeletons and some invertebrate hard parts. Phosphate is a primary nutrient in marine waters and, therefore, controls organic productivity. Phosphorites often occur as nodules, which are highly variable in size and shape; they may be several centimeters in diameter and up to a few meters in length. These nodules are usually rich in vertebrate skeletal debris, especially that of fish, and fecal material. Bedded phosphorites are commonly interbedded with limestones. Thin beds of phosphorite are rich in bones, fish scales, and fecal material. Modern phosphorites are forming in areas where cold, nutrient-rich waters rise from the ocean depths toward the surface, such as off the west coasts of Africa and North and South America.

CHERTS

Cherts, which are occasionally referred to as flint or novaculite, are not nearly as abundant as limestones; however, chert has two properties that have made it one of the most important rocks to humans. Chert has the same chemical composition as quartz but is cryptocrystalline; that is, its crystals are submicroscopic in size. Like quartz, chert is very hard and it tends to break along smooth, curved surfaces. With a little practice, one can produce a sharp edge by chipping a piece of chert with another hard object. Although early humans used numerous rock types for tools and weapons, chert was the most important for the production of arrowheads, knives, and scrapers. Consequently, chert might be considered to have contributed as much to the rise of civilization as did the development of the steam engine. In addition to chert's utility, Precambrian cherts (1.8 to 3.4 billion years old) are interbedded with important iron deposits known as banded iron formations. These cherts contain evidence of some of the earliest life to evolve on Earth.

Chert, which may be any color depending on the presence of impurities, occurs principally as spherical to oblong nodules 2 to 25 centimeters in diameter in limestones and dolostones, or as thin beds 2 to 25 centimeters deep. Nodular cherts are commonly parallel to bedding or stratification. Bedded cherts may extend laterally for great distances and are commonly interbedded with limestones and dolostones. Most geologists believe that nodular chert forms from the replacement of limestone and that bedded chert forms either by the complete replacement of carbonate-rich beds or by the diagenetic alteration of siliceous ooze.

The source of the silica may be the chemical precipitation of silicon dioxide, volcanic ash, or skeletal material. Biogenic silica is opaline, which means that it may contain up to 10 percent water. Only skeletal silica will be considered in this article. The principal biogenic sources of silica for chert are sponges, diatoms, and radiolarians. Skeletal materials from these organisms are common constituents of oceanic sediments. As this siliceous skeletal material accumulates, the opaline silica is diagenetically converted to crystalline opal and reprecipitated as bladed crystals. Continued diagenesis results in the formation of quartz chert, a mosaic of microscopic quartz crystals. The final diagenetic change often obliterates the shape or structure of the original skeletal material.

Sponges, which are abundant in most marine environments, contain as part of their supportive structure microscopic rods of opaline silica called spicules. When the sponge dies, these spicules accumulate and become part of the sediment. During burial, the opaline silica undergoes diagenesis and precipitation of crystalline opal occurs within pores in the sediment. As crystalline opal is converted into quartz chert, chemical replacement of the surrounding sediment (usually calcite) occurs, resulting in the formation of chert nodules.

Radiolarians and diatoms are microscopic organisms with disk-shaped, elongate, or spherical tests (shells) with spines and surface ornamentation composed of opaline silica. Radiolarians occur as part of the marine zooplankton, and diatoms are part of the marine and nonmarine phytoplankton. Radiolarian and diatom oozes accumulate on the deep ocean floor. In time, these beds of silica-rich ooze are diagenetically converted into thin, bedded cherts. Diatoms, in contrast, may occur in great numbers in lakes and accumulate to form diatomaceous earths, or diatomites. Diatomaceous earths have a wide variety of uses, such as in filtering agents, absorbents, and abrasives.

LIMESTONES

Limestones are the most abundant of the chemical sedimentary rocks. A minor amount of limestone forms from the inorganic precipitation of calcite from seawater or the deposition of calcite in caves and around hot springs. The majority of limestones form as the result of biological and biochemical processes that produce aragonite or calcite. Later, this material becomes part of the carbonate sediment. Once deposited, the carbonate sediment is often modified by the chemical and physical processes of diagenesis.

Numerous animal phyla, such as mollusks, brachiopods, echinoderms, bryozoans, coelenterates, and certain protozoans, produce aragonite or calcite as part of their skeletal structure. The skeletal remains of these organisms are important constituents of carbonate sediments and, eventually, of most limestones. Some marine, bottom-dwelling algae secrete aragonite or calcite. These calcareous algae represent a significant contribution to carbonate sediment. Calcareous algae called coccoliths and calcareous foraminifera are found in great numbers in the plankton of the open ocean. Accumulation of this calcareous material at the bottom of the ocean contributes greatly to the formation of chalk, a type of biogenic limestone. Modern marine organisms principally secrete aragonite; consequently, modern calcareous sediments are initially composed predominantly of aragonite. Aragonite, however, is unstable at low pressures and quickly undergoes diagenesis to calcite.

It is not uncommon to find that ancient limestones have been partially or completely transformed into dolostones; this process is known as dolomitization. Dolomitization occurs when calcium carbonate minerals are diagenetically converted into dolomite. Diagenesis may take place soon after the calcite or aragonite has been deposited or a long time after the deposition. The diagenesis is the result of the magnesium-bearing waters (seawater or percolating meteoric water) moving through the carbonate sediment or limestone. One important aspect of dolomitization is that it often leads to the formation of pores, cavities, and fissures, which enable the rock to serve as a reservoir for oil, gas, and water. Recent research suggests that much dolomitization is caused by the circulation of deep, warm fluids long after burial.

Under certain conditions, limestones are relatively easy to dissolve. In areas such as Florida or Kentucky, where limestones are abundant and there is adequate rainfall, cave systems may develop. Regions with extensive cave systems and related features are referred to as karst.

STUDY OF ROCK SAMPLES

Initially, the study of biogenic sedimentary rocks involves field investigations. Where outcrops of these rocks occur, geologists plot and map their distribution and thicknesses; note changes in the rocks' character, both vertically and horizontally; observe their association with other types of rocks; and collect samples of both the rock and any fossils that may be present. When drilling for oil, gas, or water, engineers and geologists study the same phenomena.

Samples, which are brought back to the laboratory, are analyzed with a vast array of techniques, including X-ray analysis, electron microprobe analysis, scanning electron microscopy, cathodoluminescence, and thin-section examination with a polarized microscope. Samples containing large fossils are observed with a magnifying glass or dissecting microscope. Some samples, particularly limestones, are dissolved in acid to recover both megafossils and microfossils. When samples have been ground thin enough to allow light to pass through the rock, microfossils can be observed with the help of microscopes. Scanning electron microscopy is particularly useful for observing extremely small fossils, especially radiolarians, diatoms, and coccoliths.

AREAS OF RESEARCH

In general, the chemistry of biogenic sedimentary rocks is simple; however, diagenesis can alter the structure, texture, and mineralogy of a sediment during its deposition, lithification, and burial. The analysis of diagenetic changes, particularly in limestones, is one of the most important avenues of investigation. Limestones are greatly influenced by diagenetic modifications and undergo changes in sediment size, porosity (the spaces between sediment grains), and mineralogy. Early diagenetic changes include the conversion of unstable aragonite—which is produced by most calcium-carbonate-producing organisms—into calcite and changes in the calcite's magnesium content. (Since the atoms of calcium and magnesium have similar properties, they can often substitute for each other in the crystal structure.) Dolomitization may also occur. The replacement of carbonate minerals by silica, so that chert is formed, is another important aspect of diagenesis. Stages of diagenesis can best be observed by using cathodoluminescence, which bombards a thin slice of rock with an electron beam. This process causes minerals within the rock to luminesce (emit light energy for a short interval after the energy source has been removed). The luminescences indicate various stages of diagenesis and are particularly important in studying the details of cements and crystal growth. X-ray analysis and electron microprobe techniques allow scientists to detect subtle differences in chemistries and the presences of trace elements or rare-earth elements.

Another major area of research is the classification of biogenic sedimentary rocks. The classification of limestones is of primary importance. Limestones have been classified using several criteria, but in general, classifications have focused on chemical and mineralogical composition; fabric features, such as fossils and cements; and special physical parameters, such as porosity. When examining chert, the geoscientist must ask: What was the source of the silica (volcanic ash, skeletal grains, chemical precipitate)? What was the time and rate of conversion of siliceous ooze to chert? What was the environment of deposition?

ECONOMIC AND EVOLUTIONARY SIGNIFICANCE

Biogenic sedimentary rocks are among the most important of all sedimentary rocks. Because of their biological affinities, they have preserved an important record of past life on Earth. In addition, some of these rocks were used for toolmaking by early humans; others are important resources for construction materials, fertilizers, and chemicals. Oil and natural gas, which also have biological affinities, are inseparably linked to biogenic sedimentary rocks.

Although not widespread, biogenic phosphorites are economically important, particularly to the fertilizer and chemical industries. Biogenic chert was of major importance in the past. Early humans used chert, or flint, to make tools and weapons. Biogenic limestone is the most common and widespread biogenic sedimentary rock. Economically, limestones are very important. Hydrocarbons (oil and gas) are commonly recovered from porous limestones and dolostones, and in some areas, limestones are reservoirs for groundwater supplies. Limestones often serve as host rocks for important mineral deposits, such as lead. Limestones are quarried as building stone; crushed to form construction materials, such as gravel; or processed into lime and cement. Biogenic limestones are the most important sedimentary rock containing fossils and, therefore, are the most important record of the evolution of life.

Larry E. Davis

FURTHER READING

Birch, G. F. "Phosphatic Rocks on the Western Margin of South Africa." *Journal of Sedimentary Petrology* 49 (1979): 93-100. Provides a good starting point for a review of phosphorites. Suitable for college-level readers. Available in most university libraries.

Blatt, Harvey, Robert J. Tracy, and Brent Owens. *Petrology: Igneous, Sedimentary, and Metamorphic.* 3d ed. New York: W. H. Freeman, 2005. Provides easily understood background material and has an extensive bibliography for each category of sedimentary rocks.

Boggs, Sam, Jr. *Petrology of Sedimentary Rocks.* New York: Cambridge University Press, 2009. Begins with a chapter explaining the classification of sedimentary rocks. Remaining chapters provide information on different types of sedimentary rocks. Multiple chapters describe siliciclastic rocks and discuss limestones, dolomites, and diagenesis.

Chernicoff, Stanley. *Geology: An Introduction to Physical Geology.* 4th ed. Upper Saddle River, N.J.: Prentice Hall, 2006. Reviews the scientific understanding of geology and surface processes. Includes an online address that provides regular updates on geological events around the globe.

Ham, W. E., ed. *Classification of Carbonate Rocks: A Symposium.* Tulsa, Okla.: American Association of Petroleum Geologists, 1962. This collection of ten papers is the standard reference for limestone classification. Suitable for college students. Available in most university libraries.

McBride, E. F. *Silica in Sediments: Nodular and Bedded Chert.* Tulsa, Okla.: Society of Economic Paleontologists and Mineralogists, 1979. This collection of sixteen papers provides the most complete discussion available of chert. Each paper contains an extensive bibliography. Suitable for college-level readers.

Middleton, Gerard V., ed. *Encyclopedia of Sediments and Sedimentary Rocks.* Dordrecht: Springer, 2003. Cites a vast number of scientists, also listed in the author index. Subjects range from biogenic sedimentary structures to Milankovitch cycles. An index of subjects is provided as well. Designed to cover a broad scope and a degree of detail useful to students, faculty, and geology professionals.

Oldershaw, Cally. *Rocks and Minerals.* New York: DK, 1999. This small, fifty-three-page volume offers color illustrations and is of great use to new students who may be unfamiliar with the rock and mineral types discussed in classes or textbooks.

Prothero, Donald R., and Fred Schwab. *Sedimentary Geology: An Introduction to Sedimentary Rocks and Stratigraphy*. 2d ed. New York: W. H. Freeman, 2003. A thorough treatment of most aspects of sediments and sedimentary rocks. Well illustrated with line drawings and black-and-white photographs, it also contains a comprehensive bibliography. Chapters 11 and 12 focus on carbonate rocks and limestone depositional processes and environments. Suitable for college-level readers.

Reading, H. G., ed. *Sedimentary Environments: Processes, Facies, and Stratigraphy*. Oxford: Blackwell Science, 1996. A good treatment of the study of sedimentary rocks and biogenic sedimentary environments. Suitable for the high school or college student. Well illustrated, with an index and bibliography.

Riding, Robert E., and Stanley M. Awramik. *Microbial Sediments*. Berlin: Springer-Verlag, 2010. A compilation of articles discussing various biological sources of sediment. Each article includes references. Includes subject indexing. Designed for postgraduates and professional researchers.

Tucker, Maurice E. *Sedimentary Rocks in the Field*. 4th ed. New York: John Wiley & Sons, 2011. Presents a concise account of biogenic sedimentary rocks and other sedimentary rocks. Classification of sedimentary rocks is well covered. Depositional environments are only briefly discussed. References are well selected. Suitable for undergraduates.

See also: Chemical Precipitates; Diagenesis; Hydrothermal Mineralization; Limestone; Metasomatism; Regional Metamorphism; Sedimentary Rock Classification.

BIOPYRIBOLES

Biopyriboles are important rock-forming minerals, third in abundance only to feldspars and quartz. They are especially abundant in igneous and metamorphic rocks. Important groups of biopyriboles include micas, pyroxenes, and amphiboles.

PRINCIPAL TERMS

- **amphiboles:** a group of generally dark-colored, double-chain silicates crystallizing largely in the orthorhombic or monoclinic systems and possessing good cleavage in two directions intersecting at angles of about 56 and 124 degrees
- **chain silicates:** a group of silicates characterized by joining of silica tetrahedra into linear single or double chains alternating with chains of other structures; also known as "inosilicate"
- **cleavage:** the tendency of a mineral or chemical compound to break along smooth surfaces parallel to each other and across atomic or molecular bonds of weaker strength
- **crystal system:** one of any of six crystal groups defined on the basis of length and angular relationship of the associated axes
- **micas:** a group of complex, hydrous sheet silicates crystallizing largely in the monoclinic system and possessing pearly, elastic sheets with perfect one-directional cleavage
- **monoclinic:** a crystal system possessing three axes of symmetry, generally of unequal length; two axes are inclined to each other obliquely, and the third is at right angles to the plane formed by the other two
- **orthorhombic:** referring to a crystal system possessing three axes of symmetry that are of unequal length and that intersect at right angles
- **pyroxenes:** a group of generally dark-colored, single-chain silicates crystallizing largely in the orthorhombic or monoclinic systems and possessing good cleavage in two directions intersecting at angles of about 87 and 93 degrees
- **sheet silicates:** a group of silicates characterized by the sharing of three of the four oxygen atoms in each silica tetrahedron with neighboring tetrahedra and the fourth with other atoms in adjacent structures to form flat sheets; also known as "phyllosilicates" or "layer" silicates
- **silicate:** a chemical compound or mineral whose crystal structure possesses silica tetrahedra (a structure formed by four charged oxygen atoms surrounding a charged silicon atom)

CHEMICAL STRUCTURE AND COMPOSITION

Biopyriboles include numerous but related groups of minerals. They are important constituents of both igneous rocks, whose minerals largely form as a result of cooling and crystallization of a melt (liquid), and metamorphic rocks, whose minerals are largely crystallized in a solid state at elevated temperatures or pressures. Some biopyriboles also occur as fragments in sedimentary rocks and sediments. The three most important mineral groups in biopyriboles are micas, pyroxenes, and amphiboles.

Biopyriboles are silicate minerals, meaning that they are composed of atoms of the chemical elements silicon and oxygen as well as atoms of other chemical elements. The silicon atom is surrounded by four oxygen atoms attached or bonded to it. This forms a structure known as a silica tetrahedron, which can be represented as a four-sided solid with triangular faces. This silicon-oxygen bond is very strong. These silica tetrahedra may be repeated in various ways; biopyriboles are expressed by the chain (line) and sheet (plane) structures. Pyroxenes and amphiboles are examples of chain silicates and micas of sheet silicates.

Other atoms may also occur sandwiched between oxygen atoms as chains or sheets. In some cases, some of the oxygen atoms are combined with hydrogen to form a hydroxyl group. Atoms of iron, magnesium, or other elements occur in openings bounded above and below by triangles of oxygen atoms. The enclosing oxygen atoms define a shape with eight triangular sides called an octahedron, so these sheets are called octahedral sheets. These octahedral chains or sheets alternate with the chains or sheets of silica tetrahedra. In sheet silicate groups, the sheet or plane of silica tetrahedra might alternate with aluminum bonded to its oxygens. For some sheet silicate minerals, other layers (potassium atoms bonded to oxygen or hydroxyl, for example) may be layered within. In chains, silica tetrahedra may alternate with structures of iron atoms with attached oxygens. In some groups (amphiboles), paired chains of silica tetrahedra alternate with octahedral chains.

CRYSTALS AND CRYSTALLINE MASSES

Biopyriboles are found as crystals or as crystalline masses. Crystals may be defined as inorganic or organic solids that are chemical elements or compounds. They are formed by growth and are bounded by faces or surfaces with a definite geometric relationship to one another. This relationship reflects the orderly internal arrangement of atoms and molecules. Crystalline masses are intergrowths of crystals that show incomplete expression of external faces, generally because there was either not enough space or some substance in the solution or melt inhibited crystal growth.

Crystals belong to systems, based on the relationship of three lines or axes to each other. Most micas, pyroxenes, and amphiboles fall into the orthorhombic and monoclinic systems under normal geologic conditions in the earth's crust (a very few are hexagonal or triclinic).

MICAS AND BRITTLE MICAS

The micas are one of the most common groups of rock-forming minerals. They are characterized by the following properties: the formation of thin, sheetlike crystals, which are stronger within the sheets than between them; a tendency to break in one smooth direction parallel to the sheets, a property called "cleavage," crystallization in the monoclinic system for most species; considerable elasticity; and a luster that is vitreous (glasslike) to pearly. Two of the most common members of this group are biotite and muscovite mica.

Biotite mica occurs in black, brown, or dark green flakes and is a potassium-magnesium-iron-aluminum hydroxyl silicate. The magnesium and iron atoms can substitute for each other in the octahedral layers (which alternate with a silica tetrahedron and potassium-hydroxyl layer), because charged atoms (ions) of magnesium and iron have nearly the same size as well as the same charge. The mineral occurs widely in most igneous and metamorphic rocks but is most abundant in granite, which is composed of interlocking crystals of feldspar, quartz, and, commonly, mica. It is also abundant in the coarsely foliated metamorphic rock, mica schist.

Muscovite mica occurs in clear to smoky yellowish, greenish, or reddish flakes and is a potassium-aluminum hydroxyl silicate. Aluminum occurs both in the octahedral layer and (substituting for silica) in the tetrahedral layer. Muscovite mica is especially abundant in granite and occurs in large sheets and crystals in the very coarse rock granite pegmatite. It is also prominent in mica schist and occurs microscopically in slate. Flakes of muscovite can occur as fragments in some sandstones, which are sedimentary rocks.

The brittle mica group is less common, but consists largely of hydrous calcium-bearing, iron-rich, or aluminous sheet silicates. Samples are characteristically easily broken across sheets—hence the term "brittle mica." One aluminous variety, margarite, occurs in attractive lilac or yellow crystals associated with corundum (aluminum oxide) or as veins in chlorite schist. Another common brittle mica is glauconite, a blue-green mineral abundant in some sandstones.

CHLORITES

The chlorite group is similar to the micas in that these minerals also tend to have sheetlike crystals that show cleavage parallel to the sheets and possess a pearly luster. The color of chlorite flakes is commonly dark to bright green, although it can also occur in brown, pink, purple, and colorless crystals. The potassium, sodium, and lithium that are present in mica are absent in chlorite. Its crystals are flexible but not elastic.

Chlorite may occur as fine-grained masses in clay or clay stone. The largest crystals occur in altered ultramafic rocks, such as serpentinite, and in the metamorphic rock chlorite schist. Varieties of this mineral group also may occur in association with metallic ore deposits. Chlorite differs in the arrangement and number of octahedral layers from the related serpentine and talc groups.

SERPENTINES

The serpentine mineral group is hydrous and somewhat complex. It includes magnesium-rich sheet silicates, which commonly occur as an alteration of the magnesium-iron silicates olivine and pyroxene in altered ultramafic rocks known as metaperidotites. Serpentines may also form as an alteration of olivine (forsterite) pyroxene or other minerals in marble. The various kinds vary from rather soft to moderately hard. The most common are antigorite, with a platy-massive structure, and chrysotile asbestos, which is fibrous. The fibrous structure results because the sheets that make up its structure do not match perfectly in atomic spacing, and the sheets curl into spirals or tubes. Serpentines are most commonly green, but varieties of brown, red, blue and black are known. They are economically important in some areas and occur in mountain belts such as the Alps and the Appalachian Piedmont.

TALCS

The talc group consists of sheet silicates that are rich in magnesium, aluminum, or iron. They are characterized by a very pearly luster, great softness—so they can be scratched with a fingernail—and a soapy or greasy feel; they occur in thin sheets or in scaly or radiated masses. Talc is a hydrous magnesium silicate that can form through the alteration of olivine, pyroxene, or serpentine. It is common in altered ultramafic rocks; it also occurs as talc schist and in some marbles. Pyrophyllite, the aluminum-rich analogue of talc, occurs in schists or through metamorphic alteration of aluminous rocks. The iron-rich analogue minnesotaite occurs in metamorphosed iron formations.

CLAY MINERALS

Clay minerals are of the most widespread group of sheet silicates. They include chlorite, kaolinite (china clay), smectite (also known as montmorillonite), and illite. Clays can occur as mixed sheets of chlorite, illite, and smectite.

Kaolinite is usually white and soft, and is a hydrous aluminum silicate prized for use in china. It consists of silica tetrahedron layers alternating with aluminum hydroxide octahedral layers. Smectite is unusual: It has a layer that takes up water or liquid organic molecules, and its mineral structure is therefore expandable. Illite has a structure similar to that of muscovite mica, but some of the potassium ions are replaced by hydroxyl. It is sometimes included in the mica group.

All these minerals may be important constituents of soils and sedimentary rocks (especially chlorite, illite, and smectite). Clay stones and mudstones are largely made up of clay minerals. Other members of this group occur associated with metallic ore deposits and hot springs or geyser areas.

PYROXENES

The chain silicate group includes orthopyroxenes, clinopyroxenes, pyroxenoids, orthoamphiboles, and clinoamphiboles. The pyroxene group is characterized by single silica tetrahedron chains alternating with octahedron chains and other chains with cubic structures. The octahedron chains tend to have smaller internal atoms than those of the cubic chains. Octahedron chain atoms include magnesium, iron, or aluminum, and cubic chain atoms include calcium and sodium. Sometimes the larger atoms such as calcium are called X-type, and the smaller atoms such as magnesium are called Y-type. Other letter classifications may be used by crystallographers. Pyroxenes have two directions of smooth breakage (cleavage) nearly at right angles to each other. Most pyroxenes are moderately hard and tend to be green, brown, or green-black in color.

Orthorhombic pyroxenes (also called orthopyroxenes) occur mostly in dark-colored high-temperature igneous rocks such as pyroxenites and gabbros. They range from light bronze-brown to dark green-brown in color and may have a bronzy luster. In this group, ferrous iron atoms can substitute freely for the magnesium atoms in the structure. Two important orthopyroxenes are enstatite and hypersthene.

Monoclinic pyroxenes (also called clinopyroxenes) occur in dark-colored igneous rocks and in siliceous or aluminous marbles metamorphosed at high temperatures. There are two main groups: the calcium-rich or calcic types and the sodium-rich or alkali types. Substitution of sodium and calcium atoms can occur to a certain extent between the two types. The calcic pyroxenes are usually green or brown; sodic pyroxenes may be light or bright green or, if iron-rich, blue-black to black. Common examples of calcic clinopyroxenes are diopside and augite. Magnesium and ferrous iron atoms substitute freely. Two important sodium-alkali pyroxenes are jadeite, an important carving and gem material, and aegirine.

Pyroxenoids are similar in some respects to pyroxenes; the former are also high-temperature minerals, but differ in that octahedral and cubic chains both have Y-type atoms, so that monoclinic or even triclinic structures result. They also tend to be more tabular and less blocky than are pyroxenes in their structure. In many pyroxenoids the silicate chains are twisted so the minerals do not cleave like normal pyroxenes. Wollastonite, a calcium silicate found in siliceous marbles, is an important example of this group.

AMPHIBOLES

Minerals having double silica tetrahedral chains are called amphiboles. Amphiboles are distinguished from pyroxenes, which are closely similar in color, hardness, and occurrence, by the two directions of cleavage, 124 and 56 degrees instead of close to 90 degrees, as is the case for pyroxenes. Orthorhombic amphiboles (orthoamphiboles) include the magnesium-iron amphibole anthophyllite and the aluminum-magnesium-iron amphibole gedrite. These types are restricted largely to magnesium-rich, calcium-poor metamorphosed ultramafic and mafic plutonic and volcanic rocks. Anthophyllite-gedrite varies from purple or clove-brown to yellow-brown or gray in color and may be columnar or fibrous in structure. Amphibole minerals make up one variety of asbestos. One group of clinoamphiboles, monoclinic amphiboles, is a magnesium-iron silicate similar to the anthophyllite group. The magnesium-rich end member is called cummingtonite, and the iron-rich member is known as grunerite. Cummingtonite usually occurs as fibrous or radiating crystals and is brown to gray in color. The largest group of clinoamphiboles is the calcic amphiboles. In all the calcic amphiboles, the amount of calcium exceeds that of alkalis (sodium and potassium). In the tremolite group, which is analogous to the diopside group in the clinopyroxenes, calcium and magnesium, or ferrous iron, are the major constituents of the cubic and octahedral chains, respectively, although sodium-rich tremolites occur. These amphiboles vary from colorless to green and occur mostly in marble and metamorphosed dark igneous rocks. The hornblende group of calcic amphiboles, which is somewhat analogous to the augite group of clinopyroxenes, can have a much more varied composition: Some sodium may substitute for calcium, and aluminum and ferric iron may substitute for magnesium and ferrous iron in the octahedral chains and aluminum for silica in the tetrahedral chains. Hornblendes are commonly black to dark green and occur in a wide array of igneous and metamorphic rocks.

The alkali amphiboles are rich in sodium (or, very rarely, potassium), and most range in color from dark to light blue, or violet to blue-black. Important members are the sodium-magnesium-iron-aluminum amphiboles riebeckite (blue) and glaucophane (blue to violet). Glaucophane is an important constituent of blueschist, which is formed at high pressures and is especially common in some parts of California and Japan.

Complex biopyriboles consist of combined anthophyllite-talc structures. First described from Vermont, they occur elsewhere as well. Other combinations of amphibole-talc structures are possible. Multiple chain units are characteristic of this group.

METHODS OF STUDY AND ANALYSIS

Many methods have been used to study and analyze biopyriboles. Simple physical techniques can determine color, cleavage directions, crystal form, hardness (resistance to abrasion), density, and other properties. The major biopyribole groups and the more common or distinctive kinds of micas, amphiboles, and pyroxenes can be identified with such techniques. Examination under a binocular or compound microscope can extend this process to smaller grains or crystals. Association of other minerals rich in certain elements may also be helpful in identification.

Polarizing microscopes are more powerful tools that force light to travel in a certain direction through a sample by means of polarizers, producing interference and refraction effects (bending) in the light. The chemical makeup of many pyroxenes can be studied in this way, but the more complex amphiboles and sheet silicates require more sophisticated methods.

In X-ray diffraction, X rays are generated by electron bombardment and produce multiple reflections off atomic planes in crystals, allowing for the determination of the dimensions of these planes and, therefore, the identification of the mineral. Micas, clays, serpentines, and other sheet silicates are often readily differentiated and analyzed by X-ray diffraction. Special cameras for X-ray diffraction permit the study of structure, mineral unit cell dimensions, and atomic position of the elements.

Scanning electron microscopy produces an electron photograph of the surface of a fine-grained material or small crystals. In conjunction with X-ray diffraction, this method is necessary for the unequivocal identification of fine-crystalline clay and serpentine group minerals. Transmission electron microscopy permits resolution to a few angstroms (atomic dimensions), thus allowing direct studies of mineral structure. This method is necessary for studying the detailed molecular structure of chain silicates and complex biopyriboles.

The electron microprobe, useful for chemical analysis, focuses an intense beam of electrons on some coated material (usually gold or carbon); the material then emits characteristic X rays, whose wavelength and intensity can be examined with an X-ray spectroscope. Through calculations, and with adequate corrections applied, an analysis can be produced, provided there is a mineral standard for comparison.

Differential thermal analysis uses a thermocouple method for measuring temperature differences between the material being tested and a standard material. A useful method for detailing heat-absorbing (endothermic) dehydration reactions for minerals, especially for clay minerals and sheet silicates, thermal analysis aids in identification and structural analysis.

INDUSTRIAL USES

Biopyriboles are important constituents of rocks in both the crust and the mantle of the earth. Some of the rocks containing pyroxenes and amphiboles, such as traprock and diorite, are used for road and railroad gravels, building stone, and monuments. Clay minerals, especially kaolinite and a hydrated type called halloysite, are used to make fine china and pottery, and are a constituent of ceramics, brick, drain tile, and sewer pipe. Kaolinite is also used as a filter in medical research and a filler in paper. Bentonite (smectite or montmorillonite) is used in drilling muds that support the bit and drilling apparatus in oil exploration.

Muscovite mica has been used as an electric insulating material and a material for wallpaper, lubricants, and nonconductors. Lepidolite is a source of lithium and is used in the manufacture of heat-resistant glass. Talc is highly important in the cosmetics industry. As the massive variety soapstone, talc is used for tabletops and in paint, ceramics, paper, and insecticides. Pyrophyllite, the aluminum analogue, is used for the same purposes. Serpentine has been used as an ornamental and building stone. The chrysotile variety is the main source of asbestos, which has been used in the past or fireproof fabrics and construction material. Fibrous varieties of anthophyllite (also called amosite), tremolite, and riebeckite (also called crocidolite) were also used in the past as sources of asbestos. Health considerations have largely forced the discontinuance of its manufacture and use.

Pyroxenes are not so widely used, but clear and transparent colored varieties of diopside and spodumene have been used as gemstones, and both jadeite and rhodonite are prized gem materials for carving. Spodumene is also a major source of lithium for ceramics, batteries, welding flux, fuels, and the compound lithium carbonate, used to treat persons with manic depression.

David F. Hess

FURTHER READING

Cepeda, Joseph C. *Introduction to Minerals and Rocks.* New York: Macmillan, 1994. Provides a good introduction to biopyriboles for students just beginning their studies in Earth sciences. Includes illustrations and maps.

Deer, W. A., R. A. Howie, and J. Zussman. *An Introduction to the Rock-Forming Minerals.* 2d ed. London: Pearson Education Limited, 1992. Discusses in detail the crystallography, properties, chemistry, occurrence, and origin of the pyroxenes and amphiboles, as well as micas and other sheet silicates. Suitable for college-level students.

Ferraris, Giovanni, Emil Makovicky, and Stefano Merlino. *Crystallography of Modular Materials.* New York: Oxford University Press, 2004. Contains advanced discussions of crystal structure. Includes discussion of OD structures, polytypes, and modularity. Provides a long list of references.

Klein, Cornelis, and C. S. Hurlbut, Jr. *Manual of Mineralogy.* 23d ed. New York: John Wiley & Sons, 2008. A general text of mineralogy, revised many times since James D. Dana first published the prototype in 1862. Extremely useful, it lists properties, occurrences, and uses of the amphiboles, pyroxenes, and sheet silicates, and gives background information in crystallography and descriptive mineralogy. Suitable for the high school and college reader.

Konishi, Hiromi, Reijo Alviola, and Peter R. Buseck. "2111 Biopyribole Intermediate Between Pyroxene and Amphibole: Artifact or Natural Product?" *American Mineralogist* 89 (2004): 15-19. Discusses the discovery of a naturally occurring unique pyribole chain and its laboratory synthesis. The authors comment on the probable mode of formation, using the results from these studies to interpret the formation history of geological structures.

Leake, Bernard E. "Nomenclature of Amphiboles." *American Mineralogist* 63 (November, 1978): 1023-1052. A complete but rather technical description of amphiboles and their classification. Appropriate for the college-level reader.

Oldershaw, Cally. *Rocks and Minerals*. New York: DK, 1999. This small, fifty-three-page volume is filled with color illustrations and is therefore of great use to new students who may be unfamiliar with the rock and mineral types discussed in classes or textbooks.

Robinson, George. "Amphiboles: A Closer Look." *Rocks and Minerals* (November/December, 1981). An excellent and readable summary of the properties and occurrence of this complex group. Suitable for both the high school and college reader.

Veblen, D. R. "Biopyriboles." In *McGraw-Hill Encyclopedia of Science and Technology*. 10th ed. New York: McGraw-Hill, 2007. This article provides a good overview of biopyribole formation and structure. Recommended as a prerequisite for students with limited knowledge of the topic. Provides the fundamental concepts needed to understand advanced articles on pyribole formation.

See also: Basaltic Rocks; Carbonates; Clays and Clay Minerals; Feldspars; Gem Minerals; Granitic Rocks; Hydrothermal Mineralization; Metamictization; Minerals: Physical Properties; Minerals: Structure; Nonsilicates; Orthosilicates; Radioactive Minerals.

BLUESCHISTS

Blueschists are a class of metamorphic rocks that recrystallize at depths of 10 to 30 kilometers or more in subduction zones. Blueschists are important because they contain minerals indicating that metamorphism occurred under conditions of unusually high confining pressures and low temperatures. Their presence in mountain belts is the primary criterion for recognizing ancient subduction zones.

PRINCIPAL TERMS

- **accretionary prism:** the complexly deformed rocks in a subduction zone that are scraped off the descending plate or eroded off the overriding plate
- **geotherm:** a curve on a temperature-depth graph that describes how temperature changes in the subsurface
- **lithosphere:** the outer rigid shell of the earth that forms the tectonic plates, whose movement causes earthquakes, volcanoes, and mountain building
- **metamorphism:** the alteration of the mineralogy and texture of rocks because of changes in pressure and temperature conditions or chemically active fluids
- **prograde:** metamorphic changes that occur primarily because of increasing temperature conditions
- **recrystallization:** the formation of new crystalline grains in a rock
- **retrograde:** metamorphic changes that occur primarily because of decreasing temperature conditions
- **subduction:** the process of sinking of a tectonic plate into the interior of the earth
- **tectonism:** the formation of mountains because of the deformation of the crust of the earth on a large scale
- **trench:** a long and narrow deep trough on the sea floor that forms where the ocean floor is pulled downward because of plate subduction
- **volcanic arc:** a linear or arcuate belt of volcanoes that forms at a subduction zone because of rock melting near the top of the descending plate

MINERAL CONTENT

Blueschists are a distinctive class of metamorphic rock containing one or more of the minerals lawsonite, aragonite, sodic amphibole (glaucophane), and sodic pyroxene (omphacite and jadeite plus quartz). These minerals indicate that recrystallization occurred in the temperature range of 150 to 450 degrees Celsius and at pressures of 3 to 10 kilobars or more (a kilobar corresponds to roughly 3 kilometers in depth). Blueschists of basaltic composition typically contain abundant glaucophane, a mineral that can give a rock a striking blue color. Other minerals commonly found in blueschists include quartz, mica, chlorite, garnet, pumpellyite, epidote, stilpnomelane, sphene, and rutile. The abundance of these and rarer minerals depends, as it does in all metamorphic rocks, upon rock composition, the exact pressures and temperatures of recrystallization, and the nature of chemically active fluids that have affected the rocks.

STRUCTURAL FEATURES

Blueschists in one place or another display a remarkably wide variety of structural features. In the field, many are complexly deformed, with intricate folding and refolding of compositional layering at scales of millimeters to tens of meters. Commonly, folded rocks also display a thickening and thinning of the layering, forming an interesting structure known as boudinage. Many blueschists are faulted, some so intensely that they are fragmented rocks known as breccias. When flaky minerals such as mica are lined up at the microscopic scale, a rock has a scaly foliation known as schistosity. A parallel alignment of rod-shaped amphiboles gives the rock a lineation. Blueschists typically have schistosities, and many also have lineations. The development of these features depends upon the extent to which the rocks deform while undergoing metamorphism (a geologic process known as dynamic metamorphism). Although most blueschists were so highly deformed during metamorphic recrystallization that all original features in the rocks were destroyed, some retain features from the rock's premetamorphic history, such as ripple marks, delicate fossils, or volcanic flow layering. Blueschists bearing these features were recrystallized but not highly deformed. Hence, the diversity of minerals and deformational features of the class of rocks known as blueschists is great. Many are truly blueschists, but some are neither blue in color where glaucophane is lacking nor schistose in texture when deformation was minor. These rocks are grouped with blueschists because of the presence of distinctive minerals, and by association with other blueschist rocks.

METAMORPHIC CONDITIONS OF FORMATION

Metamorphic recrystallization near 200 degrees Celsius causes anorthite (calcium-rich plagioclase) in combination with water to recrystallize as lawsonite at approximately 3 kilobars. Under the same temperature, calcite transforms into aragonite at 5 kilobars, and albite (sodium-rich plagioclase) recrystallizes as jadeite plus quartz near 7 kilobars. The breakdown of albite is one of the most distinctive indicators of blueschist metamorphism. At higher temperatures, these changes occur at higher pressures. Experiments combined with other measures of metamorphic temperature conditions indicate that most blueschists were metamorphosed at temperatures of 150 to 450 degrees Celsius and minimum confining pressures of 3 to 10 kilobars, respectively. Metamorphic pressures of 3 to 10 kilobars correspond to burial depths for recrystallization of 10 to 30 kilometers. Thirty kilometers is near the base of typical continental crust. Ultra-high-pressure blueschists containing relics of the mineral coesite (a dense mineral having the same composition as quartz that forms at extremely high confining pressures) have been found in small areas of the Alps, Norway, and China. Laboratory experiments indicate that confining pressures of 25 to 30 kilobars are required to transform quartz to coesite. Although coesite-bearing blueschists are rare, their occurrence is very important because they indicate that some blueschists recrystallized at depths of 75 to 90 kilometers—depths in the earth that are very near the base of the lithosphere.

The mineralogy of blueschists indicates that the metamorphic conditions for their formation within the earth would be equivalent to geothermal gradients of 10 to 15 degrees Celsius per kilometer depth or less. Such ratios of temperature to depth do not exist in the interior of normal lithospheric plates, because geothermal gradients are typically 25 to 35 degrees Celsius per kilometer depth. The plate tectonic setting for the generation of blueschists is thus very unusual.

SUBDUCTION

Regional terranes of blueschist extending for hundreds of kilometers in length and tens of kilometers in width are found in California, Alaska, Japan, the Alps, and New Caledonia. Smaller bodies of blueschist that are probably remnants of once-extensive terranes are found at numerous other sites around the world. Blueschists are found as fault-bounded terranes juxtaposed against deposits of unmetamorphosed sediments, igneous batholiths, or sequences of basalt, gabbro, and peridotite thought to be fragments of ocean crust (ophiolites) or other metamorphic terranes. The common feature of all occurrences is that they are regions that were probably the sites of ancient lithospheric plate convergence, a tectonic process commonly known as subduction. Subduction carries surficial rocks into the depths of the earth, where the increase in pressure and temperature causes metamorphism. Plate convergence involves localized shearing action between the descending plate and the overriding plate. As a result, blueschists and associated rocks typically undergo a complex deformational history, sometimes forming chaotic mixtures known as mélanges where deformation was particularly intense. It is of special interest that nearly all extensive blueschist terranes are of Mesozoic age or younger (less than about 250 million years). Why old blueschists are rare is a puzzle. The two most prevalent theories are that older blueschists have been destroyed by erosion and later metamorphism, or that the early earth was too hot to achieve the right combination of high pressure and low temperature.

Sites of plate subduction in the modern world are marked by ocean trenches, great earthquakes, and arcs of andesitic volcanoes. The region between the trench and volcanic arc is known as the arc-trench gap. Typically, a fore-arc basin is on the arc side of the gap and an accretionary prism is on the trench side. The fore-arc basin sits atop the overriding plate and becomes filled largely with basaltic to andesitic volcanic debris generated in the nearby arc. Fore-arc basin deposits are essentially undeformed and unmetamorphosed. In striking contrast, the accretionary prism is directly above the descending plate and consists largely of variously deformed and metamorphosed sediments that were bulldozed off the descending plate during plate convergence. Most typically, blueschists are found in the arcward parts of an accretionary prism, locally faulted directly against fore-arc basin deposits.

The unusual conditions of very low temperatures for a given depth of burial can develop within subduction shear zones because plate convergence at speeds of tens of kilometers per million years (centimeters per year) transports cold lithosphere downward faster than the earth's interior heat is conducted upward through it. As a result, after a few tens of millions of years of subduction, the front of the overriding plate cools, and the local geothermal gradients become greatly depressed. After fast plate convergence has occurred for a few tens of millions of years, temperatures less than 200 degrees Celsius at depths of 30 kilometers or more can be attained. The subduction zone metamorphism that creates blueschists is also known as high-pressure/low-temperature metamorphism.

Offscraping and Underplating

Because blueschists are found within accretionary prisms, it is important to understand how prisms grow and deform. Subduction accretion occurs by both offscraping and underplating. Offscraping is the process of trenchward growth or widening of the prism by addition at its toe of incoming sediments and seamounts (which are ocean islands such as Hawaii). It occurs by bulldozer-like action that causes the incoming pile of oceanic and trench-axis sediments to be folded and thrust-faulted. Offscraped rocks are weakly metamorphosed with the development of zeolite-group minerals and, at somewhat greater depths, the minerals prehnite and pumpellyite. Underplating is the process of addition of material to the bottom of a prism and, at greater depths, the bottom of the overlying crystalline plate. Underplating thickens and uplifts the overriding block and occurs concurrent with the shearing motions driven by the movement of the descending plate. Blueschists form in the region of underplating.

Underplating appears to be the basic process that drives both the thickening of accretionary prism and the uplift of included masses of blueschist. Underplating by itself, however, does not bring blueschists nearer to the surface. The presence of a steep trench slope (5-10 degrees) causes a prism to thin by downslope spreading, which is driven by gravity and behaves much like a glacier as it flows and thins down a mountain. Prism thinning seems to occur by a combination of normal faulting and rock flowage. Over a period of tens of millions of years, underplating-driven thickening at the base of the prism and gravity-driven thinning near the surface of the prism would slowly uplift a large terrane of blueschist near the edge of the overriding plate. The actual exposure of blueschist bedrock over a substantial area typically occurs only after subduction ceases and the top of the prism has become exposed to erosion.

The type of high-pressure/low-temperature metamorphism varies with depth. At the shallower depths of offscraping, temperatures and pressures are low, and only zeolites, prehnite, and pumpellyite develop. Surficial rocks subducted to depths of 10 to 30 kilometers and temperatures of 150 to 350 degrees Celsius are continuously metamorphosed into blueschists. At depths of 30 to 40 kilometers or more and at higher temperatures, the blueschists turn into the class of rock known as eclogite.

Prograde and Retrograde Metamorphism

The sequence of change from the zeolite to prehnite-pumpellyite to blueschist and finally to eclogite mineral assemblages is known as prograde metamorphism. Rocks in accretionary prisms commonly show all gradations of the progressive sequence. Overall, prograde metamorphism causes a general decrease in rock water content, destruction of the original minerals by recrystallization, increase in rock density, and increase in size of recrystallized crystals. At depths where the basalts and gabbros in the ocean crust (or ophiolite) at the top of the descending plate change from blueschist into eclogite, there is a large increase in the bulk density of the descending plate. This transformation decreases the buoyancy of the descending plate to such an extent that it may be the primary driving force of plate subduction and mantle convection. Most metamorphism is prograde because the loss of water or carbon dioxide prevents minerals from reverting to their original composition.

When the descending plate reaches depths of 100 to 125 kilometers, magmas are generated near its upper surface. They rise to the surface to form a volcanic arc of basaltic to andesitic composition. The presence of ultra-high-pressure blueschists directly confirms that some sediments are actually dragged down to (and returned from) the typical depths of arc magma origin. The intrusion of hot arc magmas near the surface and the eruption of volcanoes cause heating of the wall rocks, creating metamorphic rocks known as greenschists and amphibolites. This near-surface prograde metamorphism is of a low-pressure/high-temperature type. As a result, many ancient subduction zones are delineated on a regional scale by parallel belts of high-pressure/low-temperature and low-pressure/high-temperature metamorphic belts, a distinctive association known as paired metamorphic belts.

Plate convergence stops either when the relative motions between the descending and overriding plates become such that the margin becomes a transform plate margin, or when a buoyant continent or island arc is moved into a trench and "plugs up" the subduction zone. Transform plate motion occurs largely by horizontal movement along steep faults, a type of movement known as strike-slip faulting. In the process, some fault blocks rise and blueschists are eroded while others subside and blueschists become buried, reheated, and remetamorphosed as more normal geothermal conditions are reattained (a process known as retrograde metamorphism). The postsubduction destruction of blueschists by either erosion or retrograde metamorphism is the probable explanation for why most extensive terranes of blueschist are of Mesozoic age or younger.

GEOLOGICAL MAPPING OF BLUESCHISTS

Geologists study blueschists in the field, in the confines of the laboratory, and with computer modeling. Fieldwork involves going to the sites where blueschists are exposed in rock outcrops. Geological maps are made to show the field relations between blueschists and associated rocks. The first stage of geological mapping is recording on a topographic map the distribution of the major types of rocks, the orientation of bedding, and the locations and orientations of major faults and folds. Representative rock samples are collected for later laboratory study. The second stage of mapping is typically of much smaller areas. These detailed maps delineate additional variations in the types of rocks, the orientation of minor faults and folds, and associated schistosities and lineations. This stage of analysis usually provides the basis for determining the detailed movement patterns of the blueschists during subduction-zone deformation.

STUDY OF COMPONENT MINERALS

Laboratory studies of blueschists include the analysis of thin sections of the rock samples collected in the field, geochemical studies of mineral compositions, and experiments to determine the stability limits of minerals under different conditions of pressure, temperature, and fluid composition. Thin sections of the rocks are examined under polarized light with a petrographic microscope. Different minerals display different colors and other optical properties that enable their identification.

Minerals, particularly finely crystalline ones, are also identified by X-ray diffraction. The analysis of the scattering pattern of a beam of X rays focused upon the sample enables the researcher to identify minerals and—for minerals such as feldspar, pyroxene, or chlorite—to estimate their elemental composition. The elemental composition of powdered rock samples is commonly determined using X-ray fluorescence. A focused X-ray beam causes

Blueschists display a wide variety of structural features, and many are complexly deformed, with intricate folding and refolding of compositional layering at scales of millimeters to tens of meters. Seen here are examples of of boudinage (a thickening and thinning of the layering) and chevron folds in Queretaro, Mexico's El Doctor limestone. (U.S. Geological Survey)

atoms in a powder to emit other X rays whose type and intensity depend upon the types and amounts of atoms in the powder. The elemental composition of individual mineral grains is determined by analysis with the electron microprobe. A beam of high-energy electrons is focused on a 100-square-micron portion of a crystal in a highly polished thin section. As for X-ray fluorescence, the type and intensity of emitted X rays depend upon the types and amounts of atoms in a small spot in the crystal. Measurement of the composition of many spots in a traverse across a crystal enables determination of the variation in mineral composition from its core to rim, a variation known as compositional zoning. Zoning is a sensitive measure of the pressure and temperature history of the growing minerals and, hence, of both their prograde and retrograde metamorphic history.

Mass spectrometers are used to determine the isotopic ratios of the component minerals of blueschists for the calculation of the age of metamorphism. Mass spectrometers sort atoms by mass. A small amount of vaporized mineral is accelerated through a magnetic field, which deflects the atoms. More massive atoms are deflected less. Isotopic ratios of neodymium-143 to neodymium-144 and strontium-87 to strontium-86 are indicators of the geologic setting in which igneous rocks were erupted. Measurements of the ratios of oxygen-18 to oxygen-16 in coexisting minerals are indicators of the temperature of metamorphic recrystallization. The age of metamorphism for blueschists is determined from the analysis of radioactive isotopes and their daughter decay products in certain crystals. Examples are potassium-40, which decays into argon-40; rubidium-87, which decays into strontium-87; and uranium-238, which decays into lead-206. The measurement of the ratio of parent to daughter elements in either the whole rock or component minerals can be used to calculate the metamorphic age or ages of the rocks.

Experiments are conducted under controlled conditions in the laboratory to determine the stability limits and compositional relations for minerals at different pressures, temperatures, and fluid compositions. The goal is to simulate physical conditions deep in the earth. Experimental studies are also performed to determine how the ratios of oxygen isotopes in quartz and other minerals vary with different temperatures and oxygen pressure conditions. Laboratory calibration of elemental and isotopic compositions of minerals under controlled laboratory conditions is the basis for estimating the pressures and temperatures of metamorphism.

Computer simulations of the temperature conditions within subduction zones give an understanding of how temperatures change with time. Computer models that employ the principles of continuum mechanics are used to simulate the long-term tectonic deformation of an accretionary prism and the uplift of blueschist terranes. Geochemical computer models employ the principles of thermodynamics and are used to calculate the types of minerals that should develop during prograde and retrograde subduction-zone metamorphism.

Indicators of Geologic Processes

The study of blueschists is important because they are direct indicators of the geologic processes that occur deep within subduction zones that become mountain belts. Their creation indicates that abnormally cold geothermal conditions develop arcward (eastward) of the ocean trenches where rapid plate convergence occurs for tens of millions of years. Their preservation indicates that tectonic movements by faulting, folding, and rock flowage can be such that they become uplifted to near the surface while geothermal conditions remain very cold. Understanding the deformational history of blueschists is important because many of the world's largest and most destructive earthquakes occur at subduction zones at the very depths where blueschists are forming today. An understanding of how they deform and recrystallize during their downward and upward paths in ancient subduction zones will eventually provide new understanding of how destructive subduction-zone earthquakes are nucleated and, therefore, better earthquake prediction.

Subduction zones are the sites where ocean-floor-capped lithosphere plunges back into the earth to be recycled. Blueschists are direct indicators that some of the sediment on top of the descending plate has also been dragged to near the base of the lithosphere. Their presence in paired metamorphic belts is the primary way that geologists recognize ancient subduction zones. Blueschists are a key part of the geologic story of how continents grow by the addition of accretionary prisms along their edges.

Mark Cloos

FURTHER READING

Bebout, Gray E., et al., eds. *Subduction Top to Bottom.* Washington, D.C.: American Geophysical Union, 1996. Provides a fine introduction to subduction and plate tectonics. Maps, Illustrations, bibliography.

Best, Myron G. *Igneous and Metamorphic Petrology.* 2d ed. Malden, Mass.: Blackwell Science Ltd., 2003. A popular university text for undergraduate majors in geology. A well-illustrated and fairly detailed treatment of the origin, distribution, and characteristics of igneous and metamorphic rocks.

Bucher, Kurt, and Martin Frey. *Petrogenesis of Metamorphic Rocks.* 7th ed. New York: Springer-Verlag Berlin Heidelberg, 2002. Provides an excellent overview of the principles of metamorphism. Contains a section on geothermometry and geobarometry.

Cox, Allan, ed. *Plate Tectonics and Geomagnetic Reversals.* San Francisco: W. H. Freeman, 1973. Discusses the basic principles of the theory of plate tectonics and how it was developed. Chapters are introduced by short articles that discuss the importance of the following group of papers. Suitable for college-level students.

Davis, George H. *Structural Geology of Rocks and Regions.* 2d ed. New York: John Wiley & Sons, 1996. Discusses how folds, faults, and rock flowage occur. Chapter 6 covers the theory of plate tectonics and contains a short section on blueschists and their occurrence. Suitable for college-level students.

Ernst, W. G. *Earth Materials.* Englewood Cliffs, N.J.: Prentice-Hall, 1969. Discusses rocks and minerals. Sections on regional metamorphism, chemistry of metamorphic rocks, physical conditions of metamorphism, and metamorphism and the rock cycle include discussions specifically referring to blueschists. Suitable for high school students.

_____. *Subduction Zone Metamorphism.* Stroudsburg, Pa.: Dowden, Hutchinson and Ross, 1975. Covers the topic of subduction-zone metamorphism around the world. Most of the papers focus on blueschists. A series of summaries introduces groups of related papers and explains their relative importance. Appropriate for advanced college-level students.

Evans, Bernard W., and Edwin H. Brown, eds. *Blueschists and Eclogites.* Boulder, Colo.: Geological Society of America, 1986. Reports the nature of blueschists and eclogites at many sites around the world. Contains numerous pictures and diagrams of structures and mineral textures found in blueschists. Suitable for advanced college-level students.

Grapes, Rodney. *Pyrometamorphism.* 2d ed. New York: Springer, 2010. Discusses the formation of fused rocks and basaltic intrusions such as xenoliths. Heating and cooling sequences are described. The second edition includes additional references and illustrations.

Grotzinger, John, et al. *Understanding Earth.* 5th ed. New York: W. H. Freeman, 2006. One of many introductory textbooks in geology for college students. The section on plutonism and metamorphism contains a discussion of blueschists. Includes an extensive glossary of geological terms. Aimed at students at the advanced high school and freshman college levels.

Hallam, A. *A Revolution in the Earth Sciences.* New York: Oxford University Press, 1973. Discusses continental drift and the theory of plate tectonics. Blueschists are discussed in the chapter on the origin of mountain belts. Suitable for high school-level readers.

Lesnov, Felix P. *Rare Earth Elements in Ultramafic and Mafic Rocks and their Minerals.* New York: CRC Press, 2010. Discusses different chemical and mineral compositions of mafic and ultramafic rock. Compares rare-earth element composition and distribution. Writing is technical and best suited for researchers and graduate students.

Lima-de-Faria, Josae. *Structural Minerology: An Introduction.* Dordrecht, Netherlands: Kluwer, 1994. Provides a good college-level introduction to the basic concepts of crystal structure and the classification of minerals. Illustrations, extensive bibliography, index, and a table of minerals on a folded leaf.

Miyashiro, Akiho. *Metamorphism and Metamorphic Belts.* New York: Springer, 1978. An advanced textbook on metamorphic petrology that contains much discussion of blueschists, their occurrence, and their plate tectonic setting. Appropriate for advanced college-level students who want to understand the principles that control the formation of minerals in metamorphic rocks.

Perchuk, L. L., ed. *Progress in Metamorphic and Magmatic Petrology.* New York: Cambridge University Press, 1991. Although intended for the advanced reader, several of the essays serve to familiarize new students with the study of metamorphic rocks. Bibliography will lead the reader to other useful material.

Sobolev, A. V., et al. "The Amount of Recycled Crust in Sources of Mantle-Derived Melts." *Science* 316 (2007): 412-417. Discusses basaltic crust mixing with peridotitic mantle due to plate tectonics processes. Examines the chemical composition of melts.

Thompson, Alan Bruce, and Jo Laird. "Calibrations of Modal Space for Metamorphism of Mafic Schist." *American Mineralogist* 90 (2005): 843-856. Evaluates reactions of mafic rock metamorphism into greenschist, blueschist, and amphibolites-facies. Discusses the use of modal change data to determine pressure and temperature gradients of mafic schist.

Windley, B. F. *The Evolving Continents.* 3d ed. New York: John Wiley & Sons, 1995. Focuses on the origin and evolution of the continents. Several sections on blueschists are listed in a comprehensive index. Appropriate for college-level students.

See also: Batholiths; Contact Metamorphism; Granitic Rocks; Metamorphic Rock Classification; Metamorphic Textures; Metasomatism; Pelitic Schists; Regional Metamorphism; Sub-Seafloor Metamorphism.

BUILDING STONE

Building stone is any naturally occurring stone that is used for building construction. The three rock types (igneous, metamorphic, and sedimentary) are all utilized in building stone. The physical characteristics of each type of rock, such as hardness, color, and texture, determine how the stone is used.

PRINCIPAL TERMS

- **granite:** an igneous rock that is known for its hardness and durability; in modern times, it has been used on the exterior of buildings, as it is able to resist the corrosive atmospheres of urban areas
- **igneous:** rock that was formed from molten material originating near the base of the earth's crust
- **limestone:** a sedimentary rock that can be easily shaped and carved; currently, it has gained acceptance as a thin veneer
- **marble:** a metamorphic rock that has been used since Grecian times as a preferred building stone; it is known for its ability to be carved, sculptured, and polished
- **metamorphic:** rock that was formed by heat and pressure; tectonic, or mountain-building, forces of the earth's crust create and alter the mineral composition and texture of the original rock material
- **sandstone:** a sedimentary rock that is known for its durability in resisting abrasive wear; it is likely to be used for paving stone
- **sedimentary:** most commonly, rock that was formed by marine sediments in an ocean basin; it usually shows depositional features and may include fossils
- **slate:** a metamorphic rock that has a unique ability to be split into thin sheets; some slates are resistant to weathering and are thus good for exterior use

Various Forms

Building stone is any kind of rock that has supplied humankind with the material to erect monuments and edifices throughout history. The pyramids of Egypt, the temples of Greece, and the skyscrapers of the modern world have all utilized various forms of building stone. The ancients used blocks of stone stacked like building blocks, whereas the modern builder uses a thin veneer of stone anchored to the exterior of a building frame. Special physical properties of the various stone materials lend themselves to these various usages. Because of these special physical properties, only a small percentage of the

earth's rock material can be classified and utilized as building stone. Granite, an igneous rock that forms the core of the continents, is a good example. Granite is abundant, but because of mineralogical variations or structural weakness, only a small percentage of the granite areas will yield quarry blocks of suitable dimensions for building stone.

Building stone is quarried, or excavated, in most countries and then may be shipped around the world. Quite commonly, granite blocks from Brazil are shipped to Canada for sawing and finishing and then shipped to a construction site in a city in the United States. Similarly, building stone from the United States has been shipped to Italy for sawing and finishing and then shipped to England for installation on buildings. The average quarry block weighs 15 to 20 tons. These quarry blocks are sawed into slabs that vary in thickness from half of an inch to four inches. Whether the building stone is limestone, marble, granite, slate, or sandstone, the quarry block must be solid enough to yield slabs of competent rock (that is, rock capable of withstanding stress). These slabs are further cut on diamond saws to the desired shape and dimensions and then finished to the requirements of the architect or owner.

Design Criteria

Buildings constructed before or around the turn of the twentieth century were built with rectangular pieces 6 inches thick or greater. Rectangular stone forms both the supporting structure of the walls and the exterior protection of the building. Most buildings built since the 1950's, however, require that the building stone be cut into thin panels less than 3 inches thick and fastened to the exterior of the steel supporting structures. This use of stone demands that more engineering and design criteria be used to establish acceptable versus unacceptable building stone. The processes by which the architect, engineer, and geologist select and qualify a particular stone for usage help to determine the aesthetic and physical characteristics of the stone. The architect usually establishes the design criteria for the size and shape of the stone, looking at the stone for such aesthetic aspects as color and texture. The engineer and geologist then determine the suitability of the stone within the design plan.

The geologists and engineers involved with quarrying and fabricating building stone look for deposits that are uniform in texture, color, and structural integrity. The exploration of new areas requires researching and reviewing geological maps, aerial photographs, and geological libraries. Ground sleuthing or field mapping and reconnaissance are the next steps in delineating a potential quarry location. Rock sample collection, mineralogic identification, and the testing of physical properties are conducted to determine whether the stone would be suitable for the specific use. The final use of the building stone varies from a stone block size of 100 cubic feet (about 4.6 feet on a side, or 4 by 5 by 5 feet) to a rubble stone size of 1 cubic foot. Some blocks smaller than 100 cubic feet can be used to produce tiles and novelties, but for the most part the major emphasis is on quarrying larger blocks of stone.

Quarrying large blocks of stone requires geological uniformity of the rock type over an ample enough area to justify opening a quarry. This area needs to be of an adequate size to open a quarry that will yield uniform material for a time span of ten to twenty years.

COLOR AND TEXTURE

The color of the stone, so critical to the architect's design criteria, is studied by polished hand samples or test block extraction from the potential quarry area. The mineralogical content of the various granites will determine the color consistency and stability. Petrologic microscope work and thin-section studies are an important part of the search for suitable granite. Color stability is as important a factor as color uniformity; there are numerous dark limestones that, while unsuitable for exterior use, are beautiful in an interior lobby. Some minerals can make stone unsuitable. Even tiny amounts of pyrite, for example, will weather to make unsightly rust stains on the surface, releasing a sulfuric acid that attacks the stone.

Texture and mineral fabric are also important considerations with regards to the suitability of a particular stone. The use of the stone will depend on whether it has a fine or coarse texture. Interlocking mineral grains, a well-cemented matrix, and nonsoluble minerals are also important characteristics. Many fine-textured marbles are "sugary" and soft because of poorly interlocked calcite grains, whereas a similar fine-grained marble, with an interlocked mineral fabric and low porosity to surface water, may be tough and resistant to atmospheric corrosion.

SOUNDNESS AND STRENGTH

The major criterion in determining the suitability of a building stone is the soundness of the deposit. The soundness is a measure of the natural fractures and the strength of the stone. Geological conditions during the formation of various rock types will determine soundness. In a sedimentary rock, the bedding (foundation or stratification) and jointing (fracturing without displacement) will determine the block sizes capable of being produced. In an igneous and metamorphic rock, the schistosity (tendency to split along parallel planes), cleavage (tendency to split along closely spaced planes), and jointing will be contributing factors in determining the degree of soundness. In an igneous and metamorphic rock mass, the internal stresses need to be measured and understood before determining potential block sizes. In the years prior to diamond-cutting tools and hydraulic handling equipment, cleavage, jointing, and bedding were used to help pry out blocks of stone. With the development of such tools, quarry workers can cut out more rock masses for building stone.

Rock strength is another important criterion of a building stone. Test engineers use specific laboratory equipment to determine the flexural and compressive strength of each building stone. Depending on the planned usage, water absorption and abrasive wear test data are also important for knowing how a particular rock will react under certain applications. The test program may include producing test specimens of the finished product and installing them in mock-up panels.

THE BUILDING STONE INDUSTRY

Building stones have been used to produce comfortable habitations ever since cave dwellers blocked up cave entrances for protection and warmth. In the modern world, architects utilize building stone to clad the exteriors of modern buildings as a protective shield for structural steel supports. In the entrances of these same buildings, designers take polished building stone and clad interior lobbies for durability and beauty. Slate, marble, granite, limestone, and sandstone are all utilized as building stone. These materials occur throughout the world in easily exploitable deposits. From the mid-1800's to the mid-1900's, the building-stone industry in North America was both the impetus and nucleus for towns, railroads, and machine tool industries across the continent. From the 1900's through the 1930's, many technological advances in quarry drilling and finishing equipment were fostered by the building-stone industry. The 1950's through the 1970's saw many adaptations to the demands of architects and designers. Innovative technology was developed to gain competitive advantage over other producers and to satisfy the need to exploit deeper and more solid deposits of stone. This technology ranged from manual labor in the mid-1800's to the use of steam, electricity, and compressed air by the early 1900's. The equipment and tools developed in the early 1900's remained relatively unchanged until the 1950's, at which time the more sophisticated use of carbides and diamonds for cutting and a high grade of steel for drilling allowed for the more efficient excavation of stone. The use of hydraulics and electronics has also allowed for a higher degree of automation and cost savings.

In North America, the building-stone industry has shrunk from more than a thousand active quarries in the 1920's—with more than a dozen locations where quarries, mills, and shops were an integral part of the local economy—to a number in the hundreds. During the 1950's and 1960's, the building-stone industry in the United States nearly became extinct before stone became fashionable again. Yet the trend toward the increased use of stone in the late 1980's has suggested that the building-stone industry will become an important factor in local economies once again.

Lance P. Meade

FURTHER READING

Barton, William R. *Dimension Stone.* Information Circular 8391. Washington, D.C.: Government Printing Office, 1968. Describes the building-stone industry in North America before major technological advances. Appropriate for high school students.

Bates, Robert L. *Stone, Clay, Glass: How Building Materials Are Found and Used.* Hillside, N.J.: Enslow, 1987. A brief reference for junior high and advanced students.

Bates, Robert L., and Julia A. Jackson. *Our Modern Stone Age.* Los Altos, Calif.: William Kaufmann, 1982. Reference on industrial minerals. Includes discussion of building stones.

Chacon, Mark A. *Architectural Stone: Fabrication, Installation, and Selection.* New York: Wiley, 1999. Offers a clear look at stones used in the construction of buildings. Provides information on tests used to select stones, as well as the steps involved in installation.

Evans, Anthony M. *An Introduction to Economic Geology and Its Environmental Impact.* Malden, Mass.: Blackwell Science, 1997. Provides a wonderful introduction to mines and mining practices, minerals, ores, and the economic opportunities and policies surrounding these resources. Illustrations, maps, index, and bibliography.

Kourkoulis, Stavros K., ed. *Fracture and Failure of Natural Building Stones.* Dordrecht: Springer, 2006. Examines the processes involved in restoring and conserving historical stone structures. Discusses concepts of geology, physics, and archaeology. Covers mechanical properties of rocks, weathering, stress, and masonry. Discusses a number of case studies. Appropriate for those dedicated to protecting authentic material in buildings and structures.

Meade, Lance P. "Defining a Commercial Dimension Stone Marble Property." In *Twelfth Forum on the Geology of Industrial Minerals.* Atlanta: Georgia Department of Natural Resources, 1976. An objective discussion of important criteria in developing a dimension stone property. For the college-level reader.

Newman, Cathy, and Pierre Boulat. "Carrara Marble: Touchstone of Eternity." *National Geographic* 162 (July, 1982): 42-58. Photo essay on building-stone quarrying. Describes Italian quarries that provided Michelangelo with the marble for his sculptures.

Prikryl, R., ed. *Dimension Stone.* New York: A. A. Balkema Publishers, 2004. Gathers articles from the Dimension Stone Conference held in Prague in 2004. Discusses the use of stone for construction in the past and in the future. Discusses the physical properties of stones in terms of their use as construction material. Ideal for geologists, engineers, architects, and archaeologists.

Prikryl, R., and B. J. Smith, eds. *Building Stone Decay: from Diagnosis to Conservation.* London:

The Geological Society, 2007. Covers the conservation and analysis of building materials. Discusses experimental methods and tests. Various rock types, their properties, and the process of decay are examined throughout. Provides practical applications of rock decay knowledge toward conservation of original building stones.

Shadmon, Asher. *Stone: An Introduction.* 2d ed. London: Intermediate Technology, 1996. Examines the properties and features of stones, including those used as building materials. Covers the chemical and geophysical aspects of building stones, as well as restoration, conservation, and deterioration.

Smith, Mike R., ed. *Stone: Building Stone, Rock Fill, and Armourstone in Construction.* London: Geological Society, 1999. Deals with the use of stone in architecture and construction. Covers geological engineering processes. Color illustrations, maps, index, and bibliography.

See also: Aluminum Deposits; Cement; Chemical Precipitates; Coal; Diamonds; Dolomite; Earth Resources; Fertilizers; Gold and Silver; Hydroelectric Power; Hydrothermal Mineralization; Industrial Metals; Industrial Nonmetals; Iron Deposits; Manganese Nodules; Mining Processes; Oxides; Pegmatites; Platinum Group Metals; Salt Domes and Salt Diapirs; Sedimentary Mineral Deposits; Soil Chemistry; Uranium Deposits.

C

CARBONATES

Carbonate minerals are characterized by a composition that features the carbonate ion. Common carbonate minerals are divisible into calcite, aragonite, and dolomite groups.

PRINCIPAL TERMS

- **anion:** a negatively charged ion
- **aragonite:** a carbonate with the orthorhombic crystal structure of the calcium carbonate compound; it forms in marine water or under high-pressure, metamorphic conditions
- **calcite:** a carbonate with the hexagonal crystal form of the calcium carbonate compound; a common mineral found in limestone and marble
- **cation:** a positively charged ion
- **crystal:** a solid with an internally ordered arrangement of component atoms
- **divalention:** an ion with a charge of 2 because of the loss or gain of two electrons
- **dolomite:** a double carbonate that includes magnesium and has a hexagonal structure; it is abundant in ancient rocks
- **ion:** an atom that has lost or gained one or more electrons
- **isostructural:** having the same structure but a different chemistry
- **polymorphs:** different structures of the same chemical compound

CARBONATE MINERAL FORMATION

Carbonates are one among several classes of mineral, which are the materials that make up rocks. They are natural substances with a definite chemical composition and an ordered internal arrangement. Minerals can be divided into chemical groups based on their atoms and into structural groups based on the atoms' ordered arrangement.

All carbonates contain the carbonate ion as their defining anionic group. An ion is an atom that has lost or gained electrons and so has become chemically reactive. An ion that has lost electrons has a positive charge and is called a cation; an ion that has gained electrons has a negative charge and is called an anion. The number of electrons lost or gained represents the charge of the ion. The charge and the radius (size) of an ion determine how it is chemically bonded to another ion, how strong the bonding is, and the number of ions that can be coordinated to it or surround it.

The carbonate ion contains one carbon ion in the middle of a triangle formed by three oxygen ions, which occupy the corners of the triangle. The central carbon ion has a charge of +4 and the oxygen ions have a charge of -2. As a result, the carbonate ion group acts as if it were a single ion with an overall -2 charge. To form a carbonate mineral, the negative charge on the carbonate ion must be balanced by cations such as calcium, magnesium, strontium, manganese, and barium.

CARBONATE MINERAL STRUCTURE

Atoms in a mineral can be pictured as lying on imaginary planes, which are called atomic planes. These planes cut across other planes, forming definite and measurable intersection angles. The atomic planes that contain many atoms tend to develop into crystal faces. Minerals are crystalline, and the crystal faces of a mineral are related to the internal arrangement of atoms. The bonds between atoms of the same atomic plane tend to be stronger than the bonds across the atomic planes. Consequently, minerals commonly break or cleave along preferred planes, or cleavage planes, when struck by a hammer. A set of parallel cleavage planes yields one cleavage direction. Some minerals have more than one cleavage direction; others have none, since the bond strength between atoms is the same in all directions. Common carbonate minerals have three cleavage directions at oblique angles and cleave into skewed box-like shapes.

The atoms of a mineral are symmetrically related to one another. Since crystal faces are related to the internal arrangement of the atoms, these crystal faces are also symmetrically related. Scientists have found that all minerals can be grouped into thirty-two crystal classes based on their symmetry relations. A small group of atoms forms the basic building block of crystals. This building block is called a unit cell; it is an arrangement of the smallest number of atoms which, as a unit, may be repeated over and over again to form a visible crystal. The volume of a unit cell can be determined from the lengths of its lines and the angles subtended by them. Depending on the cell's shape, these imaginary lines may be parallel to its edges or may pass through opposite corners, sides, or edges. The lines are called crystallographic axes. The thirty-two crystal classes of minerals can be grouped into six crystal systems based on the shapes of their unit cells.

The common carbonates belong to two of the six: the orthorhombic and hexagonal systems. The orthorhombic crystal system is characterized by having three crystallographic axes that are unequal in length and perpendicular to one another. The unit cell is a rectangular box with unequal sides. The hexagonal system has four axes, all of which pass through a common center. Three of these axes lie on the same plane, are of equal length, and are separated from one another by 120 degrees. The fourth axis is different in length from the others and is perpendicular to them. The hexagonal unit cell is a box with 60-, 90-, and 120-degree angles.

The three common carbonates—calcite, aragonite, and dolomite—belong to three structural types. Calcite and aragonite are polymorphs of the same compound; that is, the same chemical compound occurs in different structures. Aragonite is orthorhombic and less symmetrical than calcite, which is hexagonal. Calcite and dolomite are both hexagonal, but they do not have identical structures. In calcite, calcium atomic planes lie between carbonate ion planes. In dolomite, alternating calcium planes are occupied by magnesium atoms. As a result, calcite has a higher symmetry content than dolomite, although both belong to the same crystal system. Also, dolomite shows more internal order than calcite because it requires the positioning of more different atoms in specific atomic sites.

Different chemical compounds that have identical structures are said to be isostructural. All carbonates that contain divalent cations (+2 charge) whose ionic radii are less than or equal to that of the calcium ion are isostructural to calcite, which has a hexagonal structure. Siderite (iron carbonate), magnesite (magnesium carbonate), and rhodochrosite (manganese carbonate) are isostructural to calcite.

Calcite Group

Calcite is by far the most common of all carbonate minerals. It is commonly off-white, colorless, or transparent. It may exhibit different crystal shapes, but in all cases, a careful examination will reveal the hexagonal crystal structure. It is fairly resistant to abrasion, although it can be scratched by a steel knife. Calcite fizzes and dissolves in acid because the acid breaks down the carbonate ion and releases carbon dioxide. Calcite is the main component of a rock called limestone. Most limestone is at least partly biological in origin, made of calcium carbonate secreted by marine organisms. Limestone is soluble in acid, and most surface water is slightly acidic. Many caves and some sinkholes are found in regions where the rocks are limestone, and formed because calcite is dissolved by groundwater. Sinkholes are formed when the roofs of caves collapse.

Dolomite Group

In the dolomite group, alternating calcium atomic planes are occupied by other divalent ions, such as magnesium in dolomite, iron in ankerite, and manganese in kutnahorite. Since the magnesium, iron, and manganese ions are smaller than the calcium ion, the dolomite group structure is not as highly symmetrical as the calcite structure. Consequently, although both the calcite and the dolomite group minerals are hexagonal, they belong to different classes among the thirty-two classes of crystals. Of the dolomite group minerals, the mineral dolomite is by far the most common. It is fairly common in ancient carbonate rocks called dolostones. Dolomite is white to pinkish and is similar to calcite in many ways; however, it does not fizz readily when diluted acid is dropped on it.

For a long time it was a puzzle why dolostone was so abundant, as dolomite is not forming in many places today. We now know that most dolomite begins as limestone and is converted to dolomite later. In some cases the conversion happens when fresh water mixes with salt water as the rocks are forming, but more recent research suggests that most dolomite forms when warm fluids circulating through the rocks replace calcium with magnesium.

ARAGONITE GROUP

In the aragonite group, carbonate ions do not lie on simple atomic planes. Adjacent carbonate ions are slightly out of line; also, adjacent carbonate ions face in opposite directions. Divalent cations whose ionic radii are larger than or equal to the calcium ion form carbonate minerals that are isostructural to aragonite. Of the minerals in the aragonite group, aragonite is the most common. It is generally white and elongate. It is the carbonate mineral that readily precipitates from marine water, but it is not a stable mineral, and in time it changes to the more symmetrical and hexagonal structure of calcite. Aragonite is the least symmetrical polymorph of calcium carbonate, and is found in metamorphic rocks that were formed under high pressure and comparatively low temperature.

In its most abundant form, aragonite is synthesized by aquatic organisms that are common in shallow marine environments of warm latitudes, such as the Gulf Coast. As marine organisms die, their shells, which are made of calcite, settle at the bottom. These shells may be broken into smaller fragments by browsing organisms or wave action. When buried to form limestone, the aragonite quickly recrystallizes to calcite.

IDENTIFICATION OF CARBONATE MINERALS

Carbonates are generally light-colored, soft, and easy to scratch with a knife, as compared with most other common rock-forming minerals. They form a white powder when scratched. They are harder than fingernails, however, and cannot be scratched by them. When acid is poured on carbonate powder, it fizzes, liberating carbon dioxide. Carbonates such as calcite do not even have to be powdered for the acid test, because they fizz readily. The cleavage planes and the angles between cleavages are another physical method by which carbonate minerals can be distinguished. Clear crystals such as those of calcite can produce double refraction of objects, another

Sinkholes that form by the collapse of roof rocks of near-surface caves in the limestones of warm and humid regions are a problem not only to farmers but also to homeowners. It is not unusual for a house to sink suddenly into a depression caused by a collapse into an underground cave, as this one did in 1967. (U.S. Geological Survey)

property that identifies carbonates without the aid of instruments.

Better mineral identification is done by scientists after a rock is cut to a small size, mounted on glass, and ground to a very thin section of rock (0.03 millimeters thick), which is then capable of transmitting light. The thin section is placed on a stage of a transmitted-light polarizing microscope. A lens below the stage polarizes light; it allows the transmission only of light that vibrates in one direction—east to west, for example. Another lens above the stage allows the passage only of light that vibrates in the other direction—north to south. When glass is placed on a stage and the lower polarizer is inserted across the light source, the color of the glass can be seen. When the upper polarizer is also inserted, however, the glass appears dark, because no light is transmitted. The optical properties of most minerals, including the carbonates, are different from those of glass. Other accessories are used in addition to the polarizing lenses in order to determine minerals' optical properties. Magnification by microscope permits the better determination of minerals' physical properties, such as their shape and cleavage. Special dyes can also be used to distinguish calcite from dolomite.

DETERMINATION OF CRYSTAL STRUCTURE

The crystal structure of carbonate crystals can be determined with the aid of a contact goniometer. Its simplest version is a protractor with a straight edge fastened in the middle. The goniometer is used to measure the angles between crystal faces, from which scientists can determine the crystal structure.

X-ray diffraction can ascertain the crystal structure of any substance, including carbonates. Diffraction peaks characteristic of each mineral can be displayed on a chart recorder when X rays bombard a sample. Each diffraction peak results from the reinforcement of X-ray reflections from mutually parallel atomic planes within the sample. Several diffraction peaks from one mineral indicate equivalent numbers of sets of atomic planes within the minerals. The difference in the peak heights corresponds to the density of atoms in the pertinent atomic planes. The detection device does not have to be a chart recorder; it can be a photographic paper or a digital recorder that can be appropriately interfaced to a computer for the quick identification of minerals.

PRACTICAL APPLICATIONS

Carbonates are a common group of minerals that form in environments that range from arid lands to shallow seas. After silicates, they are the most abundant minerals on the surface of the earth. Hot springs are one place where calcite precipitates. Travertine, or tufa, is a banded rock that precipitates at the mouth of springs. Caliche, deposits of carbonate that precipitate from groundwater in arid climates, is a source of serious problems to irrigation farmers. Sinkholes that form through the collapse of roof rocks of near-surface caves in the limestone of warm and humid regions are a problem not only to farmers but also to homeowners. It is not unusual for part of a house, or the whole of it, to sink suddenly into a depression caused by the collapse of an underground cave.

Carbonates are important for their regulation of the pH, or acidity content, of ocean waters. Carbonate minerals dissolve when the acid content of water is raised and precipitate when the acid content is reduced. In this way, the pH of ocean water is regulated to a steady value of 8.1.

Carbonates are also known for their industrial applications, which range from dolomite tablets to building materials such as cement and mortar. One of the finest building rocks is marble, which is composed of carbonate minerals. Marble often is delicately banded with different colors. The banding arises because the minerals are lined up in directions perpendicular to the natural pressure under which an impure limestone was metamorphosed and converted to marble. If the original limestone was pure and composed entirely of calcite, the marble that is metamorphosed from it would be white and not banded. Polished marble is used as a building material, or often as decorative stone for doors or exteriors. Polished travertine is also used as building stone, but it is placed in the interiors of buildings because of banded porous zones that can accumulate rainwater. Regular limestone is used in buildings and, most commonly, in retaining walls alongside houses and roads. Most limestone is used for cement. Cement is up to 75 percent lime or calcium oxide, created by heating limestone to drive off carbon dioxide; the rest is silica and aluminum.

The most important natural application of carbonates is in regulating the earth's climate. Carbonate rocks lock up carbon dioxide that would otherwise trap solar heat and warm the earth. Weathering and other processes also break down calcite and release carbon dioxide. The balance between carbonate formation and breakdown is critical in maintaining the earth's natural climate balance.

Habte Giorgis Churnet

FURTHER READING

Ahr, Wayne M. *Geology of Carbonate Reservoirs.* Hoboken, N.J.: John Wiley & Sons, 2008. Covers many principles of mineralogy with a focus on carbonates. Covers rock properties, petrophysical properties, stratigraphy, deposition, and diagenesis of carbonate reservoirs. A summary chapter on the geology of carbonate reservoirs ties together fundamental topics with specific field examples. Provides references and an extensive index.

Bathurst, Robin G. C. *Carbonate Sediments and Their Diagenesis.* 2d ed. New York: Elsevier, 1975. Chapter 6 discusses the chemistry and structure of the more common carbonate minerals.

Butler, James Newton. *Carbon Dioxide Equilibria and Their Applications.* Chelsea, Mich.: Lewis, 1991. Butler discusses in great detail the role of carbonates in the chemical equilibria of carbon dioxide. Includes a short but useful bibliography and an index.

Deer, W. A., R. A. Howie, and J. Zussman. *An Introduction to the Rock-Forming Minerals.* 2d ed. London: Pearson Education Limited, 1992. A work of reference useful for Earth science students. Carbonates are discussed in detail.

Klein, Cornelis, and C. S. Hurlbut, Jr. *Manual of Mineralogy.* 23d ed. New York: John Wiley & Sons, 2008. Details of carbonates are treated in chapter 10. Suitable for college-level students.

Loucks, Robert G., and J. Frederick Sarg, eds. *Carbonate Sequence Stratigraphy: Recent Developments and Applications.* Tulsa, Okla.: American Association of Petroleum Geologists, 1993. Contains several essays that address advances in the study of carbonates as they relate to stratigraphy, as well as their relevance to the petroleum industry. Bibliography and index.

Marino, Maurizio, and Massimo Santantonio. "Understanding the Geological Record of Carbonate Platform Drowning Across Rifted Tethyan Margins: Examples from the Lower Jurassic of the Apennines and Sicily (Italy)." *Sedimentary Geology* 225 (2010): 116-137. Provides an extensive overview of drowning processes and unconformities as well as specific examples to build on fundamental concepts. Although this is written to convey new research to graduate students and professionals, there is a great deal of background information to make this article accessible to undergraduates.

Mason, Brian, and L. G. Berry. *Elements of Mineralogy.* San Francisco: W. H. Freeman, 1968. An excellent and easy-to-read book on the study of minerals. Used by many colleges. Carbonates are discussed in Chapter 7.

Parker, Sybil P., ed. *McGraw-Hill Encyclopedia of the Geological Sciences.* 2d ed. New York: McGraw-Hill, 1988. Offers complete entries on all the common carbonate minerals, including aragonite, dolomite, limestone, and calcite. Written at a college level. Illustrated.

Prinz, Martin, George Harlow, and Joseph Peters, eds. *Simon and Schuster's Guide to Rocks and Minerals.* New York: Simon & Schuster, 1978. Rocks and minerals are described and illustrated with color photographs in this easy-to-read book.

Swart, Peter K., Gregor Eberli, and Judith A. McKenzie, eds. *Perspectives in Carbonate Geology.* Hoboken, N.J.: Wiley-Blackwell, 2009. A collection of papers presented at the 2005 meeting of the Geological Society of America. The papers present studies on carbonate sediments and the comparison of modern and ancient sediments. Suited for the professional geologist or graduate student.

See also: Biopyriboles; Clays and Clay Minerals; Earth Resources; Feldspars; Gem Minerals; Hydrothermal Mineralization; Metamictization; Metamorphic Rock Classification; Minerals: Physical Properties; Minerals: Structure; Non-silicates; Oxides; Radioactive Minerals.

CARBONATITES

Carbonatites are composed of carbonate minerals that appear to have formed from carbonate liquids. They typically contain many minerals of unusual composition that are seldom found elsewhere. Carbonatites have been mined for a variety of elements, including the rare-earth elements niobium and thorium. The rare-earth elements are used as phosphors in television picture tubes and in high-quality magnets in stereo systems; niobium is used in electronics and in steel to make it resist high temperatures.

PRINCIPAL TERMS

- **calcite:** a mineral composed of calcium carbonate
- **dike:** a tabular igneous rock formed by the injection of molten rock material through another solid rock
- **dolomite:** a mineral composed of calcium magnesium carbonate
- **igneous rocks:** rocks formed from liquid or molten rock material
- **ijolite:** a dark-colored silicate rock containing the minerals nepheline (sodium aluminum silicate) and pyroxene (calcium, magnesium, and iron silicate)
- **isotopes:** different atoms of the same element that have different numbers of neutrons (neutral particles) in their nuclei
- **limestone:** a sedimentary rock composed mostly of calcium carbonate formed by organisms or by calcite precipitation in warm, shallow seas
- **mineral:** a naturally occurring element or compound with a more or less definite chemical composition
- **rock:** a naturally occurring consolidated material that usually consists of two or more minerals; sometimes, as in carbonatites, rocks may consist mainly of one mineral
- **silicate mineral:** a mineral composed of silicon, oxygen, and other metals, such as iron, magnesium, potassium, and sodium

COMPOSITION OF CARBONATITES

Carbonatites are unusual igneous rocks because they are not primarily composed of silicate minerals, as are most other rocks formed from molten rock material. Instead, carbonatites are composed mostly of carbonate minerals and of minor amounts of other minerals that are rare in other rocks. The carbonate minerals composing carbonatites are usually calcite (calcium carbonate) or dolomite (calcium magnesium carbonate). The other minerals found in carbonatites often contain large concentrations of elements that rarely become concentrated enough to form these minerals in other rocks.

Carbonatites are often associated with rare silicate rocks injected at about the same time as the carbonatites. These silicate rocks are unusual, as they normally lack feldspars (calcium, sodium, and potassium aluminum silicate minerals), which are abundant in most igneous rocks formed within the earth. Instead of feldspar, many of these silicate rocks contain varied amounts of the minerals nepheline (sodium aluminum silicate) and clinopyroxene (calcium, magnesium, iron silicate) and minor amounts of minerals not commonly found in other igneous rocks. The silicate rock consisting of more or less equal amounts of nepheline and clinopyroxene is called ijolite.

OCCURRENCE OF CARBONATITES

Carbonatites often occur as small bodies of varying size and shape that cut across the surrounding rocks. Often, they occur as dikes that may be only a few feet to tens of feet wide, but they may be greater in length. An example of carbonatites occurring as dikes is found at Gem Park near Westcliffe, Colorado. Carbonatites sometimes occur as somewhat equi-dimensional bodies that are larger than those forming dikes. The Sulfide Queen carbonatite at Mountain Pass, California, for example, is about 800 meters long and 230 meters wide. Often, carbonatites contain foreign rock fragments. The cross-cutting relations and foreign rocks found within the carbonatites suggest that they form by the injection or intrusion of molten carbonate into the surrounding solid rock. The foreign rock fragments within the carbonatite could be solid rocks ripped off the walls by the moving molten carbonate material as it was injected through solid rock.

The occurrence of molten carbonate at the active volcano at Ol Doinyo Lengai in northern Tanzania, Africa, confirms that such material exists. The abundance of recently formed volcanic material composed of carbonatite at Ol Doinyo Lengai and other locations may indicate that some carbonatite can form near the surface. The carbonatite liquids at the surface may flow as lava out of the volcano or be ejected explosively into the air, similar to that of other volcanoes formed mostly from silicate liquids. Carbonatites of volcanic origin are especially abundant in Africa along a portion of the continent called the East African rift zone. In this location, Africa is slowly being ripped apart by forces within the earth. Some carbonatites that intruded below the surface as dikes may also at one time have fed volcanoes at the surface. Erosion of the extinct volcano and associated silicate rocks may have exposed dikes composed of solidified carbonatite.

COMPLEXES

The carbonatites, ijolites, and other associated igneous rocks that formed at the same place and were injected at about the same time are called complexes. These complexes have small areas of exposure at the surface; most are exposed over a surface area of only about 1 to 35 square kilometers. The Magnet Cove complex in Arkansas, for example, has about a 16-square-kilometer exposure. The Gem Park complex in Colorado is only about 6 square kilometers in area. The associated silicate rocks usually compose most of the area of the exposed complex; only a small portion is carbonatite. The overall shape of these complexes is often circular, elliptical, or oval, but departures from these shapes are common. Often, the different rock types within a complex are built of concentric zones much like the layers of an onion.

Examples of these complexes include Seabrook Lake, Canada, where the complex is roughly circular and is about 0.8 kilometer across. Its "tail," however, extends from the main circular body to about 1.2 kilometers to the south. The central core of carbonatite, composed mostly of calcite, is about 0.3 kilometer across. Smaller carbonatite dikes can be found within other rocks. The largest carbonatite body is surrounded by a dark rock containing angular blocks composed mostly of carbonate, clinopyroxene, or biotite (a dark, shiny potassium, iron, magnesium, and aluminum silicate). This latter body is surrounded by a mixture of ijolite and pyroxenite (a rock containing only pyroxene). The ijolite and pyroxenite also compose much of the tail to the south.

Another complex is located at Gem Park in Colorado. Gem Park is oval in shape, roughly 2 by 3.3 kilometers. There is no central carbonatite there. Instead, scores of small, dolomite-rich carbonatites have intruded as dikes across the silicate rocks of the complex. The main silicate rocks of the complex are pyroxenite and a feldspar-clinopyroxene rock called gabbro. The pyroxenite and gabbro form concentric rings with one another. The large amount of gabbro makes this complex unusual, as gabbros are seldom present with carbonatite in the same complex.

The complexes at Gem Park and Seabrook Lake are rather simple, as they contain very few rock types. A wide variety of minerals can be found in some complexes, resulting in a large number of rock types. The Magnet Cove complex in Arkansas, for example, has twenty-eight major rock types listed on the geologic map. (The reading accompanying the map extends the total rock types to an even greater number.) A large proportion of the rock names in geology are generated by the wide variation of minerals found in these complexes that compose merely a tiny portion of the earth's surface.

HYPOTHESES OF CARBONATITE FORMATION

Up until the 1950's, geologists believed that carbonatites were limestones that melted and were intruded as molten rock material, or that circulating waters formed carbonatites by replacing silicate minerals with carbonate minerals. Some geologists thought that carbonatites could have formed from limestones that were remobilized by a solid, plastic flow—much like the flow of toothpaste squeezed out of a tube.

A major problem with the suggestion that carbonate material could melt, however, was the apparently high melting point of pure calcite or dolomite. Few geologists could believe that the temperature within the earth was high enough to melt calcite or dolomite. Another problem concerning the belief that limestones were the source of carbonatite was that in some areas, no limestones could be found anywhere near the occurrences of carbonatites. Also, the concept of the intrusion of limestones by the plastic flow of a solid was difficult to reconcile with many observations of carbonatites, including the occurrence of foreign igneous rock fragments composed of silicate minerals within them. The absence of fossils in the "limestone" also was noted as unusual, as most limestones have abundant fossils. The lack of fossils could be explained, however, by the melting hypothesis: the fossil evidence would have been destroyed during the melting.

Experiments in furnaces at temperatures and pressures similar to those expected deep within the earth have done much to support the melting hypothesis for the igneous formation of carbonatites. Several experiments in the late 1950's showed that carbonate minerals could melt at reasonably low temperatures (about 600 degrees Celsius) if abundant carbon dioxide and water vapor coexisted with the carbonate minerals. This dispelled the notion that molten carbonate could not exist within the earth. Also, the discovery in 1960 that the Ol Doinyo Lengai volcano was extruding carbonate lavas confirmed that carbonate liquids could exist within the earth. Similar experiments at high temperature and pressure on the composition of carbon dioxide or water vapor suggested that their composition could not produce carbonatites by the replacement of silicate minerals with carbonate minerals.

These experiments, combined with field observations (including the way carbonatites cross-cut surrounding rocks and the presence of foreign rock fragments), confirmed that most carbonatites formed by the intrusion of carbonate liquids. Even so, the experiments fell short of dispelling the notion that carbonate liquids could have been derived from melted limestone. The melted limestone hypothesis met objections, because isotopic and element concentrations in carbonatites were much different from those observed in limestone. For example, the elements lanthanum and niobium are hundreds of times more concentrated in carbonatites than they are in limestone. There is no known process that can produce this magnitude of element enrichment, either by melting or leaching the carbonate liquid from the solid rock through which it moved. Such observations have caused the limestone origin of carbonatites to be rejected by most geologists.

CONTINUING STUDY OF CARBONATITE FORMATION

Scientists continue to try to understand how carbonatites form. Experiments in furnaces suggest that some carbonate liquids may separate from some silicate liquids similar to those occurring with carbonatites. This process would be like the separation of oil and water as immiscible liquids. Other experiments in furnaces

suggest that rocks more than 80 kilometers deep within the earth may melt in small amounts and produce the carbonate liquids and associated silicate liquids similar in composition to those observed in the natural rocks. Although these experiments fail to prove that carbonate liquids form in these ways, several other lines of evidence have convinced geologists that either of these possibilities could produce carbonatites in nature. For example, some possible source rocks that could melt and produce carbonatites or associated silicate rocks are sometimes carried up with lava from deep within the earth. The strontium-87 to strontium-86 ratios and the element contents of these possible source rocks have been measured and are similar to those expected for rocks that would melt and produce carbonate and silicate liquids. Strontium-87 forms from the decay of rubidium-87, and the ratio of strontium-87 to strontium-86 is a measurement of the age of the possible source rocks, and can help to identify the rocks that gave rise to the carbonatite magma.

Geologists want to understand how carbonatites form partly because of their economic importance. With this knowledge, they can design better strategies to find carbonatites or the associated silicate rocks that are as yet undiscovered. For example, carbonatites are very small targets to find on the surface of the earth, but if the silicate rocks associated with carbonatites contain abundant magnetic minerals, an aerial survey of the area can detect magnetic fields due to the rocks by using a magnetometer device. Once geologists find areas with magnetic anomalies, they can collect soil or stream samples in the area to see if any unusual minerals associated with carbonatites or the associated silicate rocks are present. They can also drill the area to see if any carbonatites are below the surface.

ECONOMIC VALUE OF CARBONATITES

The unusually high concentrations of some elements in certain minerals make carbonatites potential ores for these elements. Carbonatites have high concentrations of niobium, thorium, and the rare-earth elements of lower atomic number (such as lanthanum and cerium). Iron, titanium, copper, and manganese also have been mined from carbonatites.

Niobium has been economically extracted from the mineral pyrochlore at Fen, Norway, and at Kaiserstuhl, Germany. Niobium is used as an alloy in steel to resist high temperature; this steel is then used in gas turbines, rockets, and atomic power plants. Rare-earth elements have been mined from the carbonatite at Mountain Pass, California. The reserves of rare-earths are enormous at Mountain Pass (averaging about 7 percent) compared to other carbonatites. This carbonatite hosted the largest rare-earth mine in the United States. The mine closed for a few years, but rising demand for rare-earth elements prompted plans to reopen in 2011. Other carbonatites enriched with rare-earths occur in Malawi in Africa. The rare-earth elements are concentrated in many minerals, including perovskite, monazite, xenotime, and a variety of rare-earth carbonate minerals. The rare-earths are used as color phosphors in television picture tubes and as components in high-quality magnets used in stereo speakers and headphones. Thorium, often enriched along with the rare-earths in many carbonatites, tends to concentrate in the same minerals and deposits as do the rare-earth elements. Thorium is radioactive and has been used as a source of atomic energy. It has also been used for the manufacture of mantles for incandescent gas lights. The Oka Complex west of Montreal, Canada, has also been mined for rare-earth elements. Rising demand for rare-earths for electronics, plus dependency on China as the leading refiner of rare-earths, has spurred additional exploration for new resources.

Robert L. Cullers

FURTHER READING

Bell, Keith, ed. *Carbonatites: Genesis and Evolution.* Boston: Unwin Hyman, 1989. Papers collected from a meeting of Geological Association of Canada. Covers the formation and evolution of carbonatites. Several essays also cover research trends and recent findings. Includes illustrations and maps.

Bell, Keith, and J. Keller, eds. *Carbonatite Volcanism: Oldoinyo Lengai and the Petrogenesis of Natrocarbonatites.* New York: Springer-Verlag, 1995. Offers an in-depth study of volcanism and carbonatites in the Ol Doinyo Lengai region of Tanzania. Among the illustrations are several useful maps of the area. Bibliography and index.

Deer, W. A., R. A. Howie, and J. Zussman. *An Introduction to Rock-Forming Minerals.* 2d ed. London: Pearson Education Limited, 1992. A standard reference on mineralogy for advanced college students and above. Each chapter contains detailed descriptions of chemistry and crystal structure, usually with chemical analyses. Discussions of chemical variations in minerals are extensive. A condensation of a five-volume set originally published in the 1960's.

Faure, Gunter. *Origin of Igneous Rocks: The Isotopic Evidence.* New York: Springer, 2010. Descriptions of multiple radioactive isotope dating methods contained within this book. Principles of isotope geochemistry are explained early, making this book accessible to undergraduates. Includes data presented in diagrams, over four hundred original drawings, and a long list of references at the end.

Grotzinger, John, et al. *Understanding Earth.* 5th ed. New York: W. H. Freeman, 2006. Covers the formation and development of the earth. Written for high school students and general readers. Includes an index and a glossary of terms.

Heinrich, E. William. *The Geology of Carbonatites.* Skokie, Ill.: Rand McNally, 1966. A layperson can gain useful information from this volume, which is designed for specialists. Of special interest are sections on the history of carbonatite studies (Chapter 1), the economic aspects of carbonatites (Chapter 9), and descriptions and locations of the world's carbonatites (Chapters 11-16).

Jerram, Dougal, and Nick Petford. *The Field Description of Igneous Rocks.* 2d ed. Hoboken, N.J.: Wiley-Blackwell, 2011. Begins with a description of field skills and methodology. Contains chapters on lava flow and pyroclastic rocks. Designed for student and scientist use in the field.

Kapustin, Yuri L. *Mineralogy of Carbonatites.* Washington, D.C.: Amerind, 1980. Summarizes the vast variety of minerals that have been found in carbonatites. Much geologic jargon, but mineral collectors and readers with some geologic background will find it useful.

Larsen, Esper Signius. *Alkalic Rocks of Iron Hill, Gunnison County, Colorado.* Geological Survey Professional Paper 197-A. Washington, D.C.: Government Printing Office, 1942. Describes the wide variety of minerals and rocks at this complex containing carbonatite. The rocks are located on a geologic map. Useful for mineral collectors.

Menzies, L. A. D., and J. M. Martins. "The Jacupiranga Mine, São Paulo, Brazil." *Mineralogical Record* 15 (1984): 261-270. Provides a journal for the lay reader or mineral collector that summarizes the geologic occurrence of minerals with a minimum of technical language. It often contains beautiful photographs of minerals. An example of carbonatite minerals at a specific location.

Olsen, J. C., D. R. Shawe, L. C. Pray, and W. N. Sharp. *Rare Earth Mineral Deposits of the Mountain Pass District, San Bernardino County, California.* U.S. Geological Survey Professional Paper 261. Washington, D.C.: Government Printing Office, 1954. Describes the rare-earth carbonatite at Mountain Pass. Many detailed rock and mineral names, but a geologic map of the rock locations is included, so a mineral collector might find this source useful.

Parker, Raymond L., and William N. Sharp. *Mafic-Ultramafic Igneous Rocks and Associated Carbonatites of the Gem Park Complex, Custer and Fremont Counties, Colorado.* U.S. Geological Survey Professional Paper 649. Washington, D.C.: Government Printing Office, 1970. Gives a detailed description of the large number of minerals and the rocks found at Gem Park. Useful for a mineral collector. A layperson can read it with a dictionary of mineral and rock names. Photographs of minerals included.

Roberts, W. L., Thomas J. Campbell, and G. R. Rapp. *Encyclopedia of Minerals.* 2d ed. New York: Van Nostrand Reinhold, 1990. One of a variety of mineral references available to the layperson that gives common properties, composition, and color photographs of many minerals, including some found in carbonatites.

Sutherland, Lin. *The Volcanic Earth: Volcanoes and Plate Tectonics, Past, Present, and Future.* Sydney, Australia: University of New South Wales Press, 1995. Provides an easily understood overview of volcanic and tectonic processes, including the role of igneous rocks. Includes color maps and illustrations, as well as a bibliography.

Tuttle, O. F., and J. Gittins. *Carbonatites.* New York: Wiley-Interscience, 1966. Another fairly technical book; useful to a layperson who is not intimidated by mineral and rock names. Includes a description of the only active volcano to extrude carbonate lava, offered by the person who first descended into the volcanic vent. Includes a section on the location and description of carbonatites around the world.

Winter, John D. *An Introduction to Igneous and Metamorphic Petrology.* Upper Saddle River, N.J.: Prentice Hall, 2001. Provides a comprehensive overview of igneous and metamorphic rock formation. A basic understanding of algebra and spreadsheets is required. Discusses volcanism, metamorphism, thermodynamics, and trace elements; focuses on the theories and chemistry involved in petrology. Written for students taking a university-level course on igneous petrology, metamorphic petrology, or a combined course.

Woolley, A. R. *Alkaline Rocks and Carbonatites of the World.* Tulsa, Okla.: Geological Society, 2001. Provides information on carbonatites worldwide. Written in an encyclopedic manner, the texts organize the information by location. Contains excellent references.

See also: Andesitic Rocks; Anorthosites; Basaltic Rocks; Batholiths; Granitic Rocks; Igneous Rock Bodies; Kimberlites; Komatiites; Orbicular Rocks; Plutonic Rocks; Pyroclastic Rocks; Rocks: Physical Properties; Xenoliths.

CEMENT

Cement is a common construction material that is used to bond mineral fragments in order to produce a compact whole. The most common types of cement result from the reaction of lime and silica. These are called hydraulic cements for their ability to set and harden underwater.

PRINCIPAL TERMS

- **aggregate:** a mineral filler such as sand or gravel that, when mixed with cement paste, forms concrete
- **alumina:** sometimes called aluminum sesquioxide; a material found in clay minerals along with silica; tricalcium aluminate acts as a flux in cement manufacturing
- **clinker:** irregular lumps of fused raw materials to which gypsum is added before grinding into finely powdered cement
- **concrete:** a composite construction material that consists of aggregate particles bound by cement
- **gypsum:** a natural mineral, hydrated calcium sulfate; it helps control the setting time of cement
- **hydraulic cement:** any cement that sets and hardens under water; the most common type is known as Portland cement
- **lime:** a common name for calcium oxide; it appears in cement both in an uncombined form and combined with silica and alumina
- **silica:** silicon dioxide; it reacts with lime and alkali oxides, and is a key component in cement

TYPES OF CEMENT

By far the most common type of cement is called Portland cement, which was named for its similarity to Portland stone, a type of building stone. Portland cement consists primarily of lime, silica, and alumina. These materials are carefully ground, mixed, and heated to produce a finely powdered gray substance. In the presence of water, these ingredients react to form hydrated calcium silicates that, after setting, form a hardened product. Such a product is classified as a hydraulic cement because of its ability to set and harden under water. The hardening of concrete is not merely a matter of drying out, as the chemical reactions of cement can proceed even underwater.

While Portland cement is by far the most common type of cement, there are others. Most of these, such as high-early-strength cements, slag cements, Portland-pozzolan cements, and expansive cements, are variations on the basic Portland cement. Manufacturers can produce these specialized cements through slight variations of the basic chemical composition of Portland cement and through the use of various additives. Each of these cements is designed for a specific use, and their advantages include lower cost, higher strength, and faster setting times. Another type, high-alumina cement, is not based on Portland cement. Formed by the fusion of limestone and bauxite, high-alumina cement hardens rapidly and withstands the corrosive effects of sulfate waters (unlike basic Portland cement). Its early promise as a structural material has diminished because of a number of failures, but its ability to withstand high temperatures makes it quite useful in constructing furnaces.

Although the term "cement" is often used as a synonym for "concrete," cement is only the binding agent in concrete. Concrete is a mixture of aggregate (sand, gravel or crushed stone) and cement, which hardens to form a synthetic stone through a series of chemical reactions between the cement and the aggregate.

HYDRATION

Adding water to dry cement creates a paste that eventually hardens. The reaction of water with cement is known as the hydration process. This reaction involves much more than water molecules attaching themselves to the constituent elements of cement. Rather, the constituents are reorganized to form new, hydrated compounds. One of the first reactions involves the aluminates, particularly tricalcium aluminum oxide ($3CaO \cdot Al_2O_3$, hereafter abbreviated as C_3A). Note that the C in this formula is not the chemical symbol for carbon but shorthand for calcium oxide, or lime. Although C_3A is undesirable in cement, it forms as a by-product during the manufacturing process as the raw materials are heated. If allowed to react with water unchecked, it would lead to overly rapid setting, not allowing time for working the cement or concrete product. To avoid this difficulty, a carefully controlled amount of gypsum is added to the cement at the time of manufacture. Gypsum slows the hydration of C_3A, giving the more important calcium silicates time to react with water.

Hydration of the calcium silicates occurs more slowly than that of the aluminates but forms the basic, strength-giving structure of hardened Portland cement. Two different calcium silicates $3CaO \bullet SiO_2$ (abbreviated as C_3S) and $2CaO \bullet SiO_2$ (abbreviated as C_2S) are present in the cement. As with C, the S is not the same as the chemical symbol for sulfur. Both forms of calcium silicate react with water to produce the hydrated calcium silicate $(C_3S_2H_3)$. This product is sometimes called tobermorite because of the resemblance its molecular arrangement bears to that of the natural but rare mineral of the same name (taken from Tobermorey, Scotland). Another product of the reaction is calcium hydroxide, or $Ca(OH)_2$, which is integral to the microstructure of the hardened cement. It is important to note that as the exact ratios of the different constituents vary, so does the composition of the product. What is actually being produced, at different places and by different manufacturers, is a family of hydrated calcium silicates rather than one precise formula.

STAGES OF PRODUCTION

From the time water is added to the time the cement paste sets and fully hardens, cement goes through four general phases. Four main compounds are present during each stage: a gel of the above-mentioned hydrated compounds, crystals of calcium hydroxide, unhydrated cement, and water. The proportions of these compounds change with time; the cement gains rigidity as the percentage of hydrated calcium silicate increases with the attendant drop in the amount of free water.

These compounds eventually arrange themselves in loose, crumpled layers. For the first few minutes after adding water to cement, the two form a paste in which the cement is suspended within the water. At this stage, the cement dissolves in the water. The next stage, sometimes called the dormant period, lasts for one to four hours. During this time, the cement forms a gel and begins to set, losing pliability. The individual cement grains build a coating of hydration products, and loose, crumpled layers of hydrated calcium silicate begin to form. In the microscopic spaces between the cement grains, water is held by the surface forces of the cement particles. Larger spaces, called capillary pores, hold free water, which the cement slowly absorbs for use in the hydration process. During the third stage, which peaks about six hours after water is added, the coating of hydration products around the cement grains ruptures, exposing unhydrated cement to water and further building layers of calcium silicate. As these layers grow, they entrap water, which continues to react with the cement particles. They also contain calcium hydroxide (a by-product of the hydration of calcium silicates), which fills the larger pores and thus apparently contributes to the overall strength of the cement. The fourth stage produces the final setting and hardening of the cement. The hydration process may continue for years, and there will in all probability be a small percentage of the cement that never hydrates. Hydration of C_3S and C_2S occurs at different rates. In the first four weeks, the hydration of the C_3S contributes most to the strength of the hardening cement, and afterward the hydration of the C_2S contributes more to the cement's strength. After about a year, the hydration rates of the two compounds are roughly equal.

QUALITY MEASUREMENT

The quality of Portland cement can be measured by four principal physical characteristics: fineness, soundness, time of set, and compressive strength. The desired specifications for each may vary from one country to another and certainly will vary for the different types of cement, but certain generalizations may be made safely. The fineness of the cement plays a large role in determining the rate of hydration. The finer the cement, the larger the surface area of the cement particles and the faster and more complete the hydration. A method has recently been developed that measures particle fineness in terms of the specific surface (the surface area of cement particles measured in square centimeters per gram of cement). The two most common ways of measuring specific surface are the Wagner turbidimeter test and the Blaine air permeability test. To use the turbidimeter, a sample of cement is dispersed inside a tall glass container of kerosene (which does not react with the cement). A beam of light is then passed through the kerosene at given elevations at a specified time, and the concentration of cement is measured by a photoelectric cell. The specific surface can then be calculated from the photoelectric cell readings.

The air permeability test relies on the fact that the number and size of pores are functions of particle size and distribution. A given volume of air is drawn through a bed of cement, and the time it takes for the air to pass through the cement is used to calculate the specific surface of that sample.

Soundness is another important characteristic of cement. A sound cement is one that will not crack or disintegrate with time. Unsoundness is often caused by the delayed hydration and subsequent expansion of lime. The usual method of testing cement for soundness is in an autoclave. A small sample is placed in the autoclave after curing for twenty-four hours and is subjected to extremely high pressure for three hours. After the sample has cooled, it is measured and compared to its original length. If it has expanded less than 0.8 percent, the cement is usually considered sound. Some aggregate materials, such as chert, cause expansion and cracking as concrete sets and are unsuitable for use in concrete.

Time of setting is tested on fresh cement paste as it hardens. The ability of the paste to sustain a given weight on a needle of given diameter (usually a 300-gram load on a 10-millimeter diameter needle) can be easily correlated to its setting time.

Concrete, like other masonry products, is much stronger in compression (for example, a vertical pillar in which gravity pushes the pillar in on itself) than in tension (a horizontal beam that is bowed such that the bottom half wants to pull apart). Thus, it is most useful to study the compressive strength of concrete. The usual test is to make a two-inch cube of cement and sand (in a 1:2.75 mixture) and compress it until it breaks. Much can be learned by measuring the breaking load, type of fracturing, and other results.

USES IN CONSTRUCTION

Cement has been used for construction purposes since at least 4000 B.C.E. The Romans used a hydraulic cement based on slaked lime and volcanic ash in many of their construction projects, some of which are still standing.

One important use of cement is as mortar, or cement mixed with sand. Mortar is the substance that binds bricks, stone, and other masonry products. Cement is also the primary ingredient in grout, such as that used between tiles. The most important and common use of cement, however, is in making concrete.

Concrete is the product of mixing cement paste with a mineral aggregate. The aggregate, which acts as a filler, can be a wide variety of materials but is usually a sand or gravel. As a construction material, concrete has many advantages. First, it is inexpensive and readily available. The energy costs alone are a fraction of what they would be for a substance such as steel, and the raw materials for concrete are often available near the construction site, thus saving considerable transportation costs. Another important advantage of concrete is the ability to form it in a wide variety of shapes and sizes, quite often on the job site. Concrete is also known for its long life and low maintenance, due to its strong binding characteristics and resistance to water. Finally, concrete's high strength in compression and proven long-term performance make it a good choice for many structural components.

Concrete for structural uses comes in four major forms. Ready-mixed concrete is transported to a construction site as a cement paste and is then poured into forms to make roadways, driveways, floor slabs, foundation footings, and many other types of structural foundations. Precast concrete can be used for anything from a birdbath to wall slabs, which are cast at a concrete work and then transported to their intended site. A common example of a precast member might be the beams of a highway overpass. Reinforced concrete is any concrete to which reinforcement (usually steel rods) has been added in order to increase its strength. Prestressed concrete is a relatively new form of concrete. Developed in the 1920's, prestressed concrete is put under compression through the use of jacks or steel cables, such that a beam is always in compression, whereas an unstressed beam in the same place would experience tension.

These forms of concrete are used, often in combination, in a variety of ways. Slabs, walls, pipes, dams, spillways, and even elegant vaulted roofs (known as thinshell vaulting) are all made of concrete. Cement, especially as it is used in concrete, has played a crucial role in shaping the physical environment. Concrete is the most widely used manufactured construction material in the world. In most modern countries, the ratio of concrete consumption to steel consumption is at least ten to one. Although concrete often is taken for granted, the world is literally built upon it.

Brian J. Nichelson

FURTHER READING

Blanks, Robert F., and Henry L. Kennedy. *The Technology of Cement and Concrete*. New York: John Wiley & Sons, 1955. Offers helpful information on history, economy, and general background of the subject.

Gani, Mary S. J. *Cement and Concrete*. London: Chapman and Hall, 1997. Examines processes used to make cement and concrete, as well as the uses and applications of the materials. Appropriate for the layperson. Illustrations, index, and bibliography.

Gartner, E. M., and H. Uchikawa, eds. *Cement Technology*. Westerville, Ohio: American Ceramic Society, 1994. Examines the processes and protocol involved in the manufacture and use of cement.

Hewlette, Peter C., ed. *Lea's Chemistry of Cement and Concrete*. 4th ed. Burlington, Mass.: Butterworth-Heinemann, 2004. Revised version of Frederick Lea's classic with new information and author input. Discusses Portland cements, pozzolanic cements, and special-use cements. Covers cement admixtures and physical and chemical properties.

Mehta, P. Kumar. *Concrete: Structure, Properties, and Materials*. 2d ed. Englewood Cliffs, N.J.: Prentice-Hall, 1992. Well illustrated with tables, charts, photographs, and drawings. Definitions are abundant and clear. Contains numerous examples, all illustrated with photographs, modern-day projects, ranging from sculpture to dams. Contains information on the viscoelasticity of concrete, and the technology behind lightweight concretes.

Mindess, Sidney, J. Francis Young, and David Darwin. *Concrete*. 2d ed. Upper Saddle River, N.J.: Prentice-Hall, 2002. Covers the different cements, chemical reactions, aggregates, and all aspects of making and using concrete. Includes a brief but useful historical overview of cement. Offers a separate bibliography for each chapter.

Neville, A. M. *Properties of Concrete*. 4th ed. New York: John Wiley & Sons, 1996. Includes a lengthy discussion of cement. Provides good definitions; content is fairly technical. Covers the background on concrete and cement to prepare students for learning new techniques. Includes a detailed discussion of Portland cement, and discusses other types of cement. Includes references at the end of each chapter.

Neville, A. M., and J. J. Brooks. *Concrete Technology*. 2d ed. Upper Saddle River, N.J.: Prentice Hall, 2010. Discusses the chemistry, physical properties, and types of cement. Provides information on the role of cement in concrete constructions. Examines common challenges and problems with concrete construction.

Orchard, Dennis Frank. *Concrete Technology*. Vol. 1, Properties of Materials. 4th ed. London: Applied Science Publishers, 1979. Discusses cement for most of the book; content is challenging but gives a suitable overview for general readers. Refers to specifications and practices in both Great Britain and the United States.

Popovics, Sandor. *Strength and Related Properties of Concrete: A Quantitative Approach*. New York: John Wiley & Sons, 1998. Examines the properties and uses of cement. Accompanied by a computer disk that helps to illustrate concepts.

Portland Cement Association. *Principles of Quality Concrete*. New York: John Wiley & Sons, 1975. Designed to educate persons for employment in the concrete industry. Explains cement and concrete in the simplest terms. Provides a brief historical overview of the development of cement and concrete as well as such innovations as reinforcing and prestressing. Contains useful discussions of applications.

Taylor, Harry F. W. *Cement Chemistry*. 2d ed. London: T. Telford, 1997. Examines the chemical makeup and properties of cements and concretes, as well as chemical changes that may affect their usefulness. Emphasizes advanced chemistry and mathematics. Requires some background in the subject.

Troxell, George Earl, Harmer E. Davis, and Joe W. Kelly. *Composition and Properties of Concrete*. 2d ed. New York: McGraw-Hill, 1968. Contains a section on cement. Offers sufficient definitions of terms and other aids to allow the general reader to explore the subject.

Wilson, Alan D., and John W. Nicholson. *Acid-Base Cements: Their Biomedical and Industrial Applications*. New York: University of Cambridge, 2005. Discusses acid-base cements and the chemical reactions involved in their creation. Discusses theoretical topics as well. Emphasizes the relationship of acid-base cements with water.

See also: Aluminum Deposits; Building Stone; Coal; Diamonds; Dolomite; Earth Resources; Fertilizers; Gold and Silver; Igneous Rock Bodies; Industrial Metals; Industrial Nonmetals; Iron Deposits; Manganese Nodules; Metamorphic Rock Classification; Metamorphic Textures; Pegmatites; Platinum Group Metals; Rocks: Physical Properties; Salt Domes and Salt Diapirs; Sedimentary Mineral Deposits; Sedimentary Rock Classification; Silicates; Uranium Deposits.

CHEMICAL PRECIPITATES

Chemical precipitates are useful in nearly all human endeavors. Precipitates are materials formed by precipitation—the formation of solids from solution. Uses are numerous, including pigments and water treatment. Additionally, precipitates occur in nature to form rocks such as shale and limestone.

PRINCIPAL TERMS

- **covalent bond:** a chemical bond characterized by electron sharing
- **dye:** a colored solution
- **flocculation:** the process by which particles are released from a colloid
- **ionic bond:** a chemical bond formed from the transfer of electrons and electromagnetic attraction between the ions
- **ions:** atoms with a net charge through electron addition or loss
- **sedimentary rock:** rocks created by layering of sediment
- **semipermeable membrane:** a membrane with spaces large enough for the molecules, but not the ions, of a liquid to pass through
- **solute:** a substance that is dissolved
- **solution:** a mixture in which a solute is dissolved in a solvent
- **solvent:** a substance that dissolves another

The Solvation Process

A proper understanding of how things dissolve is needed to discuss chemical precipitates and their uses. Chemical bonds are, in general, either ionic or covalent.

Covalent bonds occur where atoms share electrons. These bonds are far stronger than ionic bonds, which occur when electrons are transferred. A covalent bond occurs when there is a large difference in electronegativity.

Electronegativity is the measure of how tightly an atom holds its electrons. Atoms with high electronegativity, such as fluorine and oxygen, hold their electrons tightly, whereas atoms of low electronegativity, such as sodium and boron, hold their electrons loosely. This is the result of properties of electron orbitals.

Atoms with eight electrons in their outermost orbital are the most stable. As a result, atoms with many electrons in their outermost orbital, such as fluorine (with seven electrons) and oxygen (with six electrons), try to gain another electron; atoms with few electrons, such as sodium (with one electron) and calcium (with two electrons), try to lose their electrons. When atoms of different electronegativity bond, the less electronegative atom gives its electrons to the more electronegative atom. Once the electrons are transferred, the atoms become ions and are then held together by electrostatic attraction.

Equally covalent bonds result in sharing; and when one is more electronegative than the other, as in water, the result is a polar molecule. In such a case, the oxygen atoms hold the electrons more tightly than the hydrogen atoms, meaning that the oxygen portion has a net negative charge. The hydrogen portion is positive. Water's shape is also of note, as it is a bent molecule. However, regardless of the shape of the molecule, the electronegative principles are the same.

Taking the example of salt dissolving in water, one can see the effect of these properties. When table salt (sodium chloride, or $NaCl$) is placed in water, it dissolves. Sodium chloride is an ionic compound. Because its bonds are weaker than the pull of water, the sodium atoms are separated from the chlorine atoms and align to the water molecule's poles. The water molecules surround the sodium and chlorine molecules, and the structure of the sodium chloride is subsumed by the water.

Dissolved systems are not static; they are under dynamic equilibrium. The solute is continuously coming in and out of solution. When the rate of precipitation exceeds the rate of dissolution, the solute comes out of solution; if the converse is true, the solute dissolves.

Many factors affect solubility. The behavior is particularly complex because it depends on the interactions of the crystal lattice. Only recently have computer models allowed for the prediction of solubility. The solubility of any given substance is related to temperature and pressure.

Also important to note is that not all soluble substances are soluble in the same solutes. The major division is between organic and inorganic solvents.

In general, organic molecules are the best solvents of organic solutes, and inorganic solvents work best on inorganic solutes.

COLLOIDS AND SUSPENSION

Solids can be contained in water by means other than through dissolving. Solids also can be contained as suspensions and as colloids.

A suspension is a system in which the particles are larger than single molecules and ions but are still small enough to remain in the solvent. A colloid is a substance in which the particles are distributed homogeneously and stably; without changes to the system, the colloid will not separate. Suspensions, in contrast, will naturally separate over time. This process is called flocculation.

Flocculation is the process wherein colloids come out of suspension in the form of clumps of particles called floc. Flocculation differs from precipitation in that the particles are suspended in a colloid rather than dissolved in a solution. The flocs can then come out of suspension.

WATER TREATMENT

Water treatment is the process by which water is cleaned for human use. The use of precipitates is critical for treatment, but much water treatment does not rely on precipitates.

First, in water treatment, waste is screened out using filters. The chunks of waste are usually disposed of in a landfill after drying. Then, after water filtration, the resulting waste is directed to a settlement tank, where the waste particles are allowed to settle. The result is called sludge, which can be refined into fertilizer. The water is then disinfected by chlorination.

The properties of precipitates can remove solutes from water. This is important because many solutes have negative effects upon water. Some solutes, such as iron, can stain objects, such as water fixtures, and can change the color of the water. Sometimes solutes are clear when drawn from a tap but turn yellow upon standing (in place without movement). Other solutes, such as hydrogen sulfide, smell like rotten eggs. Another common concern is the solute pH (potential hydrogen), which can cause corrosion. These chemical properties are caused by things dissolved in water; precipitating removes the properties from the water.

Aeration, the placing of a solute in contact with air, is a common means of precipitation. In this method, a substance dissolved in water is removed from the water by tiny bubbles passing through the solution. The increase of surface area between the solution and the air bubbles creates a space for the dissolved gases to react with the gases in the air bubbles, in this manner the air bubbles "scrub" the water clean of the solutes.

Another common precipitation process is water softening. Hard water is water with a high dissolved mineral content. Most processes aim to remove calcium, magnesium, and other metal ions from the water. One way of doing this is to add compounds to the hard water that will react with the dissolved metals to form an insoluble compound. Another process known as ion exchange works by exchanging one type of ion in a substance for another. Because the goal in water softening is to remove certain kinds of ions, the exchange of ions works fine. Typically the process works by passing water through resin beads.

Another method, often used in the American Midwest, Florida, and Texas, is lime softening, which involves adding lime to water. The water then undergoes several reactions that lead to the removal of ions through precipitation. This process is notable because the total dissolved particles decreases (with ion exchange, little to no change occurs).

Reverse osmosis is another water treatment method. Osmosis is a natural process that involves moving water (or any other liquid with different concentrations) through a semipermeable membrane. The liquids tend to move from regions of low solute concentrations to high concentrations to create chemical potential equilibrium. Reverse osmosis involves forcing a liquid with dissolved solute through the semipermeable membrane by applying a pressure greater than the osmotic pressure. This filters out the solutes. The holes in the filters are small enough to prevent the ions from passing through them.

Chelation is often used as well. In chelation, metal ions are dissolved by organic compounds. By dissolving them in the organic compound, they are removed from the water.

By all of these processes, salts and metal ions can be removed. Salts are ionic compounds that result from the neutralization of an acid with a base. Acids are acidic because they produce H^+ ions, whereas bases produce OH⁻ ions. When acids and bases come into contact, they undergo a special reaction called neutralization. In neutralization, the H^+ and OH^- ions form water, leaving the anions (negative ions) of the acid and the cations (positive ions) of the base to form an electrically neutral compound. Because it is made of ions in this manner, it is an ionic compound.

To change the pH of the water, one can add either an acid or a base to neutralize the liquid's excess. (pH is a measure of how acidic or basic something is.) According to the system, pure water is neutral with a pH of 7. Solutions with pH less than 7 are acidic and those with pH greater than 7 are basic or alkaline. pH is a logarithmic scale based on hydrogen ion activity. Improper pH can cause problems, such as pipe corrosion and harm to plants and other vegetation, so managing pH is an important part of water treatment.

PIGMENTS

A pigment is a material that provides color to something as a result of absorbing and reflecting certain wavelengths of light. Pigments are used to make many things, ranging from clothing to paint. Pigments are suspensions, but precipitation is important for manufacturing a specific kind of pigment called a lake pigment.

A lake pigment is made when a dye is precipitated with an inert binder, called a mordant. A dye is a colored solution. Dyes do not always hold well, so they need to be changed chemically to supply color. The mordant is a metallic salt. Metallic salts are compounds like sodium chloride. The colored precipitant of a reaction of a dye and a mordant, such as alumina, is usually more stable and will remain in the object being colored better than the dye alone would.

This process is ancient. In earlier times chalk, white clay, and bones were used for their calcium carbonate and calcium phosphate. Now, however, the most common salts include aluminum hydroxide, alumina (aluminum oxide), calcium sulfate, and barium sulfate. No matter the era, these mordants were all insoluble in dyes and had neutral colors. The neutral colors are needed so that the color of the precipitate is dependent upon the dye.

ROCK FORMATION

Some sedimentary rocks are formed by the process of precipitation as well. Water is a polar molecule

and a good solvent. Seawater contains many kinds of solutes, and the average liter has about 35 grams of dissolved salts.

The oceans exhibit a great deal of variance in temperature and pressure across their expanse and depths, so it is only natural that they should also display a wide variance in what and how much they can dissolve. The varying nature of seawater leads to different rock formations.

A good example of this is the formation of limestone, which is made of calcium carbonate (a soluble). In equatorial surface seawater, however, there is so much limestone dissolved in the seawater that it is practically insoluble. As a result, it tends to come out of solution to form rock. The vast majority of limestone laid down, however, comes from sea-life shells, which are composed of calcium carbonate as well. When these creatures die, they fall to the seabed and form layers of calcium carbonate. A different scenario is possible, however.

Should a creature's shell fall below a certain depth, it will dissolve. This depth is known as the carbonate (or calcite) compensation level. Under conditions that exist below a certain depth, calcium carbonate's solubility increases dramatically. The calcium compensation level depends upon many factors but is about 42 to 500 meters (138 to 1,640 feet) in the Pacific Ocean, except in the equatorial upwelling zone, where it is 5,000 m (16,404 ft). Below this level, the seabed is mainly made up of silicates; above this level is carbonates.

Additionally, under certain conditions, dissolved substances precipitate from water. Warm water often releases many precipitates, causing the mineral deposits in many salt lakes. Additionally, the deposition of certain minerals reveals to climatologists the temperature of the seas.

Another result of precipitates is manganese nodules. These are metallic spheroids that form on the ocean bottom. They may prove important in future metal production.

CEMENTATION

Sedimentary rocks are formed by minerals coming out of solution and settling on the bottom of a body of water. As this happens over time, the layers begin to compress. Water containing ions then flows between the grains and precipitates them out to form crystalline material between the grains, joining them. This can occur through groundwater or in seafloor sediments.

This process is used by humans in the making of cement. Various chemicals are mixed in an aqueous solution and then are left to dry. As the moisture content changes, the chemicals are left to precipitate out and crystallize. These crystals give concrete its strength.

Gina Hagler

FURTHER READING

Chang, Raymond. *Chemistry*. Boston: McGraw-Hill Higher Education, 2007. This comprehensive textbook is lucidly written. Discusses the chemistry of precipitation.

Crittenden, John C. *Water Treatment Principles and Design*. Hoboken, N.J.: Wiley, 2005. Written by a member of one of the leading water treatment companies, this is the definitive work on water treatment facility design.

Delamare, Francois, and Bernard Guineau. *Colors: The Story of Dyes and Pigments*. New York: H. N. Abrams, 2000. Full of pictures, this work tells the history of pigments. Also provides cultural context and a history of chemical development.

Edzwald, James K. *Water Quality and Treatment: A Handbook on Drinking Water*. New York: McGraw-Hill, 2011. A guide to water treatment methods. Addresses chemical principles, provides context, and explains water treatment.

Faulkner, Edwin B., and Russell J. Schwartz. *High Performance Pigments*. Hoboken, N.J.: Wiley, 2009. Provides details about modern high-performance pigments. Also provides details on chemistry and addresses environmental and health issues.

Lutgens, Frederick K., and Edward J. Tarbuck. *Essentials of Geology*. Boston: Prentice Hall, 2012. Clearly written, this work provides an authoritative guide to geology. Examines the formation of rock from precipitates.

See also: Crystals; Evaporites; Limestone; Manganese Nodules; Water and Ice.

CLATHRATES AND METHANE HYDRATES

Clathrates (also known as gas hydrates) are solid crystals that occur when water molecules form a dodecahedron lattice around gas molecules. A wide range of gas hydrates occur naturally, but methane hydrates are the most common. Increased scientific attention has been paid to methane clathrates, both as an important factor in climate change and as a potential energy source.

PRINCIPAL TERMS

- **carbon cycle:** a natural cycle by which carbon is absorbed by plants through the soil and air; plants in turn are eaten by animals, which exhale the carbon as carbon dioxide into the atmosphere and whose bodies decompose after death, returning carbon to the soil
- **carbon sequestration:** a chemical process whereby carbon dioxide is removed from an energy source and isolated in a secure chamber to prevent its release into the atmosphere
- **clathrate gun hypothesis:** a theory that states that the Permian extinction was caused by a sudden release of methane from gas clathrates
- **dodecahedron:** a three-dimensional geometric configuration with twelve faces
- **guest molecule:** a molecule contained inside a clathrate
- **lattice:** a cage-like shape
- **permafrost:** frozen soil found beneath ice and snow packs in polar regions

BASIC PRINCIPLES

Methane hydrates (also known as clathrates) are lattices of solid ice enveloping (but not bonding with) methane molecules. In conditions in which temperatures are low and pressure is high, water crystallizes in a twelve-sided, three-dimensional lattice (a cage-like structure) known as a dodecahedron. Hydrates are commonly found in areas in which both water and methane are available (such as the bottom of the ocean and in polar and glacial regions).

Hydrates are in and of themselves structurally unstable, but when they envelop a guest molecule (such as that of methane and other gases and nongases), they become more sound. Still, the stability of clathrates is highly dependent on the environmental conditions in which they are located. If they are exposed to lower levels of pressure and to higher temperatures, they quickly destabilize. If brought to room temperature, for example, a clathrate will, as it melts, pop and sizzle as the methane contained therein escapes. If a clathrate is exposed to an open flame, the ice of the hydrate will burn.

There are three basic types of gas hydrates, which are differentiated by their structures. For example, methane hydrates (along with carbon dioxide hydrates) are part of the structure I category, in which forty-six water molecules are organized into lattices that contain two small cavities and six large cavities. Structure II hydrates are larger, containing multiple and smaller gas molecules such as propane and isobutane. Structure H hydrates also contain multiple gas molecules, although those molecules may be larger. Both structure II and structure H hydrates are rare, however, while structure I clathrates are found with greater frequency and in larger volumes around the world.

HISTORY

In 1810, scientists Humphrey Davy and Michael Faraday, while experimenting with chlorine and water mixtures, noticed the presence of solid material forming when the liquids were just above freezing levels. For the remainder of the nineteenth century, clathrates were simply viewed as a scientific curiosity, as researchers attempted to understand the process by which an unstable water frame could form around, but not bond with, a guest molecule.

In the early twentieth century, however, prevailing attitudes toward clathrates evolved from curiosity to irritation. By the 1930's, modern civilization had developed the technology to tap into and extract natural gas and transport it through a system of pipes. E. G. Hammerschmidt, who at the time was working for the Texoma Natural Gas Company, noticed that in cold-weather environments, the company's pipelines were frequently clogged by frozen ice. However, the ice seemed to freeze at higher temperatures, which led Hammerschmidt to conclude that the frozen blockages were clathrates. Shortly thereafter, scientists began to study clathrates more closely in an effort to find ways to prevent them from forming in natural gas pipelines.

In the decades that followed, it became apparent that clathrates were naturally occurring compounds that could form not only in pipelines but also in the permafrost of polar regions and on the ocean floor. Scientists from around the world became interested in locating deposits of the substance, which had been dubbed solid natural gas. The United States, the Soviet Union, India, and other countries all launched efforts to find and access what were believed to be large deposits of clathrates around the world. By the end of the twentieth century, those countries, along with other nations, coordinated their efforts to study methane clathrates with two general perspectives in mind: the role clathrates play in the earth's natural processes and the potential that methane hydrates represent as an energy source for the future.

DISTRIBUTION OF CLATHRATES

Methane hydrates are found wherever conditions are right for their stability. They are found on the ocean floor, in regions in which temperatures are just above freezing and in which pressure is high and consistent. Clathrates also are found in permafrost, the layer of soil beneath the snow and ice in polar regions. In both types of environment, the methane hydrates are found between 150 and 2,000 meters (492 to 6,560 feet) below the surface.

The volume of methane hydrates throughout the world is significant. The total amount of clathrates in the world far surpasses the volume of conventional natural gas. According to the most recent U.S. Geological Survey review of the world's clathrate inventory, approximately 10,000 gigatons (a gigaton is the equivalent of one billion tons) of methane hydrates exist in 44 regions in which samples have been uncovered and in 113 other regions in which evidence suggests the presence of such clathrates.

A POTENTIAL ALTERNATIVE FUEL SOURCE

Scientists have become intrigued by the existence of methane hydrates in such large volume. It is therefore understandable that governments and private industry have developed an interest in pursuing clathrates as an alternative source of energy.

The challenge is extracting clathrates in a manner that is both safe and cost-effective. Governments, observing that the energy contained in methane hydrates exceeds the total volume of oil, conventional natural gas, and coal available in the world, have

launched individual and intergovernmental efforts to launch the clathrate market. Japan, for example, has developed plans to tap into clathrates as an energy source by 2016. The United States has for decades invested research dollars toward this pursuit, and a growing number of other nations are developing clathrate energy programs, too.

The dangers involved in "harvesting" methane hydrates at their remote sources are both immediate and long-term. For example, drilling for clathrates on the ocean floor can destabilize the surface, creating safety risks for workers and potentially disrupting nearby fossil-fuel drilling operations. Furthermore, scientists must find ways to store and transfer methane from these clathrates once they have been harvested. When it is released naturally, methane can break down and release its carbon components (namely, carbon dioxide, a greenhouse gas) into the atmosphere. Therefore, using computer modeling and laboratory experiments, scientists are investigating ways to utilize the energy benefits of the methane clathrates while at the same time isolating carbon (a process known as carbon sequestration). This particular aspect of research on clathrates and methane hydrates remains a major challenge and, at the same time, an important key toward the successful utilization of clathrates as a fuel source.

METHANE HYDRATES IN THE CARBON CYCLE

Clathrates have undergone an evolution in terms of scientific pursuits. The concept was at first a curiosity and thereafter developed into a problem to be solved. Today, methane hydrates are considered extremely important, warranting significant investment of money and research from both the private and public sectors.

Methane hydrates are also seen as key to the carbon cycle. The carbon cycle is a concept whereby one of the most critical elements for life on Earth is shared among the planet's many different forms of life and systems. Within this framework, carbon dioxide is absorbed by plants through the air and the soil. The plants are eaten by animals, which in turn are eaten by others (including humans). The animals also exhale carbon dioxide and, when they die, decompose and release carbon back into the soil.

Clathrates add another dimension to the carbon cycle. By capturing methane, a carbon compound, clathrates maintain a balance of the earth's methane supply. However, as stated earlier, methane hydrates are only stable under certain temperature and pressure conditions, particularly those found in permafrost and at the bottom of the ocean. If temperatures rise, the lattice can quickly destabilize, releasing its methane guest molecule. On a large scale, such events mean a rapid release of carbon into the atmosphere.

Such events have happened before in Earth's history. For example, scientists have found plentiful evidence of clathrate involvement in changes that occurred during the late Precambrian era, hundreds of millions of years ago. Studies of Precambrian rocks show signs of rapid carbon releases in areas in which clathrates had existed. These samples suggest that in light of the significant volume of carbon released during these events, clathrates played a major role in the large-scale development of plant and animal life during that era.

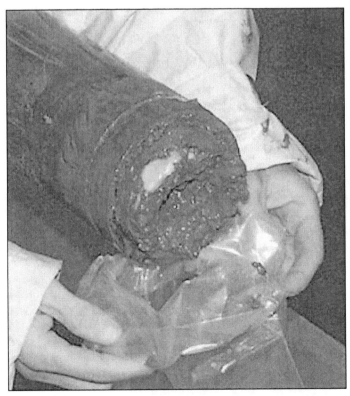

This discovery core shows white methane hydrate mixed with dark mud. The core penetrated and recovered a sample of the upper 2.1 m (5 feet) of a mud volcano 24 km (15 miles) off the southern California coast in the Santa Monica Basin. (U.S. Geological Survey)

CLATHRATES AND CLIMATE CHANGE

Methane hydrates represent a major enticement for alternative energy enthusiasts, particularly because they exist in much greater volume than other energy fossil sources and natural gas fuel sources. Additionally, if the myriad government- and privately sponsored efforts to develop ways to harvest and contain methane from these clathrates succeed, the resulting energy may be considerably cleaner than fossil fuels that produce greenhouse gases.

Separate from the issue of using clathrates as a clean energy source is the potentially significant danger that methane hydrates pose for Earth's atmosphere. As stated earlier, when destabilized naturally, methane hydrates simply release carbon dioxide (under certain conditions, those gases are released into the air). However, in their present locations (on the ocean floor and deep within the permafrost), they remain stable under precarious temperature and pressure conditions.

If these conditions are changed even slightly, the potential exists for a major release of methane into the air and the atmosphere. For example, clathrate breakdowns in the permafrost could trigger landslides, sending sediment into the sea, contributing to rising sea levels. Geological evidence from the end of the Pleistocene epoch (nearly 12,000 years ago) shows that major changes to the slopes of the earth's continents were caused by the release of clathrate gases.

Additionally, scientists believe that clathrate breakdowns in both the permafrost and on the ocean floor could significantly hasten global warming. This disastrous scenario stems from the notion that such a release, caused by rising temperatures in the ocean, would abruptly set off a chain reaction among the planet's methane hydrates and, in turn, quickly raise the planet's temperature by several degrees. Some scientists argue that such events have happened in Earth's history, most dramatically at the end of the Permian period (about 251 million years ago). Adherents to the clathrate gun hypothesis believe that a major release of methane hydrate gas, the cause of which is unknown, triggered a mass extinction of nearly 90 percent of life on Earth.

The debate concerning the validity of the clathrate gun hypothesis continues. Many scientists believe that clathrate breakdowns were not necessarily the primary cause of the Permian extinction. Nevertheless, most researchers agree that clathrates likely contributed to some degree to that event, an event from which Earth took 30 million years to recover. In light of the severity of that event, the notion that a simple elevation of a degree or two in temperature could trigger a massive breakdown in sea- and land-based gas clathrates remains a concern among global warming experts.

IMPLICATIONS AND FUTURE PROSPECTS

Clathrates, particularly methane hydrates, have evolved from a scientific curiosity to a major component in both the pursuit of alternative energy sources and the understanding of climate change. Scientists continually seek to understand the dynamics of these unusual geochemical configurations. Advances in relevant technologies, including the application of computer models and advances in thermal imaging systems, are contributing to the evolution of clathrate studies.

Although debate over the clathrate gun hypothesis continues, scientists are in agreement that methane hydrates could contribute to global warming and climate changes. However, researchers are also excited at the prospects of tapping into clathrates as an alternative energy source. To this end, scientists are pursuing technological advances to explore the safe extraction and use of methane from hydrates (including carbon sequestration practices). Their hope is to present to the modern world the use of clathrate-based energy that reduces greenhouse gas emissions and protects the planet from further warming trends.

Michael P. Auerbach

FURTHER READING

Demirbas, Ayhan. *Methane Gas Hydrates.* New York: Springer, 2010. Discusses methane hydrate within the context of its applicability as an alternative energy source, including the risks associated with such use, such as the immediate dangers posed in extraction as well as long-term risks for the environment.

Dorton, Peter, Saman Alavi, and T. K. Woo. "Free Energies of Carbon Dioxide Sequestration and Methane Recovery in Clathrate Hydrates." *Journal of Chemical Physics* 127, no. 12 (2007). Presents an analysis of the different types of clathrates and their stability in relation to the different types of gas molecules contained therein. The authors conclude that, based on simulations performed in the laboratory, methane hydrates are the most stable under certain conditions.

Giavarini, Carlo, and Keith Hester. *Gas Hydrates: Immense Energy Potential and Environmental Challenges.* New York: Springer, 2011. Discussion of the environmental risks of exploring the use of clathrates as an alternative energy source. Describes the type of methods that would be needed for extraction and transportation of methane hydrates.

Huo, Hu, et al. "Mechanical and Thermal Properties of Methane Clathrate Hydrates as an Alternative Energy Resource." *Journal of Renewable and Sustainable Energy* 3, no. 6 (2011). Presents an overview of a series of experiments conducted to understand the properties of methane clathrates. The main motive of these experiments is the eventual use of methane hydrates as a future alternative energy source.

Kurten, T., et al. "Large Methane Releases Lead to Strong Aerosol Forcing and Reduced Cloudiness." *Atmospheric Chemistry and Physics Discussions* 11, no. 3 (2011): 9057-9081. Presents an argument that a large-scale release of methane due to the destabilization of methane hydrates has the potential to significantly advance the process of global warming, a theory known as the "clathrate gun hypothesis".

Sloan, E. Dendy, Jr., and Carolyn Koh. *Clathrate Hydrates of Natural Gases.* 3d ed. London: CRC Press, 2007. Provides updated information about research conducted on methane hydrates, including discussions on the use of computer models to better understand clathrate dynamics.

See also: Crystals; Dolomite; Geologic Settings of Resources; Hydrothermal Mineralization; Unconventional Energy Resources; Water and Ice.

CLAYS AND CLAY MINERALS

Clays are fine-grained materials with unique properties, such as plastic behavior when wet. They form by the weathering of silicate rocks at the earth's surface, by diagenetic reactions, and by hydrothermal alteration. An understanding of clays is important to solving problems in petroleum geology, engineering, and environmental science.

PRINCIPAL TERMS

- **authigenic minerals:** minerals that formed in place, usually by diagenetic processes
- **cation exchange capacity:** the ability of a clay to adsorb and exchange cations, or positively charged ions, within its environment
- **chemical weathering:** a change in the chemical and mineralogical composition of rocks by means of reaction with water at the earth's surface
- **detrital minerals:** minerals that have been eroded, transported, and deposited as sediments
- **diagenesis:** the conversion of unconsolidated sediment into consolidated rock after burial by the processes of compaction, cementation, recrystallization, and replacement
- **hydrolysis:** a chemical weathering process that produces clays by the reaction of carbonic acid with aluminosilicate minerals
- **phyllosilicate:** a mineral with silica tetrahedra arranged in a sheet structure
- **shale:** a sedimentary rock with a high concentration of clays

DEFINITION AND PROPERTIES

The definition of clays varies depending on the scientific discipline or application. An engineer's definition, which is based on particle size, differs from a mineralogist's definition, which is based on crystal structure. In the broadest sense, clays are materials that have a very fine grain size (less than 0.002 millimeter) and behave plastically when wet. A more specific definition of a clay mineral is a hydrous aluminum phyllosilicate, or, more simply stated, a mineral that contains water, aluminum, and silicon, and has a layered structure. The term "clays" will be used for the broad definition, and "clay mineral" for the specific definition. Rock flour, or material that was ground to a fine powder by glaciers, would fit the definition of clays; however, it may contain minerals such as quartz and feldspars that do not fit the definition of a clay mineral. Mica is a hydrous aluminum phyllosilicate, but it often occurs as large crystals, so it does not fit the definition of clays. Certain

minerals such as zeolites and hydroxides (goethite and gibbsite) have a very fine grain size and physical properties similar to clay minerals, so they are often included with the clay minerals. Unique properties of clays, including their plastic behavior when wet and ability to adsorb water and ions in solution, can be attributed to their small crystal size, large surface area, and unique crystal structure.

CLAY MINERAL STRUCTURE

There are two basic elements of the clay mineral structure: a tetrahedral sheet and an octahedral sheet. A silicon atom surrounded by four oxygen atoms forms the basic building block of all silicate minerals, the four-sided silica tetrahedron. In phyllosilicate minerals, the silica tetrahedra are linked together by sharing the three oxygen atoms at the corners of the tetrahedra, forming a continuous sheet. The tetrahedral sheet has a negative charge and a general chemical formula of $Si_4O_{10}{}^{4-}$. The octahedral sheet consists of a sandwich of positively charged atoms (cations), like magnesium, iron, or aluminum, between sheets of oxygen and hydroxyl anions, or negatively charged ions (OH^-). Each cation is surrounded by six anions that lie at the corners of an octahedron, a shape made of eight triangles. The octahedra are linked together to form a sheet by sharing the anions on the edges of the octahedra.

There are two types of octahedral sheets: trioctahedral and dioctahedral. The prefixes refer to the fraction of octahedral occupied by cations: 2/3 for dioctahedral and 3/3 for trioctahedral. The trioctahedral sheet is composed of divalent cations such as Mg^{2+} and Fe^{2+}. For every six hydroxyl anions, the trioctahedral sheet contains three cations, resulting in a sheet where all the available octahedral sites contain a cation. The dioctahedral sheet contains trivalent cations such as Al^{3+} and Fe^{3+}. The dioctahedral sheet has only two cations per six hydroxyl anions, resulting in only two-thirds of the available octahedral sites being filled with cations. The general chemical formulas for the trioctahedral and dioctahedral sheets are Mg_3OH_6 and Al_2OH_6, respectively.

Layers in clay minerals are made of combinations of tetrahedral and octahedral sheets. The unshared oxygen of the silica tetrahedra take the place of some of the hydroxyl anions in the octahedral sheet, resulting in neutralization of the negative charges on the tetrahedral sheet. There are two types of layers: a 1:1 or T-O layer made up of one tetrahedral sheet and one octahedral sheet, and a 2:1 or T-O-T layer, which contains one octahedral sheet sandwiched between two tetrahedral sheets. Not all minerals with these layered structures are considered clay minerals.

The prototype clays have neutral layer charges, with the layers held together by weak van der Waals bonds where positive portions of one sheet are attracted to negative portions of another sheet. Kaolinite and serpentine are prototype 1:1 minerals. Kaolinite is dioctahedral and serpentine is trioctahedral. Kaolinite, a pure white clay, is the major constituent of fine porcelain. Serpentine is not generally considered a clay mineral. It can occur as chrysotile asbestos, which was widely used as insulation material before it was recognized as a health hazard. Talc and pyrophyllite are the trioctahedral and dioctahedral 2:1 prototype minerals. The waxy or slippery feel of talc (also not a clay mineral) is the result of cleavage along the weak van der Waals bonds between layers.

Because of the various types of ionic substitutions in the tetrahedral and octahedral sheets, the layers may develop a negative charge, which needs to be balanced by interlayer materials. A lower-valence cation may substitute for a higher-valence cation in the tetrahedral or octahedral sheets. Layers can also develop a negative charge if some of the sites in a trioctahedral sheet are left vacant. In the mica group of minerals, which are also not considered clays, the charge on the 2:1 layer is -1 per repeat unit and is balanced by a positively charged potassium ion that occurs between the layers. Muscovite is a dioctahedral 2:1 phyllosilicate, and biotite is trioctahedral. The perfect cleavage of micas is in the direction parallel to the tetrahedral sheets and allows them to be peeled into paper-thin sheets. Micas usually occur as larger crystals and, therefore, are not considered true clay minerals; however, illite, a common clay mineral, has a structure and chemical formula similar to muscovite, a mica. The chlorite clay minerals have an extra octahedral sheet between the 2:1 layers to balance the excess negative charges. Two groups of clay minerals known as vermiculite and smectite have negatively charged layers that are balanced by hydrated cations between the layers; a hydrated cation is a positively charged ion, such as sodium, that is surrounded by water. In smectites, this water is held very loosely between the 2:1 layers and can be easily lost or gained depending on the humidity of the environment, causing these clays to shrink or swell.

FORMATION AND DEPOSITION

Clay minerals occur in soils, sediments, sedimentary rocks, and some metamorphic rocks. Sedimentary rocks cover approximately 80 percent of the earth's surface, and shales are the most common type of sedimentary rock. Because shales are composed predominantly of clays, their abundance makes clays one of the most important constituents of the earth's surface.

Most clays form through the breakdown and weathering of minerals rich in aluminum and silicon at the earth's surface. Physical weathering is the breaking and fragmentation of rocks with no change in mineralogical or chemical composition. This process can form clay-sized particles, but it does not form clay minerals. Physical weathering, however, increases the surface area of minerals, which then favors chemical weathering. A change in the chemical and mineralogical composition of rocks by reaction with water at the earth's surface is called chemical weathering. One chemical weathering reaction that results in the formation of clays is called hydrolysis. Water and carbonic acid, a weak acid, react with aluminum silicate minerals, resulting in the production of a clay mineral plus ions in solution. (The weak acid is called carbonic acid; it forms when carbon dioxide, a common gas in the atmosphere, is dissolved in rainwater.) The type and amount of clay that forms by this reaction depend on the nature of the rock being weathered (the parent material) and the intensity of weathering.

Clays produced by weathering are eventually eroded, transported, and deposited as sediments. Most clays are transported in suspension. The brown, muddy waters of rivers are a reflection of the clays being carried in suspension. Clays are deposited and accumulate in quiet water environments, where the energy is low enough to allow the clays to settle out of suspension. Several processes enhance the deposition of clays. When fresh river water mixes with salty ocean water, the negative charges on clay surfaces are neutralized, causing them to flocculate (clump together). Clay floccules—or aggregates of clay-sized particles that behave like larger silt- or sand-sized grains—rapidly settle out of suspension. Biodeposition is a process whereby organisms ingest clays with their food; the resultant fecal pellets settle to the bottom as sand-sized particles.

TYPES OF CLAY MINERALS

There are two types of clays in sedimentary rocks: detrital and authigenic. Minerals that are transported and deposited as sediments are called detrital. Authigenic minerals form within the rocks during diagenesis, a process whereby sediments buried within the earth's crust undergo compaction and cementation into sedimentary rocks. Water expelled from the sediments during this process may react with other minerals in the sediment to form clay minerals. Kaolinite, chlorite, illite, and smectite formed by diagenesis have been observed in sandstones. Diagenesis may also result in one clay mineral being converted into another clay mineral. One reaction that commonly occurs is the alteration of smectite to form illite as a result of an increase in temperature. This reaction is important to petroleum geologists because intermediate mixed-layered illite/smectite clays form at different temperatures. The clay mineralogy of shale can be used to determine the maximum burial temperature of a rock and, in turn, help to predict the chemistry of any petroleum encountered.

Clay minerals may also form in metamorphic rocks as the result of hydrothermal alteration. Hydrothermal fluids are hot, chemically active fluids that accompany igneous intrusions. In addition to forming clay minerals as they pass through a host rock, hydrothermal fluids are responsible for producing important ore deposits such as copper ores. Economic geologists may use the distribution of hydrothermal clay minerals to locate valuable ore deposits.

IDENTIFICATION AND ANALYSIS

Clay minerals are difficult to identify and analyze because of their small crystal size. Because they can be observed neither with the naked eye nor by standard petrographic microscopes, they require the use of sophisticated equipment for their identification.

Analysis is further complicated by the fact that it is difficult to obtain pure samples of the clay minerals because they often occur as mixtures with other minerals. Before a clay mineral can be analyzed, it must be isolated from the sample by means of special physical and chemical techniques.

The tool a clay mineralogist uses most often is an X-ray diffractometer (XRD). This instrument focuses X rays onto the sample. The crystal structure of the minerals acts as a diffraction grating, and the instrument records X rays that are diffracted from the mineral. Mineralogists can determine the spacing between planes of atoms in the crystal structure by measuring the position of "reflections" produced by X rays diffracted by the mineral. Clay mineralogists prepare specially oriented samples that enhance the reflections between the layers of clay minerals called basal reflections. The basal reflections are used in determining the type of clay mineral present.

Other instruments that are used to investigate clay minerals include the scanning electron microscope (SEM) and the transmission electron microscope (TEM). The very high resolution of these microscopes allows the scientist to observe clay minerals at a very great magnification. The scientist is thus able to observe the outward crystal form of clay minerals and their texture or orientation with respect to other grains in the sample. The SEM is very helpful in distinguishing between detrital and authigenic clays in sedimentary rocks—that is, clays transported from elsewhere versus clay minerals that formed in place.

A property that is helpful in identifying clays and understanding their behavior is the cation exchange capacity (CEC). Because of their small size and unique crystal structure, clay surfaces are negatively charged and have the ability to adsorb, or attract, positive ions on their surfaces and within clays between the layers. These cations are easily exchanged with solutions. If clay that is saturated with cations is placed in a solution saturated with sodium ions, it will exchange its cations for the sodium ions. The ability of clay to adsorb and exchange cations is called the "cation exchange capacity," and depends on the type of clay mineral. Kaolinite, chlorite, and illite have relatively low CECs; smectite has a relatively high CEC. This property is especially important to soil scientists, because it controls the availability of nutrients necessary for plant growth.

INDUSTRIAL APPLICATIONS

Clays have a variety of applications in industry. They are a readily available natural resource and are relatively inexpensive. Approximately 50 million tons of clay materials worth more than $1 billion is used industrially each year. The petroleum industry uses kaolinite as a cracking catalyst in the refinement of petroleum, for triggering chemical reactions.

Sedimentary rocks that produce oil and gas by the heating of organic matter after it is buried are called source rocks. The source rock for most oil and gas is shale. By determining the type of diagenetic clay minerals present, the petroleum geologist can determine if the source rock has been heated to a temperature that is high enough to produce oil or gas. Rocks that contain oil or gas that can be easily extracted are called reservoir rocks. The best reservoir rocks are sandstones with a high porosity and permeability. Some sandstones contain clay minerals that occur between the sand grains as a cement or within the pores. Clay minerals have the potential to reduce the porosity and permeability of a reservoir. It is important to know the type and amount of clay minerals in a reservoir in order to evaluate its quality.

A knowledge of clays is also important to engineering because the concentration of clays in a soil determines it stability. Soils that contain high percentages of smectites could cause damage to the foundations of buildings because smectites swell when saturated with water and subsequently shrink when dried out.

Clays may be used as liners for sanitary landfills because the small grain size of clays allows them to be packed together very closely, forming an impermeable layer. The liner prevents toxic leachate (which forms when rainwater reacts with solid waste) from moving out of the landfill and contaminating surface water and groundwater supplies. Chemical engineers have developed what are called designer clays by altering the properties of naturally occurring clays. These clays act as catalysts in the breakdown of toxic substances to form less toxic products. Designer clays are helpful in the destruction and disposal of toxic wastes such as dioxin, and in the cleaning of existing toxic waste sites.

Clays are the basic raw material of the ceramics industry. When clays are mixed with water, they become plastic and are easily molded. A hard ceramic material is produced by firing the molded clay. In addition to the familiar pottery and dinnerware, fired clays are used in the production of brick, tiles, sewer pipes, sanitary ware pottery, kiln furniture, cement, and lightweight aggregates. Kaolinite is used as a coating on fine paper, in paints, and as a filler in plastics and rubber. Swelling clays such as smectite are used as binders in animal feed and iron ore pellets (taconite), drilling muds, industrial absorbents, and pet litter.

Annabelle M. Foos

FURTHER READING

Bergaya, F., B. K. G. Theng, and G. Lagaly, eds. *Handbook of Clay Science.* Amsterdam: Elsevier, 2006. A compilation of articles discussing various geological aspects of clay. Addresses physical properties of clay minerals, clay chemistry, as well as interactions between clay minerals and biota. The interdisciplinary coverage of this book makes it useful to anyone studying clay science.

Blatt, Harvey, Gerard Middleton, and R. Murray. *Origin of Sedimentary Rocks.* 2d ed. Englewood Cliffs, N.J.: Prentice-Hall, 1980. Covers the classification, origin, and interpretation of sedimentary rocks. The formation and classification of clays are discussed in detail in the chapter on weathering, and the chapter on mud rocks gives an excellent discussion of the distribution of clay minerals. Suitable for college-level students.

Eslinger, Eric, and David Pevear. *Clay Minerals for Petroleum Geologists and Engineers.* Tulsa, Okla.: Society of Economic Paleontologists and Mineralogists, 1988. Covers the major geologic aspects of clay mineralogy. Crystal structure, classification of clay minerals, origin by weathering and diagenesis, and distribution of clays are discussed in detail. Application of clay mineralogy to exploration and production of petroleum is also covered. The appendix contains a summary of sample preparation and X-ray diffraction analysis of clays. A suitable text for college-level students with a science background.

Klein, Cornelis, and Cornelius S. Hurlbut, Jr. *Manual of Mineralogy.* 23d ed. New York: John Wiley & Sons, 2008. A general textbook on mineralogy, covering crystallography, physical properties of minerals, and systematic mineralogy. A very good discussion on X-ray diffraction is given in Chapter 6. Phyllosilicate minerals are discussed in Chapter 11. Suitable for college-level students.

Longstaffe, F. J. *Short Course in Clays and the Resource Geologist.* Toronto: Mineralogical Association of Canada, 1981. A collection of papers by well-recognized experts in the field of clay mineralogy. The first three chapters cover the crystal structures of clay minerals and their identification by X-ray diffraction. Subsequent chapters give case histories and specific examples of applications of clay mineralogy to petroleum geology. Suitable for college-level students with a science background.

Maurizio, Galimberti, ed. *Rubber-Clay Nanocomposites.* Hoboken, N.J.: John Wiley & Sons, 2011. Provides an overview of clay minerals in section 1. It discusses various topics related to the use of clays in nanotechnology. Section 2 discusses morphology of rubber-clay nanocomposits, section 3 covers compounds containing rubber-clay composites, and the final section discusses applications. The text is designed to work as a handbook for nanotechnology researchers. Contains chapter-specific references and a comprehensive index.

Meunier, Alain. *Clays.* Germany: Springer, 2010. Covers the properties, composition, structure, thermodynamics, and origins of clay minerals. A detailed text that may seem daunting to new students, as it opens with a large amount of mathematics in the first chapter. Contains many diagrams, references, and indexing.

Paquet, Hélène, and Norbert Clauer, eds. *Soils and Sediments: Mineralogy and Geochemistry.* New York: Springer, 1997. This detailed overview of the processes of weathering and soil sedimentation and deposition includes a vast amount of information on clay minerals. Includes a bibliography and index.

Velde, B. *Introduction to Clay Minerals: Chemistry, Origins, Uses, and Environmental Significance.* New York: Chapman and Hall, 1992. This useful introductory text includes sections on the structure and classification of clay, sedimentation, and the absorption of organic pollutants into clays. Includes an extensive bibliography and index.

_____, ed. *Origin and Mineralogy of Clays: Clays and the Environment.* London: Springer-Verlag, 1995. This advanced text includes essays on such subjects such as the composition and mineralogy of clay minerals, the origin of clays by weathering and soil formation, compaction and diagenesis, and the formation of clay minerals in hydrothermal environments.

Welton, J. E. *SEM Petrology Atlas.* Tulsa, Okla.: American Association of Petroleum Geologists, 1984. An atlas of scanning electron microscope (SEM) graphs of authigenic minerals that occur in sandstones. In addition to the SEM images, the chemical composition of the minerals is given. Text is kept to a minimum; however, a summary of sample preparation and scanning electron microscope analysis is given in the introduction. Suitable for all levels.

See also: Biopyriboles; Building Stone; Carbonates; Carbonatites; Cement; Feldspars; Gem Minerals; Hydrothermal Mineralization; Metamictization; Minerals: Physical Properties; Minerals: Structure; Non-silicates; Oxides; Radioactive Minerals; Rocks: Physical Properties.

COAL

Coal is a sedimentary rock composed of altered plant debris. Its principal uses are for fueling steam power plants, as a source of coke for smelting metals, and for space heating and industrial process heating. Synthetic gas and oil are manufactured from coal on a large scale.

PRINCIPAL TERMS

- **British thermal unit (Btu):** the amount of heat required to raise the temperature of one pound of water by 1 degree Fahrenheit at the temperature of maximum density for water (39 degrees Fahrenheit or 4 degrees Celsius)
- **cellulose:** the substance forming the bulk of plant cell walls
- **fixed carbon:** the solid, burnable material remaining after water, ash, and volatiles have been removed from coal
- **humic acid:** organic matter extracted by alkalis from peat, coal, or decayed plant debris; it is black and acidic but unaffected by other acids or organic solvents
- **lignin:** a family of compounds in plant cell walls, composed of an aromatic nucleus, a side chain with three carbon atoms, and hydroxyl and methoxyl groups, and the molecule that binds cellulose fibers together
- **molecule:** the smallest entity of an element or compound retaining chemical identity with the substance in mass
- **organic molecules:** molecules of carbon compounds produced in plants or animals, plus similar artificial compounds
- **volatiles:** substances in coal that are capable of being gasified

ORGANIC MATTER

Coal is a heterogeneous mixture of large, complex, organic molecules. It is mostly carbon but contains significant amounts of hydrogen, nitrogen, sulfur, and water. Coal is derived from plant debris that accumulated as peat (plant remains in which decay and oxidation have ceased). When covered by sediments, peat begins to lose its water and more volatile organic compounds. It also compacts and progressively becomes more chemically stable. Thereafter, peat may successively alter to lignite, bituminous coal, anthracite, or graphite, as deeper burial, deformation of the earth's crust, and igneous intrusion increase temperature and pressure.

Enzymes, insects, oxygen, fungi, and bacteria convert plant debris to peat. If left unchecked, they can quickly destroy the deposit. Thus, permanent accumulation is limited to situations where oxygen is excluded and accumulated organic waste products prevent further decay. Rapid plant growth, deposition in stagnant water, and cold temperatures promote peat accumulation. Bacteria, the principal agents of decay, operate under a wide range of acidity and aeration; eventually, they remove oxygen and raise acidity so that decay stops.

Peat is a mixture of degraded plant tissue in humic acid jelly. All protoplasm, chlorophyll, and oil have decayed. Carbohydrates have been seriously attacked: First, starch, then cellulose, and finally lignin are destroyed. Epidermal tissue, seed coats, pigments, cuticles, spore and pollen coats, waxes, and resins are most durable, but they occur in relatively small amounts. Thus, peat is dominated by lignin, the most resistant carbohydrate, with an enhanced proportion of durable tissues.

TYPES OF COAL

The degree to which coal has been chemically modified after burial is called its rank. Lignites range from brown coal, which closely resembles peat but has been buried, to black or dark brown lignite, which is similar to higher-ranking coal. Lignite is partially soluble in ammonia. Its resins and waxes dissolve in organic solvents. Water content is high, and there generally is less than 78 percent carbon and more than 15 percent oxygen on an ash-free basis. Woody structure may be obvious and well preserved. Lignite yields less than 8,300 Btu per pound.

Further compression and heating progressively convert lignite to subbituminous coal. Fibrous, woody structure gradually disappears, color darkens, the coal becomes denser and harder, water content goes down, and carbon content increases. There is a pronounced decrease in alkali solubility and susceptibility to oxidation. Subbituminous coal, ranging from 8,300 to 13,000 Btu per pound, still weathers significantly and is subject to spontaneous combustion. Like lignite, subbituminous coal burns to powdery ash.

Bituminous coals range from 46 to 86 percent fixed carbon and from 11,000 to about 15,000 Btu per pound. They burn to fused or "agglomerating" ash, resist weathering, and do not spontaneously ignite. Anthracite ranges from 86 to 92 percent fixed carbon, having lost almost all water and volatiles. Additionally, it is nonagglomerating, and heating values are about 12,500 to 15,000 Btu per pound.

Unlike most rocks, which are formed from minerals, coal is said to be formed from "macerals"—recognizable varieties of plant debris. Just as minerals can occur as families of numerous specific minerals, macerals occur as "group macerals" that in turn are composed of specific materials. Coal is composed of the group macerals vitrain, a shiny black material with a glassy luster; durain, a dull black, granular material; clairain, a laminated, glossy black material; and fusain, a dull black, powdery material. Bright coal is dominated by vitrain. Banded coal, which is the most abundant, is dominated by clairain. Dull coals are mostly durain, and fusain is referred to as mineral charcoal. Microscopic study reveals that macerals are themselves composed of numerous other materials. Materials derived from woody or cortical tissues are called vitrinite or fusinite. Vitrinite is dominant in vitrain and the "bright" laminae of clairain. Fusinite characterizes durain and the "dull" laminae in clairain. Other macerals include exinite, coalified spores and plant cuticles; resinite, fossil resin and wax; sclerotinite, fungal sclerotia; and alginite, fossil algal remains. Micrinite is unidentified vegetal material. The exact chemical composition of coal is hard to determine because the large, complex, organic molecules in coal break down under attempts to separate them, and likewise in the process of analysis. The molecular composition of many derivative molecules, however, is known.

MINERAL MATTER

Mineral matter in coal includes all admixed minerals as well as inorganic elements in the coal itself. The organic elements that form the organic matter in the coal—carbon, hydrogen, oxygen, nitrogen, and sulfur—also occur in compounds, such as iron sulfide, which are part of the mineral matter. Ash is altered mineral matter that remains after the coal is burned and is not synonymous with "mineral matter." Carbonate minerals such as calcite (calcium carbonate) lose their carbon dioxide. Sulfides such as pyrite (iron sulfide) break down to yield sulfur dioxide. Clay minerals lose their water and are altered drastically in molecular structure. The minerals and inorganic elements in the coal react with one another when the coal is burned to produce an ash of mixed oxides, silicates, and glass.

Clays are the most abundant inorganic minerals in coal. Some clay is washed into the coal swamp, but much arises from chemical reactions occurring in the peat and coal during and after coalification. Sulfides generally are half as abundant as clay, with the iron sulfides—pyrite and marcasite—being the most widespread. Sulfides of zinc and lead also may be abundant. Pyrite and marcasite may originate during plant decay and coalification as hydrogen sulfide generated from organic sulfur combines with iron. Hydrogen sulfide also may result from the decay of marine organisms as areas of swamp are invaded by the sea, thus producing more pyrite. Coals associated with marine rocks generally have higher sulfur content than those coals from wholly alluvial deposits. Carbonates of calcium (calcite), iron (siderite), and magnesium (dolomite) generally are half as abundant as sulfides. Quartz is ubiquitous, ranging from small amounts to as much as one-fifth of the mineral matter. More than thirty additional minerals have been noted as abundant or common in coal.

Trace elements such as zinc, cadmium, mercury, copper, lead, arsenic, antimony, and selenium are associated with sulfides. Others, such as aluminum, titanium, potassium, sodium, zirconium, beryllium, and yttrium, are associated with mineral grains washed into the swamp. Still others find their way into the peat within plant tissues or later are concentrated from waters circulating through either the peat swamp or the coal seam. These elements include germanium, beryllium, gallium, titanium, boron, vanadium, nickel, chromium, cobalt, yttrium, copper, tin, lanthanum, and zinc. Coal ash has been a source of germanium and vanadium, and both uranium and barium have been mined from some coal seams. Selenium, gallium, zinc, and lead occurrences in coal have been investigated as possible sources of these metals, and—given the harm caused by their release in coal mining and coal combustion—the recovery of these metals in pollution control may be feasible.

CURRENT COAL FORMATION

Peats form today in two very different environments. Poor drainage in areas of recent glaciation, coupled with low temperature, facilitates peat formation in high latitudes. In warm temperate and tropical regions that are poorly drained, vigorous forests may produce peat. Good examples also can be found in coastal plains and shoreline deposits such as the Dismal Swamp and Everglades, alluvial plains such as the Mississippi Delta, and tropical alluvial plains such as the upper Amazon Basin. Most of the high-latitude peat is unlikely to be incorporated in major sedimentary accumulations and, therefore, is unlikely to become coal. Other modern peats, however, are an extension of the sort of peat accumulation that occurred in the geologic past and provide a guide to understanding coal formation.

Coal can form in two major settings. Autochthonous coal formed where plants grew and died, and was transported by rivers to deltas and coastal lagoons. The Red River of Louisiana once contained a good example of an allochthonous coal environment in the Great Raft, a vast tangle of floating vegetation that completely covered the surface of the river. The Great Raft no longer exists, as it was broken up to open the river for navigation during the nineteenth century.

Individual coal beds or seams may be very widespread or of limited extent. The Illinois #2 Coal, for example, is recognizable from western Kentucky to northeastern Oklahoma. Other coal beds cover only a few square kilometers. Coals range from a few millimeters to more than 100 meters thick. They generally are tabular but may be interrupted by filled stream channels or rolls, which are protuberances of overlying rock that apparently sagged into the coal while the original peat was still in a soft state. Thin layers of clay (splits) and cracks filled with clay (clay veins) interfere with mining and dilute mined coal with extraneous rock. Coal beds may be subhorizontal or steeply inclined, depending on deformation in the area. They also may be continuous or offset by faults—fractures of the crust along which movement has occurred. Inclination or interruption of the bed interferes with mining.

Despite a high energy content, anthracite coal is used less frequently than coals of lower rank. Because it has been metamorphosed, anthracite frequently occurs in folded rocks that make mining difficult and expensive. It also burns at such a high temperature its use requires special furnaces. If anthracite is metamorphosed still further, it becomes graphite, which is useless as a fuel but has numerous industrial uses. Contrary to popular misconception, diamonds do not form from the heating of coal. Coal is never buried deeply enough to produce diamond; the carbon in diamonds comes from the earth's mantle.

TESTING

The distribution and character of coal beds are determined by standard field geologic methods. Surface exposures are plotted on maps, and their geometric orientation is recorded. Thereafter, coal beds are projected geometrically into the subsurface. Wherever possible, their position is verified in wells and mine shafts so that the full regional extent, depth, and attitude of the coal are illustrated. If detailed information is required for mining, specially drilled test holes will be utilized to locate channel fillings and faults interrupting the coal, and to determine changes in thickness and quality. These test holes make it possible to plan mining efficiently. In addition, the relationship of the coal to rocks above and below will be investigated so that mining methods may be adjusted to potential geologic hazards, such as caving roofs and incursion of underground water.

Standardized, practical tests define quality and/or suitability for specific uses. Burning coal samples under controlled conditions at 750 degrees Celsius produces a standard American Society for Testing Materials (ASTM) ash, which defines the total ash content and its nature. Coals with low-ash content are preferred. The character of the ash—agglomerating produces a glassy clinker, and nonagglomerating produces a powder—determines the type of grate on which the coal can be burned. The heating value of the coal, figured on an ash-free basis, is determined by controlled combustion in a calorimeter: a closed "bomb" fitted with temperature sensors. The amount and quality of volatile materials are analyzed by carbonizing the coal at a standard temperature in a closed vessel and measuring the amount and kinds of substances driven out. These data are required to classify the coal according to rank and grade, and to fix its value in the market.

Analysis

More precise analytical techniques and tools are employed in research, as opposed to routine coal testing. The physical composition of the coal may be determined by microscopic study of thin sections (slices of coal mounted on glass and reduced to a thickness that allows for the transmission of light), or by examining polished surfaces with a reflecting microscope. In this way, the components of the coal may be distinguished and examined separately. More detailed study may utilize either transmission or scanning electron microscopy. Coal rank also may be determined very precisely by means of measuring reflectance—in this case, the amount of light reflected from polished vitrinite.

Mineral matter in the coal, as distinguished from ash, may be examined directly by these techniques but also may be recovered from the coal employing a low-temperature asher. The asher is an electronic device that vaporizes combustible materials without significantly raising the temperature of the sample. In this way, clays, carbonates, sulfides, and other minerals that are significantly altered by heat are delivered for study in their original state.

The elemental chemical analysis of coal provides information as to its composition at a level of limited value to coal investigators. The organic compounds in coal, however, suffer substantial alteration under most analytical techniques. The volatiles recovered by coal carbonization in the absence of oxygen may be separated by distillation, but they are not the compounds originally present in the coal. Solid organic materials may be selectively dissolved and the resultant materials subjected to organic analysis, but, again, these materials are not the ones that were present in the original coal. In spite of these limitations, coal investigators gather some conception of the original chemistry of the coal and develop information applicable to coal utilization. X-ray diffraction and other spectroscopic techniques have begun to uncover the structure of coal molecules without altering them. Concepts of coal molecular structure, however, remain rudimentary.

Fuel Source

Coal is a major source of heat energy and a significant source of organic compounds of practical use—from drugs to plastics. It fueled the Industrial Revolution and, therefore, is responsible for modern industrial society. Coal was the principal source of energy until World War I. Although its use declined thereafter under competition from oil and gas, coal is becoming increasingly important. By the end of the twentieth century, coal provided approximately 39 percent of the world's electricity. In contrast to oil and gas, the reserves of which are limited and for which production is expected to peak early in the twenty-first century, coal reserves appear to be adequate at least until the twenty-second century.

In spite of its large reserves, coal presents several significant problems. Coal is a solid fuel and inappropriate for use in domestic heating and vehicular transport. Therefore, substantial research on converting coal into liquid and gaseous fuel is needed and under way. Furthermore, coal combustion produces gaseous and solid wastes that must be managed. Sulfides are converted to metallic oxides and sulfur dioxide, which combine with water in the atmosphere to produce acid rain. This precipitation kills plants, causes respiratory difficulties, and destroys aquatic life. The carbon in coal, as well as in wood or hydrocarbons, combines with oxygen to form carbon dioxide when burned. Enough carbon dioxide has already been produced by burning wood, coal, and hydrocarbons to change the global atmospheric composition. Resultant climatic changes are feared, even though ultimate results are not yet predictable. Other deleterious elements are released, either into the atmosphere or in ash, such as lead, cadmium, arsenic, and mercury. Although these elements are emitted in small amounts, the environmental effects must be understood so that appropriate action may be taken. Mine hazards—gas and cave-ins especially—are geologically controlled. The subsidence of underground mines—in extent, timing, and ultimate cost to surface values—is geologically controlled as well. Strip-mining reclamation and the problem of mine waters entering both surface and subsurface water supplies are additional geologic concerns.

Ralph L. Langenheim, Jr.

Further Reading

Averitt, P. "Coal." In *United States Mineral Resources*, U.S. Geological Survey Professional Paper 820. Washington, D.C.: Government Printing Office, 1973. A concise account of coal formation, its rank, sulfur content, and minor elements occurring in coal. United States and world coal resources also are reviewed. Written for the informed general public.

Dalverny, Louis E. *Pyrite Leaching from Coal and Coal Waste*. Pittsburgh, Penn.: U.S. Department of Energy, 1996. Examines coal-mining policies and protocol, as well as the state of disposal procedures used by coal-mine operators. Illustrations and bibliography.

Francis, Wilfrid. *Coal: Its Formation and Composition*. London: Edward Arnold, 1961. Describes the tissues of coal-forming plants, the plants' accumulation to form peat, and formation and composition of peat, lignites, and higher-ranking coals. Reviews the composition of coal, including inorganic constituents, and discusses coal-forming processes. Moderately technical.

Freese, Barbara. *Coal: A Human History*. Cambridge, Mass.: Penguin, 2004. Appropriate for all readers, providing a history of coal's uses throughout the world. Includes information on early usage, the Industrial Revolution, mining in the past and present, and current environmental concerns.

Galloway, W. E., and D. K. Hobday. *Terrigenous Clastic Depositional Systems*. 2d ed. New York: Springer-Verlag, 1996. Discusses the character and processes of nonmarine sedimentary environments, excepting glacial environments, as related to coal deposition. Reviews the character and composition of coal as affected by depositional origin. Written for college students but intelligible to the general reader.

Gayer, Rodney A., and Jierai Peesek, eds. *European Coal Geology and Technology*. London: Geological Society, 1997. A complete examination of the technologies used in European coal mines and the coal mining industry. Illustrations, maps, index, and bibliographical references.

Heinberg, Richard. *Blackout: Coal, Climate and the Last Energy Crisis*. Gabriola Island, B.C.: 2009. Provides information on the use of coal as an energy source. Discusses both the pros and cons of coal energy, as well as the issue of limited supply. Considers coal usage around the world and concludes with possible scenarios for the future.

James, P. *The Future of Coal*. 2d ed. London: Macmillan Press, 1984. Summary account of the origin of coal,

mining methods, its use and markets, and environmental and health problems associated with coal use. Discusses coal occurrence and use on a worldwide basis. Written for the nontechnical reader.

Keefer, Robert F., and Kenneth S. Sajwan. *Trace Elements in Coal and Coal Combustion Residues*. Boca Raton, Fla.: Lewis Publishers, 1993. An examination of the chemical makeup and trace elements found in coal and coal residues. Suitable for the beginner. Includes illustrations, maps, index, and bibliographical references.

Stach, E. *Stach's Textbook of Coal Petrography*. 3d ed. Berlin-Stuttgart: Gebruder Borntraeger, 1983. Includes discussions of the origin and formation of peat and coal, a detailed account of coal's physical constituents and their origin, a discussion of trace elements in coal, the methods of coal petrography, and practical applications of coal petrography. The most comprehensive account in English. Written for professional coal geologists.

Stewart, W. N. *Paleobotany and the Evolution of Plants*. 2d ed. New York: Cambridge University Press, 2010. A well-illustrated account of the evolution of plants, including those responsible for coal. Written for beginning college-level students.

Swaine, Dalway J. *Trace Elements in Coal*. Boston: Butterworth, 1990. A chemical analysis of the trace elements and properties of coal. Although much attention is focused on chemistry, this book is suitable for the non-chemist. Illustrations help to clarify concepts. Bibliographic references.

Thomas, Larry. *Coal Geology*. Hoboken, N.J.: John Wiley & Sons, 2002. Discusses the formation of coal, occurrences, and age. Covers coal quality, physical and chemical properties, and its use as a resource. Well organized with bibliographical information, appendices, glossary, and index.

U.S. Department of Labor, Mine Safety and Health Administration. *Coal Mining*. U.S. Department of Labor, Mine Safety and Health Administration, 1997. Examines the state of mines in the United States. Considers the health risks associated with coal mining. Illustrations and maps.

Van Krevelen, D. W. *Coal: Typology-Chemistry-Physics-Constitution*. Amsterdam: Elsevier, 1961. A comprehensive treatise on the physical and chemical properties of coal, its constitution and classification, and its geology and petrology. Written at the technical level.

See also: Aluminum Deposits; Building Stone; Cement; Diamonds; Dolomite; Earth Resources; Extraterrestrial Resources; Fertilizers; Gold and Silver; Industrial Metals; Industrial Nonmetals; Iron Deposits; Manganese Nodules; Pegmatites; Platinum Group Metals; Salt Domes and Salt Diapirs; Sedimentary Mineral Deposits; Uranium Deposits.

CONTACT METAMORPHISM

Contact metamorphism is caused by the temperature rise in rocks adjacent to magmatic intrusions of local extent that penetrate relatively shallow, cold regions of the earth's crust. Many economically important metallic mineral deposits occur in contact metamorphic zones.

PRINCIPAL TERMS

- **aureole:** a ring-shaped zone of metamorphic rock surrounding a magmatic intrusion
- **contact metamorphic facies:** zones of contact metamorphic effects, each of which is characterized by a small number of indicator minerals
- **facies:** a part of a rock or group of rocks that differs from the whole formation in one or more properties, such as composition, age, or fossil content
- **hornfels:** the hard, splintery rocks formed by contact metamorphism of sediments and other rocks
- **lithology:** the general physical type of rocks or rock formations

CONTACT METAMORPHIC FACIES

Contact metamorphic rocks are moderately widespread; they occur at or near the earth's surface, where magmas of all kinds intrude low-temperature rocks. Minerals in contact metamorphic rocks are similar to those in regional metamorphic rocks of comparable metamorphic grade. Contact metamorphic effects are divided into facies (zones), each of which is characterized by a small number of concentric indicator mineral rings surrounding an intrusive rock. Nonsymmetrical zones imply special conditions, such as less (or more) chemically or thermally reactive rocks or a nonvertical intrusive body.

Development of contact metamorphic facies reflects both the history of pressure and temperature changes and the bulk-rock chemistry. Thus, by stating that a rock belongs to a particular facies, scientists convey much about the rock's history. This information is vital to exploration for metallic and industrial minerals that commonly occur in contact metamorphic aureoles and for general understanding of regional geology.

Contact metamorphic mineral facies have counterparts with regional metamorphic zones. In addition to the bulk chemical composition of a rock, temperature and pressure are two variable factors. These two factors can be independent. For example, one can find low-pressure, high-temperature facies or,

alternatively, high-pressure, low-temperature facies. With contact metamorphism, these facies occur very close to the intrusive rock. With regional metamorphism, however, the effect is widespread and may not be related to an intrusive rock. It can be hard to tell if metamorphism around deep intrusive rocks is contact or regional metamorphism, or a combination of the two.

AUREOLES

Contact metamorphic rocks are recognized by their location adjacent to igneous bodies and by evidence indicating a genetic temporal relationship. Contact metamorphic rocks are commonly massive. Granitic rocks are the most common intrusive material. The most frequent depth of the solidification of a granitic magma is 3 kilometers, corresponding to a load pressure of 800 to 2,100 bars (a bar represents approximate atmospheric pressure, or pressure under 10 meters of water). There are intrusions that solidify at a greater or shallower depth; a depth of 1 kilometer corresponds to a load pressure of 250 bars. Consequently, the load pressures effective during contact metamorphism range from 200 to 2,000 bars in most cases. In contrast, load pressures prevailing during regional metamorphism are generally greater.

When a magma intrudes into colder regions, the adjacent rocks are heated. If the heat content of the intruded magma is high and the volume of the magma is not too small, there will be a temperature rise in the bordering rock that lasts long enough to cause mineral reactions to occur. The rocks adjacent to small intrusions of dikes and sills are not metamorphosed (only baked), whereas larger plutonic rocks give rise to a distinct contact aureole of metamorphic rocks. Several zones of increasing temperature are recognized in contact aureoles.

The contact metamorphic zones surround the intrusion in generally concentric rings that approximate the shape of the intrusion. Those zones that correspond to the highest-temperature minerals are closest to the intrusion; the zone corresponding to the lowest temperature is located farthest from the intrusion. The lowest-temperature contact metamorphic zone gradually grades into unmetamorphosed country rock.

MINERAL DEVELOPMENT

Contact metamorphic rocks are characteristically massive because of lack of deformation; most are fine-grained except for a special variant called a skarn. Skarns may contain metallic mineralization in sufficient concentration that they can be worked for a profit. Such mineral concentrations are called ores.

Contact metamorphic rocks lack the schistosity or platy texture characteristic of most metamorphic rocks. The very fine-grained, splintery varieties are called hornfels. The large metamorphic gradient, decreasing from the hot intrusive contact to the unaltered country rock, gives rise to zones of metamorphic rocks differing markedly in mineral constituents. The intensity and mineral assemblage in contact metamorphic zones depend on several factors: the chemical composition of the intrusive rock; its temperature and volatile content; the composition and permeability of the host units; the structural or spatial relationship of the reactive units to intense contact effects and solutions conduits, or traps; and the pressure or depth of burial.

Argillaceous (clay rich) limestone is usually susceptible to contact metamorphism, as its diverse rock chemistry allows mineral development over a broad set of physical and chemical conditions. A typical contact metamorphic mineral assemblage formed from rocks originally of this composition is a plagioclase-garnet-epidote rock. This kind of rock frequently hosts important ore deposits. Shale usually converts to hornfels, which is characterized by the minerals cordierite, biotite, and chlorite and is essentially nonreactive and nonpermeable. This lithology usually does not host ore deposits. Porous limestone is often exceptionally susceptible to solution and to replacement by ore. Rocks formed from this material contain magnetite, garnet, and pyroxene.

Pure, massive limestone has only thin, contact metamorphic zones developed in it. Pure limestone merely recrystallizes, forming coarse-grained marble in the highest-temperature contact metamorphic zones adjacent to an intrusion. Dolomite is usually poorly mineralized and nonreactive. Siliceous dolomite, however, may act as a good host, with assemblages characterized by tremolite, diopside, serpentine, and talc. Mafic rocks, such as andesites, diabases, and diorites, can be reactive hosts capable of producing some types of ore deposits. Secondary biotite is the key alteration mineral in mafic rocks and may be associated with ore. The biotite zone may be very broad.

DIAGNOSTIC MINERALS

Contact metamorphic facies are identified by the mineral assemblages that are developed in the metamorphosed rocks. From lowest to highest temperature, the contact metamorphic facies are called albite-epidote, hornblende hornfels, pyroxene hornfels, and sanidinite.

Albite and chlorite are the contact metamorphic minerals restricted to the albite-epidote hornfels facies. Calcite, epidote, and talc also occur in this facies. Andalusite may occur in the highest-temperature part of the albite-epidote hornfels.

The amphiboles anthophyllite and cummingtonite are restricted to the hornblende hornfels facies. Muscovite is present in this facies as well as in the albite-epidote hornfels facies, and sometimes in the lowermost pyroxene hornfels facies. Grossular-andradite garnet and idocrase, sometimes with biotite and almadine garnet, are present here as well as in the pyroxene hornfels facies. Sillimanite may occur in the highest-temperature part of the facies at higher pressure; staurolite may occur in high-pressure, iron-rich rocks. Calcite is present, but not with tremolite-actinolite, epidote, or plagioclase. Under certain chemical compositions of the original rocks, other minerals that may be present include anthophyllite, cummingtonite, phlogopite, biotite, diaspore, and scapolite.

Orthoclase with andalusite or sillimanite is restricted to the pyroxene hornfels facies. Sillimanite is present here and also in the upper hornblende hornfels facies. Hyperstine and glass present in the pyroxene hornfels may also be present in the sanidinite facies. Muscovite is present only in the lowest-temperature part of the facies. For silica-deficient rocks in the pyroxene hornfels facies, dolomite, magnesite, and talc may occur only in the lowest-temperature part of the facies or at pressures of high carbon dioxide. The pyroxene hornfels facies is characterized by the breakdown of the last water-bearing minerals, muscovite and amphibole.

Diagnostic minerals of the sanidinite facies are sanidine, mullite, tridymite, and pigeonite. In silica-deficient rocks at this facies, wollastonite, grossularite, and plagioclase are present. Other minerals that may occur under special conditions are perovskite, spinel, diopside, and pseudobrookite. This is the highest-temperature facies of contact metamorphism.

STUDY AND EXPLORATION

Contact metamorphic zones are studied by standard geologic techniques, including the preparation of geologic maps through field study. Aerial photographs and satellite images are frequently used to identify contact metamorphic zones through detecting rock alteration in the contact metamorphic zone. Satellite sensors measure, analyze, and interpret electromagnetic energy reflected from the earth's surface for subsequent computer analysis. Geophysical techniques (gravity and electrical methods) are used to locate mineralized zones containing relatively heavy metallic ore minerals. The gravity contrast of these zones can be measured with surface instruments and then mapped. Some of these same minerals, particularly the ore minerals, transmit electric current in an anomalous manner. These anomalies plotted on base maps may be an additional clue to the presence of ore deposits in contact metamorphic zones.

Subsequent laboratory work involves the determination of mineral relationships by studying thin slices of rock through which light passes under a microscope (petrography). Frequently, the chemistry of entire rock samples is determined to help scientists understand the presence or absence of minerals as a guide to mineral exploration and to the composition of the original rock. Because of significant advances in laboratory instruments, detailed mineral chemistry analyses are routinely performed that determine chemical makeup of microvolumes of minerals. These determinations permit earth scientists to understand the conditions under which the contact metamorphic zones formed and to interpret the history of the contact metamorphic zone.

This information provides critical insight into the potential for metallic ore deposits within the contact metamorphic zone. Such deposits are explored by means of surveys that detect and record variations in geochemistry, gravity, and plant types and abundances. Anomalously high element concentrations in rocks, soils, and plant tissues may indicate mineral deposits that are not exposed. Many ore deposits have an anomalous gravity signature because of heavy associated silicate minerals and metallic minerals. Maps of localized variations may lead to subsurface exploration, which is conducted by drilling techniques. Continuous cylindrical samples are taken from depths of up to several thousand feet and the study of such subsurface samples leads to the evaluation of possible minable concentrations of certain metals and other elements used by modern civilizations.

ECONOMIC AND GEOLOGIC VALUE

Contact metamorphic zones frequently contain metallic mineral deposits without which modern civilization could not exist. Many significant mineral deposits worldwide occur in these zones. The metals extracted from deposits of this type include tin, tungsten, copper, molybdenum, uranium, gold, silver, and, in some cases, refractory industrial metals. Specialized surveys assess the potential of metallic ore deposits before expensive subsurface sampling by drilling.

Recognition and understanding of the contact metamorphic environment can lead to a significantly improved understanding of regional geology and the regional geologic history. In some areas, contact metamorphic zones are associated with igneous intrusives of only one geologic age. Specialized laboratory studies provide supporting data to determine the geologic history and potential for economic mineralization.

Jeffrey C. Reid

FURTHER READING

Best, Myron G. *Igneous and Metamorphic Petrology.* 2d ed. Malden, Mass.: Blackwell Science Ltd., 2003. A popular university text for undergraduate majors in geology. A well-illustrated and fairly detailed treatment of the origin, distribution, and characteristics of igneous and metamorphic rocks.

Bucher, Kurt, and Martin Frey. *Petrogenesis of Metamorphic Rocks.* 7th ed. New York: Springer-Verlag Berlin Heidelberg, 2002. Provides an excellent overview of the principles of metamorphism. Contains a section on geothermometry and geobarometry.

Deer, W. A., R. A. Howie, and J. Zussman. *An Introduction to Rock-Forming Minerals.* 2d ed. London: Pearson Education Limited, 1992. Standard references on mineralogy for advanced college students and above. Offers detailed descriptions of chemistry and crystal structure, usually with chemical analyses. Discussions of chemical variations in minerals are extensive.

Fettes, Douglas, and Jacqueline Desmons, eds. *Metamorphic Rocks: A Classification and Glossary of Terms.* New York: Cambridge University Press, 2007. Discusses the classification of metamorphic rocks. Feldspars are mentioned throughout in reference to the various types of metamorphic rocks and basic classifications.

Hyndman, D. E. *Petrology of Igneous and Metamorphic Rocks.* 2d ed. New York: McGraw-Hill, 1985. Suitable for introductory-level college education in Earth sciences. Key concepts are summarized at the ends of chapters.

Oldershaw, Cally. *Rocks and Minerals.* New York: DK, 1999. Very useful for new students who may be unfamiliar with the rock and mineral types discussed in classes or textbooks.

Perchuk, L. L., ed. *Progress in Metamorphic and Magmatic Petrology.* New York: Cambridge University Press, 1991. Although intended for the advanced reader, several of the essays will serve to familiarize the new student with the study of metamorphic rocks. In addition, the bibliography will lead the reader to other useful material.

Philpotts, Anthony, and Jay Aque. *Principles of Igneous and Metamorphic Petrology.* 2d ed. New York: Cambridge University Press, 2009. An easily accessible text for students of geology. Discusses igneous rock formations of flood basalts and calderas. Covers processes and characteristics of igneous and metamorphic rock.

Pough, F. H. *A Field Guide to Rocks and Minerals.* 5th ed. Boston: Houghton Mifflin, 1998. Provides an excellent overview of the environments in which minerals are formed and found, and simple laboratory tests that can be conducted in the field or laboratory. Offers the user a basic grasp of the subject of mineralogy. Profusely illustrated.

Sorrella, C. A. *Minerals of the World.* Racine, Wis.: Western Publishing, 1973. Presents an excellent overview of rocks and minerals. Provides information about the minerals expected in a variety of environments, including contact metamorphic zones.

Vernon, Ron H., and Geoffrey Clarke. *Principles of Metamorphic Petrology.* New York: Cambridge University Press, 2008. Discusses metamorphic processes, anatectic and nonanatectic migmatites, and formation of granite bodies. Contains a glossary, appendices, indexing, references, and color plates.

Voll, Gerhard, ed. *Equilibrium and Kinetics in Contact Metamorphism: The Ballachulish Igneous Complex and Its Aureole.* New York: Springer-Verlag, 1991. Collects a number of essays that deal with contact metamorphism in the Ballachulish Igneous Complex in the Scottish Highlands, thereby providing a useful and detailed case study of a specific region. Includes illustrations, as well as two foldout maps.

Winkler, H. G. F. *Petrogenesis of Metamorphic Rocks.* 4th ed. New York: Springer-Verlag, 1976. An introductory text at the college level. Portions may be of interest to advanced students.

Winter, John D. *An Introduction to Igneous and Metamorphic Petrology.* Upper Saddle River, N.J.: Prentice Hall, 2001. Provides a comprehensive overview of igneous and metamorphic rock formation. Discusses volcanism, metamorphism, thermodynamics, and trace elements; focuses on the theories and chemistry involved in petrology. Written for students taking a university-level course on igneous petrology, metamorphic petrology, or a combined course. A basic understanding of algebra and spreadsheets is required.

See also: Blueschists; Metamorphic Rock Classification; Metamorphic Textures; Metasomatism; Pelitic Schists; Regional Metamorphism; Sub-Seafloor Metamorphism.

CRYSTALS

Crystals are solid materials whose smallest components—molecules, ions, or atoms—are arranged in an orderly, repeating pattern that extends in all three spatial directions. There are many varieties of crystals, both natural and synthetic, and because of their varied properties, they have countless uses in technology, entertainment, and defense.

PRINCIPAL TERMS

- **boule:** a large, synthetically made single crystal with many industrial and technological applications
- **crystal growth:** the second stage of crystallization; a transfer of heat, matter, or both drives the crystal to expand from the nucleation site by adding molecules, atoms, or ions to the lattice
- **crystal habit:** the external appearance of a crystal, including its shape and visible physical features
- **crystallization:** the process of crystal formation; generally includes two steps—nucleation and growth
- **crystallography:** the scientific study of crystals and how they form
- **grain boundary:** the interface between the small crystals (crystallites) that make up a larger polycrystal material
- **lattice:** an infinite array of discrete points upon which a crystal is built; the crystal's arrangement of atoms, molecules, or ions is repeated at each lattice point
- **nucleation:** the first stage of crystallization; a new crystal forms around a nucleus, initiated by a phase change
- **periodic:** a repeating pattern
- **piezoelectricity:** the property of some crystals (and some other solids) to create an electric charge from mechanical stress or pressure
- **point group:** a group of geometric symmetries with one or more fixed points
- **unit cell:** an imaginary box that shows the three-dimensional pattern of the atoms, molecules, or ions of a crystal

AN OVERVIEW OF CRYSTALS

Crystals, both natural and synthetic, are ubiquitous. They can be found in rocks and ice, are produced naturally by some living organisms, and are produced artificially by humans for a variety of uses. Common examples of crystals include diamonds, snowflakes, and sugar, and common uses include clocks, semiconductors, and lasers.

Crystals are scientifically defined by their microscopic structure, although macroscopic descriptions are also regularly used. In scientific terms, crystals can be classified by their physical properties (the crystal lattice system, crystal systems, and crystal families) or by their chemical properties (such as types of bonds). Macroscopically, crystals are described by their "habit"—that is, their overall external appearance, shape, and features.

Crystals are abundant in nature, but not all solid materials are crystals. Most inorganic solids are actually polycrystals; they are composed of many smaller crystals (crystallites), but they themselves are not crystals because the crystallites are not organized in a periodic pattern. Other solids, such as glass and gels, are described as amorphous. They have no periodic arrangement at any level.

PHYSICAL PROPERTIES OF CRYSTALS

The microscopic structure of a crystal (the pattern of atoms, molecules, or ions) provides its scientific classification. A crystal's structure is described by a unit cell, an imaginary box showing the three-dimensional pattern of atoms. Unit cells are arranged in a lattice, and lattice parameters describe the length of the unit cell sides and the angles between them. Crystals are classified into seven different lattice systems: triclinic (the least symmetrical), monoclinic, orthorhombic, rhombohedral, tetragonal, hexagonal, and cubic (the most symmetrical and simplest).

Crystals also can be classified into seven different crystal systems, most of which correspond to the lattice systems. Crystal system classification refers to point groups (geometrical symmetries), and several of the classes contain two different lattice possibilities, explaining why the systems are not entirely the same. The triclinic, monoclinic, orthorhombic, tetragonal, and cubic classes are the same, but the trigonal crystal system contains both rhombohedral and hexagonal lattice classes. The hexagonal lattice class is also present in the hexagonal crystal system. The seven crystal systems are split further into thirty-two crystal types.

One additional classification, crystal families, is almost identical to crystal systems, except that there are six crystal families. The hexagonal family encompasses both the trigonal and the hexagonal crystal systems because they share the same lattice.

While it is important to understand that a crystal's true scientific classification comes from its microscopic structure, the macroscopic features that humans can see with unaided eyes also provide a useful classification system. One general way of describing crystals macroscopically is by referring to the clarity of definition of the crystal's flat faces. A crystal with perfectly or near-perfectly defined faces is called euhedral, while a crystal with poorly defined or undefined faces is called anhedral. Subhedral crystals are the middle ground, with moderately defined faces.

The term *crystal habit* is used to describe a crystal's overall external appearance. The crystal habit may not be entirely dictated by the microscopic structure of the crystal; other factors will affect the overall external structure, such as heat, pressure, and impurities present while the crystal grows.

There are many names for different crystal habits; the names are derived from features such as the shape of a crystal's faces, the shape of the overall crystal, or whether the crystal is actually an aggregate of multiple crystals. For example, *dodecahedral* refers to an aggregate of twelve-sided crystals, like garnet. *Fibrous* refers to very thin, fiber-like crystals, such as tremolite, which is a type of asbestos. *Rosette* refers to a rose-like aggregate of plate-shaped crystals, such as gypsum.

Aside from its physical structure, a crystal also may be characterized by physical properties. These properties include cleavage (its tendency to split along weak planes), transparency and refractive index (optical properties), and surface tension.

CHEMICAL PROPERTIES OF CRYSTALS

In addition to all of the physical classifications, crystals can be classified by their chemical properties. In particular, they can be classified according to their bonds: covalent, ionic, or molecular.

Ionic crystals are made of positive and negative ions, held together by the electrostatic attraction between them (ionic bonds). They can form when salts crystallize from a solution or solidify from a molten fluid. These crystals tend to be hard and brittle, they cleave easily along planes, and they have a high melting point. When molten, ionic crystals are good conductors of electricity; when solid, they are poor conductors. Table salt ($NaCl$) is a common example of an ionic crystal. Other examples are sodium fluoride (NaF, used for cavity prevention) and potassium chloride (KCl, a component of fertilizer).

Covalent crystals are characterized by their strong covalent bonds, where electrons are shared between two atoms. These crystals tend to have very high melting points. Diamonds are covalent crystals, as are graphite and silica. Diamonds and graphite are crystalline forms of the same solid: carbon. Chemical elements that can exist in multiple forms like this are known as allotropes. In many cases, the properties of the different forms can be different. Diamond is the hardest known solid, while graphite is extremely soft.

Molecular crystals, which are solids composed of molecules held together by weak forces, can be made of polar molecules or nonpolar molecules. Polar molecular crystals are held together by dipole-dipole attractions and by the weak London dispersion force, a type of van der Waals force; nonpolar molecular crystals are held together by just the London dispersion force. Both types have a low melting point and are generally soft, but nonpolar molecular crystals can conduct electricity in some forms, while polar molecular crystals are poor or nonconductors.

FORMATION OF CRYSTALS

Crystals form through a process called crystallization, which generally involves a crystal forming from a fluid or the dissolved materials in a fluid. (A quartz vein, for example, is formed when minerals precipitate out of a fluid flowing through a rock mass.) Often, crystallization is spurred by the cooling and solidification of a molten liquid, particularly in magmatic and metamorphic processes. In some cases, crystals can form directly from gas. A single fluid can yield many different crystals and crystalline structures because of the influence of variable factors, such as temperature and pressure.

There are two main steps to crystallization: nucleation and growth. In nucleation, a phase change is initiated, and the new crystal forms slowly around a small nucleus as the atoms or molecules begin to become correctly oriented. Nucleation can be homogeneous (without the influence of impurities) or heterogeneous (with the influence of impurities). Heterogeneous nucleation is faster than homogeneous because the foreign particles provide a scaffold upon which the crystal can grow. Indeed, when crystals are created artificially, the process can be sped up by adding small scratches to the glassware or by placing string, a rock, or small, previously made crystals (called seed crystals) into the solution. Homogeneous nucleation is fairly rare, as a large amount of energy is required to catalyze the process with no foreign surfaces present.

In the growth stage, the crystal expands from the nucleation site as ions, atoms, or molecules are added to the lattice; this step is much faster than nucleation. A perfect crystal would build slowly, but the defects and impurities naturally present in most crystallization will help speed up the process, similar to what happens in heterogeneous nucleation. The growth step involves a transfer of heat, matter, or both to drive the process.

While many crystals are created by natural geologic processes, particularly the ones found in Earth's bedrock, some are actually produced by animals. Mollusks, for example, produce calcite and aragonite. Their mantle secretes a shell of polysaccharides and glycerides, which then directs the completion of the shell by forming crystalline materials.

A variety of methods exist for creating crystals artificially, such as hydrothermal synthesis, where substances are crystallized at high temperature and pressure conditions. Another method, laser-heated pedestal growth, uses a laser beam as a heat source to grow crystals from the liquid/solid phase transition.

Industrial techniques also exist to create large synthetic single crystals called boules; these crystals are used in semiconductors and for many other purposes. The Czochralski process and the Bridgman-Stockbarger technique are two such processes. Huge single crystals also can be created naturally. The Cave of the Crystals in Mexico has many of the largest natural single crystals ever discovered, including one that is 11 meters (36 feet) long.

USES OF CRYSTALS

One of the most common uses of crystals is keeping time. Some crystals, such as quartz, are piezoelectric, meaning that electricity can accumulate in them from mechanical stress or pressure.

Because of this characteristic, quartz can be used to create a crystal oscillator, a circuit that creates an electric signal with a precise frequency based on the mechanical resonance of the crystal as it vibrates. Crystal oscillators are used in wristwatches, cell phones, radios, laboratory, equipment, and machinery such as oscilloscopes and signal generators.

The use of boules (synthetic single crystals) is widespread. Most integrated circuits contain a silicon boule, and classic semiconductors are boules or other crystalline solids. The advantage of using large single crystals over crystalline aggregates is that the single crystals do not have grain boundaries, which is the area that occurs between single crystallites in polycrystals. Grain boundaries would be detrimental to the operation of various types of equipment because they decrease the crystal's electrical and thermal conductivity and provide a weak spot that is susceptible to corrosion.

Most electronic displays, such as television screens and computer monitors, rely on liquid crystals, which have some liquid properties and some crystal properties. The key property relevant to electronic display use is that they modulate light. Because the liquid crystals are not emitting their own light, liquid crystal displays are quite energy efficient.

Crystals also have aesthetic and cultural significance, and are used in jewelry and in cultural practices such as feng shui. Also, some persons believe that crystals have healing powers.

Rachel Leah Blumenthal

FURTHER READING

Dunmur, David, and Timothy J. Sluckin. *Soap, Science, and Flat-Screen TVs: A History of Liquid Crystals.* New York: Oxford University Press, 2011. Provides a comprehensive look at the history of liquid crystals, starting with their controversial discovery in the late nineteenth century and including historical, political, and social context for their eventual acceptance and proliferation in technology.

Gavezzotti, Angelo. *Molecular Aggregation: Structural Analysis and Molecular Simulation of Crystals and Liquids.* New York: Oxford University Press, 2007. This advanced text first overviews the necessary language and concepts of molecular aggregation and then reviews studies on the processes of molecular aggregation into organic bodies, such as crystals, and their physical and chemical properties.

Khoo, Iam-Choon. *Liquid Crystals*. Hoboken, N.J.: Wiley-Interscience, 2007. Aimed at graduate students and professionals, this text provides a comprehensive look at liquid crystals—in particular, the optical properties that make them so useful to modern technology.

Sherwood, Dennis, and Jon Cooper. *Crystals, X-rays, and Proteins: Comprehensive Protein Crystallography*. New York: Oxford University Press, 2011. This textbook is aimed at biologists and biochemists. It covers the use of X-ray diffraction to determine the crystal structure of proteins and other biological molecules. It also provides a comprehensive look at the mathematics involved in protein crystallography.

Tilley, Richard. *Crystals and Crystal Structures*. Hoboken, N.J.: John Wiley, 2007. This comprehensive introductory text is perfect for students who wish to learn about crystals without necessarily specializing in the field. It covers crystal structures, classification, and symmetry extensively.

Woltman, Scott J., Gregory Philip Crawford, and Gregory D. Jay. *Liquid Crystals: Frontiers in Biomedical Applications*. Hackensack, N.J.: World Scientific, 2007. Provides an overview of advances in biomedical technology using liquid crystals, such as diagnostics, spectroscopy, lasers, and displays. It includes an introduction to liquid crystals and their optical properties.

See also: Chemical Precipitates; Clathrates and Methane Hydrates; Diamonds; Dolomite; Gem Minerals; Quartz and Silica; Silicates.

D

DESERTIFICATION

Desertification comprises a variety of natural and human processes that cause the impoverishment of ecosystems, manifested in reduced biological productivity, rapid deterioration of soil quality, and associated declines in the condition of regional human economic and social systems.

PRINCIPAL TERMS

- **albedo:** the fraction of visible light of electromagnetic radiation that is reflected by the properties of a given type of surface
- **aquifer:** a water-bearing bed of rock, sand, or gravel capable of yielding substantial quantities of water to wells or springs
- **desalinization:** the process of removing salt and minerals from seawater or from saline water occurring in aquifers beneath the land surface to render it fit for agriculture or other human use
- **moving dunes:** collections of coarse soil materials that result from wind erosion and threaten marginal vegetation and settlements as they move across deserts
- **salinization:** the accumulation of salts in the soil

Definition of Desert Conditions

Desertification—or, as some authorities prefer, desertization—is a complex set of interactions between natural and cultural forces and most often occurs in the borderlands of large, natural desert environments. The broadest conventional definition of desert conditions refers to precipitation levels: Regions that receive less than 100 millimeters of precipitation per year, for example, usually exhibit commonly recognized surface features such as flat, gravelly, or pebbly tables and accumulations of sand dunes. The long-term absence of concentrated vegetation cycles and chemical weathering results in only thin, marginally fertile soils, although in places along internal drainage systems these soils may be quite productive in the short run.

The term "desert" inevitably evokes references to land and soil, but in terms of precipitation levels and their consequences, the term can also be applied to surface regimes such as glacier fields and ice caps.

Annual precipitation in Antarctica, for example, is well below 100 millimeters. Both ocean surfaces, in the sense of annual precipitation input, and various ocean levels and seabed, in terms of the presence of nutrients analogous to soil on land, may be classified as desert or as undergoing forms of the desertification process, resulting from both natural and human-induced causes. Land deserts typically feature dispersed, perennial vegetation, some of which may be in a dormant state much of the time. Vegetation may be somewhat denser in regions where groundwater aquifers are near the surface, or may be absent entirely, as in major dune fields.

Human activities along the desert fringe, at least until recent times, were also remarkably accurate indicators of the boundaries of desert conditions. The 100-millimeter line of annual precipitation marks the effective limit of economical agriculture without the use of systematic irrigation. The presence of irrigation, therefore, is frequently a danger signal that human exploitation has ignored the limiting factors of arid environments.

Wind and Water Erosion

Desertification is primarily the result of wind and water erosion. The first step is usually the loss of diffuse, arid zone vegetation that normally is sufficient to protect the soil surface from dramatic erosion, or to intercept the wind transport of a sufficient amount of airborne sand and dust from other areas to balance the loss of local soil matter. This vegetation loss may occur from natural or human causes. The former may include long-term climatic shifts resulting in reductions in annual precipitation or in shorter-term climatic or meteorological phenomena such as abnormally long drought or erratic cloudbursts.

Once protective vegetation is lost, wind erosion can remove even the coarser materials in a soil layer in a matter of a few years. Much of this material will accumulate in moving dunes that become a threat to surviving marginal vegetation. Finer materials may be carried thousands of miles, even to other continents. Dust from the Sahara, for example, may be carried deep into Europe or even across the Atlantic Ocean to Florida and the Caribbean littoral. Dust layers in the Greenland ice cap have been traced to the Gobi Desert.

Erosion of soil matter exposes a surface of pebbles and gravels extremely sensitive to further damage, particularly damage resulting from human actions. In parts of the North African Sahara, vast areas that were scenes of armored engagements in World War II still exhibited ruts and surface disturbance from heavy-tracked vehicles nearly half a century later. Mass motorcycle races in the Mojave Desert of California have caused extensive, and what is likely to be long-lasting, surface damage.

Assessment of the desert environment in terms of average conditions often obscures the potential of water erosion to accelerate desertification. Hard stone or clay desert surfaces can become impermeable to water absorption by the action of raindrops on the finer soil materials, which seal the surface to the point where nearly all precipitation is lost by runoff. This condition not only destroys remnant vegetation beneath the surface by depriving it of water but also vastly increases the volume of water passing through internal drainage systems in periodic flash floods, which wreak havoc on surviving vegetation clinging to the banks of usually dry riverbeds.

Still another cause of desertification is accumulation of salts and alkaline substances in the soil through a process called salinization. Although this process may develop naturally along interior drainage systems, the most extreme cases result from prolonged irrigation. Virtually all water sources contain some level of dissolved minerals. In regions of poor drainage, an endemic characteristic of many arid and semiarid zones due to infrequent rainfall, these minerals accumulate over the years and eventually render the soil useless for agriculture, encouraging its abandonment to the forces of wind erosion. Thousands of square kilometers have been lost to productive agriculture through salinization in ancient agricultural centers such as the Near East and Pakistan, as well as in more recently exploited regions in California and Australia.

HUMAN EXPLOITATION OF DESERT ENVIRONMENTS

In traditional or premodern societies, human exploitation of desert environments relied upon pastoral economies, wherein grazing livestock converted plant matter for human use in the form of meat, wool, and dairy products. Herders moved their animals from region to region in a form of transhumance (the seasonal movement of herds from pasture to pasture). Social mechanisms that discourage intensive environmental exploitation and continually redistribute wealth were common in these societies. The conversion of a pastoral economy from dependence on a species such as the dromedary, a solitary and desultory grazer, to social herders and voracious grass-eaters such as sheep or goats—not to mention the issue of expanding human numbers—can destroy the equilibrium in a marginal zone in a few seasons. Sheep, for example, have been known to stimulate erosion simply because entire herds repeatedly follow the same path to a water source and destroy the vegetation and soil cover along the way.

As population density has soared, peasant populations have opened a new assault on marginal lands by cutting down the few available woody plants for firewood. Ecologists estimate that an average-sized family living in the arid borderlands of the African-Asian desert belt consumes the wood production of 1 to 3 hectares of land each year and that more than 25 million hectares per year may be denuded of trees and woody growth to provide fuel for burgeoning peasant societies.

FINDINGS FROM CLIMATOLOGY AND METEOROLOGY

The central problem in desertification research is to determine to what degree expansion of deserts results from long-term, natural fluctuations in the global environment and to what extent the phenomenon is a result of human intervention, and to distinguish the two. All the techniques that may be applied to the reconstruction of climatic history and the tracking of meteorological patterns, therefore, pertain to desertification studies.

Because of the generally brief period covered by modern meteorological records and an imperfect knowledge of the broader patterns of global climatic change, many theories have emerged that accord varying importance to natural and human-induced factors in desertification. Some climatologists propose that during the twentieth century a global swing toward greater aridity became apparent. Northern Hemisphere records suggest an expansion of polar air masses into lower latitudes with a consequent depression of moist equatorial air masses closer to the equator, so that the convergence between these systems—which roughly marks the northernmost penetration of tropical monsoons into North Africa and the Middle East—does not come as far north as it once did.

Conversely, the input of solar energy into tropical ecosystems, which drives the monsoons, could be decreasing. The GARP Atlantic Tropical Experiment (GATE), a massive, international research effort conducted in the mid-1970's to learn more about tropical environments and weather patterns, provided a knowledge base from which alarming conclusions may be drawn concerning the rapid deforestation of the tropics because of human exploitation, the consequent decrease in photosynthesis activity and increase in the albedo of the tropics, and the overall decrease in solar energy entering tropical ecosystems that results from these changes. Appreciation of the enormous importance of the tropics, as opposed to high latitudes, in understanding middle-latitude desertification was a major outcome of GATE.

SEEKING SOLUTIONS

The major objective of research into desertification is the discovery of procedures that may slow, or even reverse, what specialists in all fields agree is a relentless advance of desert frontiers. Conditions in sandy deserts with moving dunes have received particular attention in applied research because of the danger they present to settlements and installations. Careful analysis of the mineral and chemical composition as well as the physical sizes of particles, all of which may vary widely, is crucial in determining the most appropriate measures for possible fixation. Some dunes may be anchored by carefully chosen plant cover. Liquid binders and emulsions have also been tried with some success. Ironically, since intensive human exploitation is responsible for much contemporary desertification, applied research is frequently directed toward new dimensions of human adaptation to desert conditions. Results have included attempts to exploit desert plant species, inexpensive means of desalinization to utilize often plentiful sources of water in desert aquifers, and more efficient solar energy technologies; however, desert aquifers are finite and exhaustible resources.

Remote sensing techniques, using special films in aerial photography and various optical sensors in satellite observations, now provide a season-by-season record of conditions in deserts and borderlands. Detailed satellite imagery for much of the earth extends back over forty years, providing a valuable historical perspective on environmental changes. Infrared-sensitive films are capable of recording early germination of plant life, concentrations of various substances in the soil composition, and the presence of certain contaminants. Landforms and topographic configurations not readily apparent on the ground (a fact that may have a bearing on the presence of water or the beginnings of vegetation deterioration, or that may make certain areas particularly vulnerable to erosion), frequently reveal themselves through remote sensing. Small-scale mapping of vast areas and careful monitoring of conditions are keys to identifying desertification in its early stage, when reversal may still be possible.

THE SAHEL

Research on desertification in the past few decades has concentrated on the Sahel, the fragile zone on the southern margins of the Sahara where true desert conditions become transitional to the grassland environments of tropical Africa. Beginning in the 1970's, a series of calamitous droughts struck the Sahel, bringing death and misery to huge numbers of people and animals and destroying entire pastoral societies. The impact of drought in the Sahel is magnified tremendously by the fact that the greatest breadth of the African continent is precisely in this zone, so that changes in northernmost penetration of the tropical monsoons of only a few minutes of latitude can spell catastrophe over thousands of square kilometers.

Research in the Sahel demonstrates the difficulty of separating natural from human causes of desertification. In the Republic of Sudan, for example, there is well-documented evidence, even in the recent colonial period, for retreat of the boundary zone between grassland and desert in some areas several hundred miles to the south and concomitant decreases in rainfall. Yet there is also a much broader picture, derived from archaeological studies, suggesting that this pattern may extend many centuries into the past and that modern changes, therefore, cannot be entirely responsible.

Despite evidence that desertification processes have been at work for many centuries, the pressure of the recent population increase on the Sahel and other marginal areas has created an atmosphere of crisis around applied research programs, so that most are directed toward fundamental changes in local practices in agriculture and animal husbandry or toward climate modification and desert reclamation. Both strategies often have the characteristic of uncontrolled experiments. Socioeconomic intervention tends to stress traditional or small-scale economic practices, often on the partly subjective conviction that they are better adapted to local conditions.

Schemes for climate and weather modification frequently require drastic environmental change. Among the more moderate proposals for the Sahel are construction of tree shelter belts, local vegetation modification, and cloud seeding designed to force the monsoon effect northward. Other schemes demand fundamental intervention in the patterns of atmospheric circulation that broadly determine the location of major desert systems. Most of these require gross changes in surface features. The creation of huge lakes in the natural drainage basins of the Sahara, through dams, evaporation control, and exploitation of groundwater aquifers, might increase precipitation levels. Many of these basins actually held such lakes during wetter conditions during the Pleistocene. One research team has proposed paving large strips or "islands" of the Sahara with asphalt. The surface albedo of these areas would be much lower than sand or even vegetation, and their higher temperatures presumably would heat the air above them, promoting cloud formation and precipitation. Such a plan assumes sufficient moisture in the air at the outset, for cloud seeding cannot create moisture where none exists. Others have proposed periodic releases of carbon dust into the atmosphere over the Sahara to increase heat absorption and cloud formation. None of these schemes has been attempted, and all of them would be likely to generate scientific controversy given their impact on the environment and their unknown potential.

VULNERABILITY TO DESERTIFICATION

Desertification is not confined merely to marginal lands bordering on true deserts. Careless agriculture or animal husbandry can create near-desert conditions in vast areas of grassland, as in the case of the Dust Bowl phenomenon of the 1930's on the central plains of the United States. Enormous amounts of top-quality soil are lost each year to water and wind erosion resulting from inadequate farming practices.

Much of the land most vulnerable to desertification, however, lies in lower or tropical latitudes, where soil regimes and microenvironments are not nearly as resilient as they are in Europe or North America. In many of these areas, the human issues arising from desertification involve not so much a lowering of living standards as the very survival of local populations. They call for massive intervention schemes with multiple economic and social dimensions, usually on the assumption that the indigenous social fabric or the capacity to respond to emergencies of this magnitude is nonexistent.

Ronald W. Davis

FURTHER READING

Allan, J. A., ed. *The Sahara: Ecological Change and Early Economic History.* Outwell, England: Middle East and North American Studies Press, 1981. A review of ecological changes in this great desert region and human utilization of the desert margins in preclassical times. This careful reconstruction of historical environmental conditions is useful as a comparison to modern desert limits and the impact of long-term settlement.

Brooks, George E. "A Provisional Historical Schema for Western Africa Based on Seven Climate Periods, ca. 9000 B.C. to the Nineteenth Century." *Cahiers d'études africaines* 16 (1986): 46-62. Summarizes the historical evidence for climatic change in the Sahel. Includes maps projecting fluctuations of desert limits in various periods.

Bryson, Reid A., and Thomas J. Murray. *Climates of Hunger: Mankind and the World's Changing Weather.* Madison: University of Wisconsin Press, 1977. Covers climatic change and the possible extent of human-induced factors, with emphasis on deforestation and desertification around the world.

Chouhan, T. S. *Desertification in the World and Its Control.* Jodhpur, India: Scientific Publishers, 1992. Deals with desertification on a global scale but focuses on India in particular. Also deals with tactics that have been used and proposed to control the problem. Maps, index.

Climate, Drought, and Desertification. Geneva, Switzerland: World Meteorological Organization, 1997. This brief booklet, prepared by the World Meteorological Organization, deals with the climatic factors, such as drought, that lead to desertification worldwide. Includes color illustrations.

Crawford, Clifford S., and James R. Gosz. "Desert Ecosystems: Their Resources in Space and Time." *Environmental Conservation* 9 (1982): 181-195. Discusses the tensions between continuous exploitation of desert environments and the intermittent character of vital resources such as water and vegetation in a given location.

Eckholm, Erik, and Lester R. Brown. *Spreading Deserts: The Hand of Man.* New York: Worldwatch Institute, 1971. A brief, useful summary of world desertification patterns, emphasizing careless human exploitation and the interruption of natural processes that limit the spread of deserts.

Evenari, Michael, Leslie Shanan, and Nephtali Tadmor. *The Negev: The Challenge of a Desert.* 2d ed. Cambridge, Mass.: Harvard University Press, 1982. The best comprehensive study of attempts to develop settlement styles suitable to desert resources, based on a combination of archaeological research on ancient settlement patterns and the most recent findings in Earth sciences.

Geeson, N. A., C. J. Brandt, and J. B. Thornes, eds. *Mediterranean Desertification: A Mosaic of Processes and Responses.* Hoboken, N.J.: John Wiley & Sons, 2002. Begins with an overview of the Mediterranean desertification. Chapters are devoted to climatic changes and their influence on the desertification of the Mediterranean. Discusses land use, degradation, erosion, and other desertification-related processes. Followed by sections discussing changes to vegetation. Provides information on methods of study and a number of case studies.

Glantz, Michael H., ed. *Desertification: Environmental Degradation in and Around Arid Lands.* Boulder, Colo.: Westview Press, 1977. An excellent discussion of a broad selection of related phenomena, including nature and causes of desertification, theories of natural versus human-induced environmental change, case studies of desertification, experiments in weather modification, and efforts of international agencies to treat desertification as a unified, global problem. Includes an extensive bibliography.

Imeson, Anton. *Desertification, Land Degradation and Sustainability.* Hoboken, N.J.: John Wiley & Sons, 2012. Explores research from the first decade of the twenty-first century. Discusses desertification and its management and mitigation on a global and local scale. Discusses the cultural and physical impact. Well suited for undergraduates and graduate students studying desertification with a background in environmental studies, geology, ecology, or economics.

Mairota, Poala, John B. Thornes, and Nichola Geeson, ed. *Atlas of Mediterranean Environments in Europe: The Desertification Context.* New York: John Wiley & Sons, 1996. Focuses on desertification in the Mediterranean region. Includes chapters on seminatural environments and processes, socioeconomic processes and change, and field studies. Bibliography, index, maps, and plant species glossary.

Mortimore, Michael. *Adapting to Drought: Farmers, Famines and Desertification in West Africa.* New York: Cambridge University Press, 2009. Discusses research conducted in northern Nigeria on the patterns of adaptation displayed by rural communities. Describes adaptive strategies that have promoted resilience in drought-prone areas, and suggests their use in semi-arid conservation.

Oliver, F. W. "Dust-Storms in Egypt and Their Relation to the War Period, as Noted in Maryut, 1939-1945." *Geographical Journal* 106-108 (1945-1946): 26-49, 221-226. An early study of the extent to which intensive and extensive human movement and occupation of the desert may result in short-term deterioration of arid environments.

United Nations Conference on Desertification, Nairobi, Kenya. *Desertification: Its Causes and Consequences.* Elmsford, N.Y.: Pergamon Press, 1977. Includes four in-depth studies on the relationship between desertification and climate, ecological change, population, and technology, each with an extensive bibliography.

Walls, James. *Land, Man, and Sand: Desertification and Its Solution.* New York: Macmillan, 1980. An extensive collection of case studies from around the world, each illustrating facets of desertification and pragmatic, localized solutions. Contains numerous examples of how physical, biological, and social science approaches may be combined in comprehending and controlling deterioration.

Zdruli, Pandi, et al., eds. *Land Degradation and desertification: Assessment, Mitigation and Remediation.* New York: Springer, 2010. Provides background information and historical examples of land degradation. Discusses degradation and mitigation of Africa, Asia, Europe, and the Americas. Chapters are specific to the issues generated within these regions.

See also: Clays and Clay Minerals; Diagenesis; Expansive Soils; Feldspars; Land Management; Landslides and Slope Stability; Land-Use Planning; Land-Use Planning in Coastal Zones; Minerals: Physical Properties; Sedimentary Mineral Deposits; Sedimentary Rock Classification; Soil Chemistry; Soil Erosion; Soil Formation; Soil Profiles; Soil Types; Water and Ice.

DIAGENESIS

Diagenesis refers to the physical, chemical, and biological changes that sediment undergoes after it is deposited. These processes change loose sediment into sedimentary rock and occur in the upper several hundred meters of the earth's crust.

PRINCIPAL TERMS

- **bar:** a unit of pressure equal to 100 kilopascals and very nearly equal to 1 standard atmosphere
- **carbonate:** a mineral with CO_3 in its chemical formula, such as calcite ($CaCO_3$)
- **lithification:** the hardening of sediment into a rock through compaction, cementation, recrystallization, or other processes
- **pore fluids:** fluids, such as water (usually carrying dissolved minerals, gases, and hydrocarbons), in pore spaces in a rock
- **porosity:** the amount of space between the sedimentary grains in a rock or sediment
- **sediment:** loose grains of solid, particulate matter resulting from the weathering and breakdown of rocks, chemical precipitation, or secretion by organisms
- **sedimentary rock:** a rock resulting from the consolidation of loose sediment that has accumulated in flat-lying layers on the earth's surface

DISTINGUISHING DIAGENESIS FROM METAMORPHISM

Diagenesis refers to the physical, chemical, and biological processes that occur in sediment after deposition as it is buried and transformed into sedimentary rock. These processes alter the texture, porosity, fabric, structure, and mineralogy of the sediment. Through diagenesis, sand is changed into sandstone, mud is changed into shale, and carbonate sediments are changed into limestone and dolomite. The processes and degree of alteration depend in part on the initial sediment composition and the depth of burial.

As sediment is buried to increasing depths, the temperatures and pressures increase, and diagenesis grades into metamorphism. Although the exact limits separating diagenesis and metamorphism are not strictly defined, diagenesis can be considered to occur under temperatures ranging from those at the earth's surface up to nearly 300 degrees Celsius and under pressures ranging from atmospheric pressure to at least 1 kilobar (1,000 bars). These conditions occur at depths approximating 10 kilometers. Some classifications restrict the zone of diagenesis to about 0 to 1 kilometers and define zones of catagenesis (several kilometers deep, with temperatures of 50 to 150 degrees and pressures of 300 to 1,000 or 1,500 bars), metagenesis (up to about 10 kilometers deep), and finally metamorphism. Temperature is an important control on many diagenetic processes, because it influences chemical reactions such as the dissolution and precipitation of minerals, recrystallization, and authigenesis.

COMPACTION

The primary physical diagenetic process is compaction. Compaction presses sedimentary grains closer together under the load of overlying sediment, causing pore space to be decreased or eliminated and squeezing out pore fluids. In sandstones, compaction occurs by the rotation and slippage of sand grains, the breakage of brittle grains, and the bending and mashing of ductile (soft, easily deformed) grains. Brittle grains include thin shells, skeletal fragments, and feldspar grains. Ductile grains include clay or shale chips and fecal pellets. Compaction also causes some mineral grains to interpenetrate, producing irregular stylolitic contacts where the grains have intricately complex boundaries. In general, sands compact much less than muds. That is true since the average sandstone has a high percentage of hard grains, such as quartz; muds typically have a high initial water content, and water is squeezed out during compaction. The compaction of sands, however, is influenced by the nature of the sand grains present; sands with a large percentage of ductile grains are more susceptible to compaction.

CEMENTATION, AUTHIGENESIS, AND REPLACEMENT

Chemical diagenetic processes include cementation, the growth of new minerals (authigenesis), replacement, neomorphism (recrystallization and inversion), and dissolution. Cementation is the precipitation of minerals from pore fluids. These minerals glue the grains of sediment together, forming a rock. The most common cements are quartz (silica), calcite, and hematite, but such cements as clays, aragonite, dolomite, siderite, limonite, pyrite, feldspar, gypsum, anhydrite, barite, and zeolite minerals occur as well. The type of cement is controlled by the composition of the pore fluids.

Authigenesis refers to the growth of new minerals in the sediment and the transformation of one mineral into another. Some of the most common authigenic minerals in sandstones are calcite, quartz, and clay cements. Other than cements, authigenic minerals include glauconite, micas, and clay minerals. Glauconite is a green mineral that forms on the sea floor when sedimentation rates are low. Authigenic micas and clays typically form in the subsurface at higher temperatures and pressures. Often, one clay mineral is transformed into another as a result of dehydration (water loss) or chemical alteration by migrating fluids. Clays may also be formed from the alteration of feldspars or volcanic ash and rock fragments. Authigenesis also includes the alteration of iron-bearing minerals (such as biotite, amphibole, or pyroxene) to pyrite, under reducing conditions, or to iron oxide (limonite, goethite, or hematite), under oxidizing conditions.

Replacement is the molecule-by-molecule or volume-for-volume substitution of one mineral for another. Replacement generally involves the simultaneous dissolution of an original mineral and precipitation of a new mineral in its place. Fossils that were originally calcium carbonate may be replaced by different minerals, such as quartz, pyrite, or hematite. Many minerals are known as replacement minerals, including calcite, chert, dolomite, hematite, limonite, siderite, anhydrite, and glauconite. Factors controlling replacement include pH, temperature, pressure, and the chemistry of the pore fluids.

NEOMORPHISM AND DISSOLUTION

Neomorphism is a term meaning "new form"; it refers to minerals changing in size, shape, or crystal structure during diagenesis. The chemical composition of the minerals, however, remains the same. Neomorphism includes the processes of recrystallization and inversion. Recrystallization alters the size or shape of mineral grains without changing their chemical composition or crystal structure. Recrystallization can occur in any type of sedimentary rock, but it is most common among the carbonates. Limestones are commonly recrystallized during diagenesis, producing a coarsely crystalline rock in which original sedimentary textures and structures may be fully or partially obliterated. The reason that minerals recrystallize is not well understood, but it may be related to energy stored in strained crystals or to a force arising from the surface tension of curved crystal boundaries. Inversion is a process in which one mineral is changed into another with the same chemical composition but a different crystal structure. The two minerals involved are called polymorphs, meaning "multiple forms." Aragonite and calcite are polymorphs. Both have the same chemical composition, but each has a different crystal structure: Aragonite is orthorhombic and calcite is rhombohedral. Aragonite, with time, will become calcite by inversion. Inversion may occur along a migrating film of liquid, causing the simultaneous dissolution of one mineral and precipitation of its polymorph, or by solid-state transformation (switching of the positions of ions in the crystal lattice).

Dissolution refers to the dissolving and total removal of a mineral, leaving an open cavity or pore space in the rock. This pore space may persist, or it may become filled by another mineral at a later time. Some of the more soluble minerals are the carbonates and the evaporites, such as halite and gypsum. Large-scale dissolution of limestone leads to the formation of caves and caverns. Pressure solution is the dissolution of minerals under the pressure of overlying sediment. Stylolites, a common result of pressure solution, commonly occur in carbonate rocks. Stylolites are thin, dark, irregular seams with a zigzag pattern that separate mutually interpenetrating rocks. The dark material along the seam is a concentration of insoluble material such as clay, carbon, or iron oxides. Pressure solution can result in a 35 to 40 percent reduction in the thickness of carbonate rocks. The carbonate removed by pressure solution is frequently a source of carbonate cements.

Finally, minerals in solution may be deposited on existing grains. This phenomenon is very common with quartz grains. Silica in solution builds on the existing crystalline structure of the grain, causing new quartz to form. Often the sand grain develops actual crystal faces as a result. Such new mineral growths are called overgrowths.

BIOLOGICAL DIAGENETIC PROCESSES

Biological diagenetic processes occur soon after sediment is deposited and consist of the activities of organisms in and on the sediment. Bacteria are particularly important to the chemical diagenetic processes. Bacteria living in the sediment control many chemical reactions involving mineral precipitation or dissolution; they are involved in the breakdown or decomposition of organic matter (one of the steps in the formation of oil and gas), and can cause the pH of the pore fluids to increase or decrease, depending on the kinds of microorganisms, organic matter, decomposition products, and availability of oxygen. For example, in aerobic environments (those where oxygen is present), decay of organic matter generally causes decreasing pH (increasing acidity), which may lead to the dissolution of carbonate minerals such as calcite. Under anaerobic conditions, organic decay generally raises the pH and may lead to the precipitation of calcite cement. The formation of pyrite is also influenced by the activity of bacteria. Sulfate-reducing bacteria in anoxic environments change sulfate into hydrogen sulfide. If iron is present, it reacts with the hydrogen sulfide to form iron sulfides, such as pyrite.

Bioturbation is the disturbance of the sediment by burrowing (excavation into soft sediment), boring (drilling into hard sediment), the ingestion of sediment and production of fecal pellets, root penetration, and other activities of organisms. Bioturbation generally occurs shortly after deposition and causes mixing of sediment that was originally deposited as separate layers, destruction of primary sedimentary structures and fabrics, and breakdown or clumping of grains. In some cases, chemical alteration of the sediment accompanies bioturbation. For example, light-colored halos may form around burrows or roots, particularly in red or brown sediments, because of the reduction of iron.

Diagenesis may decrease or increase the porosity and permeability of the sediment. Porosity is decreased by compaction, the precipitation of cements in pore spaces, and bioturbation. Porosity is increased by dissolution. Zones of increased porosity are particularly favorable for oil and gas accumulations.

STUDY TECHNIQUES

Diagenesis is primarily studied using sedimentary petrography, which is the microscopic examination of thin sections of sedimentary rocks. Thin sections are slices of rock, typically 30 micrometers thick, bonded to glass slides, which are examined with a petrographic microscope. In this way, minerals can be identified based on their optical properties, and textural relationships can be studied, such as the size, shape, and arrangement of grains; the geometry of cements and pore spaces; the character of contacts between grains; the presence of dissolution features; and mashing or fracturing of grains. Thin sections may be enhanced, to allow easier identification of minerals, with various staining and acid etching techniques. In addition, acetate peels may be prepared from etched and stained rock surfaces for examination with the microscope.

There are a number of other techniques that can be used in conjunction with petrography to obtain more specific diagenetic data. Cathodoluminescence microscopy can provide information about the spatial distribution of trace elements in rocks. Luminescence is the emission of light from a material that has been activated or excited by some form of energy. Cathodoluminescence works by activating various parts of a polished thin section with a beam of electrons. The electron beam excites certain ions, producing luminescence. This technique can reveal small-scale textures and inhomogeneities of particles and cements through differences in their luminescence, which are related to differing concentrations of trace element ions.

Scanning electron microscopy can magnify images 70,000 times or more, permitting detailed study of extremely small particles that cannot be adequately examined using a petrographic microscope. Scanning electron microscopy involves reflections of an electron beam from a rock or mineral surface. Fine details of cements and grains may be readily observed and photographed.

X-ray diffraction is used to determine the mineralogy of sedimentary rocks, particularly fine-grained rocks such as shales. The technique is based on reflections of X rays from planes in the crystal structure of minerals. Each mineral has a characteristic crystal structure and produces a distinctive X-ray diffraction pattern consisting of peaks of different position and intensity, which are plotted on chart paper by the X-ray diffractometer.

Fluid inclusions are extremely small droplets of fluid encased within crystals or mineral grains. The fluids are a small sample of the original pore fluids from which the mineral was precipitated. By examining fluid inclusions using heating and freezing devices attached to a microscope, the geologist can determine the composition of the original pore fluids and the temperature at which the mineral was precipitated. This technique reveals that many minerals and cements were precipitated from hot, saline pore fluids.

Stable isotopes of oxygen and carbon are commonly used to determine the chemistry of the pore fluids and the temperatures under which precipitation of cements or authigenic minerals occurred. Studies of the stable isotopes of microfossils have also provided information on past climatic changes. Isotopes are different forms of elements that vary in the number of neutrons present in the nucleus; hence, the various isotopes of an element have different atomic weights. By comparing the ratios of oxygen-16 and oxygen-18 or carbon-12 and carbon-13 in minerals such as calcite, it is possible to determine whether the minerals precipitated from fresh water or marine water, or to determine the temperature of the fluid from which the mineral precipitated.

Pamela J. W. Gore

FURTHER READING

Boggs, Sam, Jr. *Petrology of Sedimentary Rocks.* New York: Cambridge University Press, 2009. Begins with a chapter explaining the classification of sedimentary rocks. Remaining chapters provide information on different types of sedimentary rocks. Describes siliciclastic rocks and discusses limestones, dolomites, and diagenesis.

_____. *Principles of Sedimentology and Stratigraphy.* 5th ed. Upper Saddle River, N.J.: Prentice Hall, 2011. Designed for undergraduate geology majors. Clearly written, comprehensive, and well illustrated. Offers a long chapter on diagenesis discussing major diagenetic processes and environment (temperatures and pressures, as well as the chemical composition of subsurface waters), major controls on diagenesis, and the major effects of diagenesis (physical, mineralogic, and chemical changes). Contains background information on other aspects of sedimentary rocks.

Jonas, E. C., and E. McBride. *Diagenesis of Sandstone and Shale: Application to Exploration for Hydrocarbons.* Austin: Department of Geological Sciences, University of Texas, Continuing Education Program, 1977. Provides clear coverage of the diagenesis of sandstones and shales. Well illustrated, with photomicrographs of thin sections showing the evidence for various diagenetic processes. Includes graphs and line drawings, which make processes easier to envision. Assumes a basic background in sedimentology but can be understood by the nonspecialist.

Larsen, Gunnar, and George V. Chilingar, eds. *Diagenesis in Sediments and Sedimentary Rocks.* New York: Elsevier, 1979. A two-volume book in the Developments in Sedimentology series, covering aspects of sedimentary geology in depth; each volume contains papers written by and for specialists in the field. Volume 1 offers an introduction to the diagenesis of sediments and rocks; subsequent chapters deal with more specialized subjects, such as the diagenesis of sandstones, coal, and carbonate rocks. Volume 2 features specialized chapters on the various phases of diagenesis, low-grade metamorphism, and the diagenesis of shales, deep-sea carbonates, and iron-rich rocks. Suitable for advanced college students.

Leeder, Mike R. *Sedimentology and Sedimentary Basins: From Turbulence to Tectonics.* 2d ed. Hoboken, N.J.: John Wiley & Sons, 2011. Covers the study of sedimentology and chemical sedimentology, and their relationship with tectonics. A good introduction to sedimentology for the layperson without much background in Earth sciences.

Mackenzie, F. T., ed. *Sediments, Diagenesis, and Sedimentary Rocks.* Amsterdam: Elsevier, 2005. Volume 7 of *Treatise on Geochemistry,* edited by H. D. Holland and K. K. Turekian. Compiles articles on such subjects as diagenesis, chemical composition of sediments, biogenic material recycling, geochemistry of sediments and cherts, and green clay minerals.

McDonald, D. A., and R. C. Surdam, eds. *Clastic Diagenesis.* Memoir 37. Tulsa, Okla.: American Association of Petroleum Geologists, 1984. Divided into three parts: basic concepts and principles of diagenesis, changes in porosity, and applications of diagenesis in the exploration and production of hydrocarbons. Articles are case histories of the diagenesis of particular rock units. Most are technical in content but some provide general overviews of specialized aspects of diagenesis. Suitable for advanced college students.

Middleton, Gerard V., ed. *Encyclopedia of Sediments and Sedimentary Rocks.* Dordrecht: Springer, 2003. Cites a vast number of scientists. Subjects range from biogenic sedimentary structures to Milankovitch cycles. Offers an index of subjects. Designed to cover a broad scope and a degree of detail useful to students, faculty, and professionals in geology.

Oldershaw, Cally. *Rocks and Minerals*. New York: DK, 1999. Useful to new students who may be unfamiliar with the rock and mineral types discussed in classes or textbooks.

Pettijohn, F. J. *Sedimentary Rocks*. 3d ed. New York: Harper & Row, 1975. Provides an introduction to the basic types of sedimentary rock and touches on various aspects of diagenesis. Of particular interest is a chapter on concretions, nodules, and other diagenetic segregations. This text explains more about secondary sedimentary structures than most other textbooks on sedimentology, and it is a useful guide for persons curious about the origin of geologic oddities and who would like a better background on the formation of sedimentary rocks. Suitable for college and advanced high school students.

Scholle, Peter A. *A Color Illustrated Guide to Carbonate Constituents, Rock Textures, Cements, and Porosities*. Memoir 27. Tulsa, Okla.: American Association of Petroleum Geologists, 1978. A superbly illustrated book on various types of carbonate rock as seen in thin section. Illustrates the major carbonate grains, along with dolomite, evaporite, silica, iron, phosphate, and glauconite minerals. Covers cements and carbonate rock textures. Provides background information on porosity and techniques for studying carbonate rocks. Provides a brief explanatory caption and geologic locality data for each photograph.

_____. *A Color Illustrated Guide to Constituents, Textures, Cements, and Porosities of Sandstones and Associated Rocks*. Memoir 28. Tulsa, Okla.: American Association of Petroleum Geologists, 1979. A color picturebook of various types of sandstone as seen in thin section. Illustrates all the major detrital sand grains, textures, cements, replacement or displacement fabrics, compaction and deformation fabrics, and porosity. Clays and shales, chert, and other types of sediment are also included. There is a brief explanatory caption and geologic locality data for each photograph. Offers basic information on sandstone classification and various techniques for studying sedimentary rocks.

Scholle, Peter A., and P. R. Schluger, eds. *Aspects of Diagenesis*. Special Publication 26. Tulsa, Okla.: Society of Economic Paleontologists and Mineralogists, 1979. First section covers the determination of diagenetic paleotemperatures; second section covers, the diagenesis of sandstones (in particular, hydrocarbon reservoirs). Includes general review articles and specific examples. Suitable for advanced college-level readers.

Sengupta, Supiya. *Introduction to Sedimentology*. Rotterdam, Vt.: A. A. Balkema, 1994. An excellent introduction to the practices, policies, and theories used within the field of sedimentology. Provides illustrations and maps to clarify concepts and processes. Index and bibliographical references included.

Tucker, M., ed. *Techniques in Sedimentology*. Boston: Blackwell Scientific Publications, 1988. Covers techniques used by sedimentologists to study diagenesis and other areas of sedimentology. Includes chapters on the collection and analysis of field data, grain-size data and interpretation, microscopical techniques, cathodoluminescence microscopy, X-ray powder diffraction, scanning electron microscopy, and chemical analysis. Explains techniques used to study diagenesis and also discusses diagenetic fabrics produced by compaction, cementation, dissolution, alteration, and replacement. Well indexed and profusely illustrated. Includes a number of excellent photomicrographs of rock thin sections illustrating diagenetic textures. For college-level students.

See also: Biogenic Sedimentary Rocks; Building Stone; Carbonates; Chemical Precipitates; Evaporites; Limestone; Sedimentary Mineral Deposits; Sedimentary Rock Classification; Siliciclastic Rocks.

DIAMONDS

Diamond is an important industrial mineral as well as the most valued of gemstones. Natural diamonds crystallize only at very high pressures and are brought to the earth's surface in kimberlite, an unusual type of igneous rock that forms in the upper mantle. Kimberlite also contains "gems" of another sort: rare pieces of the earth's deep crust and mantle.

PRINCIPAL TERMS

- **bort:** a general term for diamonds that are suitable only for industrial purposes; these diamonds are black, dark gray, brown, or green in color and usually contain many inclusions of other minerals
- **coesite:** a mineral with the same composition as quartz (silicon dioxide) but with a dense crystal structure that forms only under very high pressures
- **crystal:** a solid that possesses a definite orderly arrangement of its atoms; crystal differs from an amorphous solid such as glass, and all true minerals are crystalline solids
- **graphite:** a crystalline variety of the element carbon, characterized by its softness and ability to cleave into flakes; the carbon atoms are arranged in sheets that are weakly bonded together
- **kimberlite:** an unusual, fine-grained variety of peridotite that contains trace amounts of diamond
- **metastable:** the state of crystalline solids once they are outside of the temperature and pressure conditions under which they formed; thus, diamond forms at very high pressures within the earth but is metastable at the earth's surface
- **peridotite:** a dense, dark-green rock that is composed mainly of magnesium- and iron-rich silicates, particularly olivine; the mantle, and the ultramafic nodules derived from it, is composed of peridotite
- **ultramafic:** rocks such as peridotite that contain abundant magnesium- and iron-rich minerals

AVAILABILITY OF DIAMONDS

In the eighteenth century, virtually all the world's diamonds came from India and were hoarded by royalty; few had ever seen a diamond, let alone possessed one. Brazil became an important producer in the late 1700's, but diamonds were still unavailable to most people. In 1866, a farm boy discovered a bright pebble on the banks of the Orange River in South Africa and unknowingly started a chain of economic, social, and political upheavals that continue to this day; the stone was later determined to be a 21-carat diamond (one carat equals 200 milligrams). Within a decade, South Africa's mines would be producing 3 million carats a year to a world market; by the turn of the century, diamonds would become an important industrial commodity. Today, several South American countries also export diamonds, but South Africa remains the world's foremost diamond producer. While North America has few diamond deposits, diamonds formerly were mined in Arkansas and are currently being mined in Colorado and Canada.

Early diggings were concentrated along the Orange and Vaal rivers of South Africa, where prospectors staked small claims and shoveled into the diamondiferous gravel deposits. Fines were washed away in a rudimentary method known as wet digging, and the remaining gravel was spread out to be examined, pebble by pebble. Around 1870, a diamond was found approximately 100 kilometers from the nearest river, and prospectors began to speculate that the stones along the riverbanks did not originate there but were washed in from elsewhere. Increasing numbers of miners went into the bush to pursue diamonds, finding them in local patches of weathered rock known as yellow ground. As geologists would soon understand, the yellow ground was merely the uppermost layer of deep, funnel-shaped pipes of diamond-bearing igneous rock that had been injected into the earth's crust by ancient volcanoes. The bluish-gray rock was named kimberlite, for a nearby South African town. Diamonds are also mined from alluvial deposits in Russia and Namibia.

In some African nations, civil wars have been financed by the illicit mining and sale of diamonds, called "conflict diamonds." Frequently the diamonds are mined by slave labor. A series of international conventions requiring the registration of diamonds and strict controls on shipment has significantly curtailed the trade in conflict diamonds.

INCLUSION IN KIMBERLITE

At present, more than 90 percent of the world's diamonds are found in river gravels, beach sands, and glacial deposits of many geological ages. Only in kimberlite pipes are diamonds found in the original rock in which they were formed. Kimberlite pipes have rather inconspicuous features, seldom having diameters greater than 1 kilometer. Mining has shown them to be carrotlike bodies whose vertical dimensions far exceed the sizes of their surface outcrops. Mine shafts have penetrated about 1.5 kilometers into kimberlite pipes, but minerals contained in the kimberlite, including the coveted diamonds, suggest that the pipes extend all the way through the crust and into the earth's upper mantle to a total depth of about 250 kilometers. This is deeper than any other variety of igneous rock. Most kimberlites do not contain commercial quantities of diamonds.

Beneath the weathered yellow ground, fresh kimberlite is a hard, dark bluish-gray rock that miners call blue ground. Its texture gives strong evidence of an igneous origin, indicating that kimberlite was injected into the earth's crust as a molten liquid and then quickly solidified against cooler rocks surrounding the pipe. The major constituents of kimberlite are silicate minerals: compounds of silicon and oxygen with other metal ions. Kimberlite is a variety of peridotite ("peridot" is an ancient word for olivine), and hence its major constituent is the mineral olivine, a magnesium-iron silicate. The olivine is usually altered to the mineral serpentine, giving the rock its characteristic blue-gray color.

Exotic rocks contained within the kimberlite matrix are perhaps more interesting than the kimberlite itself. Diamonds are one such inclusion, although they constitute a minuscule proportion of the total rock. Typical diamond contents in minable kimberlite range from about 0.1 to 0.35 carat per ton; even the famous Premier Mine in South Africa has produced only about 5.5 tons of diamonds from 100 million tons of rock, which is about 0.000005 percent.

FORMATION OF DIAMONDS

Three independent lines of evidence indicate that kimberlite is formed in the upper mantle, at depths and pressures far greater than any other type of igneous rock. To calculate pressures of formation, it is usually necessary to know or assume the temperature of formation. The continental geotherm supplies the needed temperature information. At temperatures along the continental geotherm, diamond is stable only at depths greater than about 100 kilometers; at lower pressures, graphite forms instead. Diamond exists at low pressures only because it is metastable (otherwise there could be no diamonds on the earth's surface), but diamond does not form naturally at low pressures.

A second line of evidence for great pressure involves the minerals that have been trapped inside the diamonds during their growth. Diamonds sometimes contain inclusions of coesite, a mineral with the same chemical composition as quartz (silicon dioxide) but with a more compact structure that forms only at very high pressures. At temperatures along the continental geotherm, laboratory studies have shown that coesite, in turn, gives way to the silica mineral stishovite at depths greater than about 300 kilometers. Because stishovite has never been found in diamonds, the diamonds must have formed at depths of less than 300 kilometers or so. Hence, diamonds seem to have formed at depths between about 100 and 300 kilometers in the earth's upper mantle. The pressures and temperatures that have been calculated from minerals in the ultramafic nodules are in agreement with this depth range: Most of the nodules seem to have formed at depths of 100 to 250 kilometers in the upper mantle and at temperatures of about 1,100 to 1,500 degrees Celsius. Contrary to popular misconception, diamond never forms from the metamorphism of coal. Surface rocks are never buried at the depths where diamonds form, and the carbon of diamonds has been in the mantle since the formation of the earth.

PHYSICAL PROPERTIES

The unusual physical properties of diamond are a reflection of its crystalline structure. Diamond is a three-dimensional network of elemental carbon, with each carbon atom linked to four equidistant neighbors by strong covalent bonds (in which electrons are equally shared by two atoms). The dense, strongly bonded crystal structure gives diamond its extreme hardness. Another mineral made of pure carbon is graphite, the writing material in lead pencils. In graphite, sheets of carbon atoms are weakly bonded and separated by relatively large distances. Thus, graphite has a lamellar structure and is very soft.

The physical properties of diamond are remarkable in comparison to virtually all other materials. It is the hardest substance known. With a hardness of

10, it tops the Mohs scale of relative hardness and is actually about forty-two times as hard as corundum, its nearest neighbor, with a hardness of 9. Its luster falls between adamantine and greasy; it cannot be wetted by water, a property that is of great practical benefit in separating diamonds from waste rock. Diamonds vary in color from water-clear (designating the most valuable diamonds) to pale blue, yellow, deep yellow, or brown; industrial varieties are brown or grayish-black. Raw diamonds occur most often as octahedra (eight-sided polygons) or irregular shapes. Diamond can be cleaved in four directions parallel to its octahedral faces. Its refractive index of 2.42 is the highest of all gems, producing strong reflections in cut stones, and its very high dispersion (that is, its ability to separate white light into the colors of the spectrum) gives cut diamonds their "fire." Diamond is also triboelectric (electrically charged when rubbed) and fluorescent (emitting visible light when struck by ultraviolet rays).

Diamonds are separated from waste rock by first crushing the kimberlite, wetting it, then passing it over a series of greased bronze tables. Diamond cannot be wetted by water and is the only mineral that sticks to the grease, which is later scraped off as the diamonds are extracted. Another way of separating diamonds takes advantage of their fluorescent property. As the crushed rock is passed beneath ultraviolet lamps, the diamonds are spotted by photosensors, which trigger jets of compressed air that eject the diamonds into bins. Gem-quality stones of less than 1 carat are called melee. Less than 5 percent of all diamonds are suitable for cut stones of 1 carat or larger. Not all diamonds are gem quality. Black or dark diamonds full of graphite inclusions are called "carbonado"; these are suitable only for use as abrasives.

DIAMOND CUTTING AND GRADING

Diamonds have been fashioned into precious jewels for several millennia, but diamond cutting has become a major industry only during the past century, in response to worldwide demand and the abundant supply of South African diamonds. Five basic steps are involved in diamond cutting:

marking, cleaving, sawing, girdling, and faceting. The diamond is first carefully studied, sometimes for months, in order to identify its cleavage planes and to map out any inclusion-rich areas that will affect how it is to be cut. Large diamonds are usually irregular in shape and are seldom left whole. A central, master stone is commonly envisioned within the mass, and the "satellite" offcuts become fine gems in their own right. Lines for cleaving or sawing are marked in black ink, and the diamond is then sent to the cutter. If the diamond is to be cleaved, a thin groove is first established using a saw charged with diamond dust. The diamond is mounted in a dop or clamp, and a steel wedge is inserted into the groove and struck sharply with a mallet. A misdirected blow can shatter the stone: It is said that Joseph Asscher, after successfully cleaving the 3,100-carat Cullinan diamond in 1908, swooned into the arms of an attending physician. Sawing is a slower alternative to cleaving; it uses a thin, circular bronze blade that is charged with diamond dust (diamonds differ in hardness in different directions in the crystal; the cutting is done by grains that happen to have suitable orientations). Next, the diamond is girdled by placing it in a lathe and grinding it against another diamond to make it round; the "girdle" is where the upper and lower sets of facets meet. Finally, the diamond is faceted: The facets are cut and polished by clamping the stone in a holder and placing it against rotating laps that

Raw diamonds. (Geological Survey of Canada)

are charged with diamond dust. The most popular cut is the "brilliant"—a round stone with fifty-eight facets. Other cuts include the marquise (oval), emerald (rectangular), and pear. The orientations of the facets are calculated carefully to take advantage of the optical properties of the diamond.

Cut stones are graded under strict rules, using an elaborate system of four criteria: cut, color, clarity, and carat. Cut refers to the skill with which the gem has been shaped—its symmetry and reflective brilliance. Color refers to the tint of the stone: The most valuable gems are water-clear, but many fine stones are pale yellow, blue, or pink; colored diamonds are called "fancies." Clarity refers to the size, number, and locations of any inclusions that may be present. Inclusions do not necessarily degrade a stone if they are small or located in inconspicuous places. Carat refers to the weight of the stone. The largest known diamond was the Cullinan, which weighed 0.6 kilogram before it was cleaved and fashioned into an assemblage of stones that are now part of the British crown jewels.

STUDY OF DIAMONDS

Diamonds have been studied using the same analytical methods that are applied to other crystalline solids—notably X-ray diffraction—to identify their crystalline structure. Mineral inclusions in diamond (mostly coesite and garnet) have been examined with the electron microprobe, a device that employs a tiny electron beam to measure the percentages of elements that are present in a mineral. The compositions of coexisting minerals in kimberlite and ultramafic nodules have been similarly analyzed.

Igneous petrology is a subdiscipline of geology concerned with the description and origin of igneous rocks. Petrologists use all the methods listed above for individual minerals, plus larger-scale observations of entire rock masses. Minerals are assemblages of atoms, and rocks are assemblages of minerals; hence, understanding the origins of minerals, including diamond, involves an understanding of how the enclosing rock was formed. Individual rock masses such as kimberlite, ultramafic nodules, and xenoliths of the crust are glued onto glass slides, ground into thin slices, and studied under the microscope. In addition to the minerals they contain, the textures of rocks reveal much about their origins. The results of mineralogical and textural observations are then interpreted in light of even larger-scale observations, involving geological mapping on the surface and underground. Finally, all mineralogical, petrological, and geological observations must be interpreted within the constraints of geophysical data on the internal constitution of the earth.

The quest for diamonds has fostered much inquiry into the rock in which diamonds are found. Kimberlite is a very unusual type of igneous rock that forms deep in the earth's mantle, at depths up to 250 kilometers. The ultimate source of diamonds, kimberlite also contains "gems" of a different sort: pieces of the earth's deep continental crust and upper mantle that would be inaccessible by any other means. Volcanic pipes of kimberlite are therefore windows into the earth's interior, and the rock and its inclusions are avidly studied in order to learn more about the internal constitution of the earth. Diamonds contain trapped inclusions of liquids, mineral solids, and gases, mostly carbon dioxide. Analyses of these inclusions have led to a better understanding of the conditions under which diamond is formed and the volatiles that are present in the earth's mantle.

INDUSTRIAL USES

The hardness, brilliance, and fire of diamonds have made them unsurpassed as gems; the use of diamonds as industrial materials is perhaps less well known. All stones not suitable for use as gems are destined for industrial use. "Bort" refers to dense, hard, industrial diamonds, and carbonado refers to diamond that has a lower specific gravity than normal diamond. Such diamonds vary in color from off-white to black. Most industrial diamonds are used as abrasives. Crushed into various sizes, they are used for grinding wheels, grinding powders, polishing disks, drill bits, and saws. Diamond is indispensable for grinding the tungsten carbide cutting tools that have been in use since the 1930's. Industrial diamonds also are sorted for shape: Blocky stones are suited for more severe grinding operations such as rock-drill bits, and more splintery ones are reserved for grinding tungsten carbide. Gem diamonds are pretty, but industrial diamonds are absolutely essential to modern technology.

In 1955, the General Electric Company succeeded in manufacturing synthetic diamonds at low pressure. Initially more expensive than natural stones, synthetic diamonds are now widely used in grinding wheels to sharpen tungsten carbide. Synthetic diamonds are smaller but have rougher surfaces than natural stones and are manufactured in a variety of shapes and sizes for specialized abrasive applications.

Diamonds are also used as dies for drawing out the fine tungsten filaments of incandescent light bulbs, as scalpels for eye surgery, and as stereo phonograph needles. Unrivaled as a heat conductor, diamond is an important component in the miniature diodes that are used in telecommunications. Diamond has even served as a tiny instrument window on the Pioneer space probe to Venus, as it tolerates the extremes of heat and cold in outer space. A recent application of diamonds is the diamond-anvil pressure cell. Two gem-quality diamonds are used to subject samples to enormous pressure, duplicating conditions deep in the earth. The diamond windows allow the samples to be observed under a microscope during experimentation. This device, which fits in the palm of one's hand, can perform experiments that once required massive and dangerous hydraulic presses.

William R. Hackett

FURTHER READING

Barnard, Amanda. *The Diamond Formula.* Woburn, Mass.: Butterworth-Heinemann, 2000. Discusses synthetic diamonds from the gemological prospective. Covers the principles and development of diamond synthesis. Includes chapters on the subject of synthetic and natural diamonds. Easily accessible to jewelers, gemologists, and geology students.

Cox, K. G. "Kimberlite Pipes." In *Volcanoes and the Earth's Interior,* edited by Robert Decker and Barbara Decker. San Francisco: W. H. Freeman, 1982. Appropriate for high school students with a strong science background. Includes an excellent summary of the origin of kimberlite and its inclusions, including diamonds. Mostly covers the geological aspects of kimberlite and diamond.

Decker, Robert, and Barbara Decker. *Volcanoes.* 4th ed. New York: W. H. Freeman, 2005. Written for general readers. Focuses on diamonds and kimberlite pipes, described in the context of volcanoes. Includes a glossary, selected references for each chapter, and an index.

Harlow, George E., ed. *The Nature of Diamonds.* Cambridge, England: Cambridge University Press, 1998. Examines the evolution of diamonds and

diamond mines. A thorough introduction to diamonds and their properties; suitable for the layperson. Color illustrations, maps, index, and bibliography.

Hurlbut, C. S., Jr., and Robert C. Kammerling. *Gemology.* 2d ed. New York: John Wiley & Sons, 1991. A well-illustrated and complete introduction to gemstones, describing the methods of gem study. Includes a section with descriptions of minerals and other materials prized as gems. A good introduction to mineralogy for the nonscientist.

Kluge, P. F. "The Man Who Is Diamond's Best Friend." *Smithsonian* 19 (May, 1988): 72. Follows the strategy and execution of cutting a 900-carat diamond into a large stone and ten smaller stones. Color photographs document the procedure from start to finish. Appropriate for all.

Mitchell, Roger H. *Kimberlites, Orangeites, and Related Rocks.* New York: Plenum Press, 1995. Provides a good introduction to the study of kimberlites and related rocks, including an extensive bibliography that will lead the reader to additional information.

_____. *Kimberlites, Orangeites, Lamproites, Melilitites, and Minettes: A Petrographic Atlas.* Thunder Bay, Ontario: Almaz, 1997. Covers the worldwide distribution of kimberlites and other related rock types. Color illustrations and bibliography.

Newman, Renée. *Diamond Handbook.* 2d ed. Los Angeles: International Jewelry Publications, 2008. Written for jewelry professionals. Offers relevant information for other professions who use, evaluate, or characterize diamonds.

O'Neil, Paul. *Gemstones.* Alexandria, Va.: Time-Life Books, 1983. Appropriate for all readers. Offers a treasure trove of color photographs. Contains historical and geological information about many important gemstones; features outstanding chapters on diamonds. Bibliography and index.

Smith, D. G., ed. *The Cambridge Encyclopedia of Earth Sciences.* New York: Crown, 1981. Section on physics and chemistry of the earth provides a good foundation for understanding the earth's interior and hence the origin of diamonds and kimberlite. Well illustrated, carefully indexed, and suitable for college-level readers.

Tappert, Ralf, and Michelle C. Tappert. *Diamonds in Nature: A Guide to Rough Diamonds.* Berlin: Springer-Verlag, 2011. Discusses diamond's origins, morphology, color, texture, and inclusions. Discusses the natural form of diamonds; over two hundred photographs are provided to show their physical characteristics. Useful for geoscientists and gemologists.

See also: Aluminum Deposits; Biogenic Sedimentary Rocks; Building Stone; Cement; Coal; Crystals; Dolomite; Earth Resources; Fertilizers; Gem Minerals; Gold and Silver; Industrial Metals; Industrial Nonmetals; Iron Deposits; Manganese Nodules; Pegmatites; Platinum Group Metals; Salt Domes and Salt Diapirs; Sedimentary Mineral Deposits; Sedimentary Rock Classification; Uranium Deposits.

DOLOMITE

Dolomite is a sedimentary rock formed from a reaction between carbonic acid with calcium and magnesium. It forms in ancient marine environments, facilitated by chemical transformations that replace other carbonate rocks. Dolomite also can be found in limestone deposits, an important source of petroleum reservoirs because of its naturally porous structure. Dolomite is used for a variety of industrial processes, and colored dolomite is highly prized as an ornamental and decorative stone; dull varieties are often used as a component in construction materials, including concrete.

PRINCIPAL TERMS

- **anhydrous:** having little or no water in its physical structure
- **carbonate:** ionic salt formed from the interaction of carbonic acid with basic elements
- **diagenesis:** physical and chemical change occurring within a rock after solidification; results in the formation of an alternate mineral and occurs at temperatures and pressures below those required for metamorphosis
- **evaporite:** type of mineral resulting from concentration within an evaporating aqueous medium
- **fracture:** characteristic used for mineral classification based on the shape and texture of fragments produced when the mineral is subjected to force or pressure
- **luster:** characteristic used for mineral classification based on the degree to which the mineral refracts light
- **petroleum:** substance formed from the liquefied remains of fossilized organisms compressed beneath the earth's crust and often found in deposits of dolomitic limestone
- **rhombohedral:** crystal structure based on the combination of different angular lengths and resulting in a combination of rectangular shaped sections
- **salt:** ionic compound formed from a reaction between an acidic and basic substance
- **twinning:** process in crystal formation in which two separate crystals share one or more points of connection between their respective lattices

CARBONATE MINERALS

Dolomite is a carbon-rich mineral classified as a type of sedimentary rock. Sedimentary rocks are formed from the accumulation of small mineral fragments, dust, and dissolved compounds that are joined through chemical bonding. When this fragmentary sediment is buried and compressed, it solidifies.

Dolomite ($CaMg(CO_3)_2$) belongs to the class of minerals known as carbonates because they contain carbonate ions (CO_3^{2-}) within their chemical structure. Carbonate is a salt, which is a type of compound that forms from a reaction between an acid and a base. Carbonate is created from reactions involving carbonic acid and basic substances in the environment. In the case of dolomite, these basic molecules are magnesium and calcium. Carbonate minerals tend to form in aqueous environments rich in living organisms.

The most common type of carbonate mineral is calcite ($CaCO_3$), which is a component of a variety of sedimentary rocks and is present on every continent. Calcite is commonly blended with limestone in naturally occurring deposits, which are harvested and processed to create building stone.

Dolomite is similar to calcite in that it also contains atoms of calcium bonded to carbonate ions. In addition, dolomite molecules contain atoms of magnesium, and the presence of magnesium affects dolomite's color, chemical characteristics, and environmental occurrence.

Dolomite is further classified as a member of the anhydrous carbonates, which are carbonates that do not contain water molecules embedded in their crystal structure. Most anhydrous carbonates can also form into hydrated carbonates in environments where humidity and ambient moisture increase. Monohydrocalcite, for instance, is a hydrated form of calcite that is not stable at room temperature and exists only where temperatures are below freezing. Monohydrocalcite contains molecules of water bonded to carbon and calcium within the mineral matrix. Dolomite exists only in the anhydrous form, though there are hydrated carbonates that form from the combination of carbonate and magnesium.

Dolomite is closely related to ankerite, a carbonate mineral that consists of carbonate ions bonded to calcium, with magnesium, iron, and manganese integrated into the crystal lattice. Ankerite is basically dolomite in which a portion of the magnesium has been replaced by iron or manganese atoms. Ankerite and dolomite can appear in the same geologic deposits, and both minerals form from carbon- and calcium-rich aqueous solutions in which various concentrations of metallic elements contribute to the formation of dolomite in some areas and ankerite in others.

FORMATION OF DOLOMITE

Dolomite exists in two basic forms: as a pure mineral crystal called simply dolomite and as a compressed component of limestone deposits called dolomitic limestone. As a component of limestone, some dolomite is present in a variety of building materials and sculptural stones, though this dolomite is generally indistinguishable from the surrounding limestone without chemical analysis. Dolomite is a common component of many types of sedimentary rocks, often occurring in mixed beds with other carbonates.

Dolomite mineral is an evaporite, which is a type of mineral that forms primarily in aqueous, saline environments, such as salt pans, salt marshes, and salinas or salt pools. In these environments, water, which is generally derived from a nearby marine environment, is enriched in chemicals from the surrounding sediment. Solar radiation causes water to evaporate from the environment, transforming into gaseous water vapor at the water surface and moving into the atmosphere. As this occurs, the remaining material is concentrated into a mineral-rich residue; further evaporation leads to the formation of crystal deposits. In thousands of years, evaporite minerals can form into large deposits several hundred feet thick within the earth's crust.

Dolomite is common in ancient rock deposits. Until recently, geologists were uncertain about the mechanisms of dolomite formation because they could not find dolomite forming on the surface of the earth. Geologists have been trying to discover the reason for this phenomenon, often called the dolomite problem, for more than two hundred years.

Some geologists have suggested that most dolomite is first deposited as a different type of carbonate mineral, like calcite or aragonite. Exposure to warm, magnesium-rich saline waters is then believed to initiate a process called diagenesis, which results in a physical alteration of the rock's chemical structure, giving rise to dolomite. Diagenesis is similar to metamorphosis, which is the transformation of one type of mineral to another at high temperatures and pressures. By contrast to metamorphosis, diagenesis occurs at lower temperatures and pressures, such as those that commonly exist on the surface of the earth.

In 2011, Australian geologists discovered dolomite minerals within mats of red algae (*Hydrolithon onkodes*), a species common in tropical reef habitats. Red algae assist in reef building by helping to cement corals to the sea floor, thereby enabling the coral to resist the hydraulic pressure from incoming waves. This is accomplished because red algae deposit calcium around the base of the coral colonies as they metabolize food ingested from the surrounding water.

Using X-ray diffraction, geologists determined that the cells of red algae contain dolomite, magnesium, and calcite. The organism's conversion of calcite has long been understood as the mechanism behind the organism's function in cementing coral beds, but cellular dolomite is a new discovery and now believed to provide a solution to the dolomite problem.

Geologists theorize that the buildup of dolomite is caused by the spread and eventual death of red algal species on the floor of reef ecosystems. As the organisms die, their bodies decompose, leaving deposits of dolomite, calcite, and magnesite. Given this recent discovery, it appears that dolomite may actually form primarily in shallow, marine reef systems rich in red algae and coral.

After dolomite mineral has been deposited in its crystalline form, it may be buried and blended with other minerals to form dolomitic limestone. This occurs when deposits of limestone and dolomite are buried and compressed, a process called sedimentation. In thousands or millions of years, the compressed minerals form into large, solid deposits. The development and decline of marine ecosystems thereby constitute an avenue for the generation of a variety of calcium-rich minerals and rocks, including dolomite, calcite, and mixed-rock types like dolomitic limestone.

STRUCTURE OF DOLOMITE

Carbonate minerals tend to form in sheets, because carbonate molecules bond in such a way that they tend to form discrete molecular layers. In calcite, sheets of carbonate molecules alternate with sheets of calcium. In dolomite, magnesium atoms form into a single layer, alternating with layers of calcium and carbonate. Atoms of calcium and magnesium form into alternating layers because the atoms differ in size and are more stable when organized into alternating sheets, rather than in mixed layers containing both types of atoms.

Dolomite crystals typically form in a rhombohedral pattern, which is composed of rectangles of crystals layered onto one another. Dolomite also engages in "twinning," whereby two crystals share certain points of their lattice, growing from a shared connection, called a twin boundary or a composition surface. Because of twinning, the crystals often develop saddle-shaped depressions on one side of the crystal. Saddles are not present in all crystals, and some form more traditional, pointed rhombohedrons.

Luster describes the way that a substance or object interacts with incoming light. The luster of a particular crystal is related to the transparency of the materials constituting the crystal and the degree of light refraction caused by the presence of different elements within the lattice. Many dolomite crystals develop a pearly luster, which consists of translucent crystals, sometimes appearing milky with a pearlescent shine. Other dolomite crystals display a vitreous luster, which is described as glass-like and transparent, and some dolomite crystals display a dull luster, a translucent, nonreflective surface.

Dolomite forms in a variety of colors, ranging from white and off-white to green and pink. Dolomite with a yellow hue may reflect the presence of additional magnesium within the lattice. Dolomite crystals with a brownish or blackish hue result when iron becomes blended with the dolomite during formation or subsequent compaction. Pure dolomite crystals are generally pink in hue and mostly translucent, reflecting the properties of magnesium and calcium and their effect on light refraction and absorption.

Fracture is a measure of the way that crystals tend to break when subjected to force or pressure. The appearance and structure of a fractured crystal is related to characteristics involving the bonding of atoms within the crystal's chemical structure. Dolomite crystals display conchoidal fracturing, which results in fractures that follow an even plane between crystals and generally result in smooth edges. This property of dolomite is largely caused by the layering of molecules within the mineral's lattice, allowing for relatively even fractures to develop along the zones where layers meet.

USES OF DOLOMITE

Dolomite is primarily used as an ornamental stone because of its attractive crystals that are clear and colored. Smoothed and polished dolomite can therefore be used as decoration or in fabricating home furnishings and decorative stone accessories.

Dolomite also is an ingredient used in making concrete for industrial processes. Dolomitic limestone is used as a construction material and may be polished and used in a similar way to decorative marble.

In some cases dolomite is used as a precursor for deriving magnesium by subjecting the mineral to a chemical environment that separates the metals from the mineral structure. A significant amount of industrial magnesium is derived from processing dolomite.

Dolomite rock, especially that derived from ancient marine ecosystems, is known as a productive reservoir for fossil fuels. Dolomite deposits, and deposits of carbonate rock in general, tend to be porous and to fracture within the crust, leaving areas of reduced density that can serve as traps for the accumulation of petroleum and natural gas.

Petroleum may form in these same environments because of the richness of biological fauna and flora that live and die in shallow marine ecosystems. In millions of years, these decomposing organisms form into hydrocarbon residue that is buried within the surrounding carbonate rocks.

Mining dolomite quarries to obtain petroleum is a multimillion-dollar industry. Dolomitic limestone deposits are one of the leading sources for petroleum in some parts of the world. Dolomitic limestone is particularly likely to contain petroleum because it forms primarily from the chemical transformation, through biogenesis, of calcite limestone. When this occurs, the resulting rock is more porous, containing more pockets and small reservoirs in which petroleum oil can accumulate. Mining dolomitic limestone not only yields petroleum oil but also yields raw dolomite and limestone that can be further processed and sold for industrial applications.

Micah L. Issitt

FURTHER READING

Braithwaite, C. J. R., G. Rizzi, and G. Darke. "The Geometry and Petrogenesis of Dolomite

Hydrocarbon Reservoirs: Introduction." In *The Geometry and Petrogenesis of Dolomite Hydrocarbon Reservoirs*, edited by C. J. R. Braithwaite, G. Rizzi, and G. Darke. Bath, England: Geological Society, 2004. Written for advanced students of geology and professionals, this text focuses on key topics in the origin, mining, and use of dolomite.

Flugel, Erik. *Microfacies of Carbonate Rocks: Analysis, Interpretation, and Application.* 2d ed. New York: Springer, 2010. Detailed analysis of carbonate petrology and the uses of carbonate rocks in research and industrial applications. Examines the many facets of dolomite formation and structure.

Hyne, Norman J. *Nontechnical Guide to Petroleum Geology, Exploration, Drilling, and Production.* 2d ed. Tulsa, Okla.: Pennwell Books, 2001. Basic introduction to the practices and processes of the petroleum industry written for students and general readers. Contains a chapter on the industrial uses and mining of dolomite.

Monroe, James S., Reed Wicander, and Richard Hazlett. *Physical Geology.* 6th ed. Belmont, Calif.: Thompson Higher Education, 2007. General-interest text discussing processes, methodology, and theoretical principles behind modern geological research. Includes a brief discussion of dolomite chemistry, mining, and industrial uses.

Plummer, Charles C., Dianne H. Carlson, and David McGeary. *Physical Geology.* 13th ed. Columbus, Ohio: McGraw-Hill Higher Education, 2009. Introductory university text on geology that discusses the basic concepts of petrology, physical geology, and geological research. Examines dolomite structure and chemistry.

Stanley, Steven M. *Earth System History.* New York: W. H. Freeman, 2004. Detailed geological text describes the geological formation of the earth in terms of the evolution of planetary geological systems. Discusses dolomite's role in paleontological research and its geological formation.

See also: Chemical Precipitates; Clathrates and Methane Hydrates; Crystals; Evaporites; Geologic Settings of Resources; High-Pressure Minerals; Sedimentary Rock Classification; Ultra-High-Pressure Metamorphism; Unconventional Energy Resources.

E

EARTH RESOURCES

Earth's resources of metals, nonmetals, and energy undergird every modern, technological society.

PRINCIPAL TERMS

- **alloy:** a substance composed of two or more metals or of a metal and certain nonmetals
- **by-product:** a secondary or incidental product of the refining process
- **fossil fuel:** an energy source, such as coal, oil, or natural gas, which is formed from the remains of partly decayed organic matter
- **mineral:** a naturally occurring, inorganic crystalline substance with a unique chemical composition
- **nonrenewable resource:** an Earth resource that is fixed in quantity and will not be renewed within a human lifetime
- **ore deposit:** a natural accumulation of mineral matter from which the owner expects to extract a metal at a profit
- **reservoir rock:** the geologic rock layer in which oil and gas often accumulate; often sandstone or limestone
- **source rock:** the geologic rock layer in which oil and gas originate; often the rock type known as shale

RENEWABLE VERSUS NONRENEWABLE RESOURCES

All Earth resources can be subdivided into two broad categories. The first category contains the renewable resources. The word "renewable" means that these resources are replenished by nature as rapidly as humans use them up, provided good judgment is used. Renewable resources include the energy of the wind, the timber cut in a forest, and the animals used for food. Each of these Earth resources is constantly being renewed by the energy that reaches the earth's surface from the sun. As long as the sun's rays reach Earth, this pattern of replenishment can continue.

The resources in the second category are known as nonrenewable resources. These are resources that will not be renewed in a human lifetime. Only limited quantities of these resources are present in the earth's crust, and they are not replenished by natural processes operating within short periods. Examples of nonrenewable resources include coal, oil, iron, diamonds, and aluminum. While it is true that certain of these resources, such as coal and oil, are being formed within the earth's crust continuously, the processes are exceedingly slow, being measured in thousands or millions of years; the rate of use far exceeds the rate of replacement. The difference between renewable and nonrenewable resources is sometimes summed up as, "If you don't grow it, you have to mine it."

METALS

Earth resources of primary interest to the economic geologist can be divided into three categories: metals, nonmetals, and energy sources. Metals are a group of chemical elements that share a high metallic luster, or shine. Additionally, they can all be melted, conduct heat and electricity, and be pounded into thin sheets or drawn into thin wires.

Metals can be divided into two classes based on their abundance in the earth. The first class, which has been called the abundant metals, consists of those metals that individually constitute 0.1 percent or more of the earth's crust by weight. The metals in this category are iron, aluminum, manganese, magnesium, and titanium. The second class of metals, which are called the scarce metals, consists of those metals that individually constitute less than 0.1 percent of the crust. This class includes such metals as copper, lead, zinc, nickel, mercury, silver, gold, and platinum.

Certain common metals, such as steel, brass, bronze, and solder, are not pure metals but alloys—chemical mixtures of two or more metals that have characteristics of strength, durability, or corrosion resistance superior to those of the component metals. Steel, for example, is an alloy in which iron is the main constituent.

Metals are rarely found in a pure state within the earth's crust. Only the metals gold, silver, copper, platinum, and iron are ever found uncombined. All other metals are found chemically combined with additional elements to form minerals. Geologists use the term "ore deposit" to describe a rock containing metals or metal-bearing minerals from which the pure metal can be profitably extracted. Whether a rock is an ore deposit depends on a variety of factors, including how difficult it is to extract the metal from the metal-bearing mineral, how large and accessible the ore deposit is, whether valuable by-products can be obtained, and what the current price of the metal is on world markets.

NONMETALS

The second major category of Earth resources is nonmetals. The term "nonmetal" is widely employed by geologists to describe substances extracted from the earth that are neither sources of metals nor sources of energy. Nonmetals are mined and processed either because of the nonmetallic elements they contain or because they have some highly desirable physical or chemical characteristic. Some of Earth's major nonmetallic resources are fertilizers, chemicals, abrasives, gems, and building materials.

Fertilizers contain the elements nitrogen, potassium, and phosphorus. Most of the nitrogen required for fertilizer production is chemically extracted from the air, so the supply is renewable. The potassium and the phosphorus, however, come from rocks dug from the ground—potash salt layers and phosphate rocks—the supply of which is nonrenewable.

Several Earth resources provide important raw materials for the chemical industry. They include salt, which is obtained from seawater and underground beds of rock salt; sulfur, a by-product of oil production; and substances such as borax and soda ash, which are obtained from the beds of dry desert lakes. Abrasives are very hard substances used for grinding, polishing, and cleaning. They are obtained from rock and mineral substances dug out of the earth and pulverized. Gems are Earth materials that are attractive to the eye. They can be categorized as precious and semiprecious.

Building materials include the stones obtained from quarries, such as granite, sandstone, limestone, marble, and slate. There also is a high demand for crushed rock, which is used as highway roadbed and

for concrete aggregate. Sand and gravel are also used in making concrete. A number of other useful products are prepared from Earth materials: cement; plaster, from the mineral gypsum; brick and ceramics, from clay; glass, from very pure sand or sandstone rock; and asbestos, from flame-resistant mineral fibers that can be woven into fireproof cloth or mixed with other substances to make fireproof roofing shingles and floor tiles.

ENERGY SOURCES

The third major category of Earth resources is energy sources. Energy sources are frequently divided into mineral fuels, which are nonrenewable, and a second, renewable group. The first group contains coal, uranium, and oil and gas.

Crude oil, natural gas, and petroleum provided 83 percent of the United States' energy in 2009. Crude oil is a naturally occurring liquid composed of the elements hydrogen and carbon, combined into compounds known as hydrocarbons; natural gas is a gaseous form of these hydrocarbons. The oil and gas accumulate underground over long periods in source rocks and migrate into reservoir rocks, where they are trapped. Extraction is accomplished by means of drilling. Two related Earth resources are oil shale, a source rock that contains oil, and tar sands, reservoir rocks exposed at the surface.

Coal is the third most important energy source in the United States; in 2009, it met 21 percent of the nation's energy needs. Coal originates when partly decayed plant material accumulates on the floor of bogs and swamps, and is buried by overlying sediments that compact the plant material into carbon-rich rocks. The various grades of coal are lignite (brown coal), bituminous coal (soft coal), and anthracite (hard coal). Peat is partly decayed plant material that was never buried at all. Coal is mined at the surface in strip mines or mined underground. Coal, oil, and gas are sometimes referred to as "fossil fuels."

Uranium is a silver-gray metal used in nuclear reactors to produce electricity. In 2009, nuclear reactors met 9 percent of American energy needs. Within a reactor, uranium undergoes neutron-induced disintegration, producing heat. This heat is used to drive an electrical generator, just as in conventional power plants. Uranium occurs as veins or grains within a variety of rock types and is mined with standard mining methods.

There are many potential energy sources that are renewable and mostly underutilized. Foremost among these is hydroelectric power, or water power. This power is generated by means of water falling from a dammed reservoir; the force of the falling water turns the turbine of an electric generator. Barring unforeseen climate changes or silting of the reservoir behind the dam, hydroelectric power can be considered a renewable resource. Sunlight, the wind, the tides, and the steam from geysers are also renewable energy sources, as are living things, in the form of firewood and animal power. Newer applications are gasohol- and methane-fired boilers that use gases fermented from vegetation or cow manure. Renewable resources provided 7.3 percent of the energy used in 2009: 2.5 percent from hydroelectric plants, 3 percent from burning biomass, and 1.5 percent from geothermal plants and other sources. The controlled burning of garbage has tremendous potential for the future, as does the use of hydrogen from the ocean to operate "clean" nuclear fusion reactors.

STUDY OF EARTH SOURCES

Mineral deposits are quite rare in the earth's crust; either they consist of substances that are uncommon to begin with (gold, for example) or else they are composed of common substances—such as the pure sand used to make glass—that have been concentrated into workable accumulations. Much study has gone into why such concentrations exist. One valuable tool has been the plate tectonics theory, which proposes that the earth's surface is divided into a few large plates that are slowly moving with respect to one another. Intense geologic activity occurs at plate boundaries, and many mineral deposits are believed to have been formed this way.

It becomes more and more difficult to find mineral deposits, as all the easily discoverable ones have already been found. Aerial prospecting was made possible by the airplane, and it has now been replaced by satellite imaging. Most prospecting, however, is based on the search for buried deposits and utilizes indirect methods for detecting favorable underground geologic conditions. A preferred technique is seismic prospecting, in which sound waves are created underground by means of small explosive charges and are then bounced off underground rock layers in order to determine their structure. Gravitational and magnetic mapping are also powerful techniques.

Further ways in which Earth resources are studied include calculations of estimated available reserves in view of anticipated future demands; comparison of the fuel values of various energy sources; analysis of the environmental problems and types of pollution caused by the extraction, refining, and utilization of various mineral resources; and investigations of ways to conserve, recycle, or develop substitutes for mineral resources that are in short supply.

ECONOMIC EXTRACTION OF MINERAL RESOURCES

A variety of techniques have been developed for the economical extraction of mineral resources from the earth. Frequently, extraction costs are the controlling factor in whether a mineral deposit can be profitably worked. In general, extraction techniques can be divided into two groups: surface and underground methods. Surface methods are preferred whenever possible, because they are lower in cost. The traditional surface methods are quarrying, open-pit mining, and strip mining. Strip mining involves the removal of large amounts of worthless overburden so that the mineral deposit can be reached. Underground mining methods include the excavation of shafts, tunnels, and rooms. Flat-lying coal beds are now excavated by longwall mining, in which one long stretch of coal is stripped off at a time. The rock above the machine is supported by movable supports, and as the machine advances the roof in the mined-out area is allowed to subside. Fluids such as petroleum can be removed by means of drilled wells.

Mineral resources extracted from the earth rarely are ready to be sent directly to market. Generally, they require processing to separate undesirable substances from desirable ones. This processing may involve physical separation, as in the case of separating diamonds from rock pebbles of the same size, or chemical separation, which is required to remove metals from the sulfur they are combined with in certain ores. Even after the pure mineral substance is obtained, further treatment may be required, as in the smelting of iron to obtain steel or the refining of petroleum to obtain gasoline.

SUPPORT FOR MODERN TECHNOLOGY

Earth's resources of metals, nonmetals, and energy supplies support all modern technology. Houses and automobiles, televisions and refrigerators, airplanes and roads, jewelry and sandpaper, the electricity that lights a playing field and the gasoline that powers a car—a nearly endless list of goods depends on the ability to utilize or harness the resources of the earth.

A technological society relies on metals. Iron is needed for steel making, aluminum for lightweight aircraft construction, manganese for toughening steel for armor plate, and titanium for making heat-resistant parts in jet engines. Among the scarcer metals, copper is needed for electrical wiring, lead for car batteries and nuclear reactor shielding, zinc for galvanized roofing nails, nickel for stainless steel, and mercury for thermometers and silent electric switches. Silver is used for making photographic film, silverware, jewelry, and coins; gold for coins, jewelry, and dental work; and platinum for jewelry and industrial applications where corrosion resistance is essential.

Among nonmetallic Earth resources, fertilizers are used for agriculture; salt, sulfur, and soda for the chemical industry; and abrasives for sandpaper and grinding wheels. The construction industry uses cut stone, crushed rock, cement, plaster, brick, glass, and asbestos.

The energy sources obtained from the earth enable humans to perform many tasks faster than they could manually, and others that they could not perform at all. Because most machines run on electricity, the output of the energy source often must first be converted into an electric current. Automobiles, however, convert gasoline directly into power by means of the internal combustion engine. Probably the best example of the direct use of an energy source is a wind-powered sailboat—a way of using energy that has not changed in the past five thousand years.

Donald W. Lovejoy

FURTHER READING

Boyle, Godfrey, ed. *Renewable Energy.* 2d ed. New York: Oxford University Press, 2004. Provides a complete overview of renewable energy resources. Discusses solar energy, bioenergy, geothermal energy, hydroelectric energy, tidal power, wind energy, and wave energy. Covers the basic physics principles, technology, and environmental impacts. Features references and a further reading list with each chapter.

Carlson, Diane H., Charles C. Plummer, and Lisa Hammersley. *Physical Geology: Earth Revealed.* 9th ed. New York: McGraw-Hill Science/Engineering/Math, 2010. A thorough introduction to physical geology. Includes a computer disc that corresponds to chapters and topics explored. Bibliography and index.

Chiras, Daniel D., and John P. Reganold. *Natural Resource Conservation: Management for a Sustainable Future.* 10th ed. Boston: Addison Wesley, 2009. Discusses resource use and solutions regarding sustainability and conservation. Covers case studies, laws, regulations and international treaties. Discusses natural resources, geographic information systems, and remote sensing.

Craig, James R., David J. Vaughan, and Brian J. Skinner. *Earth Resources and the Environment.* 4th ed. Upper Saddle River, N.J.: Prentice Hall, 2010. Covers all Earth resources. Includes helpful line drawings, maps, tables, and charts. Features further reading and a list of principal ore minerals and production figures.

_____. *Resources of the Earth.* 4th ed. Upper Saddle River, N.J.: Prentice Hall, 2010. Includes numerous black-and-white photographs, color plates, tables, charts, maps, and line drawings. Provides an excellent overview of the metal, nonmetal, and energy resources of the earth. Covers Earth resources throughout history. Suitable for college-level readers or the interested layperson.

Cutter, Susan L., and William H. Renwick. *Exploitation Conservation Preservation: A Geographic Perspective on Natural Resource Use.* 4th ed. Hoboken, N.J.: John Wiley & Sons, 2003. Offers information on the social, economic, political, and environmental effects of resource use. Offers detailed content useful for graduate students. Presents and leaves open to discussion varying views on unresolved issues.

Davidson, Jon P., Walter E. Reed, and Paul M. Davis. *Exploring Earth: An Introduction to Physical Geology.* 2d ed. Upper Saddle River, N.J.: Prentice Hall, 2001. An excellent introduction to physical geology. Explains the composition of the earth, its history, and its state of constant change. Intended for the layperson. Includes colorful illustrations and maps.

Grotzinger, John, et al. *Understanding Earth.* 5th ed. New York: W. H. Freeman, 2006. General text on all aspects of geology. Appropriate for advanced high school and college students.

Jensen, John R. *Remote Sensing of the Environment: An Earth Resource Perspective.* 2d ed. Upper Saddle River, N.J.: Prentice Hall, 2006. Discusses the principles of remote sensing and use in geology, as well as remote sensing instruments and methodology. Covers specific resources analyzed through remote sensing, such as vegetation, water, soil and rocks, and geomorphic features. Provides practical information for use by geologists and graduate students.

Jensen, M. L., and A. M. Bateman. *Economic Mineral Deposits.* 3d ed. New York: John Wiley & Sons, 1981. Includes detailed information on different metallic and nonmetallic mineral deposits and their modes of formation. Covers the history of mineral use and the exploration and development of mineral properties. Cross-sections of individual deposits are provided. For college-level readers.

Sinding-Larsen, Richard, and Friedrich-W. Wellmer, eds. *Non-renewable Resource Issues: Geoscientific and Societal Challenges.* New York: Springer, 2012. Discusses the use, overuse, and future needs of natural resources. Elaborates on the various possible resource needs of the future. Discusses concepts of land use, sustainability, and cultural needs.

Tarbuck, Edward J., Frederick K. Lutgens, and Dennis Tasa. *Earth: An Introduction to Physical Geology.* 10th ed. Upper Saddle River, N.J.: Prentice Hall, 2010. Provides a clear picture of the earth's systems and processes. Appropriate for the high school or college reader. Includes illustrations, graphics, and a computer disc. Bibliography and index.

Tennissen, Anthony C. *The Nature of Earth Materials.* 2d ed. Englewood Cliffs, N.J.: Prentice-Hall, 1983. Contains detailed descriptions of 110 common minerals, with corresponding black-and-white photographs. Discusses the modes of formation and the classification of igneous, sedimentary, and metamorphic rocks. Includes an overview of metallic and nonmetallic mineral resources. Appropriate for college-level readers.

Thompson, Graham R. *Introduction to Physical Geology.* Fort Worth, Tex.: Saunders College Publishing, 1998. Provides an easy-to-follow examination of physical geology and the phases involved. Covers each phase of the earth's geochemical processes. Excellent for high school and college readers. Illustrations, diagrams, and bibliography included.

Warren, John K. *Evaporites: Sediment, Resources and Hydrocarbons.* Berlin: Springer-Verlag, 2006. Discusses a number of evaporitic minerals, with Chapters 4 and 6 focusing on salts. Discusses chemistry and hydrology, deposit locations, and mining. Includes many drawings of geological features.

See also: Clays and Clay Minerals; Diagenesis; Earth Science and the Environment; Hazardous Wastes; Landfills; Mining Processes; Mining Wastes; Nuclear Waste Disposal; Offshore Wells; Oil and Gas Exploration; Oil and Gas Origins; Oil Chemistry; Onshore Wells; Petroleum Reservoirs; Rocks: Physical Properties; Sedimentary Rock Classification; Strategic Resources.

EARTH SCIENCE AND THE ENVIRONMENT

An understanding of the systems of the earth, how they interact, and how they are affected by anthropogenic disturbances is essential to any study of the environment. As the model of the earth is refined and improved, it will be possible to predict the planet's response to anticipated perturbations with greater precision and confidence and adopt better public policy.

PRINCIPAL TERMS

- **El Niño Southern Oscillation:** a reversal in precipitation patterns, ocean upwelling, and thermocline geometry that is accompanied by a weakening of the trade winds; the phenomenon typically recurs every three to seven years
- **greenhouse gas:** an atmospheric gas capable of absorbing electromagnetic radiation in the infrared part of the spectrum
- **gyre:** an ocean-scale surface current that moves in a circular pattern
- **hydrologic cycle:** the cycle of water movement on the earth from ocean to land and back
- **mass wasting:** downhill movement of material by gravity, as opposed to transport by water or wind.
- **plate tectonics:** a theory that holds that the surface of the earth is divided into roughly one dozen rigid plates that move relative to one another, producing earthquakes, volcanoes, mountain belts, trenches, and many other large-scale features of the planet
- **thermohaline conveyor belt:** a system of oceanic circulation driven by the cooling and sinking of salty surface waters in the Nordic seas

GLOBAL AND SOLID EARTH SYSTEMS

The planet Earth has a large number of interconnected and interacting systems. Although the rates with which these systems respond to perturbations vary greatly, over time they all influence one another. Some of these systems are global in scale, while others are local. Many provide feedback that can either amplify or suppress a change.

The systems that have the slowest response times involve the solid earth. The surface of the earth is divided into about one dozen tectonic plates. These plates show little interior deformation over hundreds of millions of years, and yet their boundaries are constantly changing. They move across the earth at rates on the order of centimeters per year and have been doing so for at least several hundred million years. The environmental issues most closely related to this motion are the hazards of earthquakes and volcanoes; however, less obvious connections exist.

Large areas of uplift, such as the Himalayan Mountains and the Tibetan Plateau, dramatically change atmospheric circulation patterns, which in turn affect precipitation and evaporation. Rates at which tectonic plates are formed along midocean ridges vary, with far-reaching effects. The new material is hot and expanded. The time it takes to cool and contract does not depend on the rate at which it is forming, so faster growth results in larger regions underlain by expanded rock. The sea level rises as a result of this process. Some scientists have proposed a connection between volcanic processes acting along the East Pacific Rise and the occurrence of El Niño events.

Although the plate tectonic system interacts with other Earth systems and has long-term effects, it operates at such a slow rate that it is unlikely to cause significant changes to the environment during the course of a few human generations. In one hundred years, a plate may move a few meters. This is more than enough to cause tremendous earthquakes, but it is not likely to produce noticeable climate change.

Continental glaciation causes vertical movements of the earth's crust, which occur much more rapidly than plate motions. Approximately 20,000 years ago, a mass of ice about three kilometers thick covered much of Canada, the northern United States, and Scandinavia. The weight of this ice forced the crust to sink into the mantle, a process called isostasy; when the ice melted away, the crust began to rise again. As a result, some Viking fishing villages are now well above sea level.

OCEAN SYSTEMS

Much faster than solid rock systems, the circulation of the oceans is one of the most significant processes affecting the environment. It is convenient to divide this circulation into surface current circulation and thermohaline circulation (involving the effects of both temperature and salinity).

Heat from the sun warms the surface water, which then expands and floats on the cooler water beneath it. This warm body of water has a lens shape, with the center a few meters higher than the edges. As water tries to flow down the slope of the surface, Coriolis force—a result of the earth's rotation—forces it to move sideways instead. In the Northern Hemisphere, this produces a clockwise movement; in the Southern Hemisphere, it produces a counterclockwise movement. This ocean-scale circular current is called a gyre. The Gulf Stream is the northwest segment of the North Atlantic gyre.

The Gulf Stream moves 35 million cubic meters of water past Chesapeake Bay every second. The edge of the Gulf Stream forms an abrupt boundary between cool, nutrient-rich waters near the continents and the warm, nearly lifeless waters of the Sargasso Sea. Although interconnected with weather and climate in a complex way, the operation of the Gulf Stream is not the principal source of heat responsible for the mild climate of Europe. The principal source of warmth for Europe is heat transported northward by atmospheric circulation, plus heat stored in Atlantic surface waters during the summer.

This heat transfer drives the thermohaline circulation. Cold, salty Antarctic water sinks to the bottom and moves northward, eventually rising to join the surface circulation. The largest amount of bottom water is in the Pacific, warming as it circulates southward and westward. Warm water from the Indian Ocean rounds Africa into the Atlantic, moves north as the Gulf Stream, and cools and sinks in the Arctic. This circulation transports heat from regions around the equator to those around the pole and keeps the oceans mixed uniformly.

ATMOSPHERIC SYSTEMS

Atmospheric systems constitute the other major heat redistribution scheme on Earth. Because of Earth's spherical shape, the equatorial regions receive and absorb far more solar energy than upper latitudes.

As solar radiation reaches Earth, much of it is absorbed, raising the temperature of the surface. This heats the air above, which expands and rises into the atmosphere, just like a hot air balloon. The air cools as it rises, and the moisture therein condenses and precipitates, producing most of the earth's rain forests. The cooler, drier air moves to the north and south

until it eventually descends over belts of latitudes between 20 degrees north and 30 degrees north, and between 20 degrees south and 30 degrees south. Descending, this air warms up and gains moisture. At the surface, this air evaporates water effectively. Most of Earth's major deserts are in these latitudinal belts. The convection cell is completed by a return flow across the surface toward the equator. Because of the Coriolis force produced by the earth's rotation, this flow is deflected to come out of the northeast in the Northern Hemisphere and the southeast in the Southern Hemisphere. Regardless of direction, these winds are referred to as trade winds.

Another convection cell forms at somewhat higher latitudes, causing the surface winds to come out of the southwest in the Northern Hemisphere and out of the northwest in the Southern Hemisphere. These winds, called the westerlies, blow across most of the United States. As they travel over the mountains in the western states, they are forced up into higher elevations, where the air loses its water. Deserts lie to the east of these mountains, in what is called the mountains' "rain shadow."

The trade winds of both hemispheres meet along the equator and result in surface winds that tend to push the surface water of the oceans to the west. This surface water is warmer and less dense than the water beneath it, and the boundary between the two is a region of rapidly changing temperature called the thermocline. Because only warm water is pushed to the west, it piles up there, pushing the thermocline lower and lower. Under these conditions, the air just above the surface of the ocean is heated, and moisture is added to it as it travels from east to west. When it reaches the western edge of the Pacific, it rises and its water condenses. Every few years, however, the situation reverses itself. Surface winds weaken, the accumulated warm surface water flows eastward, rain comes to the deserts of Peru, and weather patterns over the entire planet are severely disrupted. The atmospheric effects of such a reversal are called the Southern Oscillation, whereas the oceanic effects are called El Niño. As they are linked and occur together, it has become common practice to refer to the phenomenon as the El Niño Southern Oscillation (ENSO).

GREENHOUSE GASES

Most of the energy that reaches the earth from the sun is electromagnetic in the visible light portion of the spectrum. A cloudless atmosphere lets this pass through easily. Upon reaching the earth's surface, some of this energy is absorbed by surface material, which then radiates the energy back, but at a much lower frequency in the infrared region, which humans detect as heat. On its way back to outer space, this energy may be absorbed by a greenhouse gas molecule. When this molecule reradiates the energy, the chances are about equal that it will be directed toward or away from the earth. In this way, a significant fraction of the infrared radiation is redirected back toward the surface of the earth.

The most important greenhouse gas is water vapor. The distribution of water vapor is uneven and subject to wide variations. It is generally assumed that the average relative humidity of the atmosphere remains constant. Because warm air contains more moisture at the same level of relative humidity than cooler air, the amount of water in the atmosphere will increase if global surface temperatures rise. Thus, water vapor is expected to enhance any effects produced by other greenhouse gases.

The greenhouse gas of greatest concern is carbon dioxide. It is well known that the trace amounts of carbon dioxide present in the atmosphere increased by 12 percent between 1958 and 1998. The amount of coal, oil, and natural gas burned in that time period is also known, and the 12 percent increase represents only about one-half of the carbon dioxide produced by this burning. Where the other half has gone remains a subject of research. Scientists are trying to determine if the observed increases in atmospheric carbon dioxide have led to changes in global surface temperatures. It appears that such temperatures rose approximately 0.5 degree Celsius between 1880 and 1980, a change well within historical rates of temperature fluctuations, leaving the significance of the increase in atmospheric carbon dioxide open to question.

Earth scientists studying gas bubbles trapped in the ice of Greenland and Antarctica have observed that carbon dioxide concentrations in the atmosphere and global temperatures changed together over the past 200,000 years. A causal relationship has not been proved, however, and the situation is complicated by biologic activity, the solubility of carbon dioxide in seawater, and many other factors that vary with temperature.

WATER AND THE SURFACE OF THE EARTH

Water evaporates from the oceans, is transported by weather systems, and falls on the land as rain and snow. Some of this water soaks into the soil, where it contributes to the chemical and physical breakdown of rocks, a process called weathering. Some water can soak deep into the ground, where it is stored as groundwater. Much of it runs across the surface, transporting loose grains of sand and silt and eroding the surface. It takes only a few tens of millions of years to erode a mountain range flat.

In addition to being transported by water, loosened surface materials can move downhill by gravity—a process called mass wasting. Slow mass wasting is called soil creep. One common sign of soil creep is trees with bent bases; young trees are tipped by creep and keep trying to grow upright. Eventually, their roots become firmly enough established to resist creep, and the trees grow straight from then on. Rapid forms of mass wasting can be extremely destructive to life and property. Slump is a slope failure where a large block of material drops but remains largely intact. Mudflows and avalanches are fast-moving quantities of debris that can cause great losses of life and property.

LOCAL SYSTEMS

The environment of Earth is maintained by the interactions of global systems. Local areas—such as a flood plain, a watershed, or an earthquake-prone region—also have interconnected systems of environmental significance. Here, too, Earth science provides essential background for policymakers.

Flooding occurs when too much water tries to move through a channel at once. Earth scientists can determine how much water a channel can hold and how long it will take various volumes of water to move down tributaries. Therefore they also can predict how high a particular flood will be and when the crest will occur. By studying geological evidence of previous floods, scientists can establish the historical frequency of floods of various sizes. Taking into account the effects of agriculture and development, they can then estimate how frequently floods of different magnitudes are likely to occur in the future. This information can be used to determine how much money to invest in flood control or how restrictive to make zoning within the flood plain.

During floods there is an overabundance of water, but at other times water can be in short supply. In many places, underground aquifers are tapped for drinking water. Earth scientists can study a watershed and evaluate the quantity, quality, and flow patterns of its underground water. They can determine how much water can be safely withdrawn without long-term detrimental effects. If pollution occurs, scientists can determine the likely paths of the pollutants, trace them using remote sensing techniques, and suggest strategies to contain or eliminate their threat.

Threats posed by earthquakes are more difficult to perceive than those posed by pollution or flooding. Large earthquakes are unlikely to strike a region twice within the span of one person's life. They may leave a record in the soils and structures of a region, and Earth scientists have been able to interpret these to extend knowledge of major earthquakes back several centuries. To detect small earthquakes, which recur frequently, Earth scientists have developed instruments called seismographs. The behavior of geologic units subjected to prolonged periods of violent shaking can be very different from their behavior under normal conditions. Earth scientists have developed theories to explain such behavior, as well as criteria that can be used to identify those materials that pose the greatest risk. By combining results from these and other areas of research, researchers have been able to estimate the risks posed by earthquakes at various locations. This information can be used to develop building codes and zoning laws to minimize the risk of death and destruction posed by major earthquakes in the future.

Significance

The environmentalists' slogan "Think globally; act locally" is an acknowledgment of the interconnectedness of many planetary processes and the importance of making local decisions based on some understanding of potential global consequences. Earth science endeavors to develop that understanding.

Earth scientists, by quantifying the many feedback loops with which systems interact with themselves and one another, can identify those natural processes that human influences are most likely to disturb. By studying the past behavior of these systems, scientists can make reasonable estimates of how they might respond to future disturbances. By informing public policy, legislation and law enforcement, and

civic leaders in general, Earth scientists can use their knowledge and insight to influence the course taken by the governments of the world.

Otto H. Muller

Further Reading

Chiras, Daniel D., and John P. Reganold. *Natural Resource Conservation: Management for a Sustainable Future*. 10th ed. Boston: Addison Wesley, 2009. Discusses resource use and solutions to issues in sustainability and conservation. Covers case studies, laws, regulations, and international treaties. Discusses natural resources in detail. Also covers geographic information systems and remote sensing.

Coch, Nicholas K. *Geohazards: Natural and Human*. Englewood Cliffs, N.J.: Prentice Hall, 1995. A general treatment of environmental problems. Provides many figures and photographs. A textbook for beginning courses for nonmajors. Emphasizes vocabulary.

Cutter, Susan L., and William H. Renwick. *Exploitation Conservation Preservation: A Geographic Perspective on Natural Resource Use*. 4th ed. Hoboken, N.J.: John Wiley & Sons, 2003. Discusses the social, economic, political, and environmental effects of resource use. Covers detailed content useful to the graduate student. Presents many varying views of unresolved issues.

Dodds, Walter K., and Matt R. Whiles. *Freshwater Ecology: Concepts and Environmental Applications of Limnology*. 2d ed. Burlington: Academic Press, 2010. Covers physical and chemical properties of water, the hydrologic cycle, nutrient cycling in water, as well as biological aspects. Written by two of the leading scientists in freshwater ecology. Appropriate for undergraduates. Includes a short summary of main topics in each chapter.

Duffield, Wendell, and John Sas. *Geothermal Energy: Clean Power from the Earth's Heat*. USGS Circular 1249. 2003. Describes historical and current geothermal energy use. Covers global geothermal applications, mining of geothermal energy, hydrothermal systems, and dry geothermal systems, and the environmental impact of geothermal energy use. Includes numerous color diagrams and images.

Fall, J.A., et al. *Long-Term Consequences of the Exxon Valdez Oil Spill for Coastal Communities of Southcentral Alaska*. Technical Report 163. Alaska Department of Fish and Game, Division of Subsistence,

Anchorage, Alaska. 2001. Covers social, economic, and environmental impact of the oil spill.

Keller, Edward A. *Environmental Geology*. 9th ed. Upper Saddle River, N.J.: Prentice Hall 2010. Emphasizes pollution but offers a broad treatment of most environmental issues. Includes examples of successes and failures with environmental incidents.

Lundgren, Lawrence W. *Environmental Geology*. 2d ed. Upper Saddle River, N.J.: Prentice Hall, 2010. Focuses on hazards and hazard mitigation. Uses case histories to illustrate key points in environmental geology. Deals mostly with regional and local issues but also provides excellent examples of how knowledge of Earth sciences can be applied beneficially.

Lutgens, Frederick K., Edward J. Tarbuck, and Dennis Tasa. *Essentials of Geology*. 11th ed. Upper Saddle River, N.J.: Prentice Hall, 2011. Covers the physical geology of the lithosphere. Examines current issues in geology with a thorough look at environmental issues. Contains many exceptional images, diagrams, and color photos.

Mackenzie, Fred T. *Our Changing Planet: An Introduction to Earth System Science and Global Environmental Change*. 4th ed. Upper Saddle River, N.J.: Prentice Hall, 1998. Provides a fairly thorough scientific foundation for the study of global change. Treatment itself is sometimes technical and dismissive of important concepts.

Murck, Barbara W., Brian J. Skinner, and Stephen C. Porter. *Dangerous Earth: An Introduction to Geologic Hazards*. New York: John Wiley & Sons, 1997. Offers a general treatment of environmental issues; includes a chapter on meteorite impacts. Contains a separate chapter on tsunamis.

Sinding-Larsen, Richard, and Friedrich-W. Wellmer, eds. *Non-renewable Resource Issues: Geoscientific and Societal Challenges*. Hoboken, N.J.: Springer, 2012. Discusses use, overuse, and future needs of natural resources. Considers possible resources needs of the future. Discusses concepts of land use, sustainability, and cultural needs.

U.S. Department of the Interior, Minerals Management Service (USDOI MMS). *Programmatic Environmental Assessment: Arctic Ocean Outer Continental Shelf Seismic Surveys*. U.S. Department of the Interior Minerals Management Service, Alaska OCS Region. 2006. Provides an overview of seismic surveys and the exploration of the Alaskan continental shelf. Content includes alternative scenarios for surveys and their evaluation. Addresses the environmental impact of such surveys.

Zektser, Igor S., et al., eds. *Geology and Ecosystems*. New York: Springer, 2010. Covers various geomorphologies and geological processes. Evaluates environmental impacts of geological research, industries, and natural dynamics. Appropriate for environmental scientists, geologists, ecologists, and similar professionals.

See also: Aluminum Deposits; Building Stone; Coal; Diamonds; Earth Resources; Earth System Science; Fertilizers; Gem Minerals; Geothermal Power; Gold and Silver; Hazardous Wastes; Hydroelectric Power; Iron Deposits; Landfills; Mining Processes; Mining Wastes; Nuclear Waste Disposal; Ocean Power; Oil and Gas Distribution; Sedimentary Mineral Deposits; Strategic Resources; Uranium Deposits.

EARTH SYSTEM SCIENCE

Earth system science is the study of four distinct but interconnected spheres within the earth's natural system. This field of study focuses on the lithosphere, the hydrosphere, the biosphere, and the atmosphere. Earth system science studies the interactions among these systems, including the effects of natural and human-made events on these spheres.

PRINCIPAL TERMS

- **greenhouse effect:** a process whereby carbon dioxide, methane, and other types of gas are released into the atmosphere, retaining the sun's heat and causing global warming
- **lithosphere:** the layer of solid and semisolid rock that forms Earth's outer crust
- **pedosphere:** the soil, rocks, and organic matter that rest on Earth's surface
- **tectonic plates:** fragmented plates that make up the lithosphere and that are in constant motion

Basic Principles

The field of Earth system science (ESS) is based on the notion that the lithosphere, the hydrosphere, the biosphere, and the atmosphere are interconnected. ESS involves the combination of a number of natural sciences, including chemistry, physics, biology, mathematics, and applied sciences, to frame and analyze Earth's various processes and systems.

ESS developed relatively recently (it was first applied in the mid-1980's) to catalog the various elements of the planet and explain how these concepts related to one another. ESS also has demonstrated relevance in the study of natural disasters, such as research on hurricanes and earthquakes. ESS has become a useful tool for examining the effects of humans on the natural environment.

Background and History

In 1983, the advisory council of the National Aeronautics and Space Administration (NASA) established the Earth System Sciences Committee. This committee was charged with researching the earth's many natural systems and processes and examining how they interact.

Scientists on the committee compared the committee's endeavor to the study of the solar system, wherein such individual concepts as gravity, solar wind, magnetism, and the planets themselves are often examined in relation to one another. A member of the committee, Moustafa Chahine of NASA's Jet Propulsion Laboratory, in a conversation with committee chair Francis Bretherton, argued that scientists had long focused on the solar system and that it was time to examine Earth's system, too.

The purpose of the development of ESS was, in part, to better understand the scope and the contributing factors of natural disasters and events. For example, volcanic activity in the lithosphere can send soot into the atmosphere and thus dramatically influence the habitats of nearby organisms. ESS has proved useful in understanding how the earth has evolved through millions of years.

The Earth System Sciences Committee also came into being as part of an effort to understand the concept of climate change and the effects that humanity was having on the planet. In 1988, the committee released a groundbreaking report, "Earth System Science: A Program for Global Change." This report included a schematic diagram that demonstrated how atmospheric, oceanic, biologic, geologic, and human factors interact, in some cases complementing one another and in other cases causing imbalances (such as climate change). The report became an important resource to which governmental and nongovernmental organizations pointed as they began to study global warming and climate change.

Elemental Interaction

ESS does not consider the four main elements of Earth's system as independent. Rather, it considers these spheres to be overlapping and interdependent. Each sphere plays a role in a cycle of biogeochemical processes.

For example, a volcano releases water vapor through the lithosphere and into the atmosphere. The water returns to Earth in the form of rain and snow. The water is collected in the hydrosphere, from which plants and animals within the biosphere consume it. The organisms of the biosphere, in turn, return the water to the lithosphere in the form of waste, where it is broken down and recycled. The process then restarts.

Because ESS encompasses such a broad range of components, the study of this type of science requires the application of many different disciplines. Scientists must take and classify biological samples, conduct chemical analyses, utilize models of physics, develop and apply mathematical models, and conduct sociological surveys to better understand the processes that occur between each of the four main areas of ESS.

THE LITHOSPHERE

One of the main components on which ESS focuses is the lithosphere, which is the layer of solid and semisolid rock that forms Earth's outer crust and uppermost mantle. The lithosphere is broken into giant tectonic plates, which are in constant motion atop the fluid asthenosphere (the softer and hotter layer just beneath the lithosphere).

When tectonic plates come into contact or release themselves from contact, the energy released along the edges of the plates (faults) causes earthquakes at the surface. Similarly, molten rock that is pushed outward through the lithosphere typically flows through fissures in the lithosphere and out through volcanoes.

Lithospheric activity implicates other systems within the ESS framework. For example, volcanic and seismic activity can significantly alter the flow of river water into and the topography of the surrounding area. In India, for example, a study of seismic and volcanic activity that occurred along a river basin through millions of years showed related changes in the terrain, forming hills and valleys and redirecting a river repeatedly.

In more dramatic fashion, a severe volcanic eruption can both level forests and kill countless people and animals. As was the case in 1980, when Mount St. Helens erupted in Washington State, however, volcanic activity also can create opportunities for life. The formerly devastated region has, since the eruption, seen more than 130 new ponds and lakes form, creating new ecosystems and habitats for wildlife.

THE HYDROSPHERE

The hydrosphere, which contains all of the planet's liquid water in oceans, lakes, rivers, ponds, and streams, plays a pivotal role in ecosystems and provides habitats for a wide range of organisms. It is central to the hydrologic cycle, whereby water is collected in surface water from groundwater sources and ice and is evaporated and transferred into the atmosphere and returned to Earth as precipitation.

For the purposes of ESS, the hydrosphere also plays a major role in the interaction among Earth's elemental spheres. The hydrosphere provides water for the biosphere, releases water vapor into the atmosphere, returns minerals, and is a vehicle for the transfer of various minerals and chemical compounds to and from the lithosphere.

THE BIOSPHERE

The biosphere is the component of the ESS model that houses the many different forms of life on Earth. The biosphere includes humans, as well as all animal and plant life. The biosphere even contains the most basic and simple forms of microscopic life, such as microbes and bacteria.

The effects that other elements of the ESS framework have on the biosphere have been well explored. The study of ecosystems, the extinction of animal species, and the damage caused to habitats and forests by natural disasters are all examples of the study of how the biosphere is affected by the other elements.

Scientists are increasingly exploring the effects of the biosphere on the other elements of Earth's system. As technology improves, for example, scientists are able to study the role microbes play in the weathering and erosion processes, breaking down base rocks and creating and cycling minerals in the lithosphere. Furthermore, biologists are examining how the gases emitted by biologic species affect the atmosphere and, through weather conditions, the hydrosphere (where precipitation is deposited).

THE ATMOSPHERE

Since the development of ESS, increased attention has been paid to the study of global warming and how human-generated emissions are contributing to this problem. Millions of years before human life began, volcanic eruptions and even an occasional asteroid impact sent large volumes of carbon dioxide and other greenhouse gases high into Earth's atmosphere, causing the earth's temperature to rise, melting ice caps and raising water levels. (Greenhouse gases are so named because they make it more difficult for heat to escape the earth, thereby keeping the planet warm; this is a process known as the greenhouse effect.) However, when carbon dioxide and other greenhouse gases are retained at the surface level, a cooling effect takes place, eventually leading to the formation of glaciers and ushering in a period known as an ice age.

The ESS model recognizes the interaction, in this case, between the atmosphere and other components of the overall terrestrial system. Recently, however, greater attention has been paid to how humans (as part of the biosphere) are, through industry and other technological activities, rapidly releasing high volumes of greenhouse emissions into the atmosphere. Most scientists believe that continued emission of such gasses will accelerate global warming, causing the ice caps to melt and increasing water levels. Such a flood could have disastrous implications for the other elements of the ESS framework, including the destruction of many of the biosphere's countless ecosystems and changing the chemical composition of the soil.

ADDITIONAL SPHERES

ESS continues to evolve as a science. Although the general consensus is that the earth's intertwining systems are localized to the hydrosphere, biosphere, atmosphere, and lithosphere, some scientists believe that other spheres should be included within this framework.

For example, some argue that the pedosphere must be considered a part of this discipline. The pedosphere is the layer of soil and organic matter that rests on the surface of the earth. The pedosphere contains a vast number of minerals and chemical compounds essential for life on Earth. For this reason, it is often referred to as the skin of the planet, and like the skin of a living organism, it is influenced heavily by its relationship with the lithosphere, atmosphere, hydrosphere, and biosphere.

Others argue that another sphere should be considered a part of ESS: the cryosphere, which encompasses the earth's frozen water (its glaciers, ice packs, icebergs, and arctic regions). The cryosphere is an important resource for the world's water supply. It also is a critical indicator of global change. When icebergs, glaciers, and other frozen bodies melt under increasing temperatures, water levels rise. For this reason, scientists concerned with global warming

carefully monitor the cryosphere for evidence of this trend.

Michael P. Auerbach

FURTHER READING

Bockheim, J. G., and A. N. Gannadiyev. "Soil-Factorial Models and Earth-System Science: A Review." *Geoderma* 159, nos. 3-4 (2010): 243-251. Includes a detailed analysis of ESS, focusing on the role the pedosphere plays in the overall network of Earth's systems.

Finley, Fred N., Younkeong Nam, and John Oughton. "Earth Systems Science: An Analytic Framework." *Science Education* 95, no. 6 (2011): 1066-1085. Discusses how ESS is rapidly replacing a long-standing view of each element as independent. Uses as an example carbon cycling. Discusses ways to teach students ESS.

Jacobson, Michael, et al. *From Biogeochemical Cycles to Global Changes*. Vol. 7 in *Earth System Science*. Maryland Heights, Mo.: Academic Press, 2000. In this volume, the authors discuss the various cycles that occur between the earth's spheres, citing the importance of studying these concepts as they pertain to global climate change.

Merritts, Dorothy, Andrew De Wet, and Kirsten Menking. *Environmental Geology: An Earth System Science Approach*. Cranbury, N.J.: W. H. Freeman, 1998. Considered the only geology textbook written from an ESS perspective. Helps students understand the connections between Earth's geological processes and its other terrestrial spheres.

Schlager, Wolfgang. "Earth's Layers, Their Cycles, and Earth System Science." *Austrian Journal of Earth Sciences* 101 (2008): 4-16. Discusses the fluid motions of the earth's layers within a context of the growing discipline of ESS.

Skinner, Brian J., and Barbara W. Murck. *The Blue Planet: An Introduction to Earth System Science*. 3d ed. Hoboken, N.J.: Wiley, 2011. Presents the atmosphere, biosphere, hydrosphere, and lithosphere in a collective light. Describes the flow of energy and the balance among these spheres.

See also: Earth Resources; Earth Science and the Environment; Geographic Information Systems; Remote Sensing.

EVAPORITES

Evaporites are sedimentary rocks that form as water evaporates from aqueous environments enriched by concentrated salts and minerals. Minerals in these environments may be derived from the surrounding sediment or from the dissolution of organic and inorganic matter within the environment. Evaporite minerals are classified according to the type of salts found within the minerals.

PRINCIPAL TERMS

- **algae:** diverse group of autotrophic organisms that live in aqueous and subaqueous environments
- **aqueous:** solution that contains water acting as a solvent or medium for reactions
- **carbonate:** salt that forms when carbonic acid reacts with a basic material
- **evaporation:** process by which molecules of water at the surface of a body of water transition to a gaseous state in response to energy derived from heat and from solar radiation
- **halide:** salt formed from the reaction between a halide ion, including chloride, bromide, and fluoride, with a basic material in the environment
- **halite:** mineral form of rock salt or common salt formed from a reaction between a chloride ion and sulfur atoms in an aqueous environment
- **playa:** dried salt lake environment known for concentrations of evaporite minerals
- **salt:** ionic compound resulting from chemical reactions forming solids through the reaction of an acidic material and a basic material
- **sedimentary rock:** one of the three basic rock types composed of solids derived from dust, debris, and fragments of material that condense and solidify into larger solid structures
- **sulfate:** salt that forms from a chemical reaction between sulfuric acid and a basic material in the environment

TYPES OF EVAPORITES

Evaporites are a type of mineral solid that forms from the accumulation of minerals and chemicals dissolved in aqueous solutions. Evaporites are generally classified as sedimentary rocks, one of the three major types of rock. Sedimentary rocks form from the accumulation of small particles and dissolved minerals that are concentrated and compacted to become solid rock.

Evaporites form in aqueous solutions only, and specifically in environments with high salt content, including salt marshes, lakes, and marine environments. As their name suggests, evaporites form as water evaporates from an aqueous solution, leaving behind elements that did not transition to gas during evaporation. Different types of evaporites may form from the same source, depending on the types of minerals dissolved in the initial solution.

Chloride evaporites are those that contain chloride ions, a negatively charged ionic form of atomic chlorine. Chloride and the acidic version of chloride, hydrochloric acid (HCl), are highly water soluble, leaving free chloride ions in aqueous solution. Chloride evaporites are also known as halides, which are the salts that derive from chemical reactions involving halide ions, such as chloride, bromide, and fluoride. Table salt, which is also called sodium chloride ($NaCl$) or halite, is one of the best-known types of chloride evaporites. Another example of the chloride-based evaporite is sylvite, which combines chloride with potassium, in the formula KCl, and forms into crystals similar to rock salt.

Sulfate evaporites contain an ionic sulfate group (SO_4^{-2}) consisting of a sulfur atom bonded with four oxygen atoms. Common sulfate evaporites include anhydrite ($CaSO_4$) and its hydrated form gypsum ($CaSO_4 \bullet 2H_2O$). Gypsum is one of the most economically important of the sulfate evaporites and is used in a variety of industrial and chemical processes.

A third major variety of evaporite rocks are the carbonates, which contain a carbonate ion (CO_3^{2-}) in their chemical structure. Common carbonate evaporites include calcite ($CaCO_3$) and dolomite ($CaMg(CO_3)_2$). Many carbonate evaporite minerals are economically important and are used for deriving minerals for industrial processes and chemical reactions. The evaporite mineral trona ($Na_3(CO_3)(HCO_3) \bullet 2H_2O$) is another common carbonate evaporite that is used in a number of industrial applications.

FORMATION OF EVAPORITE ROCKS

Evaporite minerals form in environments where aqueous solutions become enriched with dissolved minerals generally derived from rocks surrounding the body of water. Minerals become more concentrated in areas where the body of water is partially enclosed, such as in lakes and marine estuary environments.

Temporary bodies of water that form in warm or arid environments are among the most common sources for evaporite minerals. The dissolved minerals present in the aqueous solution are derived from the atmosphere, as mineral inclusions in rainwater, and from the lithosphere, as rocks within bodies of water gradually dissolve, releasing minerals into the water.

Solar radiation causes water to evaporate by exciting water molecules and thereby breaking the bonds between them. When this occurs, molecules at the top of the body of water transition to a gaseous state, called water vapor, which rises into the atmosphere. Molecules that remain solid, such as carbonate, sulfur, calcium, potassium, and phosphorus, remain in the solution as a concentrated mineral residue. Evaporite minerals form when the rate of evaporation exceeds the total rate of water inflow into the environment, including the amount of water deposited by precipitation.

At this point, the relative temperature and humidity greatly affect the types of minerals that will form as the remaining water evaporates from the solution. For instance, halite will form in environments where the ambient humidity is less than 65 percent, while environments with higher humidity will not support the growth of halite crystals. The types of evaporites found in a certain environment are therefore governed by the overall temperature and humidity of the region.

Evaporite minerals also form in permanent marine environments, such as the bottom of seas and oceans. In these environments, evaporite formations develop slowly in time as minor quantities of mineral are deposited in layers. Areas around hydrothermal vents, which are jets of superheated water at the bottom of deep marine environments, appear to be prime locations for the development of evaporite rocks because the heated water ejected from the vents contains high concentrations of dissolved minerals, which settle in the areas immediately surrounding the vents.

While deep marine environments may give rise to large deposits of evaporite materials through millions of years, evaporites form more rapidly in semimarine environments where saltwater and freshwater mix and in environments where the presence of water is temporary on more immediate geological scale. In marshes and salt lakes, for example, evaporite minerals may grow by more than 10 meters (33 feet) in thickness in one thousand years.

FORMATIONAL ENVIRONMENTS OF EVAPORITE DEPOSITS

A variety of evaporite minerals have been derived from salt flats also known as playas or sabkhas, which develop in areas where the sea level is slowly receding from the shore, generally in arid environments. Commonly, salt flats form in an area where ocean waters drain into an arid environment or where there was once a salt lake that gradually dried out.

Salt marshes, which are wetland areas surrounding marine environments, produce similar types of evaporites to those found in playas. In both salt flats and salt marshes, saltwater from the nearby ocean infiltrates the soil with the incoming tide, saturating the soil directly beneath the surface. Solar radiation causes the water to evaporate, leaving mineral deposits in the soil.

In playas and saline lakes, algae may assist in the deposition of evaporite minerals by providing a fibrous mass that helps to bind mineral crystals as they form within the soil or at the edges and bottom of the lake. As the ocean recedes and water evaporates from the system, the algal mats gradually shrink, leaving concentrated mineral deposits. The Great Salt Lake in Utah is an example of an environment where evaporite deposition is occurring. Utah's saline lake will eventually evaporate completely, leaving behind a variety of evaporite salts.

Sabkhas and salt marshes generate both gypsum and anhydrite, which form in small ponds on the surface of the marsh. Gypsum forms first, as the water recedes, but further evaporation will dehydrate the gypsum deposits, transforming them into anhydrite. An influx of rain can rehydrate anhydrite deposits, giving rise to gypsum. In this way, gypsum and anhydrite exist in a transitional flux alternating between either form depending on the relative moisture remaining in the environment. In thousands of years, thick deposits of anhydrite and gypsum can develop within the upper layers of the soil. Sabkha environments also give rise to considerable quantities of halite, which can be harvested to manufacture table salt for consumption.

Evaporite minerals also can form in shallow saltwater environments, such as salt ponds and temporary salt lakes. These environments, like the salt flats or salt marshes, give rise to halite in large quantities. The creation of artificial salt ponds, called salterns or salt evaporation ponds, allows for the distillation of salt from marine water and provides a way to harvest halite to be used in making table salt. In addition, environments of this type produce a variety of evaporite carbonates, some derived from carbonate forming from the decomposition of organisms living within the environment.

In deeper aqueous environments, such as shallow seas, marine estuaries, and other marine environments, evaporites form within the water column and settle at the bottom of the environment, where they condense into larger bodies. Gypsum forms in deeper environments as thin, needle-like rods that form primarily at the upper surface as solar radiation evaporates the layer of molecules at the surface of the water. The gypsum then gathers at the floor of the ecosystem, where it forms into fibrous mats. Halite forms in a similar manner, as do several types of carbonate evaporite minerals, developing into thin mats at the floor of the system.

Within millions of years, oceans and seas may dry out completely because of tectonic shifts changing the position of the continental crust relative to the oceans. Shallow seas that once covered portions of the intercontinental area often contain thick evaporite deposits that accumulated through millions of years of gradual evaporation and condensation as the seas shifted. The formation of ancient marine evaporite deposits is more difficult to interpret because the exact combination of climatic changes that occurred during the extended period of deposition is unknown. Geologists believe that the same basic mechanisms involved in the formation of evaporites in semi-aqueous environments also function in deeper oceanic environments, but that the process is more gradual. Ancient marine environments often contain the largest and most complex evaporite rocks, some of which may contain thousands of pounds of evaporite salts.

USES OF EVAPORITES

Halite is one of the most economically important evaporite minerals. Salt mines around the world are generally the remnants of extinct marine ecosystems or salt flats. In these environments, halite may be found in thick deposits at or beneath the earth's crust.

Halite is a fragile mineral and will dissolve in areas where humidity is high or where there is sufficient influx of water from drainage or precipitation. Halite is the only mineral that is consumed directly as a food source. Salt mining is a multibillion-dollar international industry.

Gypsum is another important evaporite mineral often used in the manufacture of construction materials. Gypsum is one of the primary ingredients in plaster and also is a component in drywall sheets, also known as gypsum boards. Gypsum forms in mixed beds containing halite, gypsum, and anhydrite.

Gypsum is generally white in color and has been used in the manufacture of gesso, a base paint used by artists to prepare canvases. Gypsum also is widely used in agriculture as an ingredient in many soil fertilizers and as a treatment for water. In some cases, gypsum may be used as a coagulant in the production of food, and it is most notably used to add stability in soy bean curd, or tofu. In this regard, gypsum provides calcium and therefore serves as a dietary supplement.

Trona is a relatively common carbonate evaporite that is derived from playas and salt pans in Africa and North America. Trona can be processed to produce sodium carbonate, which is mixed with limestone to create fertilizer, and it also can be used as a water treatment chemical. Sodium carbonate also is used in the manufacture of soaps and detergents.

Another carbonate evaporite, dolomite, is used as a decorative stone and in building construction. Crushed dolomite is used as a component in the manufacture of concrete. One of the most important uses of dolomite is as a substrate for deriving magnesium metal by subjecting the dolomite to a reductive chemical environment that results in the concentration of magnesium solids held within the rock.

Micah L. Issitt

FURTHER READING

Blatt, Harvey, Robert J. Tracy, and Brent E. Owens. *Petrology: Igneous, Sedimentary, and Metamorphic.* 3d ed. New York: W. H. Freeman, 2006. Introductory text presenting theories and research methods used to investigate petrology. Contains a discussion of evaporites in reference to sedimentary rock geology and chemistry.

Leeder, Mike. *Sedimentology and Sedimentary Basins: From Turbulence to Tectonics.* 2d ed. Malden, Mass.: Wiley-Blackwell, 2011. Geological text written for advanced students that provides detailed information on the sedimentation process and the historical development of sedimentary basins. Discusses evaporites and presents models for the deposition and evolution of evaporite geology.

Schreiber, B. C., S. Lugli, and M. Babel, eds. *Evaporites Through Space and Time.* Geological Society Special Publication 285. Bath, England: Geological Society, 2007. Collection of articles written by professional geologists and covering various aspects of evaporite origin and deposition. Details locations in which marine and terrestrial evaporites can be found and the processes involved in the formation of these deposits.

Tucker, Maurice E. *Sedimentary Petrology: An Introduction to the Origin of Sedimentary Rocks.* 3d ed. Malden, Mass.: Blackwell Science, 2001. Intermediate-level geological text covering the petrology of sedimentary rocks. Chapter 5 contains a detailed discussion of evaporites, their geological setting, and their development.

Warren, John. *Evaporites: Sediments, Resources, and Hydrocarbons.* New York: Springer, 2006. Detailed technical discussion of evaporite formation, mining, and processing for industrial applications. Written for advanced students and professional geologists.

_____. *Evaporites: Their Evolution and Economics.* Malden, Mass.: Blackwell Science, 1999. Advanced text covering the physical evolution, harvesting, and economic/industrial uses of evaporites. Detailed discussion of the difference in geologic setting and evolution between marine and terrestrial evaporite deposits.

Wicander, Reed, and James S. Monroe, eds. *Historical Geology: Evolution of Earth and Life Through Time.* 6th ed. Belmont, Calif.: Brooks/Cole, 2010. Introductory text covering geological developments in Earth's present ecosystems. Contains a brief discussion of evaporites and their use in interpreting geological history.

See also: Chemical Precipitates; Clathrates and Methane Hydrates; Dolomite; Hydrothermal Mineralization; Water and Ice.

EXPANSIVE SOILS

Expansive soils—soils that expand and contract with the gain and loss of water—cause billions of dollars in damage to houses, other lightweight structures, and pavements, exceeding the costs incurred by earthquakes and flooding.

PRINCIPAL TERMS

- **differential movement:** the unequal movement of various parts of a building or pavement in response to swelling or shrinkage of the underlying soil
- **evapotranspiration:** the movement of water from the soil to the atmosphere in response to heat, combining transpiration in plants and evaporation
- **hydration:** a process whereby, when soils become wet, water is sucked into the spaces between the particles, causing them to grow several times their original size
- **loading:** an engineering term used to describe the weight placed on the underlying soil or rock by a structure or traffic
- **moisture content:** the weight of water in the soil divided by the dry weight of the soil, expressed as a percentage
- **montmorillonites:** a group of clay minerals characterized by swelling in water; the primary agent in expansive soils
- **shrinkage:** an effect opposite to hydration, caused by evapotranspiration
- **soils and foundation engineering:** the branch of civil engineering specializing in foundation and soil subgrade design and construction
- **soil stabilization:** engineering measures designed to minimize the opportunity and/or ability of expansive soils to shrink and swell

SOIL SWELLING AND SHRINKAGE

Certain types of soil expand and contract with the gain and loss of water, primarily through seasonal changes. One meter of expansive soil has sufficient power to lift a 35-ton truck 5 centimeters. Many cracks in walls and concrete surfaces are assumed to be caused by settlement, when in fact they are produced by the lifting force of expansion.

Soils vary enormously in their ability to expand, ranging from zero in very sandy soils to highly expansive in montmorillonite clays. By contrast, kaolinite clays do not expand. Some key indicators of expansive soils are a bricklike hardness when dry, including great resistance to crushing, and stickiness and weakness when wet; also, they have a glazed surface when cut with a knife. Soil cracks and popcorn-like surface appearance are other key signs. Where soils from volcanic ash produce primarily montmorillonite clays, bentonite is formed. This extremely expansive soil prevents even vegetation from growing and is easily visible from a road, with its barren, popcornlike surface and painted-desert coloration.

The ability of montmorillonite clays to expand comes from their physical and chemical structure. Clay minerals have a molecular structure like sheets of paper. Some bond very strongly from layer to layer and have little swelling potential. Expansive clay minerals, however, have weak bonds between the layers. These layers are negatively charged and repel one another, but are bonded by positively charged ions (cations) such as sodium and calcium dissolved in the soil water. When soils rich in these minerals become wet, water is drawn into the spaces between the clay particles, a process called hydration, causing them to grow up to seven times their original size. The opposite effect, shrinkage, is caused by evapotranspiration.

In climates with great variation in seasonal moisture, especially marginally humid and semiarid lands, large changes occur in the moisture content of the soil, with subsequent swelling and shrinkage. In the midlatitudes and subtropics, winter and spring are the wet seasons, as vegetation is dormant and rain and melting snow are abundant. Summer and fall are the dry seasons, soilwise; evapotranspiration demand is high, and much water is transferred from the soil to the atmosphere. Human activity can also produce localized effects through excessive watering, leaking pipes, poor drainage around buildings, and use of trees and bushes for landscaping. It is these volume changes in the swelling soils beneath structures, caused by the gain and loss of water, that make expansive soils so damaging.

DIFFERENTIAL MOVEMENT AND VARIABLE LOADING

Expansive soils do their damage primarily through differential movement and variable loading. If the entire structure is lifted and sinks uniformly, there is little problem. Unfortunately, swelling and shrinkage are seldom uniform. Many buildings are located on sloping land, and, as a result, groundwater is more likely to be encountered on the uphill side during wet seasons. Consequently, more expansive force is exerted on that side, cracking walls and slabs. The construction season can also be a factor on sloping land, as the portion of the building pad cut into the hillside, and the portion built up with fill, tends to respond differently to changes in moisture. Even on level land, deep foundations may be stable while grade beams and slabs resting upon the surface rise and fall with the seasons.

While much attention is given to damaged buildings, damage to roads may actually have a greater impact on the public. The life of a road is measured by the total weight it carries over a period of time, divided by the strength of the roadbed. Many older roads are built directly upon the natural grade, including highly expansive soils. Because these soils are weak when wet, the heavier loading at concentrated points under truck tires causes the pavement to break down more rapidly. Once a cracking pattern is visible, a chuckhole is but a few wetting cycles away, as water can now easily reach and further weaken the soil below. Major roadbed problems may also occur near the boundaries between highly expansive and less expansive soils as a result of differential swelling. Water seepage at concentrated points beneath the roadbed also undermines the strength of the roadbed and contributes to differential movement.

Damage to structures from expansive soils can be categorized as reduced performance, economic loss, and massive failure. Massive failure is rare, but the threat of it is usually enough to demand removal of the danger. High on this list would be leaning chimneys and massively cracked concrete channels. Also, raised portions of patios, sidewalks, and driveways are hazardous to pedestrians. Reduced performance occurs when doors and windows cannot be closed because of misalignment, and unsightly cracks appear on walls and concrete slabs. Economic losses occur from such unsightly cracks when potential buyers are discouraged from purchasing property or property owners have to repair them. Also, higher heating and cooling bills result from energy losses through cracks, especially around windows and doors. For taxpayers, a little-recognized cost is incurred through more frequent road repairs.

ON-SITE GEOTECHNICAL INVESTIGATION

Fortunately, the problems caused by expansive soils are reasonably well understood by soils and foundation engineers. Typical procedures include an on-site geotechnical investigation, laboratory testing of samples obtained during the on-site investigation, and design recommendations specific to the anticipated usage of a site. The importance of sound engineering here cannot be overemphasized. The cost of such work is small compared to the cost of the project and is far cheaper than potential remedial work. Inasmuch as the soil ends up underneath the building, roadway, or other structure, it is obviously inaccessible when problems arise; it is far easier and cheaper, therefore, to compact the soil properly and install piers and drains before rather than after.

An experienced soils engineer will generally know what to expect within a given part of a city or region. Soils are notoriously diverse, however, and the only sure way to secure the accurate information needed for a sound, cost-effective design is an on-site investigation. Key information is provided by borehole samples, extracted by a drilling rig similar to that used for wells. The selection and number of borehole locations are determined from architectural layouts and the size of the project. One hole may be sufficient for a house, whereas numerous holes may be ordered for a high-rise building. Three kinds of geotechnical information are obtained directly from these holes: a description of the type and thickness of soil and rock; resistance to penetration (influenced by moisture content), which gives a crude measure of soil strength; and, if the hole is deep enough, depth to the water table.

With distressed structures, one of the most important judgments a soils engineer can make is whether remedial work is needed or will be cost-effective. Cracks often appear in new buildings and pavement; though frustrating, they may be of little consequence. By contrast, serious problems may be avoided if timely repairs are made before incipient failures, such as leaning chimneys, reach the critical point.

TESTS FOR EXPANSIVE POTENTIAL

The most common measure of expansive potential is the Atterberg limits test, run in the laboratory. Though crude, it has proved its value over many years as a predictor of expansive and nonexpansive soil characteristics. The test essentially measures the difference between the resistance of a soil sample to flow as it becomes wetter and wetter, called the liquid limit, and disintegration under rolling as it becomes drier and drier, called the plastic limit. The numbers used are moisture content percentages, obtained by computing the weight of water as a percentage of the dry weight of the soil. The difference between these two percentages is called the plasticity index, referred to as the PI. Variances among plastic limit numbers are generally small, but liquid limit numbers vary enormously. The stickier the clay, the more moisture will be required to make it flow; thus the higher the liquid limit. One characteristic of expansive soils worth remembering is their stickiness when wet. Consequently, the higher the liquid limit, the larger the PI and the greater the swell potential.

The Atterberg limits test is relatively easy and inexpensive to run. It is the simplest way to test large numbers of samples, and such tests are conducted by the thousands in soils laboratories. Where precise and detailed information is needed from key samples, however, a consolidation-swell test is run. This test is much more complex, time-consuming, and expensive, but the information gained is vital for the design of more expensive structures. The test mimics the behavior of soils under the loading applied by a building and their response to swell pressure as water is absorbed. Additional tests, such as confined and unconfined compression, are also commonly run on the more expensive projects. These test results, combined with experience and input from colleagues, form the basis for the recommended design. Good designs anticipate and avoid problems, minimize construction costs, and recommend specifications for quality-assurance testing during construction.

DESIGN ALTERNATIVES

Design alternatives include strengthening the structure, stabilizing the soil, using some combination of the first two, or anticipating movements to isolate their impact. If a slab can be strengthened so that it moves as a unit, the effects of differential movement will be mitigated. In larger buildings, reinforced grade beams are commonly suspended on piers, with space deliberately left under the grade beam to allow room for soil expansion. Major attention is given to stabilizing the soil. The goal is to compact the soil in its expanded condition, balancing the need for strength with the need to allow some room for additional expansion. A key problem is to keep the soil from drying out before it is sealed, usually by a concrete slab or asphalt. Drains are commonly installed around the perimeter and covered by concrete, asphalt, or plastic sheeting. Three types of drains are used: peripheral drains around the perimeter; interceptor drains to control subsurface water from uphill sources; and sump drains, which require a pump and are usually located under a basement floor slab.

A combination of increased strength and soil stabilization is often used. Much attention is given to the water content and density of the soil beneath the structure. When feasible, select fill (stable soil having low expansion and good binding qualities) is imported for the top layer. Lime or fly ash may be mixed with the soil to increase both its strength and its stability. This treatment is especially effective under pavement (parking areas and roadbeds). Special equipment has been invented to mix these additives into the soil efficiently. Laboratory tests are commonly run to measure the effectiveness of treatment in terms of PI reduction and strength; a common rule for lime treatment is to reduce the PI to 10 or less. Lime is the easiest to handle and its effectiveness is easiest to test, though fly ash, a by-product of coal burning in electrical power plants, is growing in use because of its low cost. The problem with fly ash is that its use requires meticulous application, beyond the abilities of many inexperienced operators. It often contains heavy metals that may contaminate water supplies. Engineering supervision of the mixing and compaction—with strict attention to moisture and time specifications—is critical to the successful use of fly ash.

Where construction cost is a key concern, it is sometimes feasible to design the structure so that slabs and interior walls can move freely without affecting the foundation or other parts of the building. Trim work can mask such movement so that aesthetic concerns are not a factor. Once structures are completed, the correction of problems is far more difficult. Corrections may range from simple drains and moisture barriers around perimeters to tearing out the pavement and redoing the soil subgrade. An accurate diagnosis is even more essential to success in this case. Consequently, skilled engineers study the problem in detail, and heroic efforts are sometimes made to obtain soil samples. The engineer's judgment of the seriousness of the problem and the remedial action that follows are especially vital. A skilled engineer can tell much from cracks and other distress patterns.

GROWING IMPORTANCE OF FOUNDATION ENGINEERING

Only since the 1970's has the severe threat represented by swelling soils been recognized by the construction industry. Several changes are responsible for this growth. First, buildings with slab-on-grade floors (concrete resting directly upon soil) have replaced basements and crawl spaces in many areas. Insofar as seasonal moisture changes are restricted to the top 1 meter and essentially disappear at a depth of 3 meters, this change created conditions far more susceptible to expansion and contraction. Second, heavier trucks and increased usage have put greater stress on roadbeds and parking areas. Once the pavement is broken, it is much easier for water to reach the soil below, and failure soon follows. Third, population growth and rising costs of land have encouraged the development of nonfarmland, which increases the odds of encountering expansive soils. Still, this extended use of land is fine, as long as the dangers are understood and taken into account in designing and planning projects.

When confronted with any hazard, one must consider whether the problem is serious enough to require expert help. Foundation engineering is a little-recognized but invaluable part of any construction program involving expansive soils. Knowledge of soil conditions prior to new construction is well worth the relatively small cost. Well-informed individuals provide an invaluable service to themselves, employers, and the organizations they serve. As part of the National Cooperative Soil Survey, the U.S. Department of Agriculture and the Soil Conservation Service provide booklets on regional surveys. In map and table format, accompanied by explanatory text, these publications present detailed information about agricultural and engineering soil conditions.

Nathan H. Meleen

FURTHER READING

Al-Rawas, Amer Ali, and Mattheus F. A. Goosen, eds. *Expansive Soils: Recent Advances in Characterization and Treatment.* New York: Taylor & Francis, 2006. Discusses classification of expansive soils. Provides information on swelling and consolidation, and the effects these processes have on buildings. Discusses how site characteristics and stabilizing agents can alter expansion processes.

Buzzi, O., S. Fityus, and D. Sheng, eds. *Unsaturated Soils.* New York: CRC Press, 2009. Discusses the dynamics of unsaturated and expansive soils. Discusses experimental studies and theoretical studies in separate volumes. Includes articles drawn from the Conference on Unsaturated Soils held in 2009. Designed to aid academics and graduate students in soil mechanics research.

Fenner, Janis L., Debora J. Hamberg, and John D. Nelson. *Building on Expansive Soils.* Fort Collins: Colorado State University, 1983. Explains the swelling mechanism. Includes a map of the United States showing the occurrence and distribution of potentially expansive materials, and gives identification techniques, design alternatives, and remedial measures. Especially recommended for builders.

Harpstead, Milo I., Thomas J. Sauer, and William F. Bennett. *Soil Science Simplified.* 4th ed. Ames: Iowa State University Press, 2001. Excellent primer for the nonspecialist. Well illustrated and suitable for high school Earth science curricula. Thorough coverage of soil topics relevant to agriculture, engineering, and geology.

Holtz, Wesley G., and Stephen S. Hart. *Home Construction on Shrinking and Swelling Soils.* Denver: Colorado Geological Survey, 1978. A short primer on the problems and possible solutions related to buildings situated on expansive soils. Includes good illustrations of typical problems, drain design, and recommended interior construction techniques.

Jochim, Candace L. *Home Landscaping and Maintenance on Swelling Soil.* Special Publication 14. Denver: Colorado Geological Survey, 1981. Prepared for the homeowner, this beautifully illustrated publication has large type and clear, simple illustrations, including a map of swell potential in Colorado, especially the Denver area. Also provides information about groundwater and how to maintain proper drainage around a house.

Katti, R. K. *Behaviour of Saturated Expansive Soil and Control Methods.* Rev. ed. Oxford: Taylor & Francis,

2002. Published under the auspices of the Central Board of Irrigation and Power in India. Focuses on soil stabilization, soil dynamics, and black cotton soil in that country. Includes an extensive bibliography and an index.

Koerner, Robert M. *Construction and Geotechnical Methods in Foundation Engineering*. New York: McGraw-Hill, 1984. Part of a series in construction engineering and geotechnical engineering, and attempts to address this mixed audience. Heavy mathematical treatment.

Legget, Robert F. *Cities and Geology*. New York: McGraw-Hill, 1973. Designed as a textbook for courses in urban or environmental geology. Highly relevant to those interested in urban planning. Uses case histories worldwide to demonstrate the effects of geology on urban growth and problem solution. An important chapter on geologic hazards is included.

Nelson, John D, and Debora J. Miller. *Expansive Soils: Problems and Practice in Foundation and Pavement Engineering*. New York: John Wiley and Sons, 1997. Nelson deals primarily with the interaction between soil and human-made structures such as building foundations and roadways.

Newson, Malcolm. *Land, Water, and Development: Sustainable and Adaptive Management of Rivers*. 3d ed. London: Routledge, 2008. Presents land-water interactions. Discusses recent research, study tools and methods, and technical issues, such as soil erosion and damming. Suited for undergraduate students and professionals. Covers concepts in managing land and water resources in the developed world.

Olson, Gerald W. *Soils and the Environment: A Guide to Soil Surveys and Their Application*. New York: Chapman & Hall, 1981. Designed for a worldwide audience and general readership. Explains how soil surveys are prepared, what the goals are, and how to understand the most important applications.

Rajapakse, Ruwan. *Geotechnical Engineering Calculations and Rules of Thumb*. Burlington, Mass.: Butterworth-Heinemann, 2008. Written for the professional geological engineer. Provides explanations of calculations and general practices of engineers building shallow and pile foundations, retaining walls, and analyzing geological features and compositions.

Sobolevsky, Dmitry Yu. *Strength of Dilating Soil and Load-Holding Capacity of Deep Foundations: Introduction to Theory and Practical Application*. Rotterdam, Netherlands: A. A. Balkema, 1995. A detailed survey of soil mechanics, soil testing, swelling soils, and foundations suitable for students who already have some background in the field.

See also: Desertification; Land Management; Landslides and Slope Stability; Land-Use Planning; Land-Use Planning in Coastal Zones; Remote Sensing; Soil Chemistry; Soil Erosion; Soil Formation; Soil Profiles; Soil Types.

EXTRATERRESTRIAL RESOURCES

During the twenty-first century, new resources may well be required to build and energize civilization on other planets and in space, as well as to enhance accelerated change in a dynamic Earth-based society.

PRINCIPAL TERMS

- **ceramics:** nonmetal compounds, such as silicates and clays, produced by firing the materials at high temperatures
- **fiber-reinforced composites:** materials produced by drawing fibers of various types through a material being cast to produce a high weight-to-strength ratio
- **hydrolysis:** the breakdown of water by energy into its constituent elements of water and hydrogen
- **off-planet:** pertaining to regions outside Earth in orbital or planetary space
- **permafrost:** a layer of soil and water ice frozen together
- **photovoltaic cell:** a device commonly made of layered silicon that produces electrical current in the presence of light; also, a solar cell
- **regolith:** that layer of soil and rock fragments just above the planetary crust
- **superconductors:** materials that pass electrical current without exhibiting any electrical resistance

CLASSIFICATION OF RESOURCES

In the largest possible sense, there will be two broad categories of resources during the twenty-first century: Earth-based resources and off-planet resources. The reason for the distinction is an economic one, based on where the materials will be used, and is driven by consideration of the gravitational field of Earth. Earth's gravitational field is very strong in comparison with extraterrestrial space, where there is little gravitational influence. It is also strong in relation to Earth's moon and Mars, where the gravitational field is much weaker than Earth's. Hence, materials required in space will be much more economical if mined from small bodies in space (such as asteroids or comets) or the moon. The same resources, if shipped into space from Earth, would cost many times more. This example is provided as an insight into the distinctive ways of looking at future resources in the developing economy of the twenty-first century. In a more systematic approach to categorization, one could fundamentally classify tomorrow's most basic resources in the same broad classes used in recent times: agricultural products, chemicals (including petroleum and derivatives), metals, ceramics, energy, wood (and derivatives), and power.

IMPACT OF SCIENTIFIC DISCOVERY

Though the same broad resource categories will be required in the future as are required today, many of their individual identities and uses will be much different. Scientists are beginning to foresee new types of materials that will have a vital use. Such an example was afforded in 1989 when two scientists announced that they thought they had discovered a revolutionary power source. One major element in their design was the rare-earth metal palladium. Until their announcement, there had been few other uses for the metal; after their announcement, the price of palladium temporarily increased as a result of its presumed importance. As new discoveries are made, new uses for existing materials, such as metals, will cause the demand for the resource to change as availability dictates.

Scientific discovery and technological development are often the key to the advancement and change of society and the resources that drive it. Examples from the past are abundant: Tungsten presently used for light filaments had few uses before the invention of the incandescent light bulb; silicon (the key element of common beach sand) assumed vital importance with the discovery of microelectronics. Probably no other single resource has had a more far-reaching impact on humankind or planet Earth than petroleum and its derivatives. The science of superconductivity produces materials that conduct electrical current without resistance. Late in 1986, this branch of physics took a revolutionary turn when it was discovered that synthetically produced materials could become superconductive at temperatures much higher than had been thought possible. The synthetic materials were described as "artificial rocks" and were made by mixing together various metal oxides under very specific conditions. As the ultimate goal of room-temperature superconductivity is approached, the materials used in producing the superconductors are likely to become highly sought-after and valuable resources.

CERAMICS AND FIBER-REINFORCED COMPOSITES

The use of ceramic materials increased dramatically in the 1980's. Ceramic materials are nonmetal compounds produced by firing at high temperatures. Ceramics can be lightweight, resilient, and heat-resistant. Materials used in the production of ceramics are clays, silicates, and calcium carbonates. Uses of ceramics include refractories (heat-shielding or heat-absorbing materials, such as those used to construct the space shuttle's heat-shielding tiles) and electrical components. They have also been used in tests as automobile engine blocks. The resources that produce such ceramics are projected to have even broader applications in the future.

Another type of product that has become quite important and is expected to play a vital role in the future is fiber-reinforced composites. Such composites are made by drawing fibers through a material being cast. When the material hardens, the fibers cast inside it make even very light products, such as aircraft wings and space vehicle fuselages, much stronger. The materials from which these composites are made (fiberglass, boron, tungsten, aluminum oxide, and carbon) will assume new importance with any increase in the use of fiber-reinforced composites.

ECONOMICS OF OFF-PLANET RESOURCES

Because the exploitation and possible settlement of space will probably be the most significant development of the twenty-first century, resources that drive that revolution will become highly valuable. The value of these space materials will be based on their origin. It will always be expensive to ship resources into space from Earth. A kilogram of aluminum mined and processed on the moon and shipped to Earth could one day cost a tenth of the amount of the same aluminum shipped to orbit from Earth because of launch-energy costs. Even under the best of circumstances, in a well-developed launch system, a liter of water launched from Earth into space will always cost thousands of times more than will a liter of water on the surface. One must always keep the gravitational interfaces in mind when considering the economics of space resources. Many studies have been conducted concerning the utilization of off-planet

resources, their availability, and their uses. Such studies have been conducted concerning the construction of moon and Mars bases and colonies, as well as the construction of enormous space colonies.

OFF-PLANET WATER AND POWER RESOURCES

The most valuable resources in space will be water and energy; the use of one is dependent on the other. Water and energy yield two other vital resources: oxygen and hydrogen. Water is the most basic of all resources necessary for human survival. Aside from the obvious purposes of direct consumption and hygiene, water will be necessary for cleaning, cooling, and producing food. Raw water can be broken down by a process called hydrolysis into elemental hydrogen and oxygen. The hydrogen may be used for energy production and the oxygen for breathing.

The mass of water is great enough that the ability to launch volumes that are sufficient to meet the needs of space resources will play a significant role in determining the use of space. Much attention has been paid to the likely sources of water off-planet, especially in deposits that can be most easily exploited. The moon may contain deposits of water ice locked up beneath the regolith. The Clementine spacecraft bounced radar signals from the lunar surface during its 1994 exploration of the moon. Radio telescopes on Earth monitoring those signals discovered that regions near the lunar poles were reflective in a way that was suggestive of the presence of water ice. The Lunar Prospector spacecraft carried a neutron spectrometer capable of detecting the energy of neutrons blasted from the lunar surface by the solar wind. It was announced on March 5, 1998, that neutrons from the polar regions carried the energy signature of having interacted with hydrogen. The most likely reservoir of hydrogen would be as a constituent of the water molecule in water ice. There may be as much as 300 million metric tons of ice particles mixed with the lunar soil of the permanently shadowed craters at each pole. This is not a huge amount of water, so future lunar colonies will have to recycle water completely. Presumably, the icy particles have been delivered to the moon over billions of years by comet impacts.

There is considerable evidence to suggest that Mars may contain vast amounts of water locked up beneath its regolith as permafrost (water ice mixed with soil). Moreover, some asteroids may contain water ice, and comets are generally considered to be mostly water ice. These resources may be captured and relocated to an orbit suitable for mining. Finally, pure water is produced as a waste product of fuel cells that react hydrogen and oxygen to produce power, but the hydrogen and oxygen of fuel cells were obtained by breaking down water in the first place.

Solar energy will be an abundant resource in space. Large reflectors can be used to concentrate heat to drive turbines and melt ores for smelting. Another source of energy is the direct conversion, by photovoltaic (solar) cells, of sunlight into electricity. Such cells may be produced off-planet by using lunar materials such as silicon, a significant constituent of lunar soil. Future space colonists may want to locate reserves of radioisotopes for nuclear power. Such reserves, if found, could be exploited with an excellent promise of safety, especially in high Earth, lunar, or solar orbit, as the isolation of the radioactive contaminants and power plant itself would be assured. Questions concerning other power sources, such as high-temperature fusion and unconventional fusion techniques, remain unresolved. Such devices, when perfected, would overcome nearly every known deterrent, allowing the fullest exploitation of space.

OFF-PLANET BUILDING MATERIAL RESOURCES

Aside from the most basic resources of survival off-planet (water, power, hydrogen, and oxygen), space colonists will require massive building-material resources. Launching millions of tons of raw building materials into space is as impractical as hauling volumes of water. A Princeton University physicist, Gerard K. O'Neill, has addressed the question of producing such quantities of building materials for construction of off-planet colonies. O'Neill and his Princeton-based Space Studies Institute have designed a mass driver, a device that could catapult "buckets" of lunar soil into lunar or Earth orbit, to be processed—in automated space factories—into sheets of aluminum, magnesium, titanium, glass, and other materials for use in construction of lunar or orbiting colonies. Lunar soil has been found to be rich enough in the necessary materials to produce such materials in space. Current studies on the moon rocks brought back from the Apollo program indicate that roughly 50,000 metric tons of aluminum sheets per year could be produced from 900,000 metric tons per year of moon ores catapulted into space by the mass drivers.

SURVEYS OF OFF-PLANET RESOURCES

Space probes have been utilized to recover information on the location of resources in space and their amounts. The Apollo mission astronauts returned samples from the lunar surface that were analyzed extensively and used in subsequent engineering and economic studies to reveal how the lunar material might be processed as building materials. One such study was performed by Gerald W. Driggers of the Southern Research Institute of Birmingham, Alabama. In Driggers' account, he estimated the cost of building an orbiting metals factory ($20 billion) and its power requirements (300 megawatts). He also estimated the final cost of the aluminum sheeting ($22 per kilogram) and compared it to the cost of the Earth-produced material ($2.2 per kilogram); he then compared it to the cost of Earth-based material launched into space ($88 per kilogram). Such economic studies will drive the capital outlays that will fund and justify such substantial expenditures.

The resources of Mars were surveyed at two landing sites in 1976 by the United States' Viking landers. The Viking landers were not designed to return samples directly from Mars but were sent with a device called an X-ray fluorescence spectrometer, which would analyze the Martian soil for its inorganic materials. No probe has ever taken a direct observation of an asteroid, although many scientists are convinced that the Martian moons Phobos and Deimos are captured asteroids. Flyby studies of these bodies indicate that their densities are low enough that they could be made up of some water ice in their interiors. In 1986, several spacecraft flew past Halley's comet as it approached the sun, returning spectacular, closeup photos of the body and taking measurements of its density. These flybys have supported earlier speculation that comets are largely water ices covered over by exceptionally dark, carbonaceous material.

DEVELOPMENT OF ENERGY PRODUCTION

The survivability and quality of every human's life is directly affected by the resources available to each individual. From water and power to food and building materials, humankind has been in a constant struggle to improve the availability and quality of resources while reducing the magnitude of the struggle to obtain them. The historical propensity of civilization, if not its fundamental purpose, is to ease that struggle and increase the availability of the resource base while constantly improving its superiority and basic usefulness.

The resources of the future will follow this trend, with momentous advances being made in the most elemental domain: energy. Energy production directly affects the acquisition of all other resources. Improved ways of generating power will directly influence food production for all people, but the most profound influence will affect those populations in regions fixed on the edge of continual famine. High-temperature superconductivity would change the basis of electronic devices, resulting in more efficient and cheaper electrical power production, storage, transmission, and use. The exploitation of this science would enable supercomputers that operate at unprecedented speed. Transportation on Earth and in space would be revolutionized.

Dennis Chamberland

FURTHER READING

Arnopoulos, Paris. *Cosmopolitics: Public Policy of Outer Space.* Toronto: Guernica, 1998. Offers a thorough description of space exploration and industrialization, and the laws and legislation surrounding these fields.

Cadogan, Peter. *The Moon: Our Sister Planet.* New York: Cambridge University Press, 1981. Discusses the discoveries arising from the exploration of the moon during the 1960's. Details the resources discovered from close analysis of the Apollo moon rocks. Speculates about water on the moon and other possible moon-based resources. Well illustrated.

Chiras, Daniel D., and John P. Reganold. *Natural Resource Conservation: Management for a Sustainable Future.* 10th ed. Boston: Addison Wesley, 2009. Discusses resource use and solutions concerning issues in sustainability and conservation. Covers case studies, laws, regulations, and international treaties. Discusses natural resources in detail. Covers geographic information systems and remote sensing.

Cooper, Henry S. F., Jr. *The Search for Life on Mars: Evolution of an Idea.* New York: Holt, Rinehart and Winston, 1980. A superlative account of the search for life on Mars. Relates the instruments used in all the Viking lander science, including a discussion of the device that measured the constituents of Martian soil. An excellent work for understanding

how scientists comprehend the details of another planet's resources.

Fridleifsson, Gudmundur O., and Wilfred A. Elders. "The Iceland Deep Drilling Project: A Search for Deep Unconventional Geothermal Resources." *Geothermics* 34 (2005): 269-285. Written for readers with a background in engineering or geophysics. Provides a good overview of common issues in deep-drilling and focuses the article on the benefits of harnessing geothermal power.

Gulkis, Samuel, D. S. Stetson, Ellen Renee Stofan, et al., eds. *Mission to the Solar System: Exploration and Discovery.* Pasadena, Calif.: National Aeronautics and Space Administration, Jet Propulsion Laboratory, 1998. Deals with space exploration and the possibility of resource exploration on other planets. Provides a look into current practices, as well as future potential in space exploration.

Gupta, Chiranjib, and Harvinderpal Singh. *Uranium Resource Processing: Secondary Resources.* Berlin: Springer-Verlag, 2010. Discusses the topics of metallurgy, mineralogy, and leaching as they apply to uranium deposits, as well as the recovery of uranium from phosphates and other resources. Includes a further reading list and a complete subject index.

Hazen, Robert M. *The Breakthrough: The Race for the Superconductor.* New York: Summit Books, 1988. Features the discovery of high-temperature superconductors in 1987. Relates what resources were used in the construction of the "artificial rocks" that were made superconductive. Also describes the revolutionary changes that such discoveries would bring.

Heppenheimer, T. A. *Colonies in Space.* Harrisburg, Pa.: Stackpole Books, 1977. Depicts what resources will be required to settle space en masse. Names and enumerates studies that have independently addressed the issue in sufficient particulars that the reader will understand the complexity and degree of effort required to make this next great "leap" of humankind.

Letcher, Trevor M., ed. *Future Energy: Improved, Sustainable and Clean.* Amsterdam: Elsevier, 2008. Begins with the future of fossil fuels and nuclear power. Discusses renewable energy supplies, such as solar, wind, hydroelectric and geothermal. Focuses on potential energy sources and currently underutilized energy sources. Contains a section covering new concepts in energy consumption and technology.

Miles, Frank, and Nicholas Booth. *Race to Mars: The Harper and Row Mars Flight Atlas.* New York: Harper & Row, 1988. Specifies the details for a flight to Mars and its ultimate settlement. Discusses the possibilities of Mars's resources, such as water, availability of building materials, and power sources. A historical accounting of the difficulties in finding sufficient resources to set up colonies off-planet and the efforts required to establish a foothold on new frontiers.

O'Neill, Gerard K. *2081: A Hopeful View of the Human Future.* New York: Simon & Schuster, 1981. Discusses the facts concerning the delivery of raw materials not only to the space and planetary colonies but also to Earth from space. Discusses the "capture" of asteroids and the use of moon-based mass drivers to deliver raw materials for use in space. Illustrated; written for all readers.

Schmidt, Victor. *Planet Earth and the New Geoscience.* 4th ed. Dubuque, Iowa: Kendall/Hunt, 2003. Produced through the University External Studies Program at the University of Pittsburgh in coordination with the television series *Planet Earth.* Provides an overview of Earth sciences, mineral resources, and astronomy, and their future possibilities.

Sinding-Larsen, Richard, and Friedrich-W. Wellmer, eds. *Non-renewable Resource Issues: Geoscientific and Societal Challenges.* New York: Springer, 2012. Discusses the use, overuse, and future needs of natural resources. Offers hypotheticals of what resources will be needed in the future. Reviews concepts of land use, sustainability, and cultural needs.

See also: Desertification; Earth Resources; Earth Science and the Environment; Hazardous Wastes; Landfills; Land-Use Planning; Land-Use Planning in Coastal Zones; Mining Processes; Mining Wastes; Nuclear Waste Disposal; Soil Chemistry; Soil Formation; Soil Types; Strategic Resources.

F

FELDSPARS

Feldspars are the most abundant group of minerals within the earth's crust. There are many varieties of feldspar, distinguished by variations in chemistry and crystal structure. Although feldspars have some economic uses, their principal importance lies in their role as rock-forming minerals.

PRINCIPAL TERMS

- **crystal:** a material with a regular, repeating atomic structure
- **igneous rocks:** rocks formed from the molten state; they may be erupted on the surface (volcanic) or harden within the earth's crust (plutonic or intrusive)
- **ion:** an atom that has gained or lost electrons and thereby acquired an electric charge; most atoms in minerals are ions
- **metamorphic rocks:** rocks formed by the effects of heat, pressure, or chemical reactions on other rocks
- **polarized light:** light whose waves vibrate or oscillate in a single plane
- **sedimentary rocks:** rocks formed on the earth's surface from materials derived by the breakdown of previously existing rocks
- **silicate:** a mineral containing silica tetrahedra, which may be separate from one another or joined into larger units by sharing their corner oxygen atoms
- **silica tetrahedron:** the fundamental molecular unit of silica; a silicon atom bonded to four adjacent oxygen atoms in a three-sided pyramid arrangement

COMPOSITION AND PROPERTIES

Considered as a group, the feldspars are the most abundant minerals in the earth's crust. They form in igneous, metamorphic, and sedimentary rocks and are among the principal repositories for sodium, calcium, potassium, and aluminum in the crust. The feldspars are thus compounds of aluminum, oxygen, and silicon, together with one or more of the elements sodium, potassium, and calcium. They form two principal series: potassium feldspars and plagioclase feldspars. There are a few rare barium feldspars as well. The feldspars are a more complex group than this summary suggests at first glance, as they undergo subtle but important changes in crystal structure depending on temperature and pressure. Also, there are several distinctive mixtures of potassium and plagioclase feldspars.

The physical properties of all the feldspars are similar. They all have a Mohs scale hardness of 6 (they can scratch most glass, but cannot scratch quartz). Their densities are all in the range of 2.6-2.75 grams per cubic centimeter, or about the average density of most common rock-forming minerals. Feldspars are usually, though not always, light in color. They are usually translucent in thin splinters but in rare cases can be transparent. All the feldspars have good cleavage, or a tendency to split easily along smooth planes dictated by the atomic structure of the mineral. They cleave along two perpendicular or nearly perpendicular planes.

CRYSTAL STRUCTURE

Most of the feldspars crystallize in the triclinic crystal system; a few feldspars are monoclinic. Crystals are classified according to their atomic arrangements. The fundamental atomic unit that makes up any crystalline material can be pictured as fitting inside an imaginary box, or unit cell, with parallel sides. Unit cells stack together to form a crystal, and the shape of the crystal reflects the shape of its unit cell. In triclinic crystals, the angles between the faces or edges of the unit cell are never right angles. In monoclinic crystals, two pairs of faces of the unit cell are perpendicular, but the third is not. A monoclinic feldspar unit cell looks like a carton with no top or bottom, sheared slightly out of shape so that its outline is an oblique parallelogram instead of a rectangle.

One feature of the crystal structures of the feldspars is especially notable. The feldspar minerals have a pronounced tendency to exhibit distinctive kinds of crystal twinning, or abrupt changes in crystal growth patterns. The growth of a crystal can be pictured as stacking planes of atoms on one another in a specific pattern. There are often many equally possible ways to stack one plane on the next. When a crystal has been built up according to one stacking pattern and then begins following a different pattern, there is an abrupt change in the atomic structure of the mineral. This changeover of atomic structure is called twinning. Sometimes the results are visible to the unaided eye, and the mineral appears to consist of two crystals stuck to one another or penetrating each other. In other cases, the results of twinning may only be visible through the microscope. The twinning of feldspars is a valuable clue to the geologist, because twinning often makes it easy to distinguish feldspars that are otherwise very similar.

Feldspars belong to the tectosilicate group: silicate minerals in which silica tetrahedra link to form three-dimensional networks. The silica tetrahedra in feldspars link to form zigzag chains, and the chains in turn are linked to adjacent chains to create a continuous network. The aluminum in feldspars occupies the centers of some of the tetrahedra in place of silicon, and the potassium, sodium, or calcium occupy the open spaces between the chains.

POTASSIUM FELDSPARS

There are three important potassium feldspars: microcline, orthoclase, and sanidine. Microcline is perhaps the most familiar potassium feldspar, because it is the normal potassium feldspar in granitic rocks and the principal potassium feldspar in metamorphic rocks. Microcline can be any light color but is most often white or pink. The familiar pink color of many granites is caused by microcline, which is colored pink by microscopic plates of the iron oxide mineral hematite within the feldspar. Amazonite is a distinctive variety of microcline, unusual for its bright green color. Amazonite is a minor gemstone.

The green color is of uncertain origin. It has been attributed to small amounts of the elements rubidium or lead, or to changes in the crystal structure of the microcline because of natural radioactivity in the surrounding rocks. Colors in minerals frequently have complex causes and are often the result of very tiny amounts of impurities. It is common for a given color to have several different causes. Orthoclase is a variety of potassium feldspar that forms at somewhat lower pressure than does microcline. It occurs chiefly in granitic rocks that form near the earth's surface and cool quickly, and also in volcanic rocks. Sanidine is a high-temperature potassium feldspar found in some volcanic rocks. All three varieties of potassium feldspar are distinguished by differences in their crystal structure, particularly as seen under the microscope.

PLAGIOCLASE FELDSPARS

The plagioclase series of feldspars is one of the best natural illustrations of a solid solution. Just as solutions exist in liquids, a solid solution is a blend of two or more distinct materials on the atomic scale. Metallic alloys are other familiar examples of solid solutions. Solid solutions differ from chemical compounds in that the components can have variable proportions. They differ from simple mixtures (for

Feldspar is one of the two most abundant mineral types in the earth's crust; here its typical two-directional, 90-degree cleavage can be seen. (U.S. Geological Survey)

example, salt and pepper) in that the components are interchangeable on the atomic level. These concepts are important to understand because feldspars include both solid solutions and mixtures.

The plagioclase series consists of a solid solution of two components, or end members: albite and anorthite. The proportions of aluminum and silicon are different in anorthite because calcium ions in minerals normally have a +2 electric charge, compared with the +1 of sodium. Therefore, an aluminum ion (+3) must substitute for silicon (+4) to compensate for the extra charge on the calcium ion. The plagioclase series is subdivided into six members, depending on the relative amounts of albite and anorthite. In increasing order of anorthite content, the plagioclase feldspars are albite, oligoclase, andesine, labradorite, bytownite, and anorthite. The plagioclase feldspars become somewhat denser and usually darker in color with increasing anorthite content.

Albite contains less than 10 percent of the anorthite component and forms in sodium-rich environments. Albite forms in marine sedimentary rocks as a cementing mineral, and forms in marine volcanic rocks when sodium from seawater replaces calcium in their plagioclase feldspars. Albite can also form during the metamorphism of sodium-rich rocks. Albite is rare in igneous rocks because igneous rocks are rarely so rich in sodium and poor in calcium that the mineral would form. Oligoclase contains 10 to 30 percent of the anorthite component and is very common because it is the normal plagioclase feldspar in granitic rocks. Andesine contains 30 to 50 percent anorthite and is common in igneous rocks that are less silica-rich than granite, such as diorite. Labradorite contains 50 to 70 percent anorthite and is the principal plagioclase feldspar in silica-poor igneous rocks such as basalt or gabbro. Labradorite is also the principal feldspar in rare igneous rocks called anorthosite, which consist mostly of plagioclase. Anorthosite is a common rock type of the moon. Terrestrial anorthosites, rare as they are, are commonly used as ornamental building stones. They generally are dark gray with large feldspar crystals a centimeter or more across, and show attractive bright bluish reflections from cleavage planes within the feldspar. Bytownite, with 70 to 90 percent anorthite, is perhaps the rarest plagioclase. It occurs most often in very silica-poor igneous rocks. Anorthite is any plagioclase with 90 percent or more of the anorthite end

member (or less than 10 percent albite). Anorthite most often forms through the metamorphism of rocks rich in calcium and aluminum but very poor in sodium, such as clay-rich limestones.

OTHER FELDSPARS

To some extent, potassium feldspars and sodium-rich plagioclases also form a solid solution. Feldspars containing roughly equal parts potassium feldspar and albite are called anorthoclase. In addition to pure feldspar minerals, there exist a wide variety of mixtures of feldspars. At high temperatures, potassium feldspars and plagioclases coexist in solid solution much more readily than at low temperatures. Crystals that formed a stable solid solution when an igneous or metamorphic rock formed often become unstable when the rock cools. As the feldspar cools, plagioclase and potassium feldspar often separate. The final result is a patched or streaked network of plagioclase and potassium feldspar filaments interlocked with one another. This texture is easily visible to the unaided eye. When the feldspar consists mostly of potassium feldspar enclosing small amounts of plagioclase, the mixture is called perthite. Perthite is very common; almost any large microcline crystal will exhibit perthitic texture, with the pink microcline enclosing milky filaments of plagioclase. Less often, plagioclase is the dominant feldspar, enclosing small inclusions of potassium feldspar. Such a mixture is called antiperthite.

There are very few feldspar minerals other than the potassium and plagioclase feldspars. The openings in the atomic structure of feldspar are so large that only very large ions can be held there. Magnesium and iron ions are too small, so there are no magnesium or iron feldspars. Lithium, although chemically similar to sodium and potassium, is also too small to form feldspars. Cesium and rubidium feldspars have been made artificially; some feldspars are rich in cesium and rubidium, but no special names have been assigned to these minerals. Barium feldspars, such as celsian, do exist in nature. Hyalophane can be considered a solid solution of celsian, albite, and potassium feldspar. Banalsite is another barium feldspar. The rare mineral buddingtonite forms when ammonia-rich volcanic solutions alter plagioclase, replacing sodium and calcium with ammonia.

FELDSPAR-LIKE MINERALS

A few feldspars and feldspar-like minerals form when other small ions substitute for aluminum or silicon within the silica tetrahedra. Feldspar-like minerals that form in this manner include reedmergnerite, eudidymite, danburite, and hurlbutite. All are uncommon.

A few minerals that are geologically akin to the feldspars deserve mention. The feldspathoids have feldspar-like chemical compositions and form when rocks are too poor in silica to form feldspars. They are usually softer than feldspars and with quite different crystal forms. The scapolite minerals are essentially plagioclase feldspars that include molecules of sodium chloride (halite or table salt), calcium sulfate (gypsum), or calcium carbonate (calcite) within their atomic structures. It has been suggested that scapolite may be a common mineral on Mars, formed when plagioclase absorbed carbon dioxide from the planet's atmosphere, and that much of the planet's original carbon dioxide may now be locked up in scapolite.

TECHNIQUES FOR IDENTIFYING FELDSPARS

A wide range of techniques has been developed to probe the structure of minerals with polarized light. The two most obvious features of minerals in polarized light are interference color and extinction. When light enters a mineral, it splits into two beams polarized in perpendicular planes. The orientation of the planes is closely related to the crystal structure of the mineral. The two light beams travel at different speeds through the mineral. When the beams emerge, they recombine into a single light beam whose direction of vibration is usually different from the original direction. Some light passes through the second polarizer. The resulting interference color is determined strictly by the amount the two beams of light separated within the mineral. If the specimen is rotated (petrographic microscopes are normally equipped with rotating specimen stages), it will black out, or undergo extinction, at intervals of 90 degrees. Extinction occurs when the vibration directions of light in the mineral match those of the polarizing filters. In these positions, light leaving the mineral experiences no change in vibration direction and is blocked by the second polarizing filter.

Under the microscope, in normal illumination, feldspars are colorless and similar in appearance to quartz. They can often be distinguished from quartz by cleavage (which appears as straight, parallel cracks) and by a dusty appearance caused by chemical alteration. Quartz is almost immune to chemical alteration. Between crossed polarizers, quartz and feldspar display similar interference colors, but the twinning habits of the feldspars usually make it easy to distinguish them. Because the crystal structure changes abruptly when twinning occurs, twinned crystals are obvious as a result of abrupt changes in the optical properties. A crystal that looks like a single entity in normal illumination appears as distinct regions with different interference color or extinction between crossed polarizers.

Geologists sometimes use staining techniques to identify feldspars. The most common method involves etching the specimen with hydrofluoric acid (an extremely hazardous material) and applying a series of chemicals that stain potassium feldspars yellow and plagioclase pink.

For probing the atomic structure of feldspars in detail, many highly sophisticated techniques are in use. The chemical composition of feldspars can be determined for even tiny specimens by bombarding the specimen with electrons (electron microprobe) or with X rays (X-ray fluorescence) and measuring the energies of radiation given off by the specimen. The atomic arrangement of feldspars is determined by X-ray diffraction, in which X rays are reflected off atoms within the specimen. The amount of X rays reflected in different directions can be used to determine the geometric arrangement of atoms within the crystal.

GEOLOGIC SIGNIFICANCE

Feldspars have some economic value, but their principal importance lies in their role as major building blocks of the earth. Because they are such tremendously important reservoirs of common elements, feldspars are key minerals in classifying rocks, and the composition of feldspar minerals in a rock is a powerful clue to its origin and history.

Feldspars are also geologically significant for their role in radiometric dating. Both the potassium-argon and rubidium-strontium dating methods rely on elements that are found in feldspar, either as principal ingredients (potassium) or as common trace elements (rubidium and strontium).

ECONOMIC VALUE

Feldspars have minor use as gemstones. Amazonite is a green variety sometimes used as a gem, and moonstone is translucent feldspar with microscopic inclusions that give it a milky appearance. Aventurine is a clear feldspar with tiny plates of other minerals that impart it with a sparkly appearance. Some feldspar-rich rocks called anorthosite are used as an ornamental building stone.

The most important uses of feldspar are less glamorous. Feldspar is the principal ingredient of porcelain. Indirectly, feldspar is the source of aluminum. Rocks that are rich in feldspar and poor in iron are broken down by tropical weathering so that all but the aluminum is dissolved away. The final result is a mixture of aluminum minerals called bauxite, the principal ore of aluminum. In temperate climates, weathering of feldspar releases potassium, an essential plant nutrient.

Steven I. Dutch

FURTHER READING

Blackburn, William H., and William H. Dennen. *Principles of Mineralogy*. 2d ed. Dubuque, Iowa: Wm. C. Brown, 1993. A college-level mineralogy text with chapters on mineralogical theory and methods, plus descriptions of common minerals. Provides a lengthy section on the feldspars. Chapter 12 is useful for its good coverage of modern analytical techniques used in the study of minerals.

Deer, W. A., R. A. Howie, and J. Zussman. *An Introduction to the Rock-Forming Minerals*. 2d ed. London: Pearson Education Limited, 1992. The chapters on feldspars contain detailed descriptions of chemistry, crystal structure, and identification techniques.

Fettes, Douglas, and Jacqueline Desmons, eds. *Metamorphic Rocks: A Classification and Glossary of Terms*. New York: Cambridge University Press, 2007. Discusses the classification of metamorphic rocks. Feldspars are mentioned throughout the book in reference to the various types of metamorphic rocks and basic classifications.

Klein, Cornelis, and Cornelius S. Hurlbut, Jr. *Manual of Mineralogy*. 23d ed. New York: John Wiley & Sons, 2008. One of the most widely used college mineralogy texts. Contains chapters on most mineralogical methods, plus descriptions of common minerals. Particularly good for its attractive illustrations of atomic structure of minerals, its description of optical properties and the petrographic microscope, and its survey of research on the causes of color in minerals.

O'Donoghue, Michael. *The Encyclopedia of Minerals and Gemstones*. London: Orbis, 1976. Summarizes basic geological, crystallographic, and optical concepts that relate to gem minerals. Aimed at a general audience. Second half offers descriptions of about one thousand minerals. Liberally illustrated with large, attractive color photographs.

Parsons, Ian, ed. *Feldspars and Their Reactions*. Dordrecht, Netherlands: Kluwer, 1994. A collection of papers presented at the North Atlantic Treaty Organization (NATO) Advanced Study Institute on Feldspars and Their Reactions in Edinburgh, Scotland, in 1993. Although many of the discussions may be too technically advanced for the high school student, the bibliography is useful for anyone seeking information on feldspars and geochemistry.

Roberts, Willard Lincoln, Thomas J. Campbell, and George Robert Rapp, Jr. *Encyclopedia of Minerals*. 2d ed. New York: Van Nostrand Reinhold, 1990. A summary of the physical properties of 2,200 minerals. Notable for its almost complete collection of color photographs. Its alphabetic arrangement simplifies locating mineral descriptions, but the lack of cross-reference lists of minerals by chemistry, crystal structure, or other properties is a serious drawback for students who wish to compare minerals.

Wood, Elizabeth A. *Crystals and Light*. 2d ed. Princeton, N.J.: Van Nostrand Reinhold, 1977. A short book aimed at beginning college students that describes many of the optical phenomena encountered in the study of minerals. Describes basic concepts of crystallography and the behavior of polarized light in simple language.

See also: Biopyriboles; Carbonates; Cement; Clays and Clay Minerals; Diagenesis; Expansive Soils; Gem Minerals; Hydrothermal Mineralization; Igneous Rock Bodies; Industrial Nonmetals; Landfills; Metamictization; Minerals: Physical Properties; Minerals: Structure; Non-silicates; Oil and Gas Origins; Oxides; Radioactive Minerals; Soil Chemistry; Soil Formation.

FERTILIZERS

Fertilizers increase food, feed, and fiber production by providing the minerals needed by plants. Organic refuse, gases, rock phosphate, and salts from ancient seas are the chief sources of materials for fertilizers.

PRINCIPAL TERMS

- **compound:** a chemical combination of elements with distinct properties that may differ from the elements from which it formed
- **element:** a pure substance that cannot be broken down into anything simpler by ordinary chemical means
- **grade:** the percentage by weight of a nutrient present in a fertilizer, expressed as a standard element or compound
- **leach:** to dissolve from the soil
- **macronutrient:** a substance that is needed by plants or animals in large quantities; nitrogen, phosphorus, and potassium are macronutrients
- **micronutrient:** a substance that is needed by plants or animals in very small quantities
- **mineral:** a naturally occurring substance with definite chemical and physical properties
- **nutrient:** a substance required for the optimal functioning of a plant or animal; foods, vitamins, and minerals essential for life processes
- **pH:** a term used to describe the hydrogen ion activity of a system; a solution of pH 0 to 7 is acid, pH 7 is neutral, pH 7 to 14 is alkaline
- **sedimentary:** formed from material that has been deposited from solution or eroded from previously existing rocks

TYPES AND GRADES

The major nutrients needed by plants are carbon, oxygen, hydrogen, nitrogen, sulfur, and phosphorus. They form 95 percent of a plant's dry weight. Carbon and oxygen can be taken in directly from the air. Hydrogen comes from water absorbed by plant roots. Nitrogen, although present in the atmosphere, is not taken in by plants in the gaseous form. It must come from soluble nitrogen-containing compounds in the soil or be produced by bacteria that live in the roots of some plants. The remaining 5 percent of plant dry weight is composed of phosphorus, potassium, calcium, magnesium, silicon, sodium, sulfur, and chloride. Selenium, iron, boron, molybdenum, zinc, and manganese are necessary to most plants in very small amounts. Silicon has little biological use but is present in water occupied by plants and typically is sequestered in silica grains within the plant. Because plants cannot take in nitrogen directly from the air, and because nitrates are easily leached from soils, nitrogen fertilizers are the most commonly used fertilizers. Phosphorus and potassium are commonly added to soil by fertilizers as well.

Two types of fertilizers may be used. Organic fertilizers, which are produced by plants and animals, include bone meal, blood meal, humus, peat, domestic sewage sludge, and manure. Guano, the accumulated wastes of birds or bats, is also an organic fertilizer. Inorganic fertilizers begin with atmospheric gases, natural gas, or deposits of sedimentary rock. The rocks precipitate from mineral-rich coastal waters or form when shallow lakes and seas evaporate. Inorganic materials are chemically processed to yield commonly used fertilizers such as liquefied ammonia, urea, and superphosphates.

The price and recommended use of commercial fertilizers depend on their grade. Both organic and inorganic fertilizers are labeled with a grade. Fertilizer grade is usually written as three numbers, such as 10-16-10 or 10-0-10. Each number tells the percentage by weight of one of the three macronutrients. They are always given alphabetically: nitrogen, phosphorus, potassium. The first number represents total nitrogen (in the form of elemental nitrogen) in percent by weight. Plants use the second macronutrient, phosphorus, when it is soluble in ammonium citrate. Usable phosphate is calculated as phosphorus pentoxide (P_2O_5). Potassium, the third major nutrient, is calculated as water-soluble potassium oxide. A fertilizer with a grade of 10-5-10 would contain 10 percent by weight of nitrogen, 5 percent by weight of available phosphorus, and 10 percent by weight of water-soluble potassium. A fertilizer may contain more of a nutrient than is stated on the label. If chemical tests show that the nutrient is in a form not usable by plants, the amount is not used in calculating the grade.

Nitrogenous Fertilizers

Ammonia is the principal nitrogenous fertilizer. Pure ammonia is a colorless, pungent, and irritating gas. As inorganic fertilizer, it becomes usable to plants when liquefied and injected into the soil. Most fertilizers sold are ammonia or a compound derived from ammonia. Anhydrous (without water) ammonia is made by combining atmospheric nitrogen and natural gas.

Urea contains 45 to 46 percent nitrogen. The nitrogen in urea must be converted to the ammonium ion by soil enzymes and held in humus or clay to be useful to plants. Urea can be applied as pellets or in a water solution. Chemical manufacturing plants synthesize urea from ammonia and carbon dioxide. Another solid source of nitrogen is ammonium nitrate. It contains 33.5 percent nitrogen and includes both ammonium and nitrate forms of nitrogen. The nitrate can be used directly by plants, but it is easily leached. Ammonium fixed onto clays becomes available to plant roots. Ammonium nitrate is synthesized from ammonia and nitric acid. Natural deposits occur in Chile. Ammonium sulfate is made from coke-oven gases. It is 21 percent nitrogen as the ammonium ion, NH_3^+. Ammonium sulfate supplies sulfur as well as nitrogen and is often used on wet soils.

Urea or other solid fertilizers often are coated with sulfur or compounded from materials of different solubilities. The sulfur coating slows the solution of the fertilizer granule. When materials of different solubilities are blended, some will dissolve immediately, while others take longer to dissolve. These fertilizers release nitrogen continuously and save on the costs of application when supplements are needed over a long period of time. Nurseries and turf farms prefer continuous-release fertilizers.

Phosphate Fertilizers

A second macronutrient supplied by fertilizers is phosphorus. Phosphate fertilizers are available in both inorganic and organic forms. Rocks rich in phosphorus are mined in the southeastern and western United States. Rock phosphate fertilizers contain about 4 to 5 percent equivalent of phosphorus pentoxide. The rock is merely ground into particles small enough to be applied to the soil. Bone meal, the phosphorus fertilizer that has been in use the longest time, contains about the same amount of available phosphorus as rock phosphate. Bones from slaughterhouses provide the raw materials for bone meal.

Rock phosphate can be enriched in several ways. When mixed with sulfuric acid, it becomes superphosphate with 16 to 20 percent phosphorus pentoxide. Mixtures with phosphoric acid yield triple superphosphate, with a 0-45-0 grade. Other phosphate fertilizers are diammonium phosphate (46 to 53 percent phosphorus pentoxide and 18 to 21 percent nitrogen), monomonium phosphate (48 percent phosphorus pentoxide and 11 percent nitrogen), and nitric phosphate (20 percent phosphorus pentoxide and 20 percent nitrogen). Nitric phosphates are made by treating phosphate rock with nitric acid and phosphoric acid. The addition of muriate of potash (potassium chloride) to nitric phosphates gives a 10-10-10 grade fertilizer containing all three basic nutrients. Polyphosphate fertilizers, another inorganic source of phosphorus, contain high amounts of phosphorus pentoxide. They form soluble combinations with iron and aluminum. These chemicals, called complexes, can be used by plants. Other phosphate fertilizers may form insoluble compounds in soils with high iron and aluminum content. Thus, the phosphate is not available to plants.

Phosphate rock can be mined in Florida, Tennessee, North Carolina, Idaho, Wyoming, and Montana. North Africa and the region west of the Ural Mountains also have large deposits. In the United States, geoclinal (West Coast) deposits form in deep areas at the margins of continents. The geoclinal deposits of the northern Rocky Mountains extend over hundreds of square kilometers. Platform (East Coast) deposits were formed closer to shore where phosphate-rich waters warmed on approaching land. Weathered or residual deposits form when calcium-rich rocks are leached and less soluble phosphate remains. The "brown-rock" deposits of Tennessee are of this type. The phosphate sometimes goes into solution and is deposited lower in the soil profile. The Pliocene Bone Valley formation of Florida contains phosphatic pebbles and nodules reworked from the underlying Miocene Hawthorn formation. Phosphatic guano occurs on some Pacific islands where old deposits of bird excreta have lost their nitrogen through decomposition. Guano that has decomposed and been leached by percolating waters may reach 32 percent phosphorus pentoxide.

Five nations—the United States, China, Jordan, Morocco, and South Africa—have 90 percent of the world's phosphate rock deposits, which are likely to decline in availability toward the end of the twenty-first century. This event, dubbed "peak phosphorus," will have profound effects on global food production. Fortunately, phosphorus supplies can be extended by better application of fertilizers to minimize waste, and by recycling of phosphorus that would otherwise be discarded as sewage.

POTASSIUM FERTILIZERS

The third nutrient commonly supplied in fertilizers is potassium. The most widely used potassium supplements contain muriate of potash, sulfate of potash, or sulfates of potassium and magnesium. These are all inorganic salts. Muriate of potash, chemically potassium chloride, is the most commonly used potassium source, but due to its chlorine content muriate of potash can be toxic to some plants. Potatoes, tobacco, and avocados are sensitive to excess chlorine. Sulfate of potash contains 48 to 50 percent available potassium oxide. Sulfate of potash-magnesia is about 18 percent magnesium oxide and 22 percent potassium oxide. Other potassium fertilizers include potassium nitrate and potassium polyphosphate.

Potassium salts are mined in Canada, New Mexico, Utah, and elsewhere from deposits left behind by the withdrawal or evaporation of ancient shallow seas. Brines from the Great Salt Lake are evaporated to yield magnesium and potassium chlorides. Potash salts are the last salts to come out of solution when seawater evaporates. This fact, combined with their solubility in water, makes potash salts more rare than gypsum and rock salt. The largest mineral deposits of potash salts occur in the Devonian muskeg or Prairie formation of western Canada, North Dakota, and Montana. The Saskatchewan deposits occur nearest the surface and are, for that reason, the leading production sites. The Permian Zechstein evaporites of England, Germany, and Poland are equally famous. Guano derived from the wastes of birds or bats is an organic source of potassium nitrate fertilizer.

SPECIALTY FERTILIZERS

Specialty fertilizers supply nutrients that may be depleted by fertilizer use, soil management practices, irrigation, or heavy crop production. Sulfur deficiencies can be alleviated by using ammonium or potassium sulfate fertilizers. The use of ammonia fertilizers, leaching, or excessive moisture may lead to acidic soils. Dolomitic limestone adds calcium and magnesium, and corrects soil pH. Some ammonia fertilizers include recommendations for sweetening soils that may become acidic because of the production of hydrogen ions. Boron is supplied by borax. Borates, from which borax is made, occur in bedded deposits beneath old playas, brines of saline lakes, and hot springs or fumaroles. Glass particles added to fertilizers provide boron.

Micronutrient metals, such as molybdenum, iron, zinc, manganese, and copper, can be added to inorganic fertilizer as needed. Soil tests, chemical analysis of crop samples, and crop performance help indicate such needs. Molybdenum provided by sodium or ammonium molybdates can be added in small quantities to other fertilizers or mixed with water and used to soak seed. Iron, zinc, and copper sulfates supply soluble metals.

The formation of water-insoluble chelates may lead to shortages of micronutrient metals. Chelates are organic chemicals that bond to a metal to keep it from forming a precipitated salt. Helpful chelates are soluble in water. Manure contains water-insoluble chelates that may tie up metals. Excessive use of manure as a fertilizer, therefore, can cause deficiency of metals.

Commercial solid fertilizers combine several types of ingredients. The "carrier" contains the nutrient element. Nitrates, ammonium sulfates, and ammonium carbonate carry nitrogen. Phosphates carry phosphorus. Potassium can be carried with nitrogen in potassium nitrate. The second ingredient of many fertilizers is a conditioner. Conditioners prevent caking and assure good flow. Vermiculite and organic wastes condition by absorbing water. Diatomaceous earth, oils, plastic, and waxes maintain flow. Often, inert substances are selected so that reactions will not affect the solubility of the carrier. Neutralizers, such as ground dolomitic limestone, may be added to counteract acidity. Fillers, such as sand, bring the product up to a standard weight. Special additives—including micronutrients, fungicides, herbicides, and insecticides—help save on labor costs or prevent unwanted plants from absorbing nutrients intended for the crop. Fertilizers are often blended for the needs of specific soils growing specific crops.

AGRICULTURAL AND ENVIRONMENTAL CONCERNS

Most current research on fertilizers is related to more efficient agricultural use and assessing and reducing the effects of fertilizer on the environment. Coated fertilizers and combination fertilizers with timed release of nutrients are being developed. Studies show that for some crops, applying fertilizer as a liquid and at certain stages of plant growth is preferable to solid application. Micronutrients have been found to be of importance in the budding and setting of fruit.

Areas of environmental concern include the assessment of fertilizer's impact on the groundwater and health of developing nations. Using locally available materials, such as greensand, manure, chitin from crustacean shells, or phosphate rock, may boost Third World agricultural production without incurring foreign debt or environmental damage. In the western United States, excessive use of fertilizers combined with irrigation from salty river waters has left some soils too salty to grow certain crops. Research continues on the use of low-salt fertilizers and salt-tolerant crop varieties. Finally, research on the use of sewage sludge as a fertilizer and conditioner is important. The accumulation and uptake of heavy metals may be prevented by the addition of fungi, which concentrate metals and prevent them from entering crop plants. Fertilizer use becomes more efficient as researchers learn more about plant metabolism and the role of specific nutrients at different stages of the plant life cycle.

A soil test can reveal the need for fertilizers. Agricultural specialists may then suggest the best type of fertilizer for a particular soil and crop. Some nutrient carriers, such as muriate of potash, may harm salt-sensitive crops such as tobacco and potatoes. Crops such as trees or turf grass may require a continuous-release form of fertilizer. Carefully planned fertilization avoids waste and minimizes hazards to the applicator and the environment.

Dorothy Fay Simms

FURTHER READING

Bates, Robert L. *Geology of the Industrial Rocks and Minerals.* Mineola, N.Y.: Dover, 1960. Provides information on the occurrence and use of nonmetallic minerals, including mineral fertilizers.

Birkeland, P. W. *Soils and Geomorphology.* 3d ed. New York: Oxford University Press, 1999. One of the best books written on soils from the perspective of a geologist. Provides many examples from around the world. Covers the newly adopted soil profile description symbols. Discusses soil taxonomy in the early chapters. Suitable for university-level readers; early chapters are easily understood by high school readers.

Buchel, K. H., H.-H. Moretto, and P. Woditsch. *Industrial Inorganic Chemistry.* 2d ed. Weinheim: Wiley–VCH, 2000. Discusses hydrogen, sulfur, and halogen compounds, as well as fertilizer minerals such as phosphorus, nitrogen, and potassium. Metal compounds discussed include alkaline earth metals, aluminum, chromium, silicon, and manganese. Covers uses of these materials as nuclear energy, pigments, fillers, and other products.

Buol, S. W., F. D. Hole, and R. J. McCracken. *Soil Genesis and Classification.* 5th ed. Ames: Iowa State University Press, 2003. A classic soil science text. Contains an update on the new soil horizon symbol nomenclature. Provides an in-depth and fairly easy discussion of soil taxonomy classification. Suitable for university-level readers. Familiarity with Birkeland and Singer and Munns will assist readers approaching this text.

Colwell, J. D. *Estimating Fertilizer Requirements: A Quantitative Approach.* Wallingford, England: CAB International, 1994. Provides mathematical and statistical models to test and estimate the properties of fertilizers and soils. Appropriate for students with some familiarity with the subject.

Donahue, Roy L., Raymond W. Miller, and John C. Shickluna. *Soils: An Introduction to Soils and Plant Growth.* 6th ed. Upper Saddle River, N.J.: Prentice-Hall, 1990. Includes chapters on soil chemistry, fertilizers, and plant nutrition.

Foth, Henry D., and Boyd G. Ellis. *Soil Fertility.* 2d ed. Boca Raton, Fla.: CRC Lewis, 1997. A good introduction to soil and fertilizers, this book describes soil chemistry, fertilizers, and plant nutrition, as well as ways to test for fertility.

Govett, G. J. S., and M. H. Govett, eds. *World Mineral Supplies: Assessment and Perspective.* New York: Elsevier, 1976. A compilation of the location and availability of industrial and agricultural minerals.

Gowariker, Vasant, et al. *The Fertilizer Encyclopedia.* Hoboken, N.J.: John Wiley & Sons, 2009. Provides over four thousand terms related to fertilizers and soils. Covers topics of physical and chemical properties of soil, physiology, manufacturing, economics, and soil fertility.

Havlin, John L., et al. *Soil Fertility and Fertilizers*. 7th ed. Upper Saddle River, N.J.: Prentice Hall, 2004. Less technical than previous versions and more accessible to readers looking for an introduction to soil fertility. Useful for soil professionals and crop consultants. Discusses nutrient and soil management with minimal environmental impact.

Jensen, Mead L., and Alan M. Bateman. *Economic Mineral Deposits*. 3d ed. New York: John Wiley & Sons, 1981. Discusses the world's best-known mineral deposits. Discusses principles of mineral deposit study. Includes chapters on nonmetallic mineral deposits, such as fertilizer minerals.

Pettijohn, F. J. *Sedimentary Rocks*. 3d ed. New York: Harper & Row, 1975. A thorough discussion of the origin, classification, and uses of sedimentary rocks.

Sides, Susan. "Grow Powder." *The Mother Earth News* 114 (November/December, 1988). Discusses organic, unprocessed, and minimally processed fertilizers.

Sopher, Charles D., and Jack V. Baird. *Soils and Soil Management*. 3d ed. Reston, Va.: Reston Publishing, 1986. A concise description of the major fertilizers with some crop recommendations.

See also: Aluminum Deposits; Building Stone; Carbonates; Carbonatites; Cement; Chemical Precipitates; Coal; Diamonds; Dolomite; Gold and Silver; Industrial Metals; Industrial Nonmetals; Iron Deposits; Limestone; Manganese Nodules; Pegmatites; Platinum Group Metals; Salt Domes and Salt Diapirs; Sedimentary Mineral Deposits; Sedimentary Rock Classification; Uranium Deposits.

G

GEM MINERALS

A gem mineral is any mineral species that yields a gem upon cutting and/or polishing. Gem minerals have value based on their potential to produce gems. Although gems have had various uses in the past, including special powers attributed by folklore, their principal uses have been personal adornment (as jewels), a mode of investment, and a symbol of wealth and power.

PRINCIPAL TERMS

- **gem:** a cut and polished stone that possesses durability, rarity, and beauty necessary for use in jewelry and, therefore, of value
- **gemstone:** any rock, mineral, or natural material that has the potential for use as personal adornment or ornament
- **inclusion:** a foreign substance enclosed within a mineral; often very small mineral grains and cavities filled with liquid or gas; a cavity with liquid, a gas bubble, and a crystal is called a three-phase inclusion
- **mineral:** a naturally occurring inorganic substance with a characteristic chemical composition and atomic structure, manifested in its external geometry and other physical properties
- **mineral species:** a mineralogic division in which all the varieties in any one species have the same basic physical and chemical properties
- **mineral variety:** a division of a mineral species based upon color, type of optical phenomenon, or other distinguishing characteristics of appearance
- **rough:** gem mineral material of suitable quality to be used for fashioning gemstones
- **synthetic mineral:** a human-made reproduction of the structure, composition, and properties of a particular mineral

PHYSICAL PROPERTIES

Gem minerals are those mineral species that have yielded the material from which specimens have been fashioned into gems. Gems and therefore gem mineral varieties must have the same properties. These properties are beauty (color, phenomenon, or clarity), durability (hardness and toughness), and value (rarity, demand, and tradition). Of the approximately five thousand mineral species known, only about ninety of these produce material in a quality and quantity suitable for gems, and this quantity of suitable material makes up generally only a very small portion of the gem mineral found. Often the quality of this gem material is such that the material is given a gem varietal name or mineral varietal name based upon color or phenomenon. A different name is given for each color variation; hence, there is a large number of gem names compared with the gem mineral names. Two or more different gem minerals may have the same gem name, as in the case of jade and the minerals nephrite (actinolite and tremolite) and jadeite. Finally, new gem varieties are found from time to time. For example, tanzanite was found in the 1960's and tsavorite during the 1970's.

BEAUTY

The beauty of a gem may be inherent in the gem mineral species—such as clarity, color, or phenomena (stars, eyes, and the like)—or it may be brought out as it is fashioned by cutting and polishing. Faceting is the cutting of a stone to add faces (facets) or flat surfaces, generally with a regular geometric form. Because faceting enhances the brilliance of a stone by causing reflections of both front and back facets, those mineral varieties having superior clarity became important gem varieties. The faceting of a clear gem mineral produces a color phenomenon called dispersion. This is the ability of a particular gem mineral to break light up into a rainbow of component colors, as does a prism. This property is commonly known as "fire" and is well known in diamonds.

Color is a very important property of gems. There are many gem varieties of gem minerals based solely on this property. For example, the named color varieties of beryl include aquamarine (blue-green), chrysolite (yellow-green), emerald (intense green), heliodor (brownish-green), and morganite (pink or orange). Clarity of color determines much of the beauty and value of gemstones. A "play of color" is produced by various combinations of reflection, refraction, diffraction, and interference phenomena; it is well known from such gems as opal (opalescence) and labradorite (labradorescence). Some stones are selected entirely for their color. These may be opaque or nearly so and may or may not take a high polish. They include such stones as turquoise, jade, and many varieties of quartz, such as jasper, bloodstone, agate, and carnelian.

Some stones have inclusions that produce aesthetic effects, such as stars, eyes, spangling, or even pictures. Stars and eyes are produced by minute aligned voids or needles of minerals like rutile. If they are in one direction, they produce eyes, as in chrysoberyl and tourmaline. If they are in a hexagonal or in other star-shaped patterns, they produce stars, as in star rubies, sapphires, diopside, and garnet. Spangling results from variously oriented small inclusions of reflective minerals within the gem mineral, such as mica, rutile, or tourmaline.

DURABILITY

Durability of a gem mineral is its resistance to scratching or breaking, a necessity for an owner of jewelry because one does not want to damage a gem and thus markedly reduce its value simply by wearing it.

Hardness is the ability of a gem mineral to resist scratching. Mineralogists and gemologists rank minerals against a hardness scale of common minerals, called the Mohs hardness scale. From hardest to softest, the scale is 10 (diamond), 9 (corundum), 8 (topaz), 7 (quartz), 6 (orthoclase feldspar), 5 (apatite), 4 (fluorite), 3 (calcite), 2 (gypsum), and 1 (talc). The hardness of 7, the hardness of quartz, a very common mineral, is very important in the selection of gems or gem minerals. Quartz is a very common mineral, which means a stone with a hardness of less than 7 becomes easily scratched in our environment and therefore lacks durability. Most of our well-known gem minerals have a hardness greater than or equal to 7.

Resistance to breaking is referred to as tenacity. Most minerals are brittle, but native gold is malleable, ductile, and sectile.

Minerals break in two major ways called cleavage and fracture. Cleavage is the breaking of minerals parallel to more weakly bonded atomic planes found within the mineral and produces flat reflecting surfaces. Fracture is the random breaking of a mineral across atomic planes. Fracture along curved surfaces is called conchoidal fracture. Tenacity is directly related to the bond strength between adjacent atoms. Cleavage thus reduces tenacity, as in easily cleaved topaz, while an intergrown mat of fine crystals increases tenacity of a specimen, as in the case of the jades nephrite and jadeite. Cleaving was probably the first way to facet gems and still is very important in the shaping of rough diamonds before cutting and polishing.

VALUE

The value of a gemstone or a particular gem is based upon the interaction of many factors that can be divided into four major groups: beauty, rarity, quality of fashioning, and demand. Beauty includes such properties as color, quality of color, and clarity (freedom of inclusions, fractures, or cloudiness); it also includes phenomena such as quality of stars, eyes, or play of color. Rarity involves not only the rarity of the variety of the gem mineral but also the rarity of that quality of stone being used. For example, size influences the value because large stones are rarer than small stones; perfection influences the value because high-quality stones are less common than flawed stones. Rarity of the gem mineral can increase value or decrease value. Very common varieties of quartz, for example, may have less value than similar qualities and sizes of corundum because corundum is less common. Some gems, however, are so rare that people have not heard of them, and there is no demand except by connoisseurs and hobbyists, which in turn decreases value. Quality of fashioning includes degrees of perfection of symmetry and customizing of facet junction angles to the properties of the gem mineral species, which maximizes brilliance, color, and dispersion, and fineness or degree of polishing. Demand includes such issues as fashion, fads, and engagement ring and birthstone traditions. These associations, in general, increase value. Good or bad luck attributed to gemstones may increase or decrease their value, respectively.

The terms "precious" and "semiprecious" used with a gemstone name are misleading in assessing the value of the stone. "Precious" has been historically applied to gems such as diamonds, rubies, sapphires, opals, and emeralds, while "semiprecious" has been applied to such gems as aquamarine, tourmaline, chrysoberyl, citrine, topaz, and amethyst. Almost any gem mineral is found in a wide variety of qualities and therefore so are gems; thus, there are high-quality stones called semiprecious that are often more valuable than low-quality stones called precious. Unfortunately, the term "precious" has also been applied all too often to low-quality diamond, ruby, sapphire, opal, or emerald in order to deceive the buyer.

FORMATION OF GEM MINERALS

Gem minerals originate in the same way as all other minerals. They occur in four broad categories: igneous deposits (formed from molten rock—lava and magma), hydrothermal and pneumatolytic deposits (formed from hot water and steam, respectively), metamorphic deposits (formed by heat, pressure, and chemically active fluids with crystallization in the solid state), and sedimentary deposits (formed by processes of weathering, erosion, and deposition). Corundum is an example of a gem mineral that may be found under several conditions. A well-known igneous gem mineral formed from the cooling of molten rock is olivine (peridot), which is found in lava. Gem minerals found in plutonic rock (molten rock cooled within the earth) are diamond, various garnets, zircon, corundum, spinel, labradorite, orthoclase, and albite-oligoclase. Well-known gem minerals formed by hydrothermal and pneumatolitic processes are varieties of quartz (amethyst, citrine, crystal, smoky), varieties of beryl (aquamarine, morganite, emerald), various kinds of tourmaline, topaz, spodumene, and microcline. These minerals are commonly found in bodies or rock called pegmatites, or veins. Metamorphic rocks containing gem minerals form in the roots of mountain chains and/or adjacent to igneous bodies, where there are high pressures and temperatures. Well-known gem minerals formed here include corundum, kyanite, lazurite (lapis lazuli), various garnets, jade (both nephrite and jadeite), spinel, beryl, and chrysoberyl. Gem minerals found in sedimentary deposits may be divided into two major divisions: placer deposits (gravels that contain heavy minerals resistant to weathering and left by erosion and deposition of streams) and primary sedimentary minerals grown in the sediments. Placer gem minerals include diamond, corundum, spinel, tourmaline, topaz, various garnets, zircon, and chrysoberyl and may be associated with placer gold. Primary sedimentary minerals are mainly varieties of quartz (agate, chalcedony, onyx, sard, carnelian, petrified wood, heliotrope, bloodstone, jasper). Gem minerals may also result from weathering processes. A fine example is turquoise, which is produced by the weathering of copper-bearing hydrothermal veins under arid conditions.

Garnet, one of many gemstones, forms equidimensional crystals with twelve to thirty-six faces. (U.S. Geological Survey)

TOOLS OF THE GEMOLOGIST

The binocular microscope is one of the most useful tools to the gemologist, whether for making identifications, grading, or preparing a stone for lapidary work. These microscopes have relatively low magnification, but the twin eyepieces allow the gemologist to view the stone in three dimensions. Other magnifying instruments used include the hand lens and loupe, which are mostly used for preliminary inspections of gems. Properties of minerals that can be determined for identification purposes with the binocular microscope include hardness—by noting polish quality or wear characteristics—and crystallography, which can be determined from traces of cleavage on the unpolished girdle of the stone, noting single or double refraction, orientation of inclusions, and other qualities. Properties of minerals that can be determined for the grading of gems include clarity, quality of cutting and polishing, symmetry, and color. Properties of minerals that can be determined for lapidary purposes include crystallographic orientation, cleavage directions, and location of color variations and inclusions. Most gems are considered flawless when no inclusions, fractures, surface blemishes, or cutting flaws are visible to the naked eye.

The refractometer is also one of the most useful tools of the gemologist for purposes of gem identification. This technique makes use of the fact that each mineral species has a distinctive and characteristic index of refraction, or indices of refraction. Different minerals may have one, two, or three indices of refraction. A refractometer such as the Rayner or Duplex refractometer can measure the indices of refraction of a gem by use of the critical angle of light for that mineral, which is dependent upon the mineral's refractive index or indices. The index or indices of refraction are measured from flat facets of cut gems or crystal faces; an average of the indices may be obtained from a curved face or from any polished surface of a fine-grained aggregate, such as jade or chalcedony. Tables of refractive indices allow gemologists to compare results of their testing with those of known gem materials.

The polariscope is a simple yet very valuable instrument. It is a light source with two Polaroid plates mounted above and below the gem. The polariscope is used to distinguish between crystalline double refractive material (minerals whose optical properties vary with direction in the mineral) and singly refractive crystalline material or amorphous materials, and to identify doubly refractive fine-grained aggregates, such as jades and chalcedony. For example, ruby is doubly refractive and will change appearance when rotated in polarized light, whereas a similarly colored garnet will not.

The spectroscope may on occasion be a useful instrument. The hand-held types are difficult to use, and the table models are very expensive. They are generally not necessary for the identification of stones but are useful in confirming or determining the presence of metallic ions like iron or chromium. Light traveling through the mineral is separated into its color components in a rainbow of color by prism or diffraction. There will be dark lines on this rainbow, corresponding to light frequencies absorbed by atoms in the mineral. Genuine ruby will show absorption lines for chromium; red garnet does not.

ROLE IN ECONOMICS AND TECHNOLOGY

The study of gems or gem minerals interests the connoisseur, hobbyist, investor, and those whose business is the gem trade. Two reasons for a large public ownership of gemstones in Western society are the traditional diamond engagement ring in North America and the European use of the diamond as an investment. It is particularly in this market that most of the fraudulent practices in the gem and gem mineral trade exist. Investing in gems and gem minerals has been common for several reasons, including short-term and long-term gains, and as a hedge against inflation. Gemstones have also been valued for their portability, especially in unstable political situations. As with any investment, however, there is an element of risk. The degree of risk may vary with the gem mineral; for example, finding a large new deposit drops the price. Cornering of the market by a particular group forces the price up or down for some political or economic reason. Diamond pricing is probably the most structured and stable because it has been controlled by the De Beers Central Selling Organization (CSO) in London for the last hundred years. Nevertheless, prices for a top-quality, flawless one-carat diamond (round brilliant) fluctuated from more than $7,000 in 1977 through $50,000 to $60,000 in 1980 down to about $12,000 in 1985 to more than $17,000 in 1988 in a very volatile world market. Many of the African nations where diamonds are mined are prone to civil war, and diamonds are frequently used to fund those wars. Such diamonds, called "conflict diamonds," are a major humanitarian problem because they are often mined with forced labor and battles are frequently fought for control of mining areas. Many diamonds are now engraved by laser with serial numbers that can be used to verify they did not originate in conflict zones.

Because of their perfection, gems are used by scientists for research concerned with crystals and crystallization processes. They also have many technological applications, from ruby (corundum) lasers to quartz crystals in citizen-band radios and telephone communications equipment. Because of the large demand for these minerals, synthetic substitutes are generally utilized today for their technological applications. Mineral species, of which the gem-quality varieties may make up only a very small quantity, may also have other major non-gem-related industrial uses. The mineral diamond, for example, has use as a gem not only in transparent, attractively colored, and colorless varieties but also in translucent, nearly opaque, or highly included stones, which are not pleasing to the eye yet have important industrial cutting and abrasive uses. Much high-technology grinding, cutting, and polishing is based upon diamonds, from the bits that drill oil wells and cut metals and the saws that cut stone to the fine abrasives used to polish gems and lenses. The mineral corundum, the second hardest mineral known, not only produces specimens of ruby and sapphire but is also extensively used as an abrasive. Some of the abrasive uses, such as coarse grinding with emery, have decreased because of the introduction of relatively inexpensive artificial diamonds, but other uses, such as polishing with alumina, have increased. The mineral hematite is best known as the chief ore of iron, but it is also used for black gems and has become more fashionable in recent years. The jewelry trade knows it best for yet another use: jewelers' rouge, a finely powdered hematite used to polish metals and some gems. The gem mineral quartz, with many gem varieties—amethyst, citrine, smoky, rose, crystal, agate, carnelian, bloodstone—has many nongem uses, including glassmaking and the production of abrasives (sandpaper) and sand for sandboxes, beaches, and concrete.

Charles I. Frye

FURTHER READING

Baurer, Max. *Precious Stones.* 2 vols. Mineola, N.Y.: Dover, 1968. The original was published in German in 1896, went through several editions, and was translated by L. J. Spencer of the British Museum in 1904. Most of the information is still up-to-date, though some noteworthy gems have been found since publication, as well as new localities and new technologies developed. Written for both the hobbyist and the gemologist. The original German and English editions are collector's items.

Cavey, Christopher. *Gems and Jewels.* London: Studio, 1992. Cavey sets out not only to provide information about gems and jewels but also to correct many of the misconceptions concerning their study. Although the book includes many illustrations, only a few of them are in color.

Desautels, Paul E. *The Gem Kingdom.* New York: Random House, 1970. Not intended to be an exhaustive study of gems or gem minerals but instead touches on most aspects of the gem world using the Smithsonian Institution Gem and Mineral Collections for illustration. Might be considered more useful for gem appreciation than gemology. Excellent photographs and a lively text. For anyone, such as collectors, interested in gems, particularly those found at the Smithsonian.

Gubelin, E. J. *Internal World of Gemstones.* 3d ed. Santa Monica, Calif.: Gemological Institute of America, 1983. A lovely book of exquisite photomicrographs. Good information on the internal nature of gemstones and a classification of inclusions. Very useful to the gemologist or the jeweler wishing information concerning inclusions in gemstones. Also excellent for the hobbyist and collector.

Hall, Cally. *Gemstones.* 2d ed. London: Darling Kindersley, 2002. Introductory overview of gemstones and their properties that includes numerous color illustrations and an index.

Hurlbut, Cornelius S., Jr., and Robert C. Kammerling. *Gemology.* 2d ed. New York: John Wiley & Sons, 1991. A commonly available textbook on gemology, written for college-level students.

Klein, Cornelis, and Cornelius S. Hurlbut, Jr. *Manual of Mineralogy.* 23d ed. New York: John Wiley & Sons, 2008. An introductory college-level text on mineralogy. Discusses the physical and chemical properties of minerals and describes the most common minerals and their varieties, including gem mineral varieties. Contains more than 500 mineral name entries in the mineral index and describes in detail about 150 mineral species.

Kunz, George. *The Curious Love of Precious Stones.* Mineola, N.Y.: Dover, 1971. A delightful book about the mythology surrounding gems. Written at the high school level, this volume provides a wealth of information on the folklore of gems and was first published in 1913 by J. B. Lippincott. The original edition is a collector's item.

Liddicoat, Richard T., Jr. *Handbook of Gem Identification.* 12th ed. Santa Monica, Calif.: Gemological Institute of America, 1993. The best all-around text written for the gemologist, jeweler, or layperson who wishes to identify gemstones. Gems are listed in alphabetical order. Many tables of gem properties are included, and gem-testing equipment is described. Synthetics, imitations, and fakes are discussed, and information is frequently updated with new editions. High school level.

Manutchehr-Danai, Mohsen. *Dictionary of Gems and Gemology.* 3d ed. Germany: Springer-Verlag, 2005. At over one thousand pages long, this reference provides vital information for anyone entering the world of gemology. Contains 25,000 entries and thousands of figures, charts, and diagrams.

O'Donoghue, Michael. *Gems.* 6th ed. London: Robert Hale, 2008. Provides information on the properties and classification of gems as well as their applications, natural sources, and properties of synthetic gems. This text is designed to provide information for gemology students, jewelers, and gem collectors.

Rygle, Kathy J., and Stephen F. Pedersen. *The Treasure Hunter's Gem and Mineral Guides to the U.S.A.* 5th ed. Woodstock, Vt.: GemStone Press, 2008. Four-volume set provides information on locating and collecting gems, precious stones, and minerals from all regions of the United States. Content includes mining techniques, equipment, and safety. Locations are listed by state with information the minerals found there and public access. Although these volumes are specific to the United States there is useful information throughout on techniques, tips, and trivia.

Shipley, Robert M. *Dictionary of Gems and Gemology.* 6th ed. Santa Monica, Calif.: Gemological Institute of America, 1974. An international reference source for the gemological profession. Defines the vocabulary associated with the gem trade, including gem and gem mineral names, equipment names, and names associated with synthetics and other imitations. The glossary has more than four thousand entries.

Sinkankas, John. *Gemstones of North America.* Tucson, Ariz.: Geoscience Press, 1997. Recommended for those who wish to know more about North American gem mineral locations. Written for the hobbyist, at the high school reading level.

Webster, Robert F. G. A. *Gems: Their Sources, Descriptions, and Identification.* 5th ed. Boston: Butterworth-Heinemann, 1994. A technical college-level text that is an important reference for the gemologist or the senior jeweler. Better for sources and descriptions than for identification. Essential for the gemologist.

See also: Aluminum Deposits; Biopyriboles; Building Stone; Carbonates; Clays and Clay Minerals; Crystals; Diagenesis; Diamonds; Feldspars; Hydrothermal Mineralization; Metamictization; Minerals: Physical Properties; Minerals: Structure; Non-silicates; Oxides; Pegmatites; Radioactive Minerals; Silicates.

GEOGRAPHIC INFORMATION SYSTEMS

Geographic information systems (GIS) capture, manage, manipulate, and present geographical data. GIS involves a combination of hardware, software, and collated data to analyze conditions, trends, and events in the natural environment. It may be used to map and track natural resource, seismological, volcanic, meteorological, and climatological conditions around the world, and is an invaluable tool for the analysis of regional and global systems.

PRINCIPAL TERMS

- **computer-aided design:** software that generates high-resolution images that may be manipulated by the user
- **open source:** software in which the source code may be accessed by the public and modified without the application of copyright laws
- **optimization:** aspect of a software program that enables the user to modify it to conform to the user's needs
- **raster:** cell-based, high-resolution image that assigns data on surface conditions
- **rendering:** manipulated GIS image that shows data overlain on the original map image
- **spatial:** pertaining to space and geography
- **vector:** spatial point used in GIS to indicate surface features such as structures, bodies of water, and roadways

BASIC PRINCIPLES

Geographic information systems (GIS) are integrated devices used to map natural events, trends, and conditions. GIS employs a number of different and separate hardware and software tools, such as satellite and aerial sensors and cameras, which are specialized computer databases and other systems.

GIS software employs a geographic reference, such as a digitized map, to study a given region. Depending on the type and complexity of the scientific pursuit, the GIS then applies one or more layers, applying the compiled data to that map. The resulting image (or rendering) provides a composite of the area being studied according to the respective scientist's thesis.

Three general areas make up an integrated GIS program. The first is a series of numerical algorithms, which enable researchers to assign data to values on the geographic reference. The second is a table of statistics. In GIS research, this table is frequently extensive, comprising a large volume of data. The third is optimization: In the application of GIS to a specific area of focus, the user or users must be able to modify the software appropriately to conform to the research.

GIS is utilized in a number of scientific disciplines and fields, including the social sciences (such as sociology, political science, and history), engineering, epidemiology, and the natural sciences. For the purposes of Earth science, GIS has a wide range of applications of value. Scientists using this technology can map changes in the earth's topography, follow climate shifts, analyze water resources, and study the impact of volcanic eruptions.

GIS focuses on the collection of data from a specific geographical location. This information, when combined with data from other locations, can help scientists create a composite of a particular concept. For example, GIS is used in the long-term study of the melting of the world's glaciers. It also can be used to map the destruction caused by volcanic eruption or to monitor pollution in a given area.

HISTORY

GIS has its roots in cartography. Indeed, scientists maintain that GIS could not exist without knowledge of the way maps are (and have been) created. One of the earliest examples of GIS comes from the middle of the nineteenth century, when British physician John Snow began investigating the spread of cholera in London.

Snow hypothesized that an outbreak in 1854 was caused by a bacterium in contaminated water. He drew a map of London and, using data from the hospitals in the area, plotted on the map the locations in which cholera deaths had occurred. His map revealed that most victims lived near a particular water pump. Upon his recommendation, that pump was shut down, leading to an end to the outbreak.

GIS development continued much in the same manner as it did in the mid-nineteenth century. Before the advent of computer technology, one of the more popular methods employed by scientists when studying natural events and trends in a geographic location was to use an existing map of a region. When analyzing data, scientists would simply lay a clear plastic sheet over the map, showing the points of interest. GIS continued to develop along these lines; even today, many GIS systems require the use of computer-aided design (CAD) software, a system that scans high-resolution hard copies of existing maps. The images derived from these scans are used as the basis for overlapping images suited for a given study.

As a computer-based system, GIS was largely developed for governments. However, by the start of the twenty-first century, the private market for GIS software and hardware had developed quickly. By 2000, the private sector spent between $15 billion and $20 billion on GIS. Furthermore, the use of GIS programs has grown worldwide, with software and technology available and affordable for even the most austere budgets.

RASTER AND VECTOR DATA

Two types of reference data are utilized on a GIS rendering, in addition to a separate data table. The first of these reference data is a set of indicators known as vectors. Vector data appear in three general manifestations: polygon, line, and point.

On the image, a polygon (a multisided geometric shape) represents a city or other boundary (such as a body of water or forest region). A line, or arc, depicts linear features, such as streets, rivers, railways, and trails. The third form of vector is the point. This type of data identifies singular points of interest, such as buildings and bridges. Vectors simplify locations on a given rendering, reducing clutter on the image. The user can obtain an enhanced view with the vectors, which can show more detail. Adjacent structures or physical features may come into view with the zoom.

The second type of reference data in a GIS image is the raster. A raster focuses on surfaces rather than features. They are, in essence, high-resolution images (such as aerial and satellite photographs) comprised of a grid of pixels. Rasters appear in two forms: continuous and discrete. Continuous rasters vary consistently with no clear boundaries. Examples of this type of cell include surface temperature and geographical elevation. Discrete rasters, however, have boundaries with attached categories, classes, and descriptions. Some examples of discrete rasters are data for population density and political boundaries.

Advantages and disadvantages come with the use of rasters and vectors. Vectors, for example, use less of a computer system's memory because of their relative simplicity and low volume of data. However, because of their simplicity, vectors can be easily adjusted for scale. Rasters, on the other hand, use a complex system of values, allowing for a number of different operations, including mathematical calculations. Then again, the high resolution and complexity of rasters require the user's computer system to contain a significantly larger memory to perform any spatial analysis.

The ease with which vector data are managed in GIS analyses has made vector analysis a preferred choice among many researchers. Software developers are, as a result, introducing GIS programs that can convert reference data from raster to vector formats. Still, a large number of GIS programs have various raster-vector conversion programs, enabling the user to access and manipulate a range of maps and images.

SPATIAL ANALYSIS

GIS accesses and analyzes spatial (or geographic) data. In spatial analysis, the data for each map layer are provided in an adjoining table. Each data point is assigned a value, which is typically performed through numerical algorithms and statistics. Analysis considers the relationships among each point. In many cases, these points lie within a larger feature (such as a building or geographic area). Spatial analysis also may require assessing the relationships among two or more points in different geographic locations.

The data provided on a given area may not automatically correspond to a predesigned map, or it may be anticipated that the map will change in light of certain conditions. For example, satellite-based thermal images may not delineate a mountain or a structure. Additionally, a scan of a shrinking rainforest may require multiple images and points of reference to demonstrate a chronological change. In such cases, it is important to generate an appropriate map. In these situations, GIS software typically calls upon modeling and forecasting programs to create a base image for spatial analysis.

GIS-generated spatial analysis helps scientists collate and utilize the attributes of a given reference point (or series thereof). Such data help researchers develop clear illustrations of a given geographic area in a far more comprehensive manner than that which is uncovered simply by flying over or driving through a target area.

In 2011, for example, scientists in Israel conducted research on the availability of drinking water in nearby Gaza Strip. The project entailed the acquisition of data on rainfall, population increases, and the number of reservoirs available in Gaza, and the generation of a useful map to serve as the basis for the spatial analysis. Using GIS, this project created an analytical framework that identified the areas of highest consumption from groundwater reservoirs. The models produced by this GIS application also provided accurate short-term forecasts for groundwater usage in this area.

MAPPING TOPOGRAPHY

Until the mid- to late twentieth century, scientists were limited in their ability to study the earth's topography. Although aerial photographs and ground-based analysis prove effective in some areas, scientists were until this period unable to study a wide range of locations and trends in the natural environment.

One such location is the ocean floor. Although hidden from plain sight, the ocean floor contains fish habitats and provides indications of the impact of humanity on the natural environment. To better survey the suboceanic environment, scientists are increasingly calling upon GIS to map the ocean floor.

For example, scientists at Oregon State University have utilized GIS to create a surficial geologic habitat map. This map is seen as critical for the region and its fishing industry, which has been in crisis because of concerns of depleted fish populations. Using a series of different mapping software that employs sonar data and density information, and seismic imaging and other databases and systems, researchers can now effectively map the floor off the coast of Oregon with three-dimensional images and other relevant applications.

In another example, GIS technology was employed to study the bedrock of Oklahoma. Here, researchers used a digital elevation model to map the sandstone and shale that compose that state's bedrock. Understanding the geology of this area is important for that state, with implications in its agricultural and civil engineering sectors.

NATURAL RESOURCES

Applying GIS to the study of the earth's natural resources has been evolving since the late 1970s. However, this endeavor has seen increased success

with the development of computer modeling and satellite-based technologies. In rural areas of North America and Pakistan, for example, GIS has proved effective in mapping forest resources. The use of such technology is particularly useful for studying the decline of forest cover from overdevelopment and unsustainable logging practices.

In an era in which sustainable development and protection of natural resources, such as drinking water and agriculture, are of increasing importance, GIS is a highly effective vehicle for analyzing trends and, ultimately, creating environmentally friendly policies. In developing countries and regions, this issue is particularly critical. For example, the area around Turkey's Tuz Lake (the second-largest lake in Turkey and one of the largest salt lakes in the world) is considered an important region for agriculture and for many different forms of animal life.

In 2010, scientists began to develop a strategy for land use in the area, considering the area's delicate ecological balance and such elements as soil and water quality, climate, and land suitability. Using GIS mapping, researchers created an overlapping series of maps of the region accounting for each of these ecological factors. The result of the study was a comprehensive land-use framework that outlined sustainable agricultural use, effective waste management, and the protection of the lake area's wildlife.

SOFTWARE

GIS involves integrated and open-source software technologies. Among these systems is CAD software.

CAD is used to design a high-resolution image that can be manipulated by zooming in and out, adding overlapping images, and viewing segments of an image from multiple angles. Although many CAD systems are built into appropriate hardware, a growing number of CAD software programs may be obtained independently, including through the Internet.

In addition to CAD, GIS requires the integration of spatial indexing software. This type of system enables spatial data to be stored, modified, and analyzed within the database. Spatial indexing software is used to collate statistics and data and assign them to an appropriate vector and raster. Spatial indexing and development software must be compatible with the CAD program in use, as the resulting reference data (vectors and rasters) will be overlain on the main composite image. Like CAD, this software is now available on the Web. Furthermore, because GIS-based studies commonly feature a large amount of data, it is important that the integrated system in use includes a high-volume database application.

In addition to Web-based software, another innovation useful to GIS researchers is mobile GIS applications. Because the memory storage of smartphones, iPads, and other mobile systems is ever-increasing, researchers can now access and utilize GIS software from their mobile devices.

IMPLICATIONS AND FUTURE PROSPECTS

GIS has evolved so extensively that it is being applied across a broad spectrum of scientific disciplines. Social scientists use GIS to map and analyze demographics and residential populations. Civil engineers employ GIS to assess the viability of constructing roads, bridges, and other structures. Public health professionals use it not only to track diseases but also to analyze potential health threats and benefits for populations.

The evolution of GIS also has been aided by the parallel evolution of sensory hardware. In terms of Earth science, advances in thermal imaging, radar, seismography, and other technologies add a number of additional resources to spatial analysis. Furthermore, the accessibility of the global positioning system and other satellite-borne hardware systems now enable scientists to study a geographic location with ever-greater detail. Many of these advanced systems can "see" through weather, foliage, and even topographical features, all of which were previously considered obstacles to spatial analysis.

As with other scientific disciplines, it is likely that GIS will continue to have relevance for researchers in the field of Earth science. The private sector market for GIS continues to grow because of its applicability to so many fields. Government and nongovernmental research on such issues as sustainable development, climate change, and natural resource management will only continue to drive this interest.

Michael P. Auerbach

FURTHER READING

Bolstad, Paul. *GIS Fundamentals: A First Text on Geographic Information Systems*. St. Paul, Minn.: Eider Press, 2007. Describes the many scientific applications of computer-based tools for spatial analysis. In addition to providing a comprehensive overview of GIS and its applications, discusses future trends in GIS.

Longley, Paul A., et al. *Geographic Information Systems and Science*. 3d ed. Hoboken, N.J.: Wiley, 2010. A comprehensive, authoritative text that introduces the major components of GIS and explains how they are applied in the twenty-first century. The authors describe the many issues and challenges facing GIS and how GIS is designed to address such obstacles.

Ormsby, Tim, et al. *Getting to Know ArcGIS Desktop*. 2d ed. Redlands, Calif.: ESRI Press, 2010. A tutorial on ESRI's ArcGIS product and a textbook on how GIS is integrated and applied. Describes how spatial analysis is conducted using this particular system, providing information that helps any individual seeking to understand the many uses for GIS.

Pick, James B. "Geographic Information Systems: A Tutorial and Introduction." *Communications of AIS* 14 (2004): 307-331. Provides an overview of GIS's basic structure and concepts. Discusses the application of state-of-the art technologies to GIS usage, including wireless and Web-based access.

Sei-Ichi, Saitoh, et al. "Some Operational Uses of Satellite Remote Sensing and Marine GIS for Sustainable Fisheries and Aquaculture." *ICES Journal of Marine Science/Journal du Conseil* 68, no. 4 (2011): 687-695. Describes how GIS may be applied to the study of the ocean. Specifically discusses how data from satellite-based remote-sensing systems are utilized in GIS analysis to study current and ideal, sustainable fishing activity and overall climate changes.

Thrall, Grant Ian. "MapInfo Professional 8.5: Internet-Enabled GIS." *Geospatial Solutions* 16, no. 9 (2006): 36-39. Introduces an innovative, Web-based GIS system and describes the advantages of Web-based software for GIS users, including easy access to a wide range of integrated services and bundled data programs.

Wade, Timothy G., et al. "A Comparison of Vector and Raster GIS Methods for Calculating Landscape Metrics Used in Environmental Assessments." *Photogrammetric Engineering and Remote Sensing* 69, no. 12 (2003): 1399-1405. An overview of vector- and raster-based GIS analysis conducted for the purposes of environmental studies. Using vector and raster studies focusing on watersheds in the mid-Atlantic region of the United States, explores the viability of each data reference approach for conducting environmental assessments.

See also: Earth Science and the Environment; Earth System Science; Geologic Settings of Resources; Land Management; Land-Use Planning; Land-Use Planning in Coastal Zones; Remote Sensing.

GEOLOGIC SETTINGS OF RESOURCES

Buried organic matter accumulates in porous and fractured sediment, forming reservoirs of hydrocarbons that can be harvested for commercial use. Petroleum and natural gas are hydrocarbon substances that form from the accumulation, decomposition, pressurization, and heating of organic matter. Minerals and metallic ores are other resources that are harvested from the earth's crust by mining. Deep-sea hydrothermal environments also include metallic ores and hydrocarbon reserves that are later available at the continental surface because of the shifting of ocean systems.

PRINCIPAL TERMS

- **carbonate:** material that contains carbon as a key structural or chemical ingredient
- **evaporite:** type of mineral or rock that results from the accumulation of saline sediments in environments where an aqueous solution is evaporating from the surface
- **hydrocarbon:** organic compounds consisting of hydrogen and carbon molecules
- **hydrothermal vent:** chamber or fissure in the oceanic crust that releases water into the ocean that has been heated by geochemical processes arising from magmatic substances beneath Earth's surface
- **metallic ore:** sedimentary rock containing minerals rich in metals and often used in the harvest and derivation of metal for industrial processes
- **petroleum:** liquid or semiliquid hydrocarbon substance that results from the decay of organic material within the earth's surface as it is heated and pressurized by geochemical forces
- **sediment:** material consisting of fine portions of rocks and other minerals dissolved from larger rocks and carried within a gaseous or liquid fluid
- **seep:** an area where water, gas, petroleum, or another fluid emerges from subterranean chambers onto the surface of the earth
- **seismic waves:** low-frequency acoustic energy that travels through solid rock and other structures
- **tectonic movement:** gradual movement of plates, which are portions of the earth's crust and lithosphere connected at deep sedimentary cores and that move as units

Hydrocarbon Reservoirs and Traps

Petroleum and natural gas are hydrocarbon-rich fuel sources harvested from geological sediment. Hydrocarbon fuels form as organic matter decays and sinks through the upper layers of the earth's crust, infiltrating the rocks beneath.

Hydrocarbon fuels form only under certain specific conditions. First, organic matter must be buried during decomposition, preferably by soft sedimentary material such as sand and mud. After burial, the carbon residue of decomposing organisms is subjected to increasing pressure, while gases and heat emanating from decomposition raise the temperature within these pockets of buried material. As this occurs, the resulting petroleum and gas begin rising to the surface because they are less dense than the overlying medium.

Petroleum that cannot return to the surface because it is sealed by the overlying rock forms a petroleum trap. Traps are the source of most of the world's petroleum and result from tectonic movement of the earth's crust, which causes physical deformations in the rock surrounding a developing petroleum reservoir. The most common petroleum trap is formed when tectonic movement causes the sediment to buckle and fold. The underlying petroleum moves toward the surface because it is more buoyant than the underlying rock but cannot reach the surface because the overlying sediment is impermeable to the oil.

Hydrocarbon Exploration

Hydrocarbon exploration involves a variety of methods to locate petroleum and other hydrocarbon fuel reservoirs. In some cases, petroleum or natural gas may break through to the surface of the crust under natural conditions, creating a seep, which is a rupture in the earth's surface that slowly releases oil or natural gas, or both. When a seep is located, engineers can utilize equipment to contain and further develop the seep, extracting hydrocarbon fuel from the connected reservoir.

Engineers also locate reservoirs by using seismic surveys. Seismic technology involves using machines that can propagate seismic waves, which are waves of acoustic energy that can travel through solid rock. Seismic waves are usually propagated on a low frequency, as these types of force waves are more suitable to traveling through solid materials and will travel significant distances before diminishing. As seismic waves move through the sediment, they are also reflected back to the source of the wave as they collide with materials under the surface. By measuring and analyzing the reflection of seismic waves, engineers can determine the location of pockets within the crust that might contain certain types of materials. This method is therefore used to find hydrocarbon traps that can then be explored and harvested by mining.

Geologists can increase their chances of finding new reservoirs by analyzing existing harvesting operations. For instance, a number of productive reservoirs are contained within dolomitic limestone deposits, a sedimentary rock that forms from ancient marine environments. The shallow marine environments containing dolomitic limestone also support the formation of petroleum reservoirs because environmental conditions favor the burial and compression of large quantities of organic matter. In addition, dolomitic limestone is a porous rock that lends itself to fracturing, thereby reducing density and providing space for the development of a hydrocarbon trap. Given this, engineers and geologists can seek rocks and minerals indicative of dolomitic limestone deposits or other types of rock that develop in similar environments; engineers can then concentrate exploration efforts in these areas, increasing the chance of finding productive reservoirs.

UNCONVENTIONAL PETROLEUM RESOURCES

Conventional oil traps and reserves compose a finite resource that will be exhausted given current trends in human hydrocarbon consumption. Engineers have begun looking into alternative methods for finding and harvesting additional petroleum and natural gas. Unconventional petroleum exploration is too cost prohibitive and technically complex to constitute a significant alternative to conventional sources, but research is ongoing to find new ways of developing alternative systems to supplement existing petroleum sources.

Oil sands are deposits of petroleum blended with find sedimentary material. Harvesting petroleum from oil sands requires a complex and expensive process that separates and removes the hydrocarbons contained within the surrounding sediment.

The process is vastly inferior to removing petroleum from conventional traps because it requires a greater initial investment of energy to power the purification process and uses a large amount of water, which becomes polluted during the process.

Oil shale is another potential source of hydrocarbons. Oil shale comprises collected and extracted petroleum contained within porous shale rock, a type of sedimentary rock that contains hydrocarbons tightly bound in the structure of the rocks. Through combustion, the petroleum within shale can be extracted, though combustion is far less efficient and produces greater pollution than standard methods of petroleum harvesting.

MINERAL AND ORE RESOURCES

Minerals used in industrial and chemical processes are harvested from a variety of geological settings. Among the most common types of mineral resources are metals, including copper, manganese, iron, magnesium, platinum, and gold.

These minerals are harvested through mining and by chemically processing various rocks to remove and purify metallic deposits. The precious metals, including gold, platinum, and silver, are mined and refined for use in the construction of cosmetic items and jewelry; they also have other applications. Nonprecious metals, such as iron, lead, and nickel, are commonly used in industrial processes and in the production of consumer items.

Most minerals and ores are harvested by mining, which is the process of removing sediment, rock, and other materials from the earth's crust. Surface mining is used to harvest sediment from the upper layers of the crust and surface, while subsurface mining utilizes tunnels extending into the lower layers of the crust to harvest materials that are found at greater depths. Various metallic ores and other minerals are found at both the surface and lower depths, so both types of mining are important for retrieving minerals and ore.

Many metallic ore deposits, like petroleum deposits, result from ancient marine ecosystems where metallic compounds gathered at the bottom of a sea or ocean and were subsequently concentrated as the water receded. Iron, for instance, is harvested largely from banded iron deposits, which are rocks that formed in prehistoric seas, such as the Western Interior Seaway, which once covered large portions of North America.

Banded iron rocks consist of layers of iron-rich minerals, like magnetite, alternating with layers of other, nonmetallic sedimentary rock. Most banded iron deposits are found in sediment that was once part of prehistoric oceans. Geologists believe that similar iron-rich rock deposits are forming on the ocean floor in modern seas. Much of the iron used in construction and other industrial applications is refined and processed from banded iron ore deposits.

As mentioned, dolomite and dolomitic limestone are often found associated with petroleum reserves. These carbonate rocks are also a source of other minerals used in industrial applications. Dolomite, for instance, contains high levels of magnesium bound into the mineral structure of the rock. Chemical processing can separate this magnesium, which can then be processed to yield magnesium metal for a variety of applications.

Many dolomite deposits also contain a layer of bauxite, an aluminum-rich sedimentary rock that is used to provide aluminum for industrial and manufacturing processes. Bauxite provides the majority of the world's aluminum and is generally harvested by strip mining, a type of surface mining that strips the top layer of sediment from the crust. Bauxite forms in sedimentary environments, where aluminum becomes blended with various silicon-rich rocks, like granite and shale. To remove and process the aluminum within bauxite, engineers subject samples of bauxite to a heated, acidic environment, thereby melting and separating the aluminum in the rock.

Copper and lead, two of the most useful metals for industrial applications, are generally harvested by strip mining areas with high concentrations of metallic ore. Lead is most commonly derived from galena, a naturally occurring form of lead sulfide, whereas copper is extracted from chalcocite, an ore that forms from copper sulfide. In both cases, rocks are harvested through strip mining operations and then subjected to intense heat and chemical environments that melt and separate the metals in the ore.

RESOURCES AND HYDROTHERMAL VENTS

Many metallic ores derive from ongoing marine geologic processes that lead to the buildup of metallic ore through millions of years. Hydrothermal vents, which are pockets of trapped water that develop under the surface of ocean sediment, are thought to be important in the formation of many metallic rocks.

The water and gas trapped within the vents are heated by volcanic material and become enriched with compounds dissolved from surrounding rock. Eventually, this heated water erupts to the surface, producing a jet of superheated solution. Water leaking into the hydrothermal chamber from elsewhere fuels the system, leading to jets that may erupt multiple times each day for thousands of years.

As water erupts from the vent's funnel, it meets the cold water of the surrounding oceans, facilitating chemical reactions that cause minerals and metals to precipitate and fall to the surface surrounding the vent. In millions of years, waves of deposited minerals and metals develop into thick deposits that may contain many kilograms of metallic ore. Seafloor spreading then distributes these mineral deposits around the ocean. As the oceans shift and shrink in response to tectonic movements, portions of former oceanic sediment are revealed at the surface. Beneath the top layer of sediment, the earth's crust often contains sediment that was once part of marine environments, which may include sediment derived from hydrothermal vent systems.

Deep ocean environments also are rich in mineral resources but are difficult to explore because of the pressure and other environmental factors that make it difficult for divers or diving technology to operate at these depths. One example can be found in the manganese nodules that cover much of the ocean floor around hydrothermal vent environments.

Manganese nodules develop over thousands of years from the accumulation of manganese-rich sediment released from the vents. The chemical nature of this metallic substance and the activities of organisms maintain these nodules at the surface of the ocean, despite the deposition of further sediment. Though there is no system for efficiently harvesting manganese from deep oceanic environments, the deposits represent a large, untapped reserve of manganese that may be harvested in the future.

Micah L. Issitt

FURTHER READING

Evans, Anthony M. *An Introduction to Economic Geology and Its Environmental Impact.* Malden, Mass.: Blackwell Science, 1997. Introductory geological textbook discussing the economic uses of minerals and geologic resources and the environmental impact of harvesting geologic materials worldwide. Contains detailed discussions of mining in the petroleum industry and the harvest of coal and metallic minerals and rocks.

Hyne, Norman J. *Nontechnical Guide to Petroleum Geology, Exploration, Drilling, and Production.* 2d ed. Tulsa, Okla.: Pennwell, 2001. General introduction to petroleum mining and processing. Contains references to numerous locations utilized in the production of petroleum and their geological history.

Monroe, James S., Reed Wicander, and Richard Hazlett. *Physical Geology.* 6th ed. Belmont Calif.: Thompson Higher Education, 2007. General text in geologic sciences written for the introductory student of Earth science. Contains references to geologic settings commonly used to harvest petroleum, natural gas, coal, and other mineral resources.

Plummer, Charles C., Diane H. Carlson, and David McGeary. *Physical Geology.* 13th ed. Columbus, Ohio: McGraw-Hill Higher Education, 2009. Introductory text covers the basic theories, research practices, and principles of geology. Contains information on petroleum mining and processing and references to numerous other types of mineral resources and their industrial uses.

Pohl, Walter L. *Economic Geology: Principles and Practice.* Hoboken, N.J.: Wiley-Blackwell, 2011. Advanced textbook outlining the economic use and development of geologic research. Contains discussions of coal, petroleum, and natural gas harvesting and processing and detailed discussions of the harvesting and processing of metallic minerals.

Thomas, Larry. *Coal Geology.* Hoboken, N.J.: John Wiley & Sons, 2002. Detailed technical textbook covering all major aspects of coal formation, geologic structure, geologic setting, and its mining and usage in the production of energy. Chapter 7 contains a detailed discussion of the global setting of coal resources and the processes used to harvest coal in various environments.

Wenk, Hans-Rudolf, and Andrei Bulakh. *Minerals: Their Constitution and Origin.* New York: Cambridge University Press, 2004. Intermediate-level textbook covering the basic structure, formation, and mining of minerals. Includes a discussion of economically important minerals and their functions in industrial applications.

See also: Aluminum Deposits; Chemical Precipitates; Dolomite; Evaporites; High-Pressure Minerals; Hydrothermal Mineralization; Iron Deposits; Manganese Nodules; Oil and Gas Distribution; Unconventional Energy Resources; Water and Ice.

GEOTHERMAL POWER

Geothermal power, having its source in the earth's internal heat, offers a form of energy used in areas of the United States and other countries. Although limited by current technology, the earth's heat as a power source offers immense resources, high versatility, and cost-effectiveness.

PRINCIPAL TERMS

- **binary cycle:** the process whereby hot water in the primary cycle gives up heat in a heat exchanger; a fluid such as isobutane in the secondary cycle absorbs heat, is pressurized, and drives a turbine generator
- **direct or single flash cycle:** the process whereby hot water under great pressure is brought to the surface and is allowed to turn, or "flash," to steam, driving an electrical turbine generator
- **double flash cycle:** the process whereby two flash vessels are employed in cascade, each operating a turbine to extract more power
- **hydraulic fracturing:** the underground splitting of rocks by hydraulic or water pressure as a means of increasing the permeability of a formation
- **hydrostatic pressure:** pressure within a fluid at rest, exerted at a specific point
- **hyperthermal field:** a region having a thermal gradient many times greater than that found in nonthermal, or normal, areas
- **permeable formation:** a rock formation that, through interconnected pore spaces or fractures, is capable of transmitting fluids
- **thermal gradient:** the increase of temperature with depth below the earth's surface, expressed as degrees Celsius per kilometer; the average is 25 to 30 degrees Celsius per kilometer

GEOTHERMAL PHENOMENA

Geothermal power, as evidenced by hot springs, steam vents, geysers, and volcanic activity, is the earth's inner heat escaping through faults and weak spots in the crust. Temperatures below the surface vary from one location to another and depend upon the depth and temperature of the heat source and ability of the subsurface rocks to conduct heat. Generally, geothermal water is not sufficiently hot to produce electricity but still can be used in a variety of industrial and agricultural applications.

Most of the heat energy escapes through the crust at lithospheric plate boundaries near young volcanic centers, such as those in Hawaii, Alaska, and the western United States. In the latter region, most of the geothermal hot spots are in Nevada, which has more than nine hundred hot springs and wells. This region is the Basin and Range geologic province, with a thin continental crust traversed by a system of north-south faults, allowing an easy path for hot water to percolate to the surface. By way of comparison, the greater part of the earth's surface is nonthermal, having temperature gradients from 10 to 40 degrees Celsius per kilometer of depth. It is important to keep in mind, however, that even in nonthermal areas there is a great amount of heat several kilometers below the surface. The hot rocks at these depths, although not permeable, could be accessible except for the cost of extracting heat of this grade. The presence of a geothermal field is not always marked by the surface thermal activity. Excellent fields have been detected in places completely devoid of any surface thermal manifestations.

GEOTHERMAL ENERGY RESOURCES

Geothermal heat currently can be commercially exploited only in regions where hot pore fluids circulate within permeable formations, and can be reached by cost-effective drilling. The geothermal energy resources now in use or under experimentation are dry steam, hot water convection systems, geopressured systems, and hot dry rocks. For temperatures of 200 degrees Celsius or higher, the most economical and easily used source is dry steam—so called because the incredibly hot pure vapor lacks droplets—but this accounts for only 0.5 percent of all U.S. geothermal resources. The hot water convection systems that make up 10 percent of the world's geothermal resources are heated by magma sources near the surface, thereby transferring heat to the surface rocks. Water that is confined within the rock layers is heated above the boiling point, but pressure from surrounding rocks prevents the water from becoming steam. Water heated by molten magma may reach temperatures of 300 degrees Celsius or more.

Geopressured fields do not fall within the general classification of geothermal fields, as they occur in non-thermal areas. These fields occur at depths up to 6,000 meters, where temperatures range from less than 93 to more than 150 degrees Celsius. The hot water is pressurized in excess of the hydrostatic values encountered at that depth; that is, the water is sealed in the rocks and compressed. These very high pressures are thought to be caused by gradual subsidence along faults of rock that has trapped pockets of water below and between alternating layers of sandstone and shale. Geopressurized fields produce three types of energy: thermal, as a result of the fluid temperatures; hydraulic, as a result of the high excess fluid pressure; and chemical, as a result of the caloric value of the methane gas dissolved in the water. The immense energy contained in geopressurized fields is not at this time fully attainable because of their great depths below the surface and the undeveloped technology necessary to exploit this source.

All of these geothermal systems utilize processes of groundwater circulation whereby drillers merely need to construct a well to reach the water table. If nature does not oblige, drillers attempt to inject water in the ground and create fractures in the rock strata using explosives. The fractures, in turn, provide pathways for the flowing water to pick up heat from the surrounding rock. Hot dry rocks are found at moderate depths by proven methods, but a far greater number lie at depths beyond which drilling is not cost-effective. Many believe that the large-scale extraction of energy from hot dry rock could have significant long-range payoffs. Considering regions of thermal gradients of 40 degrees Celsius or more for each kilometer of depth, it is estimated that these heat sources could provide up to ten times the heat energy of all coal deposits in the United States.

A small-scale geothermal resource is heat stored near the surface, which can be extracted by a heat pump for heating a building during the winter. The same system usually can be reversed for cooling in the summer. Heat pumps are most suitable for climates with moderate winters. In this case, the source is mostly solar heat absorbed by the earth's surface, not heat from the deep interior of the earth.

GEOTHERMAL FACILITIES

The direct flash system is used at the Geysers, the world's largest geothermal facility, located 140 kilometers north of San Francisco. In 1999, the Calpine Corporation, which operates the Geysers facility, announced its capacity to be at more than 1,000 megawatts. Commercial geothermal power has been produced at the Geysers since 1960. Steam at these facilities is first allowed to pass through centrifugal separators, which remove rock particles that may damage turbine blades. The steam travels to the turbines by means of insulated pipes.

The United States' largest geothermal resource is probably located in the Imperial Valley of California. The valley itself is flat but covers six known geothermal resources that have thermal fluid temperatures ranging from 120 to 330 degrees Celsius, circulating in layers of porous rock between 1,200 and 6,000 meters below the surface. In 1997, fourteen plants with a combined capacity of more than 400 megawatts were in operation in the Imperial Valley.

A 10-megawatt demonstration plant using the direct flash method is located in Brawley, California. Water at temperatures above 200 degrees Celsius is pumped to the surface under pressure and routed through a series of pathways, which reduces the pressure and allows some of the liquid to vaporize or "flash boil." The steam, constituting 20 percent of the fluid, is expanded through a standard steam turbine. A problem with the Brawley geothermal fluid is that it contains 15 percent total dissolved solids, which can choke a three-inch pipe diameter to half an inch after 100 hours of operation. Mineral buildup, or scaling, has been a major engineering challenge at the first hot-water power plants.

FACILITIES UTILIZING BINARY SYSTEMS

Located less than 2 kilometers away, another demonstration plant will be the first commercial facility employing a binary cycle system. In the binary system, pressure prevents the water from flashing to steam. Instead, the hot brine is allowed to flow through a heat exchanger that is surrounded by a heat exchanging fluid (commonly isobutane, which vaporizes at -11 Celsius). The fluid expands and under high pressure drives a turbine. By keeping the brine under high pressure when it leaves the turbine at a temperature of 70 degrees Celsius, engineers believe that they can avoid the scaling problems that clog pipes. The brine is pumped to a well injection station 4 kilometers away, on the edge of the Heber geothermal reservoir. It is believed that brine returned to the ground will be geologically recycled and available for future use.

The Imperial Magma test facility in the East Mesa, California, geothermal field is a binary system that uses brine at a low to moderate temperature with two hydrocarbon working fluids, each passing through a heat exchanger. Heat from the East Mesa brine first passes hot brine at 180 degrees Celsius to vaporize isobutane, which will run a 12-megawatt turbine. The brine cools and moves on to a second-stage heat exchanger, which vaporizes the propane at a lower temperature and rotates a 2-megawatt turbine generator. The second turbine, operating at a considerably lower temperature, increases the efficiency of the system.

The appeal of the binary system is that it gains higher efficiency with moderately low-temperature brines. For high-temperature geothermal fields, engineers believe that using working fluids with lower boiling points will permit engineers to extract energy from the fluids as the brine temperature drops over time. This will help extend the production life of some fields, since most must produce for thirty to thirty-five years to be economical. Another appeal is that a well-designed binary power plant can operate with virtually no emissions. Geothermal systems at low temperatures operate at a lower efficiency than conventional power plants, which typically operate at high temperatures.

GEOTHERMAL EXPLORATION

Geothermal exploration involves the teamwork of specialists versed in a variety of disciplines. Geothermal explorers in general will have several objectives in mind: They must find likely locations of geothermal fields and decide whether the field located has a sufficient source of heat and whether the field is steam- or water-dominated. Of the sources of exploitable geothermal energy, the high-temperature fields are the most promising commercially. These fields are almost always located in young orogenic regions or mountain belts where there has been recent volcanism.

The task of the team's geologist is to construct as accurately as possible a model of the thermal region's geological structure and to predict promising drilling sites. This is accomplished to a great extent by surface mapping and the study of the tilt of rock outcrops. Much of the model is deduced by direct observations, which are not generally possible below the surface. By studying hot-spring deposits, the geologist

can estimate approximate subsurface temperatures.

The function of the hydrologist is to work closely with the geologist and to determine the paths that water will follow underground through geologic strata and within the boundaries of the geologist's model. The hydrogeologist studies the gradients, porosities, and permeabilities of the various geological formations and may be able to offer a reasonable explanation for the thermal fluids reaching permeable zones in the field and, from that point, how they escape to the surface.

The geophysicist's task in geothermal exploration is to determine as accurately as possible the physical properties of the subsurface and to detect anomalies. A geothermal field with large volumes of steam and hot water in permeable rocks will likely appear anomalous when compared to surrounding regions. The geophysicist will use the thermometer and its various forms, including thermocouples and thermistors, to deduce temperature gradients, heat-flow rates, and local hot spots. Electrical resistivity measurements may be taken by placing electrodes into the ground and measuring the voltages between them. Differences in resistance of rocks can be attributed not only to the physical differences in rocks but also to the presence of steam or electrically conductive thermal waters. Additional techniques employed may be gravity and seismic measurements, which can delineate variations in the densities of rock strata and the presence of faults for the migration of pore fluids.

The function of the geochemist is to analyze the chemistry of natural thermal discharges. If the fluids are hot, chemical equilibrium will be achieved rapidly; the chemical nature of the discharged fluids is a reflection of the temperature achieved at equilibrium. The geochemist will look for the presence of silica and magnesium and at ratios of sodium to potassium as indicators of deep reservoir temperatures. The lower the sodium-potassium ratio, the higher the fluid temperature. The geochemist also is able to detect valuable minerals in the thermal fluids that are of interest to industrial geothermal developers.

The aim of this preliminary exploration is to choose promising locations for exploratory well drilling. If the exploration team believes that a useful field exists, then drilling is used to locate zones of permeable rocks saturated with hot thermal fluids. Drilling expense and time are saved if the exploration drilling bore samples are of small diameter;

larger bores may follow once the potential of the field has been established.

VALUE AS ENERGY SOURCE

Geothermal energy offers significant savings: 25 to 55 percent of conventional thermal cost, and 30 to 35 percent of nuclear fuel expense. The construction of geothermal power plants can yield significant savings, as the geothermal energy is already present within the earth and does not need to be generated. When used along with other forms of energy production, geothermal power can help reduce the overall cost of energy. The use of geothermal power is becoming increasingly important on a worldwide scale; in fact, in some twenty countries geothermal exploration is actively pursued, with more than sixty countries involved in geothermal power development. Nevertheless, some predictions are that geothermal energy is unlikely ever to meet more than 10 percent of energy needs. The key lies in the extent to which humans can harness the earth's heat, particularly the deep heat sources that are at the limits of current technology.

The ultimate goal is universal heat mining; that is, utilizing the earth's heat wherever it is needed, even in the nonthermal areas. Meeting this goal could require drilling 6.5 to 9 kilometers for power generators operating at thermal efficiencies of 15 to 20 percent. Included in heat mining is the direct tapping of the magma pockets of active volcanoes; because of the very high temperatures encountered, this is a formidable task for conventional drilling even at shallow depths. Drilling into magma presents no danger of triggering a volcanic eruption, because the drill hole is far smaller than the size of a volcanic vent. In fact, a geothermal project in Iceland briefly spurted lava in 2009.

Geothermal energy has several limitations. Such gases as methane and carbon dioxide do not contribute to energy production by flashing and must be removed in some geothermal fields. The thermodynamic efficiency is always low as well. Thermodynamic efficiency is the temperature drop in a process divided by the initial temperature, as measured in degrees kelvin, or degrees from absolute zero (-273 degrees Celsius). If steam has a temperature of 400 kelvins (127 degrees Celsius) and 300 kelvins (27 degrees Celsius) after flashing, the thermodynamic

efficiency is $100/400 = 25$ percent. The low efficiency is the inevitable result of using steam at relatively low temperatures. After engineering losses are taken into account, most geothermal plants achieve efficiencies of only a few percent.

In addition, the fluids from hydrothermal fields accompanied by gases and certain water-soluble chemicals can be potentially hazardous to the machinery components of a geothermal facility as well as to the environment. Some deposits, such as silica and iron sulfide, build up in discharge channels on turbine blades and pipes, which reduces efficiency. Carbon dioxide and hydrogen sulfide in solution may form large corrosion pits on the pipe walls. The discharge of hydrogen sulfide into the atmosphere at geothermal power plants may have an adverse effect on crops, river life, and electrical equipment. Fortunately, scrubber systems can clean hydrogen sulfide and other gases from the effluent, and sometimes those gases can even be converted into useful products, such as fertilizer. Also, well-designed geothermal plants release only a few percent of the carbon dioxide and sulfur compounds produced by coal-fired plants and oil-fired plants. As stated above, closed binary cycle plants can operate with virtually no emissions.

By the end of the 1990's, the installed capacity of geothermal electrical power plants was about 6,000 megawatts worldwide, including 2,200 megawatts in the United States. The United States uses an additional 1,000 megawatts for direct heating. This places geothermal energy third in the United States among renewable energy sources (0.5 percent of all energy used), following hydroelectricity (4.5 percent) and biomass (4.3 percent), and ahead of solar power (0.1 percent) and wind power (0.05 percent). Most of the people in Iceland and more than 500,000 people in France use geothermal heat in homes and public buildings.

Michael Broyles

FURTHER READING

Albu, Marius, David Banks, Harriet Nash, et al., eds. *Mineral and Thermal Groundwater Resources.* London: Chapman and Hall, 1997. Considers groundwater and its relationship to geothermal processes. Suitable for college-level students with some background in Earth science. Contains bibliographical references and index.

Anderson, Greg M., and David A. Crerar. *Thermodynamics in Geochemistry: The Equilibrium Model*. New York: Oxford University Press, 1993. An exploration of geochemistry and its relationship to thermodynamics and geothermometry. A thorough, somewhat technical resource. Recommended for readers with a background in chemistry and Earth science.

Armstead, H. Christopher. *Geothermal Energy*. 2d ed. Bristol, England: Arrowsmith, 1983. A comprehensive textbook covering the historical applications of Earth's heat, the nature and occurrence of geothermal fields, and exploration techniques. Discusses methods of electric power generation from the geothermal energy of specific regions, along with the comparative costs in each case. Suitable for the general student of geothermal energy.

_____, ed. *Geothermal Energy: Review of Research and Development*. Lanham, Md.: UNIPUB, 1973. Discusses geothermal energy exploration and utilization for a general audience. An excellent supplement to the author's 1983 text. Softcover format with many diagrams, especially of power plant machinery, including noise silencer construction, pipeline expansion arrangements, and circulating water systems.

Boyle, Godfrey, ed. *Renewable Energy*. 2d ed. New York: Oxford University Press, 2004. Provides a complete overview of renewable energy resources. Discusses solar energy, bioenergy, geothermal energy, hydroelectric energy, tidal power, wind energy, and wave energy. Discusses basic physics principles, technology, and environmental impact. Chapters include references and a further reading list, which makes this book an excellent starting point.

Di Pippo, Ronald. *Geothermal Power Plants*. 2d ed. Burlington, Mass.: Butterworth-Heinemann, 2008. Discusses geothermal geology and exploration, geothermal well drilling, and the many uses of geothermal energy. Examines geothermal physics and analyses, and includes case studies. Covers environmental impacts of geothermal power. Well organized with a summary and references following each chapter, seven appendices, and indexing.

Duffield, Wendell, and John Sas. *Geothermal Energy; Clean Power from the Earth's Heat*. USGS Circular 1249. 2003. Begins with a description of historical and current geothermal energy use. Includes discussions of global geothermal applications, mining of geothermal energy, hydrothermal systems and dry geothermal systems, and the environmental impact of geothermal energy use. Provides numerous color diagrams and images.

Gregory, Snyder A., Clive R. Neal, and W. Gary Ernst, eds. *Planetary Petrology and Geochemistry*. Columbia, Md.: Geological Society of North America, 1999. Provides an excellent overview of the field of geochemistry and its principles and applications. Appropriate for college students.

Grotzinger, John, et al. *Understanding Earth*. 5th ed. New York: W. H. Freeman, 2006. Contains some discussion of the structure and composition of the common rock-forming minerals. Discusses the relationship of igneous and metamorphic petrology to the general principles that form the basis of modern plate tectonic theory. Suitable for advanced high school and college students.

Harsh, K., and Sukanta Roy Gupta. *Geothermal Energy: An Alternative Resource for the 21st Century*. Boston: Elsevier Science, 2006. Written for the layperson. Explains recent developments in the exploration of geothermal resources, methods for study and exploitation, and technology. Provides a basic overview of geothermal energy in the earth and summarizes types of geothermal systems.

Hodgson, Susan F. *A Geysers Album: Five Eras of Geothermal History*. Sacramento: California Department of Conservation, Division of Oil, Gas and Geothermal Resources, 1997. Provides an easy understanding of geothermal history in the United States. Includes useful graphics, illustrations, and maps as well.

Johnson, T. "Hot-Water Power from the Earth." *Popular Science* 222 (January, 1983): 70. Discusses the direct flash and binary cycle systems, and provides several case histories from the California region.

Kerr, Richard A. "Hot Dry Rock: Problems, Promise." *Science* 238 (November, 1987): 1226-1229. A discussion on tapping the potentially enormous heat reserves in rocks too dry to yield steam or hot water. Deals with the latest research on hydraulic fracturing in deep wells as a means of creating pathways for water injected to circulate in deep hot rocks. Presents case histories in Los Alamos, New Mexico, and in Cornwall, England. Suitable for the general science reader.

Kruger, Paul, and Carel Otte, eds. *Geothermal Energy: Resources, Production, Stimulation*. Stanford, Calif.: Stanford University Press, 1973. Offers a good assessment of U.S. resources in terms of producible geothermal energy and elaborates on the characteristic problems of utilization. Deals with corrosion and scaling of machinery, and presents a complete chemical analysis of geothermal waters from several sources. A valuable reference for the geothermal student.

Letcher, Trevor M., ed. *Future Energy: Improved, Sustainable and Clean.* Amsterdam: Elsevier, 2008. Discusses the future of fossil fuels and nuclear power. Reviews renewable energy supplies, such as solar, wind, hydroelectric and geothermal resources. Focuses on potential energy sources and currently underutilized energy sources. Covers new concepts in energy consumption and technology.

Plate, Erich J., et al., eds. *Buoyant Convection in Geophysical Flows.* Boston: Kluwer Academic Publishers, 1998. Provides good descriptions and discussions of heat convection theories, geophysics, and heat flows. Offers illustrations and a bibliography.

Sacher, Hubert; and Rene Schiemann. "When Do Deep Drilling Geothermal Projects Make Good Economic Sense?" *Renewable Energy Focus* 11 (2010): 30-31. Provides technical aspects of deep-drilling projects and economic feasibility.

See also: Earth Resources; Gold and Silver; Hydroelectric Power; Hydrothermal Mineralization; Nuclear Power; Nuclear Waste Disposal; Ocean Power; Oxides; Radioactive Minerals; Sedimentary Mineral Deposits; Solar Power; Unconventional Energy Resources; Wind Power.

GOLD AND SILVER

Silver and gold have played important roles in human economies, industries, and finances for thousands of years. They have served widely in applications as diverse as medicine and jewelry, and remain two of the most useful and sought-after metals.

PRINCIPAL TERMS

- **amalgam:** an alloy of mercury and another metal; gold and silver amalgams occur naturally and have been synthesized for a variety of uses
- **electrum:** a term commonly used to designate any alloy of gold and silver containing 50 to 80 weight percent gold
- **fineness:** a measure of the purity of gold or silver expressed as the weight proportion of these metals in an alloy; gold fineness considers only the relative proportions of gold and silver present, whereas silver fineness considers the proportion of silver to all other metals present
- **karat:** a unit of measure of the purity of gold (abbreviated "k"); pure gold is 24 karat
- **lode deposit:** a primary deposit, generally a vein, formed by the filling of a fissure with minerals precipitated from a hydrothermal solution
- **placer deposit:** a mass of sand, gravel, or soil resulting from the weathering of mineralized rocks that contains grains of gold, tin, platinum, or other valuable minerals derived from the original rock
- **troy ounce:** approximately 31 grams; there is about 14.5 troy ounces per pound, compared to the 16 ounces in everyday (avoirdupois) ounces.

NATURAL FORMS

Gold, a deep yellow metal, is chemical element number 79, with an atomic weight of 196.967. Natural gold consists of a single isotope, gold-107, and in pure form exhibits a density of 19.3 grams per cubic centimeter. Silver, a brilliant white metal, is chemical element number 47, with an atomic weight of 107.87. Natural silver consists of two isotopes, silver-107 (51.4 percent) and silver-109 (48.6 percent), and in pure form exhibits a density of 10.5 grams per cubic centimeter. Both metals are very malleable, and this property has led to the common use of gold in gilding, a process by which extremely thin films are applied to surfaces of metal, ceramic, wood, or other materials for decorative purposes. Gold is so malleable that it can be pounded into translucent films so thin that more than 300,000 of them would be required to form a pile 1 inch (2.5 centimeters) high (the sheets are about 100 atoms thick, in other words).

Gold and silver occur in a variety of minerals. Native gold and electrum, the gold-silver alloy, are the most commonly occurring natural forms of gold in ore deposits. Although laboratory studies indicate that all compositions of gold-silver alloys may be synthesized, naturally occurring gold and electrum compositions are nearly always between 50 and 95 weight percent gold. The most commonly encountered compositions contain about 92 percent gold.

ECONOMIC VALUE

Although gold and silver are commonly thought of as precious metals, they differ greatly in their chemical properties and economic value. Gold is a true noble metal; that is, it is almost totally resistant to chemical attack and dissolution. Gold forms few chemical compounds and minerals. By comparison, silver is less noble, meaning that it is considerably easier to dissolve and forms several compounds and minerals. Gold is about sixty times more costly than silver, though relative values fluctuate; this difference is partly because of gold's superior chemical properties but mostly because of its rarity relative to silver. These differences affect the geochemical and geological behavior of these metals as well as the intensity of the mining and mineral exploration activities directed at their recovery.

The average abundance of gold in the earth's crust is about 0.004 part per million, and the average abundance of silver is 0.08 part per million, about twenty times that of gold. These amounts of gold and silver are far too small to pay for the extraction of these metals from common rock. In some places in the earth's crust, however, hydrothermal solutions have circulated, collected some of the gold and silver from the rocks, and transported them to sites of deposition. The rocks at these sites contain gold and silver values many times greater than those of average rocks and are sought out as ore deposits. In addition, the weathering of gold deposits and transportation of gold particles by streams creates placer deposits, where gold can be extracted from sand and gravel by simple techniques. In order for gold deposits to be mined economically, the concentrations of the gold generally must be on the order of 1 to 3 parts per million (about 0.1 troy ounce per ton), but some of the richest deposits contain ores with 8 to 10 parts per million (about 0.2 troy ounce per ton). Because silver is worth much less, deposits must generally contain 150 to 300 parts per million (5 to 10 troy ounces per ton) if they are mined solely for silver. Much silver, however, is extracted as a by-product from copper, lead, and zinc ores in which the silver is present in concentrations of 1 to 50 parts per million.

Because economic concentrations of gold are so low, very sensitive chemical, analytical techniques must be used to determine the amount of gold present. One of the oldest but still widely used is fire assaying. In this method, lead oxide is mixed with a powdered ore sample and melted until the lead oxide is converted to metallic lead that picks up the gold and silver and settles to the bottom of the crucible. The lead, gold, and silver bottom is melted in a porous cup where the lead is again converted to lead oxide that is absorbed by the cup, leaving a gold and silver mixture known as doré metal. The doré metal is weighed and the silver is dissolved by nitric acid. The remaining pure gold is reweighed to determine the amount of gold.

MINING AND EXTRACTION

When economic gold-bearing deposits are found, the method of mining the ore and extracting the metals (some silver is always present) varies, depending on the location of the deposit and its mineralogy. If the gold occurs as large enough grains, it is readily concentrated by gravity techniques, and the high density of the gold (15 to 19 grams per cubic centimeter) allows it to be physically separated from other minerals that have much lower densities. The gold pans and sluices used by early miners and some mines today are small-scale examples of gravity methods. Conversely, if the gold is present in very small, sometimes submicroscopic grains, chemical solvent techniques are utilized. The most commonly employed method applies a sodium cyanide solution that dissolves the gold. The gold is removed from the solution by activated charcoal, and the solution is re-used. The gold is redissolved from the charcoal and then electrochemically precipitated on steel wool. The steel wool is mixed with silica, borax, and niter

and melted to produce a doré metal of the gold and silver and a slag with all the impurities. The doré metal is sent to a refinery to separate the gold and silver.

Silver-rich ores usually contain much higher concentrations of silver-bearing minerals than do gold mines. After grinding the ores to 0.01 to 0.05 inch (0.2 to 1.0 millimeter), the silver minerals are separated by flotation techniques in which small bubbles generated in soaplike solutions are used to pick up and concentrate the small grains. The concentrated materials are shipped to refineries for chemical separation of the metals and sulfur.

The principal gold- and silver-producing areas in the United States are the Carlin district, Nevada, and the Coeur d'Alene district, Idaho, respectively. South Africa has the largest gold deposit in the world, called the Witwatersrand district. This district is thought to contain at least 760 million troy ounces (24,000 metric tons) of gold and produces about 22 million troy ounces (700 metric tons) of gold annually. The Real de Angeles Mine in Mexico is the world's largest producer of silver. This district produces about 13 million troy ounces of silver annually. Gold and silver are produced in many other areas of the world as well. In addition, significant amounts of gold and silver are recovered as by-products during the processing of the ores of other metals, such as copper.

HYDROTHERMAL SYSTEMS

There are many types of hydrothermal systems that produce deposits of gold and silver. Many of these hydrothermal systems are ancient analogues of modern geothermal systems such as the one at Yellowstone National Park, Wyoming. Geologists deduce the nature of the hydrothermal system that produced a particular deposit by studying its size, shape, mineralogy, ore grade, and tectonic setting. The size and shape of the ore body are determined from maps of the surface outcrop and from geological cross-sections of the subsurface extent of the ore body, prepared by consulting drill core logs and maps of outcrops in underground workings. Observations of hand samples, microscopic observation (transmitted and reflected light), X-ray diffraction analysis, and electron microprobe analysis are common processes used to determine the mineralogy of the ore (mineralized rock from which metals can be economically extracted), the gangue (minerals that have no economic value and that were deposited by the hydrothermal solutions), the country rock (the unmineralized rock surrounding the deposit), and the alteration zone (country rock that was changed chemically or mineralogically by reactions with the hydrothermal solutions). The types, compositions, and associations of these minerals are used to infer the chemical conditions at the time of the ore deposit's formation. Fluid inclusions, or small bubbles of the hydrothermal solution trapped in growing mineral grains, are commonly observed features that are particularly useful in determining the nature of the fluids from which the ore minerals precipitated. The gold and silver content of the rocks (ore grade) is determined by chemical analysis of systematically selected samples from various parts of the ore body. Some ore bodies show sharp cutoffs of grade between the mineralized rock and the country rock. Others show a gradual decrease from the mineralized zone to the background levels of the country rock. The tectonic (deformational crust) and structural setting of the ore body give evidence of the general geologic framework for the ore-forming process. Nearly all hydrothermal ore deposits are associated with tectonically active areas, where the rate of heat flow from the earth's interior to the surface is quite high. It is this heat flow that increases the temperatures of the hydrothermal solutions and drives them through the rocks. Midocean ridges and rises and island arcs associated with subduction zones (where the edge of one crustal plate descends below the edge of another) are recognized as tectonically active areas that have high rates of heat flow and are thus settings for the potential formation of hydrothermal ore deposits.

Weathering breaks down rocks both chemically and physically. The products of weathering are ions that are carried away by solutions and solids that are eroded from the surface by rainwater. When hydrothermal ore deposits weather, most of the ore minerals are destroyed by oxidation, and their components are carried away in solution. If these solutions penetrate downward into the outcrop, they sometimes encounter more reducing conditions at depth, and the ore elements reprecipitate, or separate out again, to form an enriched zone. This process is called secondary enrichment. The grade of gold and, particularly, silver deposits can be improved by this process. Most of the time, however, gold is simply too unreactive to be destroyed by chemical weathering

and instead concentrates in the soil horizon to form an eluvial placer. As the soil erodes, the gold grains are carried into streams and rivers where they are transported until they encounter areas of slower water velocity. There, they are dropped by the stream and accumulate, because these grains have a much greater density than common silicates. Gold and electrum grains have densities ranging from 15.5 to 19.3 grams per square centimeter, whereas common silicate minerals have densities ranging from 2.6 to 3.0 grams per square centimeter. Thus, the common silicates simply are washed away, leaving the gold and electrum grains behind. The Witwatersrand deposits are paleoplacer deposits; that is, they are placer deposits that formed about 2 billion years ago when rivers and streams carrying gold grains flowed into a large basin. As the waters of these rivers and streams entered the basin, their velocity slowed and they deposited their load of gold-bearing gravels and sands. These deposits were eventually covered by other rocks and became lithified, or changed to stone. The very large extent and relatively high grade of these deposits (8 to 10 parts per million) make the Witwatersrand district the largest gold district in the world.

One of the more interesting problems surrounding the formation of silver and lode gold deposits is determining how these elements were transported by aqueous solutions. Gold is a noble metal, meaning that it is very resistant to the natural chemical attacks that would dissolve it. Silver is not as noble as gold and therefore tends to form compounds (minerals) with a wide variety of elements. These silver minerals, as well as native silver, have low solubilities. The solubility of gold and silver and their minerals does seem to increase with temperature; however, even at very high temperatures, their solubility in pure water is much too low to account for the amounts of the metals transported into ore deposits. The most likely explanation for the ability of hydrothermal solutions to transport these metals is that they react with other ions in the solution, such as chloride, bisulfide, and perhaps others, to form very stable complexes. The formation of complexes increases the stability of the gold or silver species in the aqueous solution and therefore makes gold and silver soluble enough to account for their transport to form ore deposits. When solutions containing these complexes are cooled, mixed with solutions containing lower concentrations of complexing ions, oxidized by reactions with other minerals, or boiled to remove chloride or bisulfide as vapors, the complexes tend to break down, lowering the solubility of gold and silver and causing these elements to precipitate in metallic form or as constituents of minerals.

CONTINUING DEMAND

Gold and silver were among the earliest metals used by humans. They were employed as amulets and jewelry as early as 5000 B.C.E., and their use as money and in coinage had begun in Asia Minor and Greece by about 600 B.C.E. Although these practices continue today, gold and silver have found many additional uses and have diverged in their principal applications. In the United States, gold today is used primarily in jewelry and in the arts (about 55 percent of the total usage); other major uses include solid-state electronic devices (34 percent), dental supplies (12 percent), and investment products (0.1 percent). In contrast, silver is primarily an industrial metal, with 43 percent being used in photography and 35 percent being used in electrical products. Other important silver uses include jewelry (6 percent), sterling ware (6 percent), and coinage (4 percent). Gold and silver coins were widely minted in many countries, especially in the 1700's and 1800's, but were generally dropped from use in the early 1900's. The United States minted gold coins from 1849 until 1933 and silver coins from 1794 until 1964. In 1964, the price of silver for photography exceeded the value of the silver in coins, and silver coins were discontinued. Since 1986, the United States has minted gold and silver bullion coins that do not have assigned denominations but rather contain specified amounts of gold or silver and vary in value as metal market prices fluctuate.

Throughout history, the quest for gold and silver has played a very important role in the development and expansion of human civilization, culture, and enterprise. By the time of the Egyptian empire, humankind had already developed sophisticated mining and metallurgical processes for producing precious metals. The wealth and prosperity of Athens was based in large part upon the silver production of the mines at Laurion in southeastern Attica (Greece), and the Roman Empire flourished as a result of gold and silver obtained from the Iberian Peninsula. Christopher Columbus's encounter with gold-bearing Native Americans on his first voyage provided a powerful stimulus for the Spanish to explore and exploit the New World. The new gold and silver stocks (181 metric tons of gold and 18,000 metric tons of silver) shipped from the New World to Europe from 1500 to 1650 produced a revitalization of the economic system that helped fuel the later stages of the Renaissance. Paradoxically, Spain, glutted by such money, became impoverished as a result. The search for gold and silver also played a major role in the development of the United States, especially the gold rush of 1849 that lured thousands to California. Although less known, the first gold rush in the United States was in the Carolinas and Georgia in the 1830's, and concerned rocks quite similar to those of the Sierra Nevada foothills of California. Subsequent gold discoveries in the Black Hills of South Dakota in 1874 and in the Klondike and Yukon in 1896 drew additional treasure hunters. The quest for gold also led to the opening of Australia in 1851 and South Africa in 1886. Currently, gold exploration and gold rushes in Brazil have produced an invasion of much of the tropical rain forests.

Gold in its native state. (U.S. Geological Survey)

Gold and silver are expensive metals, and for this reason there is a constant search for less expensive substitutes. Many pieces of jewelry and electronic products can be made by alloying other metals with gold and silver, or by merely coating base metals. Other products can be redesigned to maintain their utility while using smaller amounts of gold. In many cases, less expensive palladium and silver can be substituted for gold. Although more expensive, platinum is sometimes used instead of gold for coins, jewelry, and electrical products. Aluminum and rhodium are less expensive substitutes for silver in mirrors. Surgical plates, pins, and sutures can be made of tantalum instead of silver. Silver tableware can be replaced by stainless steel products. Video cameras, silverless black-and-white film, and xerography have reduced the silver demand of copying and photography. Even with these substitutions, it is unlikely that the demand for these metals will diminish.

J. Donald Rimstidt and James R. Craig

FURTHER READING

Boyle, Robert W. *The Geochemistry of Silver and Its Deposits.* Geological Survey of Canada Bulletin 160. Ottawa: Queen's Printer, 1968. A general semi-technical presentation of the chemical behavior of silver in various geological environments and of the minerals that contain silver. Describes many of the world's major silver deposits and techniques employed in prospecting for silver.

_____. *Gold: History and Genesis of Deposits.* New York: Van Nostrand Reinhold, 1987. Provides a fine overview of the history of gold and its mode of formation from primitive to modern times. Presents annotations on articles covering the general geochemistry of gold and various theories on the mechanisms by which deposits form. Intended for the scientist or the serious layperson.

Brooks, Robert R., ed. *Noble Metals and Biological Systems: Their Role in Medicine, Mineral Exploration, and The Environment.* Boca Raton, Fla.: CRC Press, 1992. Explores not only the economic importance of precious metals but also the importance of the roles they play in biogeochemical and physiological research and applications. Slightly technical at times; appropriate for someone with a background in the field.

Cotton, Simon. *Chemistry of Precious Metals.* London: Blackie Academic and Professional, 1997. Illustrates the procedures and protocols used to determine the chemical makeup and properties of precious metals. Illustrations, bibliography, and index.

Evans, Anthony M. *An Introduction to Economic Geology and Its Environmental Impact.* Malden, Mass.: Blackwell Science, 1997. Provides a wonderful introduction to mines and mining practices, minerals, ores, and the economic opportunities and policies surrounding these resources. Illustrations, maps, index, and bibliography.

Gasparrini, Claudia. *Gold and Other Precious Metals: From Ore to Market.* New York: Springer-Verlag, 1993. Offers a step-by-step account of the procedures used in mining precious metals. Covers all aspects of the field, from mining the ores to the metals' economic value in the marketplace. A good book for someone without a background in the field.

Gee, George E. *Recovering Precious Metals.* Palm Springs, Calif.: Wexford College Press, 2002. Focuses on gold and silver recovery and recycling processes. Discusses many precious metal sources, including silver from photography solutions, and mirror-maker's solutions. Covers the separation of gold and silver from platinum.

Gray, Theodore. *The Elements: A Visual Exploration of Every Known Atom in the Universe.* New York: Black Dog & Leventhal Publishers, 2009. An accessible overview of the periodic table. Written for the general public. Provides a useful introduction of each element to high school students and the layperson. Offers excellent images, complete with index.

Laznicka, Peter. *Giant Metallic Deposits.* 2d ed. Berlin: Springer-Verlag, 2010. Discusses the location, extractions, and future use of many metals. Organized by topic rather than by metal. Discusses geodynamics that result in large metal ore deposits and the composition of metals within deposits. Discusses hydrothermal deposits, metamorphic associations, and sedimentary associations.

Marsden, John O., and C. Iain House. *The Chemistry of Gold Extraction.* 2d ed. Littleton, Colo.: Society for Mining, Metallurgy and Exploration, 2006. Discusses the history of gold extraction, ore deposits, hydrometallurgy, leaching, solution purification, and effluent treatment provide theory and application to the gold processing industries.

Background in geology or chemistry is helpful but not necessary.

St. John, Jeffrey. *Noble Metals*. Alexandria, Va.: Time-Life Books, 1984. An excellent general treatment of gold, silver, platinum, and other precious metals. Discusses the geological origin of these metals, methods of exploration, means of extraction, and uses. Develops a useful historical perspective on the role of these metals in society. Spectacular illustrations of natural grains, nuggets, and crystals, as well as manufactured precious metal objects, jewelry, and coins.

U.S. Geological Survey. *Mineral Facts and Problems*. Bulletin 675. Washington, D.C.: Government Printing Office, 1985.

_____. *Yearbook*. Washington, D.C.: Government Printing Office, issued annually. Chapters on gold and silver provide concise discussions and many data tables on uses, resources, technology, and production relationships of gold and silver.

Watkins, Tom H. *Gold and Silver in the West: The Illustrated History of an American Dream*. Palo Alto, Calif.: American West, 1971. An abundantly illustrated historical account of the search for and discovery of gold and silver in North America. Traces the paths westward and provides a wonderfully readable and accurate history of those who sought gold and silver and the boomtowns they built. Written for the layperson.

White, Benjamin. *Silver: Its History and Romance*. Detroit, Mich.: Tower Books, 1971. Gives the history of silver use and mining, dating back to prehistoric times. Contains a fair amount of production and monetary statistics, as well as moderately detailed historical accounts of important silver-producing regions. Chapters deal with the various uses of silver in jewelry and coins of various cultures.

See also: Aluminum Deposits; Biogenic Sedimentary Rocks; Building Stone; Carbonates; Cement; Chemical Precipitates; Coal; Diagenesis; Diamonds; Dolomite; Earth Resources; Fertilizers; Industrial Metals; Industrial Nonmetals; Iron Deposits; Land Management; Manganese Nodules; Minerals: Physical Properties; Pegmatites; Platinum Group Metals; Salt Domes and Salt Diapirs; Sedimentary Mineral Deposits; Soil Chemistry; Soil Profiles; Uranium Deposits.

GRANITIC ROCKS

Granitic rocks are coarse-grained igneous rocks consisting mainly of quartz, sodic plagioclase, and alkali feldspar, with various accessory minerals. These rock types occur primarily as large intrusive bodies that have solidified from magma at great depths. Granitic rocks can also occur to a lesser degree as a result of metamorphism, a process referred to as granitization.

PRINCIPAL TERMS

- **aphanitic:** a textural term that applies to an igneous rock composed of crystals that are microscopic in size
- **crystal:** a solid made up of a regular periodic arrangement of atoms
- **crystallization:** the formation and growth of a crystalline solid from a liquid or gas
- **granitization:** the process of converting rock into granite; it is thought to occur when hot, ion-rich fluids migrate through a rock and chemically alter its composition
- **isotopes:** atoms of the same element with identical numbers of protons but different numbers of neutrons, thus giving them a different mass
- **magma:** a body of molten rock typically found at great depths, including any dissolved gases and crystals
- **migmatite:** a rock exhibiting both igneous and metamorphic characteristics, which forms when light-colored silicate minerals melt and crystallize, while the dark silicate minerals remain solid
- **phaneritic:** a textural term that applies to an igneous rock composed of crystals that are macroscopic in size, ranging from about 1 to over 5 millimeters in diameter
- **pluton:** a structure that results from the emplacement and crystallization of magma beneath the surface of the earth
- **porphyritic:** a texture characteristic of an igneous rock in which macroscopic crystals are embedded in a fine phaneritic or aphanitic matrix

GRANITIC ROCK OCCURRENCE

The term "granitic rocks" generally refers to the whole range of plutonic rocks that contain at least 10 percent quartz. They are the main component of continental shields and also occur as great compound batholiths in folded geosynclinal belts. Granitic rocks are so widespread, and their occurrence and relation to the tectonic environment are so varied, that generalizations often obscure their complexity. Major granitic complexes are, in general, a continental phenomenon occurring in the form of batholiths and migmatite complexes.

When large masses of magma solidify deep below the ground surface, they form igneous rocks that exhibit a coarse-grained texture described as phaneritic. These rocks have the appearance of being composed of a mass of intergrown crystals large enough to be identified with the unaided eye. A large mass of magma situated at depth may require tens of thousands or even millions of years to solidify. Because phaneritic rocks form deep within the crust, their exposure at ground surface reflects regional uplift and erosion, which has removed the overlying rocks that once surrounded the now-solidified magma chamber.

Along continental margins, belts of granitic rocks developed as batholiths composed of hundreds of individual plutons. The formation of batholithic volumes of granitic magma generally appears to require continental settings. Some of the more prominent batholiths in North America are the Coast Range, Boulder-Idaho, Sierra Nevada, and Baja California. The largest are more than 1,500 kilometers long and 200 kilometers wide, and have a composite structure. The Sierra Nevada, for example, is composed of about 200 plutons separated by many smaller plutons, some only a few kilometers wide.

GRANITIC ROCK CLASSIFICATION

As with other rock types, granitic rocks are classified on the basis of both mineral composition and fabric or texture. The mineral makeup of an igneous rock is ultimately determined by the chemical composition of the magma from which it crystallized. Feldspar-bearing phaneritic rocks containing conspicuous quartz (greater than 10 percent in total volume) in addition to large amounts of feldspar can be designated as granitic rocks. This nonspecific term is useful where the type of feldspar is not recognizable because of alteration or weathering, for purposes of quick reconnaissance field studies, or for general discussion.

Granitic rocks consist of two general groups of minerals: essential minerals and accessory minerals. Essential minerals are those required for the rock to be assigned a specific name on a classification scheme. Essential minerals in most granitic rocks are quartz, sodic plagioclase with 30 to 50 percent calcic plagioclase, and potassium-rich alkali feldspars (either orthoclase or microcline). Accessory minerals include biotite, muscovite, hornblende, and pyroxene.

When an initial phase of slow cooling and crystallization at great depths is followed by more rapid cooling at shallower depths or at the surface, porphyritic texture develops, as is evident in the presence of large crystals enveloped in a finer-grained matrix or groundmass. The presence of porphyritic texture is evidence that crystallization occurs over a range of temperatures, and magmas are commonly emplaced or erupted as mixtures of liquid and early-formed crystals.

Classification of granitic rocks can be based either on the bulk chemical composition or on the mineral composition. Because oxygen is ubiquitous in rocks, chemical compositions are given as the weight percentages of oxides rather than individual elements, whereas mineral composition units are in approximate percentages in total volume. The mineral composition of granitic rocks provides a reliable basis for classification. Specific granitic rock types are defined on the basis of the presence of quartz and the ratio of potassium feldspar and plagioclase. The presence of other accessory minerals may be indicated in the form of a modifier (such as biotite granite).

GRANITIC ROCK GROUPS

Granitic rocks include true granites, tonalite, granodiorite, quartz monzonite granite, soda granite, and vein rocks of pegmatite and aplite. The mineralogy of these varies, and the distinction between different granitic rocks can be gradational, but all are defined by the presence of quartz. Tonalite has mostly plagioclase feldspar and very little potassium feldspar. Granodiorite is composed predominantly of plagioclase feldspar, with subordinate potassium feldspar

and biotite, hornblende, or both as accessory minerals. Quartz monzonite is composed of approximately equal amounts of potassium and plagioclase feldspars, with biotite, hornblende, or both as accessory minerals. Granite is composed predominantly of potassium feldspar with subordinate plagioclase feldspar and biotite alone or with hornblende or muscovite as accessory minerals. Soda granite is composed predominantly of very sodium-rich plagioclase feldspar, with small amounts of sodium-bearing pyroxene or amphibole.

Pegmatite is a very coarse-grained complex rock in terms of mineralogy. Structurally, pegmatites occur as dikes associated with large plutonic rock masses. Dikes are tabular-shaped intrusive features that cut through the surrounding rock. The large crystals are inferred to reflect crystallization in a water-rich environment. Most pegmatites are simple in terms of mineralogy, consisting primarily of quartz and alkali feldspar, with lesser amounts of muscovite, tourmaline, and garnet; they are referred to as simple pegmatites. Other pegmatites can be very complex and

A large granite boulder shaped by exfoliation, in the Kaweah Basin, Sequoia National Park, California. (U.S. Geological Survey)

contain other elements that slowly crystallize from residual, deeply seated magma bodies. High concentrations of these residual elements can result in the formation of minerals such as topaz, beryl, and minerals rich in rare-earth elements, as well as quartz and feldspar. Aplite is a very fine-grained, light-colored granitic rock that also occurs as a dike and consists of quartz, albite, potassium feldspar, and muscovite, with almandine garnet as an occasional accessory mineral.

Other rocks that are occasionally grouped with granitic rocks are migmatites. Migmatites, or mixed rocks, are heterogeneous granitic rocks that, on a large scale, occur within regions of high-grade metamorphism or as broad migmatite-rich zones bordering major plutons. Migmatites appear as alternating light and dark bands. The light-colored bands are broadly granitic in mineralogy and chemistry, while the darker bands are clearly metamorphic. Migmatites are believed to reflect the partial melting of metamorphic rocks in which the portions rich in quartz and feldspar melt first.

Granitic rocks are classified by the relative proportions of quartz, plagioclase, and potassium feldspar, and can also be classified by their aluminum content. Peraluminous granites have excess aluminum present after the formation of feldspar, and thus contain aluminum-rich minerals like sillimanite or garnet. Peralkaline granites run out of aluminum before using up their sodium and potassium and have alkali-rich minerals like sodium amphibole. Granitic rocks can further be divided into two groups, S-types and I-types, according to whether they were derived from predominantly sedimentary or igneous sources, respectively. Enormous volumes of I-type granites, which constitute most plutons, occur along continental margins overriding subducting oceanic material.

INCLUSIONS

A common feature of granitic rocks, notably in granodiorite to dioritic plutons, is the presence of inclusions or rock fragments that differ in fabric and/or composition from the main pluton itself. The term "inclusions" indicates that they originated in different ways. Foreign rock inclusions, called xenoliths, include blocks of wall rocks that have been mechanically incorporated into the magma body. This process is referred to as stoping. Some mafic inclusions are early-formed crystals precipitated from the magma itself after segregation along the margin of the pluton, which cools first. These inclusions are called antoliths. Other inclusions may reflect clots of solidified mantle-derived magma that ascended into the granitic source region, or residual material that accompanied the magma during its ascent.

PETROGRAPHIC STUDY

Granitic magmas can be derived from a number of sources, notably the melting of continental crust, the melting of subducted oceanic crust or mantle, and differentiation. The problem facing the geologist is to decide which tectonic environment led to the formation of the granite. To accomplish this objective, both petrographic and geochemical information is used.

The standard method of mineral identification and study of textural features and crystal relationships is the use of rock thin sections. A thin section is an oriented wafer-thin portion of rock 0.03 millimeter in thickness that is mounted on a glass slide. The rock thin section is so thin that even normally opaque minerals transmit light and reveal mineral content, abundance and association, grain size, structure, and texture. The thin section also provides a permanent record of a given rock that may be filed for future reference.

Thin sections are studied with the use of a petrographic microscope, which is a modification of the conventional compound microscope commonly used in laboratories. The modifications that render the petrographic microscope suitable to study the optical behavior of transparent crystalline substances are a rotating stage, an upper polarizer or lower polarizer, and a Bertrand lens, which together are used to survey the optical properties of the mineral over a wide range of angles at once. With a magnification that ranges from about 30 to 500 times, the petrographic microscope allows one to examine the optical behavior of transparent crystalline substances or, in this case, crystals that make up granitic rocks.

The study of granitic rocks may be greatly facilitated by various staining techniques of both hand specimens and thin sections. Staining is employed occasionally to distinguish potassium feldspar from plagioclase and quartz, the three main mineral constituents of granitic rocks. A flat surface on the rock is produced by sawing and then polishing. The rock surface is etched using hydrofluoric acid, an extremely hazardous material that requires great care in handling. This step is followed by a water rinse and immersion in a solution of sodium cobaltinitrate. The potassium feldspars will then turn bright yellow. After rinsing with water and covering the surface with rhodizonate reagent, the plagioclase becomes brick red in color. Staining techniques are available for other minerals, including certain accessory minerals such as cordierite, anorthoclase, and feldspathoids.

Measurement of the relative amounts of various mineral components of a rock is called modal analysis. The relative area occupied by the individual minerals is estimated or measured on a flat surface (on a flat-sawed surface, on a flat outcrop surface, or in thin section) and then related to the relative volume. Caution must be used, because the relative area occupied by any mineral species on a particular planar surface is not always equal to the modal (volume) percentage of that mineral on the rock mass, especially if mineral grains are clumped or strongly aligned.

GEOCHEMICAL STUDY

When a rock specimen is crushed to a homogeneous powder and chemically analyzed, the bulk chemical composition of the rock is derived. Chemical analyses are normally expressed as oxides of the respective elements because of the overwhelming abundance of oxygen. Analysis of granitic rocks shows them to be typically rich in silica potassium and sodium, with lesser amounts of basic oxides such as magnesium, iron, and calcium oxides. Magnesium, iron, and calcium oxides are present in higher abundance in basalts, which contain plagioclase feldspar, pyroxene, and olivine.

Isotopes such as those of strontium, oxygen, and lead can also be used as tools in evaluating granitic magma sources, such as a mantle origin and crustal melting of certain rock types. Some accessory minerals reflect the trace element content of the magma and thus the possible nature of their source. Much research is focused on chemical tracers. Tracers help distinguish the source region of a granitic magma, such as lower-crustal igneous or sedimentary rock or mantle material.

INDUSTRIAL APPLICATIONS

Granitic rocks have been used as dimension stones for many years. Dimension stones are blocks of rock with roughly even surfaces of specified shape and size used for the foundation and facing of expensive buildings. When crushed, granitic rocks can be used as aggregate in the cement and lime industry. In addition to these uses, granitic rocks are valued because of their geographic association with gold. Gold ores are found in close proximity to the contacts of the granitic bodies within both granitic rocks and surrounding rocks.

Pegmatites can also be very valuable. Simple pegmatites are exploited for large volumes of quartz and feldspar, used in the glass and ceramic industries. Complex pegmatites can also be a source of gem minerals, including tourmaline, beryl, topaz, and chrysoberyl. In spite of varied mineral composition, relatively small size, and unpredictable occurrence, pegmatites constitute the world's main source of high-grade feldspar, electrical-grade mica (used for supporting heater elements in toasters), certain metals (including beryllium, lithium, niobium, and tantalum), and some piezoelectric quartz, although most industrial quartz is now grown synthetically rather than mined.

Stephen M. Testa

FURTHER READING

Carmichael, Ian S., Francis J. Turner, and John Verhoogen. *Igneous Petrology.* New York: McGraw-Hill, 1974. A well-known reference presenting both the mineralogical and geochemical diversity of granitic rocks and their respective geologic settings. Chapter 2, "Classification and Variety of Igneous Rocks," and Chapter 12, "Rocks of Continental Plutonic Provinces," are particularly recommended.

Chappell, B. W. and A. J. R. White. "Two Contrasting Granite Types: 25 Years Later." *Australian Journal of Earth Sciences* 48 (2001): 489-499. Discusses I-type and S-type granites found in the Lachlan Fold Belt. The original study was reviewed and updated with more recent research.

Deer, W. A., R. A. Howie, and J. Zussman. *An Introduction to Rock-Forming Minerals.* 2d ed. London: Pearson Education Limited, 1992. A standard reference on mineralogy for advanced college students and above. Each chapter contains detailed descriptions of chemistry and crystal structure, usually with chemical analyses. Discussions of chemical variations in minerals are extensive. A condensation of a five-volume set originally published in the 1960's.

Grotzinger, John, et al. *Understanding Earth.* 5th ed. New York: W. H. Freeman, 2006. This comprehensive physical geology text covers the formation and development of the earth. Written for high school students and general readers. Includes an index and a glossary of terms.

Frost, B. Ronald, et al. "A Geochemical Classification for Granitic Rocks." *Journal of Petrology* 42 (2001): 2033-2048. Discusses the chemical makeup of granitic rock. The authors classify granitic rock based on Fe-number, the MALI, and the ASI. Further subdivisions are also discussed. Requires a working knowledge of chemistry.

Hutchison, Charles S. *Laboratory Handbook of Petrographic Techniques.* New York: John Wiley & Sons, 1974. Stresses the practical aspects of laboratory and petrographic methods and techniques.

Klein, Cornelis, and Cornelius S. Hurlbut, Jr. *Manual of Mineralogy.* 23d ed. New York: John Wiley & Sons, 2008. An introductory college-level text on mineralogy. Discusses the physical and chemical properties of minerals and describes the most common minerals and their varieties, including gem mineral varieties. Contains more than 500 mineral name entries in the mineral index and describes in detail about 150 mineral species.

Pellant, Chris. *Smithsonian Handbooks: Rocks and Minerals.* New York: Dorling Kindersley, 2002. An excellent resource for identifying minerals and rocks. Contains colorful images and diagrams, glossary, and index.

Perch, L. L., ed. *Progress in Metamorphic and Magmatic Petrology.* New York: Cambridge University Press, 1991. Although intended for the advanced reader, several of the essays in this multiauthored volume will serve to familiarize the new student with the study of igneous rocks. In addition, the bibliography will lead the reader to other useful material.

Phillips, William Revell. *Mineral Optics: Principles and Techniques.* San Francisco: W. H. Freeman, 1971. A standard textbook discussing mineral optic theory and the petrographic microscope and its use in the study of minerals and rocks.

Ramo, O. T. *Granitic Systems.* Amsterdam: Elsevier, 2005. A compilation of articles discussing granite-related topics. Discusses plate tectonics and other processes, mineralogy, and geochemistry.

Smith, David G., ed. *The Cambridge Encyclopedia of Earth Sciences.* New York: Crown, 1981. Chapter 5, "Earth Materials: Minerals and Rocks," gives a good discussion of the processes involved in the formation of granitic rocks and a description of the mineral and chemical composition of granitic rocks. A well-illustrated and carefully indexed reference volume.

Sutherland, Lin. *The Volcanic Earth: Volcanoes and Plate Tectonics, Past, Present, and Future.* Sydney, Australia: University of New South Wales Press, 1995. Provides an easily understood overview of volcanic and tectonic processes, including the role of igneous rocks. Includes color maps and illustrations, as well as a bibliography.

See also: Andesitic Rocks; Anorthosites; Basaltic Rocks; Batholiths; Carbonates; Carbonatites; Igneous Rock Bodies; Kimberlites; Komatiites; Minerals: Structure; Orbicular Rocks; Plutonic Rocks; Pyroclastic Rocks; Rocks: Physical Properties; Xenoliths.

H

HAZARDOUS WASTES

Hazardous wastes include a broad spectrum of chemicals, produced in large quantities by industrial societies, that are dangerous to the health of the environment. These wastes often reach the environment, where reactions in soils and groundwater may work to reduce the hazard.

PRINCIPAL TERMS

- **biodegradation:** biological processes that result in the breakdown of a complex chemical into simpler building blocks
- **halogen:** one of a group of chemical elements including chlorine, fluorine, and bromine
- **heavy metal:** one of a group of chemical elements including mercury, zinc, lead, and cadmium
- **hydrocarbons:** chemicals composed chiefly of the elements hydrogen and carbon; the term is usually applied to crude oil or natural gas and their by-products
- **xenobiotic:** refers to a chemical that is foreign to the natural environment; an artificial chemical that may not be biodegradable

HAZARDOUS BY-PRODUCTS OF MANUFACTURING PROCESSES

Broadly defined, hazardous wastes include chemical, biological, or radioactive substances that pose a threat to the environment. Small amounts of radioactive wastes are rendered harmless by dilution, while larger amounts are held in isolation. Biological wastes may be incinerated, and some chemical wastes, such as nerve gas, are turned into harmless by-products by incinerating them at sufficiently high temperatures. The largest volume of hazardous wastes are the raw materials, the by-products, or the end-products of modern manufacturing processes. In that context, the elements and chemicals are all useful and necessary. When they end up in places where they do not belong—such as soils, groundwater, lakes, rivers, the ocean, and the air—and in higher than acceptable concentrations, they become hazardous wastes. For example, heavy metals, such as lead and chromium, are used in many chemical processes. They may enter the environment from the time they are mined from

the ground until they are used in a chemical process. They become hazardous wastes when their concentrations in the environment are sufficient to harm plants and animals.

Many hazardous waste chemicals are derived from petroleum. Petroleum is a complex mixture of chemicals, called hydrocarbons, composed mainly of carbon and hydrogen atoms. During refining and processing, the various petroleum hydrocarbons in the mix are separated; the carbon and hydrogen atoms are rearranged; and new elements, such as chlorine, bromine, and nitrogen, are introduced into the hydrocarbon molecules. In this way, naturally occurring chemicals, such as those that make up gasoline, are separated and concentrated, and new, synthetic chemicals are produced. These synthetic chemicals, and products derived from them, enable the production of several types of plastics (such as PVC pipes and polyethylene bags), nonstick surfaces, synthetic fibers in clothes and carpets, and many pharmaceuticals. Many of the synthetic chemicals that are used to make useful products—and the end-products themselves—may be hazardous wastes in the environment. For example, polychlorinated biphenyls (PCBs) are synthetic halogenated hydrocarbons, which are useful as heat-insulating materials in electrical transformers. When the transformer is scrapped, PCBs become a waste that must be discarded with care. Wastes from nuclear reactors and medical waste containing radioactive materials are a special kind of hazardous waste and are not discussed in this article.

HAZARDOUS WASTE RELEASES

In time, most environments can "cleanse" themselves of accidental spills or discharges of natural hazardous wastes by diluting, reacting with, or destroying the wastes. Several metals will be removed from water by attaching to sediment or soil mineral particles. Although the metals are still in the environment, they are in a form that makes them less mobile and less likely to be absorbed by plants and eaten by animals.

Hydrocarbon spills and discharges—from crude oil blown from a damaged offshore oil rig to leaks of refined gasoline from the underground storage tanks at filling stations—are probably the most common type of hazardous waste release. Hydrocarbon spills are also a formidable cleanup task for the environment. A major factor in the natural cleansing process is that the hydrocarbon molecules represent sources of energy. The stored energy is tapped in each molecule by burning the molecules in combustion engines. Bacteria and fungi in the ocean, lakes, and soils can also tap the energy in hydrocarbon molecules and use them as food. Some of the molecules in the complex mixtures are easier to use than others, but given enough time and the right conditions (such as a plentiful supply of oxygen for the primary biological process of respiration), most of the hydrocarbons can be broken down, or biodegraded, to carbon dioxide and water.

Synthetic chemical spills are much more difficult for the environment to handle without help because the amounts are often larger than the environment can handle, and the time for natural remediation is very long. The chemicals, being human-made, are new to the environment, and the natural mechanisms to biodegrade them into simple, harmless chemicals are not immediately available. For this reason, synthetic chemicals may remain in the environment for longer periods of time. As a result, there is more time for contact between the synthetic chemicals and plants and animals. There is also more opportunity for the plant or animal to absorb or eat harmful quantities of the chemical. Once inside an organism, a xenobiotic chemical will tend to accumulate. Very few biochemical mechanisms are available to the organism to break down the molecule so, as more molecules are ingested, the concentration of the chemical increases in the organism's tissue. An equal problem is that many inert materials eventually break down into tiny particles that organisms mistake for food. These particles can clog their digestive tracts or cause death from malnutrition.

TRANSPORTATION OF HAZARDOUS WASTES

Scientists studying hazardous wastes in the environment must answer three important questions.

First, how is the waste transported? Second, what is the concentration of the waste during transport? Finally, can processes in the natural environment reduce the distance the waste is transported?

Most hazardous wastes are transported either in air or in water; some wastes can be transported both ways. Because plants and animals use water and air in their daily lives, organisms absorb wastes as part of the normal life process. To determine how much of a given hazardous waste will be transported in solution (dissolved in water) or in air, a scientist determines the solubility or volatility of the waste. Solubility is a measure of how much of a chemical will dissolve in water. It is said that oil and water do not mix, which suggests that substances such as crude oil are not soluble in water. Yet crude oil is a complex mixture of chemicals, some of which can dissolve in water. Once dissolved, they may be carried away in solution to a river or a drinking-water well.

CONCENTRATION OF WASTES DURING TRANSPORT

Just as some of the chemicals in a crude oil spill are soluble in water, some of the chemicals are volatile. The smell of gasoline (a refined product of crude oil) vapors at a filling station indicates that some chemicals volatilize and are transported in the atmosphere. In the open atmosphere, air is moving so rapidly that under most normal weather conditions volatile hazardous chemical concentrations never get large enough to cause any concern. When weather conditions stagnate over large open areas, such as the Los Angeles basin, or in small areas, such as a cellar, the concentrations may become large enough to cause illness and death.

In an area where a hazardous waste spill is suspected, scientists will take samples of the local soil, water, and air. The samples will be analyzed for the soluble and volatile components of the waste. Surface water samples may be collected from lakes or streams. Groundwater samples may be collected from existing water wells or specially constructed monitoring wells. Air samples for volatile chemical analysis may be collected in restricted areas, such as basements, by using a vacuum pump to draw a certain amount of air through an activated carbon filter. The hazardous chemicals are absorbed by the carbon, and the sample is taken to a laboratory for analysis. Sometimes volatile hazardous chemicals can be identified in the soil atmosphere (the air filling the spaces between soil particles above the water table). Investigators pound a hollow pipe into the ground and, using a similar vacuum pump and carbon filter, draw soil air up through the pipe and over the absorbent carbon.

SORPTION AND BIODEGRADATION

Natural processes in soils and groundwater often act to destroy the chemicals in a hazardous waste. For example, a waste's rate of movement through the environment can be slowed by natural means. Some of the chemicals and elements in hazardous wastes have low solubility in water. These are called hydrophobic ("afraid of water"), as they attach themselves to the surface of a soil mineral grain or piece of dead plant material rather than stay in solution. Sorption, a natural process involving the removal of chemicals from solution by attachment to solid surfaces, slows a waste's rate of migration from the spill site. Some of the heavy metals and many of the hydrocarbons are hydrophobic. Sorption studies in the laboratory with soil materials and pure chemicals help to explain this complex natural phenomenon.

Microscopic organisms, such as bacteria and fungi, living in the soil and groundwater can limit the spread of a hazardous waste spill by consuming some of the chemicals. Hydrocarbons—common chemicals in a hazardous waste spill—can be an energy source for organisms. Biodegradation of hazardous chemicals is an important natural mechanism for limiting the transport of the chemical and ultimately cleaning up the wastes. Unfortunately, the microorganisms that consume the most hazardous chemicals are not common in most environments. The more common microbes, when they are in the environment, are often overwhelmed by the sudden influx of food represented by the spill. Other chemicals (such as oxygen gas) necessary for them to destroy the hazardous chemicals are quickly depleted, and the rate of destruction rapidly diminishes.

Lowry Landfill in Arapahoe County, Colorado, 1978: Garbage and hazardous wastes mix with groundwater. (U.S. Geological Survey, photo by N. Gaggiani)

Scientists hope to exploit the natural biodegradation of hazardous chemicals in two ways. One way is to stimulate microbial activity in the area of a spill. This usually means providing a fresh supply of oxygen and other nutrients necessary for the organisms to work at full speed. The other method is to grow microorganisms in the laboratory. When a spill occurs, the microorganism or group of microorganisms is brought from the lab to the spill site. Once there, it may be injected into the ground, or the water contaminated with waste may be pumped to the surface and into tanks containing the microbes. The biologically treated water is filtered before being discharged, to keep the microorganisms in the treatment tanks. The latter technique seems to be more effective because it is easier for scientists to provide the optimal conditions for biodegradation in the controlled treatment tanks. Advances in genetic engineering may make it possible to create organisms specifically designed to deal with hazardous materials.

ECONOMIC IMPACT

The chemicals that constitute hazardous wastes can be found in kitchens, workshops, garages, or garden sheds. Hazardous chemicals are present in household cleaners, gasoline and oil, paints, varnishes, thinners, herbicides, and insecticides. Handling hazardous chemicals has become part of our way of life. Through carelessness and ignorance, hazardous chemicals become hazardous wastes when they are released into the environment. When they are released, their movement can become uncontrolled, and their concentration in air or water sufficiently high to damage the environment. Most of the chemicals that represent hazardous wastes at one time were useful. It is only when they are accidentally spilled or disposed of improperly that they become hazardous wastes.

The government spends a significant amount of money to clean up hazardous waste sites and to prevent future spills. Billions of dollars will be spent as part of the Comprehensive Environmental Response, Compensation and Liability Act (CERCLA), also known as Superfund, to find and clean up uncontrolled waste sites. More is being spent through the Resource Conservation and Recovery Act (RCRA) to limit or eliminate hazardous chemical releases to the environment, particularly from the factories that use and produce large quantities of the chemicals.

Banks and savings and loans, which hold the mortgages on all types of properties—from large industrial complexes to private homes—are now concerned about hazardous wastes. If buried or spilled hazardous wastes are found on a piece of property, the expense of a cleanup may be so high that the property occupant will declare bankruptcy. The holder of the property mortgage may then be liable for the cleanup. Before committing themselves to finance a land purchase, banks now require a preliminary hazard assessment of the property. Such an assessment usually includes an examination of historical records, indicating the uses of the land by previous owners, and a site survey to identify any obvious signs of problems.

Richard W. Arnseth

FURTHER READING

Brown, Michael. *Laying Waste: The Poisoning of America by Toxic Chemicals.* New York: Pantheon, 1980. Follows the events surrounding the discovery of hazardous wastes at Love Canal. Follows that story with case histories of hazardous waste's effects on communities around the country. Accessible to the general reader. Offers a good index but no bibliography.

Byrne, John, and Steven M. Hoffman, eds. *Governing the Atom: The Politics of Risk.* New Brunswick, N.J.: Transaction, 1996. Volume from the Energy and Environmental Policy Series researches the environmental and social aspects of the nuclear industry. Considers government policies that have been implemented to monitor safety and accident prevention measures.

Dadd, Debra L. *The Nontoxic Home and Office: Protecting Yourself and Your Family from Everyday Toxics and Health Hazards.* Los Angeles: Jeremy P. Tarcher, 1992. Discusses the toxic chemicals found in common household products and their potential health hazard. Gives insight into the wide range of common products that become hazardous wastes when discarded. Provides information on nontoxic alternative products. Suitable for the general reader.

Gay, Kathlyn. *Silent Killers: Radon and Other Hazards.* New York: Franklin Watts, 1988. Covers a range of hazardous chemicals, from Agent Orange to lead-based paint to radioactive soil, used as fill for home construction. Offers small bits of technical descriptions between larger discussions of the political, economic, and social impacts of hazardous wastes. Written in a nontechnical style with a bibliography of books and magazine articles. Recommended for the general reader.

Hackman, Christian L., E. Ellsworth Hackman III, and Matthew E. Hackman. *Hazardous Waste Operations and Emergency Response Manual and Desk Reference.* New York: McGraw-Hill, 2002. Provides practical application of hazardous material regulations and legislature. Discusses federal agencies and other organizations related to hazardous waste management. Designed to guide individuals through risk assessment, safety protocol development and implementation, and emergency response. Appropriate for corporate safety and health managers. A valuable resource for management and professionals responsible for OSHA compliance.

Hawkes, Nigel. *Toxic Waste and Recycling.* New York: Gloucester, 1988. Profusely illustrated introduction to toxic waste problems for young readers. Emphasizes recycling as a solution. A good introduction to the scope of the problem and potential solutions for junior high students.

LaGrega, Michael D., Philip L. Buckingham, and Jeffrey C. Evans. *Hazardous Waste Management.* 2d ed. Long Grove, Ill.: Waveland Press, 2001. Covers the fundamentals of hazardous waste, from the technical definition and chemical and physical properties, to federal regulations. Discusses current preventive practices, as well as disposal methods and risk assessment. Appropriate for advanced undergraduate and graduate students with some background in engineering.

Moeller, Dade W. *Environmental Health.* 3d ed. Boston: President and Fellows of Harvard College, 2005. Covers such topics as toxicology, drinking water, electromagnetic radiation, economics, and environmental law and monitoring. Contains references and indexing. Differentiates among clinical medicine, public health, and environmental health issues. Includes many tables and figures.

O'Leary, Philip R., Patrick W. Walsh, and Robert K. Ham. "Managing Solid Waste." *Scientific American* (December, 1988): 36. Discusses the mounting solid waste disposal problem related to dwindling landfill space and the restriction on siting new landfills. Discusses alternatives to landfill disposal: recycling, volume reduction techniques such as incineration, landfill design, and energy from garbage. Suitable for the general reader.

Parker, Sybill P., ed. *McGraw-Hill Concise Encyclopedia of Environmental Science.* 6th ed. New York: McGraw-Hill, 2009. Covers industrial wastes, soil, groundwater, and solid waste management. Sections are written in a nontechnical style, with a few illustrations and tables of facts. Includes short bibliographies for more technical reading.

Pichtel, John. *Waste Management Practices: Municipal, Hazardous, and Industrial.* Boca Raton, Fla.: CRC Press, 2005. Discusses various types of waste, the history of waste management, laws and regulations, as well as sampling, chemical properties, collection, transport, and processing of municipal solid wastes. Covers technical components of recycling and composting. Also discusses landfills. Includes references, suggested readings, questions, and a sizeable index.

Rosenfeld, Paul E., and Lydia G. H. Feng. *Risk of Hazardous Wastes.* Burlington, Mass.: Elsevier, 2011. Provides case studies and numerous other real-life examples of waste management issues. Chapters are separated by industry and include references. Discusses current treatment, disposal, and transport practices. Focuses on toxicology, pesticides, environmental toxins, and human exposure issues, along with risk assessment, regulations, and accountability. Includes numerous appendices with tables of a number of OSHA, NIOSH, ACGIH, ATSDR, WHO, and EPA limits and regulatory values.

Turnberg, Wayne L. *Biohazardous Waste: Risk Assessment, Policy, and Management.* New York: John Wiley & Sons, 1996. A complete account of the policies, laws, and legislation surrounding medical, infectious, and hazardous waste management. Provides an analysis of the risks involved.

Wagner, Kathryn D., et al., eds. *Environmental Management in Healthcare Facilities.* Philadelphia: W. B. Saunders, 1998. Provides an examination of the procedures which health-care facilities abide in regard to the disposal of medical, infectious, and hazardous waste, and the precautions they take to prevent environmental pollution.

See also: Earth Resources; Earth Science and the Environment; Landfills; Mining Processes; Mining Wastes; Nuclear Power; Nuclear Waste Disposal; Petroleum Reservoirs; Solar Power; Strategic Resources.

HIGH-PRESSURE MINERALS

Minerals are formed from a number of factors, including pressure. Some minerals are formed in the low to moderate pressures of the earth's outer layers. Other minerals are produced under great pressure. One of the best-known minerals formed in such conditions is the diamond, which is transformed from deep in the earth's mantle and brought to the surface through a number of geologic processes.

PRINCIPAL TERMS

- **deformation:** geologic process whereby the shape and size of rocks and minerals are altered because of the application of pressure, temperature, or chemical forces
- **equilibrium:** geologic state in which temperature and pressure, along with other environmental conditions, are in balance
- **hydrostatic:** state of fluid pressure equilibrium
- **mantle:** layer of hot, solid rock between the earth's core and outer crust
- **metamorphism:** geologic process whereby igneous or sedimentary rocks are changed compositionally under high pressure
- **sedimentary basin:** lower area in the earth's crust in which sedimentary rocks and minerals accumulate
- **subduction:** geologic process whereby a tectonic plate is forced below another plate

BASIC PRINCIPLES

The earth has myriad minerals, each of which is formed under particular circumstances. Some minerals are formed in the earth's outer crust, combining the eroded elements and compounds produced when water dissolves rocks on the planet's surface. Other minerals are formed in cooling magma as it moves outward toward the earth's surface. Still others are formed in the mantle (the layer of hot, solid rock that rests between the earth's core and outer crust), which is located about 28 meters (90 miles) below the planet's surface.

In the last of these scenarios, minerals are produced under high amounts of pressure. This pressure comes from downward movement of the planet's tectonic plates (massive slabs of rock that are in constant motion beneath the earth's surface). The plates themselves are under pressure from gravity and the weight of oceans and glaciers, or are forced beneath another plate (a process known as subduction).

Minerals also may be produced under the intense pressure between tectonic plates. Here, other minerals and rock that have been pushed outward from the mantle may be caught between two contacting plates. In these cases, the pressure causes the rocks to change compositionally to form new minerals (a process known as metamorphism).

Still other types of minerals are formed not in the mantle but deep beneath the earth's crust in sedimentary basins. Here, a number of environmental factors create different pressures that, in turn, cause varying types of mineral configurations.

MINERAL FORMATION UNDER PRESSURE

A mineral is an amalgamation of basic elements, such as carbon, copper, iron, and oxygen. Some minerals comprise a single element. Minerals are solid in texture, formed in nature with crystalline structures. Each element contains atoms that bond with other atoms through the electronic charges that draw together the atoms' respective electrons and protons. Thousands of minerals exist, developed under a variety of degrees of pressure and temperatures. Glass and ice are minerals, too, although ice ceases to be a mineral when it melts.

Two major natural forces—temperature and pressure—are significant in the formation of minerals. Temperature, for example, causes atoms to vibrate and to distance themselves from one another, causing instability in the crystalline structure and, as temperature increases, liquefaction. Pressure draws atoms together, creating more dense crystalline structures.

For a mineral to coalesce from the liquid and semisolid materials emanating from Earth's core and mantle, high degrees of pressure (measured in atmospheres, the equivalent of approximately 14.7 pounds per square inch) are often essential. For example, muscovite (also known as mica) and feldspar—minerals that commonly form igneous rocks (rocks that form from cooling magma)—depend upon high levels of pressure for such bonding to occur.

CONDITIONS IN THE MANTLE

In many cases, high-pressure minerals are formed under the intense pressure and temperatures of the mantle. The mantle appears in two distinct parts: upper and lower. Conditions in each part (including pressure and temperature) are somewhat different. Therefore, different high-pressure minerals may be found in these two areas.

In the upper mantle (on which the earth's tectonic plates rest) the temperature of the rock is at or near a melting point. Scientists believe this area produces the magma that is pushed out through volcanoes. Upper mantle minerals form under pressure from subducting plates above it. Scientists attempting to discern the types of minerals that are present in the upper mantle typically rely on examinations of material ejected from erupting volcanoes. One mineral, iron (which is found in the earth's core), is typically disregarded in this pursuit. However, other minerals formed under high levels of pressure in the upper mantle include peridotite, eclogite, and diamonds.

The pressure in the lower mantle is far greater than found in the upper mantle. Scientists link the pressure in which a mineral is produced to its overall strength. Hence, minerals produced in the lower mantle tend to have greater strength than those in the upper mantle. Lower mantle minerals, such as perovskite, possess a much greater level of strength (including a resistance to higher temperatures) than upper mantle minerals such as peridotite.

DIAMONDS

One of the best-known high-pressure minerals is the diamond. One of the hardest materials on Earth, diamond is also among the most sought-after minerals. A common misperception is that diamonds are formed from coal. In reality, the process by which diamonds are formed provides an excellent illustration of the environment in which many high-pressure minerals are formed.

Diamonds are formed in a number of environments. The notion that diamonds are made from coal comes from the knowledge that diamonds, like a large percentage of minerals, consist of carbon. Diamonds that are formed in the earth's lower mantle are derived from carbon that was deposited deep in the interior when the planet was still forming. The areas in which diamonds are formed are known as diamond stability zones, wherein the pressure and temperature is constant. Diamonds are dislodged from these zones during volcanic eruptions and are then carried into sedimentary basins from which they are mined.

The lower mantle is not the only place where diamond stability zones exist. Diamonds may be formed farther outward in the upper mantle. Here, however, the carbon from which they are formed comes from rocks such as limestone and marble (and organic debris, such as plants) that have been subducted by oceanic plates. Diamond stability zones are formed in these environments as oceanic plates are subducted under continental plates, creating extremely high pressure on the upper mantle (where the diamonds are formed). The churning effects of subduction move the diamonds, along with other rocks in the upper mantle, from stability zones and either out toward the surface or deeper into the mantle.

METAMORPHISM

A mineral's assemblage and texture may be altered significantly because of pressure. In other words, higher levels of pressure may in essence significantly change the composition of the rock or mineral. This process is known as metamorphism. As pressure increases, so, too, does the rate of change in the mineral (through a concept called prograde metamorphism). Scientists believe there also exists a rare form of metamorphism in which, as pressure is decreased, a mineral's assemblage may change. This latter manifestation is known as retrograde metamorphism.

Pressure that induces metamorphism is not of a singular nature; a number of different types of pressure can influence how a mineral is formed. For example, the shapes and crystalline structures of certain minerals are frequently altered because of varying degrees of pressure emanating from one side of the mineral or another. This differential stress can cause deformations such as rounding or disintegration into sheets. In other cases, the stress is uniform: It is applied in equal proportions from all around the mineral.

PLATE TECTONICS AND MINERAL FORMATION

The theory of plate tectonics (the notion that the earth's lithosphere, located just below the outer crust, comprises a series of giant stone slabs that are in constant motion) reveals a great deal about how high-pressure minerals are formed and re-formed.

The concept demonstrates how molten material flows outward from the mantle toward the earth's outer crust, contributing to the movement of the plates (which are also moving because of gravitational forces). Within the plate tectonics framework, thousands of the earth's minerals (including high-pressure minerals) are created.

According to plate tectonics, basalt (an iron- and magnesium-rich form of rock that is low in density and is therefore commonly found in volcanic lava) flows out through the lithosphere and toward the plates. Contained in the basalt are the elements and minerals essential for the formation of other minerals. As the basalt arrives at the lithosphere, some of the minerals are crystallized, depending on the elements present and on environmental conditions (which include pressure).

In time, however, the minerals that have become conjoined with the rest of a tectonic plate eventually start to descend, largely because of the strong forces of gravity pushing down on the plates. As the plate descends, the minerals again undergo increasing levels of pressure, causing mineral deformation and metamorphosis.

An example of this process may be found in a study of the mountains of the Czech Republic. Geologists studying the geodynamic processes of this region found large concentrations of high-pressure mineral grains. These minerals provide evidence of the thickening of the plate suggested in plate tectonics. The study also revealed a diverse array of pressure environments. (Various concentrations of certain high-pressure minerals, such as eclogite and peridotite, found in the research illustrate that the region is in a dynamic state.) In other words, when minerals are formed under certain high-pressure conditions, they may be metamorphosed when the pressure equilibrium is upset as the plate is pushed down by gravity.

SUBDUCTION

A key process in the formation of minerals under high grades of pressure is subduction. In this phenomenon, tectonic plates are pushed down because of the presence of glaciers or oceans above them or because of pressure exerted upon them through contact with other plates. This process causes intense downward pressure on the mantle. It also causes extreme pressure in the areas (the subduction zones) between the plates.

Scientists can study the minerals formed in these high-pressure environments after the minerals are forced into subduction zones and out through volcanic activity. The configuration of the rocks containing these minerals often provides evidence of repeated metamorphism, demonstrating that high-pressure minerals are formed in a dynamic environment. For example, mineralogists and geologists studied a sample of eclogite found in an area of central China in which the South China block, a tectonic plate, is deeply subducted. The sample showed multiple decompression textures, revealing that the minerals had been formed, transferred, and re-formed under the high pressure of the subduction zone.

HIGH-PRESSURE FORMATION IN SEDIMENTARY BASINS

As suggested earlier, minerals formed under great degrees of pressure are not necessarily located in the mantle. Indeed, a number of conditions exist in which minerals are formed and re-formed closer to the outer crust. For example, studies of sedimentary basins (lower areas in the earth's crust in which sedimentary rocks and minerals accumulate) show that high levels of pressure exist in certain areas, facilitating the formation of certain mineral deposits.

Geologists are increasingly focused on the presence of water and other liquids in a number of sedimentary basins. These liquid deposits, contained in pockets (or pores), exert pressure on the surrounding sediment. This liquid pressure may be in a state of equilibrium, a condition known as hydrostatic pressure. The fluid pressure is great enough to deform existing rocks and minerals and to form new ones.

If the pore is influenced by increases in temperature, increased compaction from additional sediment, or other factors (including volcanic activity), the fluids may experience a state of overpressure. Even human-caused compaction, such as exploratory mining for fuels and gems, can increase pressure on a pore. When the pore reaches an overpressure point, it may explode violently. The same fluid pressure that can form minerals in sedimentary basins also can cause large amounts of damage and fatalities.

Michael P. Auerbach

FURTHER READING

Bolin, Cong, ed. *Ultrahigh-Pressure Metamorphic Rocks in the Dabieshan-Sulu Region of China.* New York: Springer, 1997. Comprehensive look at the ultrahigh levels of pressure that are present beneath the earth's crust and how they affect mineral formation. The editor specifically looks at a region in China, which has one of the world's largest exposed areas of high-pressure metamorphosed rock.

Chen, Ren-Xu, Yong-Fei Zheng, and Bing Gong. "Mineral Hydrogen Isotopes and Water Contents in Ultrahigh-Pressure Metabasite and Metagranite: Constraints on Fluid Flow During Continental Subduction-Zone Metamorphism." *Chemical Geology* 281, nos. 1/2 (2011): 103-124. Discusses the involvement of water and fluid pressure in the creation, deformation, and metamorphism of minerals in the Sulu area of China. Cites these minerals in terms of the implications of fluid pressure for miners and drillers.

Duffy, Simon, E. Ohtani, and D. C. Rubie, eds. *New Developments in High-Pressure Mineral Physics and Applications to the Earth's Interior.* Atlanta: Elsevier, 2005. Provides a detailed study of the earth's geophysical systems and how scientists measure them. Discusses the high-pressure-producing forces that form minerals in the earth's deep interior.

Feldman, V. I., L. V. Sazonova, and E. A. Kozlov. "High-Pressure Polymorphic Modifications in Minerals in the Products of Impact Metamorphism of Polymineral Rocks." *AIP Conference Proceedings* 849, no. 1 (2006): 444-450. Examines a series of laboratory experiments designed to re-create the different types of high-level pressure that fosters and changes the structures of the minerals olivine and pyroxene. The experiment used shock waves, seeking to reproduce the pressures created by asteroids and other space debris when they impact the earth's outer crust.

Liu, Yi-Can, et al. "Ultrahigh-Pressure Metamorphism and Multistage Exhumation of Eclogite of the Luotian Dome, North Dabie Complex Zone (Central China): Evidence from Mineral Inclusions and Decompression Textures." *Journal of Asian Earth Sciences* 42, no. 4 (2011): 607-617. Examines samples of eclogite, the product of the subducted South China block. The sample is the product of multiple subduction processes.

Pellant, Chris. *Smithsonian Handbooks: Rocks and Minerals.* London: DK Adult, 2002. Provides extensive photographs and descriptions of more than five hundred types of rocks and minerals, including those produced in high-pressure environments.

See also: Crystals; Diamonds; Hydrothermal Mineralization; Igneous Rock Bodies; Minerals: Physical Properties; Minerals: Structure; Pyroclastic Flows; Ultra-High-Pressure Metamorphism.

HYDROELECTRIC POWER

Hydroelectric power is the most commonly used renewable energy resource. It can be stored in the form of impounded water and is relatively nonpolluting.

PRINCIPAL TERMS

- **electricity:** a flow of subatomic charged particles called electrons used as an energy source
- **energy:** the capacity for doing work; power (usually measured in kilowatts) multiplied by the duration (usually expressed in hours, sometimes in days)
- **flow rate:** the amount of water that passes a reference point in a specific amount of time (liters per second)
- **generator:** a machine that converts the mechanical energy of the turbine into electrical energy
- **head:** the vertical height that water falls or the distance between the water level of the reservoir above and the turbine below
- **kilowatt:** 1,000 watts; a unit of measuring electric power
- **megawatt:** 1 million watts; a unit of measuring electric power
- **penstock:** the tube that carries water from the reservoir to the turbine
- **power:** the rate at which energy is transferred or produced
- **pumped hydro:** a storage technique that utilizes surplus electricity to pump water into an elevated storage pond to be released later when more electricity is needed
- **turbine:** a device used to convert the energy of flowing water into the spinning motion of the turbine's shaft; it does this by directing the flowing water against the blades mounted on the rotating shaft

SUPPLY AND TRANSMISSION

Even though running water has been used by people for centuries to turn the wheels of gristmills and sawmills, it was not until the end of the nineteenth century that waterpower began to be employed to generate electricity. Hydroelectric projects can be as small as a waterwheel supplying energy to a single household or as large as a system of dams and storage projects that supply electricity to many cities and millions of people. Electric energy, generated by water-powered turbines, is transported to houses, factories, mills, and other sites of consumption along high-voltage transmission lines that may extend for more than 1,500 kilometers. These transmission lines are either alternating current (AC), the type of electricity used in houses, or direct current (DC), the type of electricity used in batteries, and can deliver hundreds of megawatts of power. In the United States, agreements between states, regions, and Canada have created a network of transmission lines that allows the flow of electricity from one part of North America to another. This ability to transport electricity from one place to another was one of the driving forces behind the relocation of factories and mills from along the rivers to locations adjacent to sources of raw materials. The mobility of electricity has also allowed for the growth of numerous cities located away from sources of energy.

The easiest way to harness a river for the purpose of generating hydroelectric power is to construct a dam across a river and funnel the water through a turbine that creates electricity. A dam with a large reservoir of water behind it is best for generating electricity, because both the amount of water in the river and the demand for electricity vary throughout the year. For example, in the Columbia River system in the Pacific Northwest, the river reaches peak flow during the spring snowmelt, but demand for electricity is greatest during the late winter (for heating homes) and summer (for air conditioning). The ability to store large volumes of water behind each of the dams in the system allows electric utilities to meet summer and winter demands, "storing" the water until electricity is needed. Water from spring runoff is stored and released to generate electricity to power air conditioners in the summer and heat homes in the winter.

Large storage dams also allow a utility to increase or decrease electric generation to match the demand. Electrical demand in the morning is met by releasing extra water. At night, when the demand is less, water is either kept in the reservoir or passed through the turbines using a process called spinning. Spinning is the method in which water passes through a turbine but no electricity is produced. In a matter of seconds, the spinning turbine can be engaged and electricity produced.

PUMPED HYDRO

Another form of stored hydroelectric power is pumped storage. During the period of midnight to six in the morning, when energy demand is at its lowest, hydroelectric projects must maintain a minimum outflow of water for navigation, agriculture, mining interests, recreational interests, fish breeding, and water quality. A utility may choose to use the electricity produced by minimum stream flow regulations to pump water into a storage pond. Then, during the day, when energy demand increases, the pumped water is released and electricity is generated.

During the late 1980's, pumped hydro required 1.3 to 1.4 kilowatt-hours for every kilowatt-hour produced. That may not seem economic, but because the water is pumped using surplus energy (energy that cannot be saved), the utility is able to postpone generation until demand is present. Another consideration is the difference in the cost of energy at peak and nonpeak hours. Electricity sold during peak demand may be several times the cost of electricity sold during nonpeak hours. Some utilities have even pumped water into excavated caverns. Hydroelectric power produced by the pumped-hydro plants amounts to thousands of megawatts.

WATER FLOW RATE AND HEAD

Hydroelectric power is produced by converting the potential power of natural stream flow into energy. This is commonly done by employing falling water to turn turbines, which then drive generators and produce electricity. The amount of electric power produced is dependent upon the flow rate of the water and the head.

The flow rate of water is the volume of water that moves past a point during a specific period of time. The quantity of water is commonly measured by first determining the cross-sectional area (width and depth) of the river; second, the speed of the water is measured by defining a reference length of river to monitor, dropping a float at the upper end of the reference length, and recording the length of time the float takes to travel down the reference length of the river. Then, if one calculates the volume of water (length × width × depth) divided by the time the float

took to travel down the reference length of the river, the result is volume per time, or flow rate.

The head of a stream is the vertical height through which the water falls. The head measurement of a hydroelectric project is the elevation difference between where the water enters the intake pipe, or penstock, and the turbine below. When waterwheels are employed to produce electricity, the head measurement is the total distance that water falls to the waterwheel. As with all energy conversions, friction results in some loss. The type of turbine or waterwheel utilized can also contribute to greater or lesser losses of energy. The theoretical maximum power available in kilowatts is equal to the head, measured in meters, multiplied by the flow rate, measured in cubic meters per second, multiplied by 9.8. Efficiencies (actual energy produced divided by the amount of energy available in the flowing water) for hydropower plants (turbines, waterwheels) vary from a high of 97 percent, claimed by manufacturers of large turbines, to less than 25 percent for some waterwheels.

WATERWHEELS

Two devices used to convert the potential energy of water to mechanical energy are the waterwheel and the water turbine. The type of waterwheel or turbine used is dependent upon the flow and head. The ideal situation is high head (more than 18 meters) and high flow, but it is feasible to produce electric energy with any combination of high head and low flow, or low head and high flow.

Waterwheels are the simplest machines employed to generate hydroelectric power. The central shaft of the waterwheel, which in the past was directly connected to a grindstone, is hooked to a generator to produce electricity. Efficiency has been claimed by waterwheel manufacturers to be around 90 percent, but the usual efficiency ranges from 60 percent down to around 20 percent. The most efficient is the overshot wheel, in which water falls onto the top of the wheel and turns it. The wheel is suspended over the tailwater (the water on the downstream side of the wheel) and is not resting in the water but suspended above the water surface. This type of waterwheel requires at least a 2-meter head of water.

Three other types of waterwheels are able to operate at lower heads than does the overshot wheel, but all three are costly to construct. The first type, the low and high breast wheels, are turned by water striking the wheel at a point one-third to over one-half of the height of the wheel. The "low" or "high" defines the level at which the water enters the wheel. The second type, the undershot wheel, is probably the oldest style presently in use. The wheel is turned by water running under the wheel. Although this type of waterwheel has an efficiency of less than 25 percent, it can operate with less than a third of a meter head. The third low-head waterwheel is the Poncelet wheel, which is an improved undershot that rests just at the water level and depends upon the velocity of the water to turn the wheel. Because this wheel forces the water through narrow openings on the wheel, it is suitable for heads of less than 2 meters, but it is easily damaged by debris carried in the water.

Because the waterwheel rotates at a slow rate, the gear box, which transfers the rotation energy to the turbine, is a very costly, complex collection of gears. This expense is a major disadvantage. Another disadvantage is the large size of a waterwheel. Given the large amount of time and material involved and the low efficiency, the rate of monetary return is low. Overshot wheels, however, have the advantage of being able to operate with fluctuating water flows better than do water turbines. A second advantage is that once set up, and unlike low-head water turbines, an overshot wheel requires little repair and is not damaged by grit or clogged by leaves.

TURBINES

There are two types of hydraulic turbines: impulse turbines, which utilize water under normal atmospheric pressure, and reaction turbines, which use water under pressure to drive them. Impulse turbines are driven solely by the impact of water against the turbine blades. The Pelton impulse wheel was designed in 1880 and is the crossover from waterwheels to turbines. The Pelton wheel is composed of a disk with buckets attached to the outside of the wheel. The wheel requires a head of at least 18 meters (roughly 1.8 times atmospheric pressure) but can operate under low flow rates because the water is forced under its own pressure through a nozzle to strike the buckets. The water striking the buckets causes the wheel to spin. Because operating efficiencies are commonly over 80 percent, this wheel is still a favorite of many small utilities in North America. The turgo impulse wheel represents an improvement over the Pelton. The water jet

is aimed at the buckets at a low angle, thus allowing the stream of water to strike several buckets at once. That results in higher efficiencies with smaller wheels and lower flow rates than those needed for the Pelton. The turgo has an efficiency reported over 80 percent and is suited for use with heads greater than 10 meters. The cross-flow turbine is a drum-shaped impulse turbine with blades fixed along its outer edge. The drum design allows water to pass over the blades twice: once from the outside to the inside, then (after entering the drum) back outside again. The net result is up to 88 percent efficiency in large units and the ability to operate with heads as low as 1 meter. The cross-flow turbine is in widespread use around the world. It is simple to operate and largely self-cleaning.

Reaction turbines of several types are normally used in the largest hydroelectric projects; a single unit at Grand Coulee Dam can produce 825 megawatts. Reaction turbines exploit not only the impact of water but also the drop in water pressure as it passes through the turbine. Reaction turbines work by placing the whole runner (which is what is left of the "wheel" and resembles all blades set into a central shaft) into the flow of water. The water is carried to the turbines from the reservoir by a long tube called a penstock. The penstock can be more than 10 meters in diameter and tens of meters long. A propeller turbine is a reaction-type turbine that resembles a boat propeller in a tube. This type of turbine may be set either horizontally or vertically, depending on the design of the system. The Kaplan turbine is a turbine with adjustable blades on the propeller to allow operation at different flow rates. The water pressure in this system must be constant or the runner will become unbalanced. Very large hydroelectric plants usually install the Francis turbine. This type of turbine is designed to be set up and adjusted for the specific site. It can be used with a head of 2 meters and has an efficiency rating greater than 80 percent. The turbine spins as water is introduced just above the runner and directed onto the blades, causing the blades to rotate. The water then falls through and out a draft tube. A complicated mechanical governor is often used to guide the water around the runner.

COSTS AND BENEFITS

Hydroelectric power is the most developed renewable energy resource. It supplied about 20 percent of the world's electricity as of 1999. Where available, hydroelectric power likely is less expensive than alternatives. Its primary environmental impacts come from flooding land behind a dam and blocking a river's flow; flooding destroys habitat and displaces the population of the flooded area. Mercury levels in impounded water rise somewhat for about twenty years in a newly flooded area but return to normal as the mercury is leached from the flooded soil. Carbon dioxide and sulfur dioxide are released to the atmosphere as newly submerged vegetation decays, but even so hydroelectric plants emit only about 4 percent as much greenhouse gases as natural-gas-fired or oil-fired power plants, and about 3 percent as much as coal-fired power plants.

During the late 1980's, government and private industry studies were still identifying many regions in the world that could be developed with hydroelectric power plants. The available technology allowed for construction of dams that would permit unhindered fish migration, coexistence of fish hatcheries and hydroelectric projects, and maintenance of natural fish and wildlife populations. Concerns about commercial and sport fishing populations have resulted in the close monitoring of hydroelectric power plant operation by biologists to ensure fish survival.

A major obstacle to the development of new hydroelectric power plants is the large amount of paperwork involved. The cost of environmental impact studies can exceed the cost of actual construction. Dams must be licensed by the federal government and meet hundreds of county and state regulations. The amount of water allowed to flow downstream is regulated to ensure that agriculture, sport and commercial fisheries, recreation, environmental considerations, and Native American water rights are satisfied. These competing interests imply that no single user of a given river will determine the quantity and conditions by which water moves through the dams.

Susan D. Owen

FURTHER READING

Alward, Ron, Sherry Elisenbart, and John Volkman. *Micro-Hydro Power: Reviewing an Old Concept.* Butte, Mont.: National Center for Appropriate Technology, 1979. Delineates all the components of a hydroelectric system with detailed but easy-to-understand pictures. Includes pictures of turbines used on larger hydroelectric projects. Contains an international list of manufacturers and suppliers of hydroelectric system components, as well as an excellent bibliography.

Andrews, John, and Nick Jelley. *Energy Science: Principles, Technologies, and Impact.* New York: Oxford University Press, 2007. Discusses environmental and socioeconomic impacts and covers principles of energy consumption, including information on the generation, storage, and transmission of energy. Many examples require a strong mathematics or engineering background. Appropriate for undergraduates and professionals with interest in energy resources.

Basson, M. S., et al., eds. *Probabilistic Management of Water Resources and Hydropower Systems.* Highlands Ranch, Colo.: Water Resources Publications, 1994. Supplies the reader with mathematical models used to assess management of water supply, watersheds, and hydroelectric plants. Includes illustrations and bibliography.

Boyle, Godfrey, ed. *Renewable Energy.* 2d ed. New York: Oxford University Press, 2004. Provides a complete overview of renewable energy resources. Discusses solar energy, bioenergy, geothermal energy, hydroelectric energy, tidal power, wind energy, and wave energy. Discusses basic physics principles, technology, and environmental impact. Includes references and a further reading list, which makes this book an excellent starting point. Advanced technical details are limited.

Freeze, R. Allan, and John A. Cherry. *Groundwater.* Englewood Cliffs, N.J.: Prentice-Hall, 1979. Considered the leading text on groundwater hydrology. Presents the subject in an interdisciplinary manner, with practical sampling methods and tests. (It is important to understand the relationships between surface and subsurface water systems before understanding hydroelectric systems.)

Jiandong, Tong, et al. *Mini Hydropower.* New York: John Wiley & Sons, 1997. Examines the earth's renewable energy resources, water supply management, and the operations of hydroelectric plants. Includes illustrations and maps, index and bibliographical references.

Letcher, Trevor M., ed. *Future Energy: Improved, Sustainable and Clean.* Amsterdam: Elsevier, 2008. Begins with the future of fossil fuels and nuclear power. Discusses renewable energy supplies, such as solar, wind, hydroelectric, and geothermal. Focuses on potential energy sources and currently underutilized energy sources for the future. Covers new concepts in energy consumption and technology.

Majot, Juliette, et al., eds. *Beyond Big Dams: A New Approach to Energy Sector and Watershed Planning.* Berkeley, Calif.: International Rivers Network, 1997. Essays and case studies focused on the environmental and social aspects of dams, hydroelectric plants, and electric power production, as well as the government policies surrounding these processes. Suitable for the layperson.

McGuigan, Dermott. *Harnessing Water Power for Home Energy.* Pownal, Vt.: Garden Way Publishing, 1978. Explains how to build any type of small to microscale hydroelectric facility. Lists manufacturers in the United States and the United Kingdom, as well as the 1989 cost of the equipment.

Palmer, Tim. *Endangered Rivers and the Conservation Movement.* Berkeley: University of California Press, 1986. Examines the "flip side" of hydroelectric power: the river. Details the conservation battles to preserve rivers in their natural states. Based on hundreds of interviews and reviewed by a number of environmentalists and politicians. Easy, enjoyable reading.

Sullivan, Charles W. *Small-Hydropower Development: The Process, Pitfalls, and Experience.* 4 vols. Palo Alto, Calif.: Electric Power Research Institute, 1985-1986. Explains hydroelectric power plants, how to determine where to place them, cost, regulations, environmental impact, and a number of other related topics. Completed under contract with the U.S. Department of Energy. Suggests computer programs for organizing data. Detailed enough for any group wanting to build a hydroelectric plant and simple enough to read and understand on a nontechnical level.

United States Bonneville Power Administration. *Columbia River Power for the People: History of Policies of the Bonneville Power Administration.* Portland, Ore.: Author, 1980. Provides a good description of the development of one of the largest hydroelectric systems in the world. Documents the harnessing and development of the Columbia River and the politics involved. Helps the reader to understand the social, economic, and cultural forces that must be pacified in order to create a successful hydroelectric system.

United States Federal Power Commission. *Hydroelectric Power Resources of the United States, Developed and Undeveloped.* Washington, D.C.: Government Printing Office, 1992. Reports hydroelectric power resources, both developed and undeveloped, as of 1976. Lists the state, project owner, river, developed and undeveloped generation capacity, and gross static head of all projects licensed by the federal government.

See also: Earth Resources; Geothermal Power; Hydrothermal Mineralization; Nuclear Power; Ocean Power; Solar Power; Strategic Resources; Tidal Forces; Unconventional Energy Resources; Wind Power.

HYDROTHERMAL MINERALIZATION

Hydrothermal mineralization occurs when mineral particulates precipitate from water. This process is closely tied to tectonics and volcanic activity, which create the heat and convection necessary for the transportation and deposition of minerals. Hydrothermal mineralization can occur in submarine or terrestrial environments. Hydrothermally created mineral deposits are economically important, and they provide data that help geologists to understand Earth processes.

PRINCIPAL TERMS

- **black smoker:** a columnar structure formed deep in the ocean when sulfides precipitate from hydrothermal fluids
- **crust:** the outermost layer of the earth; crust may be continental or oceanic
- **divergent boundary:** an area where two tectonic plates are spreading apart instead of crashing together or sliding against each other
- **fumarole:** a gas- and steam-producing vent at volcanic craters
- **hot spot:** a column of magma that rises from the mantle, remaining in one place as the lithospheric plate moves over it; also known as a mantle plume
- **hydrothermal vent:** a fissure through which magmatically heated fluids escape to Earth's surface
- **isotopes:** atoms of an element that contain the same number of protons but different numbers of neutrons
- **magma:** hot fluid or semifluid from beneath the earth's crust; later forms igneous rock
- **mantle:** a hot layer of the earth below the crust, asthenosphere, and lithosphere; surrounds the inner core of the planet
- **silicate:** a compound containing silica; includes mica, feldspar, and quartz
- **sulfide:** a compound made with sulfur; an iron sulfide is a compound of iron and sulfur

Hydrothermal Mineralization:
The Big Picture

Liquids frequently contain minerals, and when the liquids are heated, the minerals tend to fall from solution (precipitate). An everyday example of this is the mineral layer that sometimes forms on the bottom of tea kettles or in coffee makers.

In geologic terms, hydrothermal mineralization means that water is heated by magmas or hot rocks, and as the water travels, it leaves behind the minerals it was holding in solution. Other factors that affect precipitation are chemical reactions, pressure, oxygen levels, acidity, and rate of cooling.

To understand how water becomes heated and chemically active, it is necessary to understand wider geologic forces. Humans are familiar with the earth's crust because it is the layer of inhabitation, but beneath the crust are the lithosphere (composed of tectonic plates), the asthenosphere, and the mantle, which is hot. Convection causes the plates to move. Sites where two tectonic plates crash together are called convergent boundaries. Sites where they are moving apart are called divergent boundaries.

Movement of tectonic plates creates seismic activity and facilitates the release of heat and hot magma from the earth's interior. Evidence of this process comes in the form of volcanoes. Hydrothermal activity occurs at divergent boundaries (often midocean ridges) because new, hot material comes to the surface as the plates spread apart. Hydrothermal processes also occur at subduction zones (associated with convergent boundaries), where surface crust slides under another plate as they collide and descends to an area of increased heat and pressure. The rock in the crust then melts and is less dense than surrounding material, allowing the rock to again rise to the surface.

Because hydrothermal activity is associated with plate boundaries and hot spots (localized areas where a column of magma rises from the mantle), one can infer that hydrothermal mineralization occurs at divergent boundaries such as the East African Rift, the Mid-Atlantic Ridge, the East Pacific Rise, and the Red Sea Rift. Hydrothermal mineralization also occurs at subduction zones, such as the ones along the west coast of the Americas. Hydrothermal activity also occurs at oceanic hot spots such as the Hawaiian Islands and at continental hot spots such as the Yellowstone National Park region of the United States.

HYDROTHERMAL PROCESSES AND MINERAL DEPOSITION

Hydrothermal vents—fissures in which water heated by magmatic activity rises and escapes to the surface—can occur in submarine or terrestrial environments. The types of mineral deposits associated with these vents depend upon the composition of the nearby rock and magma. Magma from divergent plate boundaries is made of a high percentage of iron and magnesium, but at convergent plate boundaries, the magmas tend to be higher in silicon and oxygen because they are mixing with continental sediments rich in these elements.

At submarine vents, ocean water heated to temperatures nearing 350 degrees Celsius (662 degrees Fahrenheit) rises, leaching metals from the basaltic rocks it passes through. The metals later precipitate from the water, creating mineral deposits. These deposits can occur in the form of black smokers, which are tower-like structures built from iron-rich sulfides. White smokers, which have lower temperatures, contain calcium sulfate and silica deposits. Some silver, lead, zinc, and copper sulfides are produced near submarine hydrothermal vents, too.

Deep ocean deposits are neither large enough nor accessible enough to mine economically. However, in the future, resource scarcity and improved mining technology may make these deposits more attractive to exploit. Some exploitable ore deposits can be traced to ancient seafloor spreading centers that have moved great distances in time and have been lifted by plate collision. An example can be seen in the island nation Cyprus, where copper and other minerals, formed eons ago, are now near the surface.

Silicates, including feldspar, quartz, and mica, are not as economically valuable as metal deposits, but they are also formed by hydrothermal processes. Continental hydrothermal deposits are often rich in silica because igneous rocks that make up continental crust are high in silica. Silica-rich igneous rock that cools and solidifies underground is called granite, but silica-rich igneous rock that cools at the surface is known as rhyolite. These rock compositions are important because hydrothermal fluids pass through the rock and pick up minerals from them.

Hydrothermal fluids seep into existing spaces and precipitate minerals called cavity-filling deposits. Sometimes, however, the hydrothermal fluids can react with the rocks they pass through and alter the rock to form a deposit made from both hydrothermal precipitates and host-rock material. This process is called hydrothermal replacement.

Terrestrial hydrothermal vents occur in several forms associated with volcanic activity. Examples can be seen in and around Yellowstone National Park, which is located at a hot spot. A fumarole is a type of gas- and steam-producing vent that occurs in volcanic craters or along their sides. Deposits form when the vent's gases cool. The Yellowstone area is rich in rhyolite, so the hydrothermal fluids there are silica-rich. Hot springs or geysers frequently contain silicon dioxide and calcium carbonate. Precipitation of these minerals can be aided by the biologic activity of organisms that live in and around hot springs.

Other continental hydrothermal deposits were formed in ancient subduction zones where metals were liquefied as they made their way down toward the mantle then rose into fractures in the crust. As the fluids flowed through, such metals as copper, molybdenum, silver, gold, lead, and zinc were left behind. These metals are called porphyry metal deposits because they occur in porphyritic igneous rock associated with hydrothermal activity. Deposits of this type have been found in North America, and are being mined.

READING THE EARTH

Even when hydrothermal deposits have no economic value, they have scientific value. Because of the association of seafloor spreading centers with hydrothermal activity and the creation of mineral deposits, scientists can use deposits and rock samples from and near oceanic ridges to learn about cycles of spreading, subduction, and sediment transport. In addition, because humans cannot descend into the mantle to gather data about it, the material brought to the crustal layer through geologic events is a valuable extrapolative tool.

Hot hydrothermal fluids emerge from point sources (vents) and begin to spread out and cool as they mix with cold ocean water. Scientists can track the movement of these fluids by using a method called isotopic tracing, which uses isotopes (forms of an element with the same number of protons but differing numbers of neutrons) to provide information about the vertical and horizontal movement of water. Isotopes also can be used to determine the age of particular rock samples or the conditions under which they formed. This is helpful in establishing patterns in tectonic plate movement or the rate of seafloor spreading.

Because many of the hydrothermal deposits and magmas at seafloor spreading centers contain iron, which can be magnetized, they also provide a record of Earth's magnetism. The needle of a compass will invariably indicate north because that is where the earth's magnetic pole is located. This has not always been true. Earth's magnetic poles reverse approximately every few hundred thousand years. The process occurs gradually and does not follow strict cycles, but because seafloor spreading occurs with magmatic activity, and because magma becomes magnetized at a specific point during cooling, scientists can construct a relatively precise timeline of the earth's magnetism based on the magnetism of rocks and deposits near seafloor spreading centers.

One of the other benefits of studying mineral deposits at seafloor spreading centers is to compare magnetic and other qualities of the deposits on each side of the ridge. Minerals on one side of the ridge should have a matching counterpart on the other side of the ridge. By looking for similarities, scientists can see plate movement outward from a spreading center, but they also can track horizontal movement, such as the type of movement at transform fault boundaries.

J. D. Ho

FURTHER READING

Chernicoff, Stanley, and Donna Whitney. *Geology: An Introduction to Physical Geology.* Upper Saddle River, N.J.: Prentice Hall, 2007. Introductory descriptions of basic tectonic processes. The chapter on igneous rocks covers the formation of rock from magma as well as crystallization, composition, geographic distribution, and types of rock.

Hibbard, Malcolm J. *Mineralogy: A Geologist's Point of View.* New York: McGraw-Hill, 2002. Covers many aspects of mineralogy but is particularly useful for its easy-to-understand descriptions of crystallization in magmas and precipitation of minerals from hydrothermal solutions.

Jerram, Dougal, and Nick Petford. *The Field Description of Igneous Rocks.* Hoboken, N.J.: Wiley-Blackwell, 2011. Although more of a rock guide, this book provides background information on formation processes and on the composition of rock types. Includes locations where one can find rocks formed by hydrothermal activity. Also describes physical properties and classification of rock types.

Keller, Edward A. *Introduction to Environmental Geology.* Upper Saddle River, N.J.: Prentice Hall, 2012. Contains several useful chapters with clear and concise descriptions of all aspects of mineral formation, from the movement of tectonic plates and the interior structure of the earth to the molecular structure of rock types and the distribution and extraction of mineral resources.

Libes, Susan M. *Introduction to Marine Biogeochemistry.* Burlington, Mass.: Academic Press, 2009. Focuses on ocean systems. Chapter 19 examines mineral deposits created by hydrothermal activity. Explicates the geology of seafloor spreading, subduction zones, and hot spots; chemical reactions in deep-sea systems; and the biology and ecology of hydrothermal vents.

Morgan, Lisa A. *Integrated Geoscience Studies in the Greater Yellowstone Area: Volcanic, Tectonic, and Hydrothermal Processes in the Yellowstone Geoecosystem.* Reston, Va.: U.S. Geological Survey, 2007. Much of this publication is technical, but Chapters A and O provide more accessible coverage of the geology of the Yellowstone area and a useful case study of continental hot spots; information on tectonics, ore, and rock formation; and a discussion of hydrothermal activity.

See also: Aluminum Deposits; Chemical Precipitates; Crystals; Geologic Settings of Resources; High-Pressure Minerals; Igneous Rock Bodies; Iron Deposits; Orbicular Rocks; Quartz and Silica; Silicates; Sulfur; Ultra-High-Pressure Metamorphism; Ultrapotassic Rocks; Uranium Deposits.

IGNEOUS ROCK BODIES

Igneous rock bodies are formations of rock created when magma generated within the earth's mantle cools and solidifies into solid rock strata. Magma forms when a portion of the mantle is disrupted and decompressed, which causes pockets of the material to liquefy. Igneous rock is the most common type of rock in the crust and is a common component in construction and other industrial applications.

PRINCIPAL TERMS

- **aphanitic:** igneous rocks with fine grains that can be distinguished only with visual aid
- **crystalline:** solid structure composed of atoms arranged in a regular, repeating pattern to form a repeating three-dimensional structure
- **extrusion:** igneous rocks that are formed after liquid lava cools on the surface of the earth
- **feldspar:** family of silicon-based minerals that occur in igneous rocks; formed from silica combined with calcium, potassium, and sodium in various concentrations
- **felsic:** igneous rocks mainly comprising silica blended with feldspar and composed largely of such lighter elements as oxygen, sodium, potassium, and aluminum
- **intrusion:** liquid rock that forms beneath the earth's surface
- **mafic:** igneous rocks that are rich in magnesium and iron and contain less feldspar
- **pahoehoe:** rope-like structures formed when a flow of lava cools on the earth's surface
- **phaneritic:** igneous rocks with large grains that can be seen without a microscope
- **pluton:** a body of igneous rock formed by intrusion

ORIGIN OF IGNEOUS ROCKS

Igneous rocks are crystalline or glassy rock bodies that form when superheated rock originating in the earth's mantle rises toward the surface and cools to form solid structures. While it remains under the surface, this superheated rock is known as magma, and when it breaks through the crust to the surface of the earth, it becomes lava. The classification of igneous rocks depends largely on whether the rock solidified from magma or from lava.

Magma is a combination of solid, liquid, and gas molecules and is composed of a variety of elements, including oxygen, aluminum, silicon, sodium, potassium, magnesium, and iron. The two most important components of magma are water and silica, which is a molecule composed of silicon and oxygen. The proportion of silica and water within a deposit of magma determines its overall grain and texture and therefore determines the types of solids that will develop if the magma cools.

The surface of the earth, known as the crust, is formed of solid, low-density rock that floats on top of a layer of higher-density rock called the mantle. The mantle layer is partially solid, but because it is subject to greater pressure, it is more ductile. The mantle is constantly moving in response to heat rising from the inner layers of the earth.

Below the mantle is a layer of liquid rock known as the outer core, which remains liquefied because it is subjected to intense pressure and is continually heated by the radioactive decay of elements contained within. Heat rising from the inner core causes currents to develop within the mantle; the crust of the earth moves along the path of these currents.

The crust and the upper layers of the mantle are divided into a series of structures known as tectonic plates, each of which moves as a single unit. The movement of tectonic plates and mantle generates pressure and friction that cause portions of the mantle to liquefy into magma. Most magma forms at depths ranging from 10 to 200 kilometers (km), or 6 to 124 miles (mi) below the surface.

Some magmas form along subduction zones, which are areas in which two of the earth's plates are driven together by convection currents. As the plates collide, a portion of one plate is driven under the leading portion of the opposing plate, which disrupts both the crust and the upper mantle. This leads to reduced density among the molecules of mantle rocks. Once freed from their dense compaction, the molecules of the mantle rock are excited by heat and pressure and begin to vibrate, causing a portion of the rock to morph into the liquid phase, becoming magma. Magma that forms along subduction zones is generally known as granitic or felsic magma because it is formed from melted rock derived from both the crust and the upper mantle.

Magma also forms in areas known as rifts, where two continental plates diverge. As the plates move apart, convection currents from deep within the mantle push upward, decompressing a portion of the upper mantle and leading the rock to become liquid. This type of magma is known as basaltic or mafic magma and is primarily composed of rock from the upper mantle.

Magma that forms away from the edges of tectonic plates is called intraplate magma. This type of magma results from what geologists call a hot spot, an area in which a portion of the mantle becomes destabilized from convection currents rising from the inner core, forming a plume of liquid rock that rises to the surface because it is less dense than the rock layers of the upper mantle or crust. The portion of the crust overlying the hot spot will continue to float along the mantle; through millions of years, this can lead to the formation of a chain of volcanoes breaking through the crust.

CLASSIFICATION OF IGNEOUS ROCKS

Extrusive deposits are those that form when lava erupts through the crust and then solidifies into solid rock on the surface. Extrusive deposits typically involve volcanic eruptions and may occur either at areas where tectonic plates collide, at rift zones, or where hot spots form within the body of a tectonic plate.

By contrast, intrusive igneous rocks form under the surface of the earth, sometimes taking thousands or millions of years to solidify. Intrusive igneous formations are hidden within the crust until erosion or some other disruption of crustal rock exposes the igneous deposits at the surface.

Igneous rocks are classified according to both texture and chemical composition. The texture or grain of an igneous rock takes into account the thickness of individual grains, the presence of glass within the rock, and whether the rock contains hollow cavities

or pores. In terms of chemical composition, geologists often use the proportion of silica as the basis for classifying igneous rocks.

All igneous rocks are composed primarily of silica, which is the main mineral component of the earth's crust. Silica combined with calcium, potassium, and sodium may form feldspar. Feldspar occurs in many varieties and is generally classified according to the dominant types of minerals mixed in with the silica base. Therefore, feldspar types are calcium dominant, potassium dominant, and sodium dominant. Igneous rocks containing 50 to 70 percent silica blended with potassium-dominant feldspar are called felsic rocks, whereas mafic rocks contain only 40 to 50 percent silica and have a high proportion of iron or magnesium. Feldspar that occurs in mafic rocks is typically of the calcium-dominant variety, though mafic rocks contain much lower levels of feldspar than do felsic rocks.

Igneous rocks with large grains, visible without a microscope, are called phaneritic rocks, while aphanitic rocks have grains that typically cannot be seen except under a microscope. The texture of the rocks depends on the specifics of the cooling and solidification process. Phaneritic rocks result from slow cooling processes and are therefore characteristic of intrusive rock formations, in which the magma had cooled beneath the earth's surface. Aphanitic rocks form from rapid cooling and are therefore characteristic of extrusive rock formations, such as when lava cools in contact with air or water after an eruption.

Under special conditions that lead to extremely rapid cooling, lava will form into glassy rocks that contain no minerals and no crystalline structure. The black glass known as obsidian is one example of a glassy igneous rock that forms when lava undergoes a rapid cooling process.

Another type of igneous rock, called porphyritic rock, combines aphanitic and phaneritic characteristics. Porphyritic rocks appear as fine-grained aphanitic rocks with large crystals embedded within their matrix. These crystals result when the rock experiences a mixed solidification regime including alternating periods of rapid and slow cooling.

Other igneous rocks are classified as vesicular because the rock contains large holes or pores that result from bubbles of gas that formed while the rock was cooling. After the rock solidifies, these pores remain in the rock's structure.

EXTRUSIVE IGNEOUS BODIES

Igneous formations that develop from lava are usually aphanitic in texture because the rapid cooling of the rock causes the formation of small crystals. Rapid cooling also may result in the formation of glass deposits, like obsidian.

Among the most common types of deposits resulting from lava flows are basalts, which are a type of aphanitic rock, rich in calcium and containing low levels of larger mineral deposits. Basalt is composed primarily of plagioclase, a type of feldspar that combines silica with aluminum, calcium, and sodium. Basalt is one of the most common types of igneous rocks and is a major component of the earth's crust. Geologists use the term *flood basalts* to describe large deposits of basalt sediment that cover vast areas of terrestrial or submarine environments after large-scale volcanic eruptions.

Lava flows that occur in marine environments may result in pillow lavas, which are bulbous, spherical formations named for their resemblance to pillows. Lava flows that cool on the surface of the earth often form a variety of structures. Rapidly cooled surface lava may become frozen in the shape of a flowing liquid, leaving behind ropey structures called pahoehoe, a Hawaiian term for these formations, which are common on the Hawaiian Islands. In some cases, a lava flow may cool in such a way that the outer surface of the lava cools while the inner portion remains liquid. As the liquid lava flows from this solid shell, it leaves behind a hollow structure called a lava tube. Large lava tubes may combine to form a lava cave system with subterranean caverns stretching for miles.

INTRUSIVE IGNEOUS BODIES

Intrusive rock bodies occur when pockets of magma infiltrate preexisting rock, known as country rock. These formations are called plutons and are generally classified according to their position relative to the surrounding bed of country rock.

Tabular plutons form when magma is forced into cracks between layers of country rock. These formations are named for the sheet-like or table-like forms they take as they squeeze into the country rock. Tabular plutons that form parallel to the existing sediment are called sills, which tend to be thin, reaching less than 50 meters (164 feet) in thickness.

Many sills are connected to dikes, which are plutons that form in a concordant direction to the sediment of the country rock such that they extend at a horizontal angle through the bedrock. Dikes also tend to be relatively thin, rarely reaching more than 20 meters (66 feet). As the dikes extend horizontally, they may form one or more sills at different levels of the sediment. Dikes and sills often result from a flow of magma deeper in the mantle.

In some cases, a large sill may form such that the central portion of the sill bubbles up toward the surface. These formations, called laccoliths, may grow to such a size that they push the crust into a mountain-sized mound. The Henry Mountains in Utah were formed by an ancient laccolith that developed under the crust.

Larger, irregularly shaped plutons are sometimes grouped together as massive plutons. A stock is a type of massive pluton that results from an ancient magma chamber deep within the crust and upper mantle. Stocks are usually about the size of a large mountain and are often connected to one or more volcanoes at the surface. The magma body that forms the basis of the stock may feed active volcanoes and may also feed dikes and connected sills leading from the body of the magma chamber toward the surface. Geologists usually restrict stocks to formations that are less than 100 kilometers (62 miles) across.

The largest igneous formations are called batholiths, which result from several magma chambers that have joined; batholiths can extend for thousands of square kilometers just under the surface of the crust. Batholiths may take millions of years to cool and solidify and may be renewed by infusions of newly generated magma from within the mantle.

Batholiths are often connected to collections of stocks, dikes, and sills in massive networks that extend toward the surface of the crust. The Sierra Nevada in California are formed from a massive batholith. The interior of many mountains contain batholith material that cooled during the mountains' formation. Because of a batholith's size, erosion does not extend to its lower level, making it impossible to know exactly where the lower limit of the batholith resides. Some of the larger batholiths may extend through most of the crust into the upper mantle.

In some cases, intrusive rock bodies may form close to the surface, leading to a formation that has characteristics between those of rocks that form on the surface and those that form at deeper depths. These types of igneous formations are called hypabyssal and result in rocks that cannot be classified as either plutons or volcanic rocks. Hypabyssal rocks are usually porphyritic in structure, containing a medium-grained sediment punctuated by inclusions of various minerals and other types of rock. Hypabyssal rocks are rocks that did not cool as quickly as those rocks that reached the surface; hypabyssal rocks took less time to cool than rocks characteristic of batholiths, stocks, and other deep plutons. This intermediate cooling period leads to the structural variations observed in the rocks.

USES FOR IGNEOUS ROCK

Because igneous rocks vary widely in texture, color, and other characteristics, they are essential to many industrial processes. One of the most common types of igneous rock is granite, which is the main component of many intrusive deposits, including batholiths and stocks found within mountains and under much of the earth's crust.

Granite is one of the most common types of basement rock, which is the layer of rock that occurs under the uppermost surface layer of the earth's crust. Because of its thickness and stability, granite is commonly used as a sculptural and building material, especially in the manufacture of headstones and statues.

Basalt is another common type of igneous rock that typically forms from extrusive lava flows. Large basalt deposits are formed in deep-sea rifts where the continents diverge and magma is forced to the surface. Because it is rapidly cooled at the surface, basalt is a finely grained stone, largely devoid of intrusive minerals. The mineral composition of basalt tends to result in relatively dark colored stone.

Basalt underlies more of the earth's surface than any other type of rock, and its abundance has made it a popular stone for construction. Basalt is often ground and used as a base for concrete and other types of composite stones in building and sculpture.

Pumice is another kind of volcanic rock that forms when lava is rapidly ejected from the mouth of a volcano. As the rock cools, minerals within the lava begin to depressurize, leaving hollow pores and pockets within the rock. Pumice is an important component of insulating stone used in many construction projects. The open pores trap air, which creates a buffer zone mediating the loss of heat from the internal environment. In addition, ground pumice is commonly used as an abrasive additive to cleaners and hand soaps. Shaped pumice stones are sometimes used as a cosmetic aid to rub away callused and dead skin from the body.

Some types of crystallized igneous rocks are used in the manufacture of jewelry because of their attractive appearance. Igneous glass formations, like obsidian and basaltic glass, are often polished and used to make jewelry. A number of gemstones are found within igneous rock formations, including both diamonds and emeralds. For this reason, mining of igneous formations is important to the jewelry industry. Many types of precious metals are found within igneous deposits, too, and the deposits from dikes and stocks can become an important source of precious metals.

Micah L. Issitt

FURTHER READING

Duff, Peter McLaren Donald, and Arthur Holmes, eds. *Holmes' Principles of Physical Geology*. 4th ed. New York: Chapman & Hall, 1998. Introduction to geological theories, research, and methodology. Chapter 5 describes igneous rock formation.

Grotzinger, John, Frank Press, and Thomas H. Jordan. *Understanding Earth*. New York: W. H. Freeman, 2009. Text covering concepts in chemistry, geology, and physics and the integration of these principles in Earth's formation and development. Contains detailed descriptions of issues surrounding igneous rock formation and uses in industry.

Monroe, James S., Reed Wicander, and Richard Hazlett. *Physical Geology*. 6th ed. Belmont, Calif.: Thompson Higher Education, 2007. General text covering aspects of physical geology, including methodology, methods, and environmental issues surrounding mineralogical formation and harvest.

Montgomery, Carla W. *Environmental Geology*. 4th ed. Columbus, Ohio: McGraw-Hill, 2008. Introductory level text on environmental geology. Includes an overview of the environmental impact of mining and climate change on Earth's lithosphere.

Stanley, Steven M. *Earth System History*. New York: W. H. Freeman, 2004. Discusses rock formations in relation to Earth's historical development. Chapter 6 covers igneous rocks.

Woodhead, James A., ed. *Geology.* 2 vols. Pasadena, Calif.: Salem Press, 1999. Basic introductory text to geological sciences. Contains a detailed description of igneous rock bodies and their formation.

See also: Crystals; Feldspars; High-Pressure Minerals; Pyroclastic Flows; Quartz and Silica; Rocks: Physical Properties; Silicates; Ultra-High-Pressure Metamorphism.

INDUSTRIAL METALS

Metals have played a major role not only in human survival but also in the high standard of living that most cultures enjoy today. Humans have developed an understanding of the geologic conditions under which minerals form in nature and have learned to prospect for and produce those minerals from which metals are derived. A use has been found for virtually every metallic element.

PRINCIPAL TERMS

- **basic:** a term to describe dark-colored, iron- and magnesium-rich igneous rocks that crystallize at high temperatures, such as basalt
- **by-product:** a mineral or metal that is mined or produced in addition to the major metal of interest
- **granite:** a light-colored, coarse-grained igneous rock that crystallizes at relatively low temperatures; it is rich in quartz, feldspar, and mica
- **hydrothermal solution:** a watery fluid, rich in dissolved ions, that is the last stage in the crystallization of a magma
- **laterite:** a deep red soil, rich in iron and aluminum oxides and formed by intense chemical weathering in a humid tropical climate
- **nodule:** a chemically precipitated and spherical to irregularly shaped mass of rock
- **ore mineral:** any mineral that can be mined and refined for its metal content at a profit
- **pegmatite:** a very coarse-grained igneous rock that forms late in the crystallization of a magma; its overall composition is usually granitic, but it is also enriched in many rare elements and gem minerals
- **placer:** a surface mineral deposit formed by the settling of heavy mineral particles from a water current, usually along a stream channel or a beach; gold, tin, and diamonds often occur in this manner
- **primary:** a term to describe minerals that crystallize at the time that the enclosing rock is formed; hydrothermal vein minerals are examples
- **refractory:** a term to describe minerals or manufactured materials that resist breakdown; most silicate minerals and furnace brick are examples
- **reserve:** that part of the mineral resource base that can be extracted profitably with existing technology and under current economic conditions
- **resource:** a naturally occurring substance, the form of which allows it to be extracted economically

- **secondary:** a mineral formed later than the enclosing rock, either by metamorphism or by weathering and transport; placers are examples
- **vein:** a mineral-filled fault or fracture in rock; veins represent late crystallization, most commonly in association with granite

MAJOR METALS

For convenience, industrial metals may be divided into three groups: the major nonferrous metals, the ferroalloys, and the minor metals. The major nonferrous metals are aluminum, copper, tin, lead, and zinc. Copper has been used longer than any metal except gold. It occurs in at least 160 different minerals, of which chalcopyrite is the most abundant ore. The world's most important source of copper is in large masses of granite rock known as porphyry copper bodies, which are found throughout the western United States but are especially numerous in southern Arizona. Hydrothermal solutions deposited copper-bearing minerals in openings and cracks throughout these masses, and later weathering concentrated the ores beneath the surface in amounts that could be recovered economically. Copper also occurs in sedimentary rocks. Such deposits account for more than one-fourth of the world's copper reserves, with the best examples in the Zambian-Zairean copper belt of Africa. A future resource is the copper in manganese nodules that cover large portions of the ocean floor, especially in the North Pacific. Chile, the United States, and Canada, in that order, are the world's leading producers of copper.

Like copper, tin has long been used by humans. The principal ore mineral of tin is the oxide cassiterite, "tinstone." Primary tin-bearing deposits include granite pegmatites and hydrothermal veins. Much more important from a commercial standpoint, however, are secondary stream placer deposits. Although domestic mine production is negligible, the United States is the world's leading producer of recycled tin. World leaders in the mining of tin are Malaysia, Brazil, Indonesia, Thailand, China, and Bolivia.

Zinc and lead now rank just behind copper and aluminum as essential nonferrous (not containing iron) metals in modern industry. The most important geologic occurrences of lead and zinc are within stratified layers of metamorphic or carbonate rocks. The deposits at Ducktown, Tennessee, and Franklin, New Jersey, are in metamorphic rocks, while those in the upper Mississippi Valley, southeastern Missouri (the Viburnum Trend), and the Tri-State mineral district (Missouri-Oklahoma-Kansas) are found in carbonate rocks. This latter type of occurrence is now referred to as Mississippi Valley-type deposits. The principal ore minerals for lead and zinc are the sulfides galena and sphalerite (zinc blende), respectively. These minerals were probably emplaced in the host rocks by later hydrothermal solutions. The United States is a major producer of lead, with most coming from the Viburnum Trend of Missouri. Recycled lead is also very important in the United States, accounting for more production than mining. Other major lead-producing countries are Australia, Canada, Peru, and Mexico. Canada, Australia, Peru, Mexico, and the United States are among the world's leading producers of zinc.

Aluminum is in a class by itself because it is not mined from hydrothermal deposits in rocks but from tropical soils. Deep weathering of aluminum-rich rocks, typically igneous rocks rich in feldspar, removes all soluble elements and leaves a residue containing mostly aluminum, iron, and silica. This ore is called bauxite.

FERROALLOYS

The ferroalloys are a group of metals whose chief economic use is for alloying with iron in the production of carbon and various specialty steels. Manganese is the most important of the ferroalloys, and it occurs principally in sandstone deposits. Chemically precipitated nodules on the ocean floor contain up to 20 percent manganese and represent an important potential resource.

The elements chromium, nickel, titanium, vanadium, and cobalt most commonly occur in basic igneous rocks, such as gabbro. Examples of such deposits are the Stillwater complex of Montana and the Bushveld complex of South Africa. The Bushveld is the largest such body in the world and holds most of the world's chromium reserves. Nickel ranks second in importance to manganese among the ferroalloys.

Copper ore. (U.S. Geological Survey)

The large igneous deposits at Sudbury, Ontario, Canada, have been the world's major supplier. The most important economic occurrences of titanium are in secondary placer deposits, notably the rutile beach sands of Australia and the ilmenite sands of northern Florida. In addition to its occurrence in dark igneous rocks, vanadium is found as a weathering product in uranium-bearing sandstones such as those of the Colorado Plateau region of the United States. Vanadium is also produced from the residues of petroleum refining and the processing of phosphate rock. While the primary occurrence of cobalt is in dark igneous rocks, it also is produced from laterites and as a by-product of the sedimentary copper deposits of the African copper belt.

Molybdenum, tungsten, niobium, and tantalum occur commonly in quartz veins and granite pegmatites. Most molybdenum production comes from the very large, but low-grade, porphyry deposits at Climax, Colorado, and as a by-product of the porphyry copper deposits at Bingham Canyon, Utah. Production of the minerals scheelite and wolframite from quartz veins accounts for more than one-half of the world's production of tungsten. Because of their rarity, there is no mining operation solely for the production of either niobium (also known as columbium) or tantalum. Both are always produced as by-products of the mining of other metals.

The United States is self-sufficient and a net exporter only with respect to molybdenum. It is almost totally dependent on foreign sources for manganese, chromium, nickel, cobalt, niobium, and tantalum, and to a lesser degree vanadium, tungsten, and titanium.

MINOR METALS

There are a number of metallic elements that are rare in nature. Most of these have a primary origin in hydrothermal veins or granite pegmatites and are produced either from such deposits (antimony, cadmium, beryllium, bismuth, arsenic, lithium, cesium, rubidium, and mercury) or from placer deposits that resulted from the weathering and erosion of the pegmatites (zirconium, hafnium, thorium, and the rare-earth elements). The sole exception is magnesium, which is produced mostly from seawater and natural brines in wells and lakes, as at Great Salt Lake in Utah.

Most elements in this group are produced exclusively as by-products of the mining and processing of other metals. Antimony, in the mineral stibnite, is closely associated with lead ores in Mississippi Valley-type deposits, and most antimony is produced as a by-product of the mining and processing of lead, copper, and silver ores. China is the world's leading producer of antimony. Cadmium is a trace element that is similar to zinc and therefore substitutes for zinc in some minerals. All cadmium is produced from the mineral sphalerite as a by-product of zinc production. Large but untapped potential resources of cadmium exist in the zinc-bearing coal deposits of the central United States. There are no specific ore minerals of bismuth, and virtually all production is from the processing of residues from lead smelting. Mercury, or quicksilver, is the only metal that is a liquid at ordinary temperatures. It is produced from the sulfide mineral cinnabar and is marketed in steel "flasks," each flask containing 76 pounds of liquid mercury. The mines at Almaden, Spain, have been the leading

producers. Magnesium is the lightest metal and is especially strong for its weight. It is an abundant metal, both in the earth's crust and in seawater. Lithium, cesium, and rubidium are often classified as the rare alkali metals, the abundant alkali metals being sodium and potassium. Like magnesium, lithium is abundant enough in brines and evaporite deposits to be processed economically from these sources. Rare-earth elements, or lanthanide elements, are a group of chemically similar elements, of which cerium is the most abundant and widely used. Thorium is a heavy metal, the parent element of a series of radioactive decay products that end in a stable isotope of lead. It is not an abundant element, but it is widespread, occurring in veins, placers, and sedimentary rocks.

The United States is well endowed with some of these minor metals, including beryllium, lithium, magnesium, and rare-earth elements. In contrast, it must import virtually all of its supplies of antimony, cadmium, bismuth, mercury, arsenic, and rubidium, and one-half of its annual consumption of zirconium and thorium ores and products.

EXPLORATION FOR METALS

It is a fair assumption that almost all the large and easily accessible metallic ore bodies have been discovered. The focus of the years ahead will be the detection of deposits in more remote localities, concealed deposits, or lower-grade deposits, as well as extending the useful life of existing deposits. While research provides more sophisticated tools, the basic exploration approach remains the same. Four main prospecting techniques are employed: geological, geochemical, geophysical, and direct.

Geological exploration generally involves plotting the locations of rock types, faults, folds, fractures, and areas of mineralization on base maps. Examples of deposits that have been discovered by simple surface exploration and mapping are as follows: chromite, which is resistant to weathering and "crops out"; manganese-bearing minerals, which oxidize to a black color; and molybdenite, with a characteristic silver color. Because ore minerals are known to be associated with rocks formed in certain geologic environments, the focus of study today is on understanding such associations as clues to locating mineral deposits.

Geochemical exploration consists of chemically analyzing soil, rock, stream, and vegetation samples.

Concentrations of metallic elements in the surface environment are assumed to be representative of similar concentrations in the rocks below. Areas of low mineral potential can be eliminated, while targets for further study and testing can be outlined. In some instances, it is necessary to trace metals back through the surface environment to their points of origin. Research is being done not only on the movement and concentration of economically important metallic elements, but also on the elements that are often associated with them but that nature, for various reasons, can more easily move and concentrate. Cobalt, for example, is mobile, moving through rocks and sediments easily. It may be possible to trace this element back to its source and in the process locate other associated metals.

Geophysical techniques range from large-scale reconnaissance surveys to detailed local analysis. These techniques detect contrasts in physical properties between the ore bodies and the surrounding host rocks. Airborne magnetic and electromagnetic surveys are useful for rapid coverage of remote and inaccessible terrain, and of areas where ore bodies are covered by glacial sediments. Radiometric surveys can be used to detect concentrations of radioisotopes and have been effective in locating deposits of thorium, zirconium, and vanadium-bearing minerals. Other applications of geophysical techniques include seismic and gravity studies to determine the thickness of overburden, airborne infrared imagery to detect residual heat in igneous deposits, light reflectance of vegetation, side-looking radar, and aerial and satellite photography. In general, airborne reconnaissance techniques are followed by detailed geological, geochemical, or geophysical ground surveys.

The final stage in the exploration process is the direct stage. Here, a prospect is directly sampled by drill, pit, trench, or mine to determine its potential. This step is the most expensive. It is the purpose of geological, geochemical, and geophysical prospecting to narrow the possibilities, lower the odds, and thereby reduce the final cost by selecting the most promising prospects for direct sampling.

USES OF MAJOR AND MINOR METALS

The great value of copper derives from its high thermal and electrical conductivity, corrosion resistance, ductility, and strength. It alloys easily, especially with zinc to form brass, and with tin and zinc to form bronze. Principal uses of tin are plating on cans and containers, in solder, in bronze, and with nickel in superconducting alloys. Restrictions on the use of lead in pipes and solder should cause an increase in tin consumption, as a lead replacement. Bottle and can deposit laws, enacted in a number of states, will increase the use of scrap (recycled) tin. The principal use of lead is in automobile batteries, but cable sheathing, type metal, and ammunition are other important applications. Zinc uses include galvanizing for iron and steel, die castings, and brass and bronze.

The minor metals have a variety of industrial applications, both as metals and in chemical compounds. Antimony is used primarily as a fire retardant, but it is also alloyed with lead for corrosion resistance and as a hardening agent. This last characteristic is important for military ordnance and cable sheathing. Cadmium is used in the electroplating of steel for corrosion resistance, in solar cells, and as an orange pigment. Beryllium is alloyed with aluminum and copper to provide strength and fatigue resistance, and is widely used in the aerospace industry. It is a "nonsparking" metal that can be used in electrical equipment. Oxides of beryllium are used in lasers and ceramics, and as refractories and insulators. The principal use of bismuth is in the pharmaceutical industry. It soothes digestive disorders and heals wounds. Salts of bismuth are widely used in cosmetics because of their smoothness. Bismuth metal lowers the melting points of alloys so that they will melt in a hot room. This allows them to be used in automatic water sprinkler systems and in safety plugs and fuses.

Mercury, because it is a liquid at room temperature, has applications in thermometers, electrical switches, fluorescent lights, and, with rubidium, in vapor lamps. It also is used in insecticides and fungicides, and has medical and dental applications. Because of its toxicity, such uses as thermometers are being phased out. Magnesium is alloyed in aircraft with aluminum to reduce weight and at the same time provide high rigidity and greater strength. The metal also burns at low temperature, which makes it suitable for flash bulbs, fireworks, flares, and incendiary bombs. The largest use of magnesium is in the oxide magnesia for refractory bricks. Arsenic is mostly used in chemical compounds as a wood preservative and in insecticides and herbicides. Lithium-based greases have wide applications in aircraft, the military, and the marine environment, as they retain their lubricating properties over a wide temperature range and

INDUSTRIAL NONMETALS

Nonmetallic Earth resources consist of fertilizer minerals, raw materials for the chemical industry, abrasives, gemstones, and building materials for the construction industry. They provide the necessary base for a technological society.

PRINCIPAL TERMS

- **asbestosis:** deterioration of the lungs caused by the inhalation of very fine particles of asbestos dust
- **catalyst:** a chemical substance that speeds up a chemical reaction without being permanently affected by that reaction
- **guano:** fossilized bird excrement, found in great abundance on some coasts or islands
- **metal:** a shiny element or alloy that conducts heat and electricity; metals are both malleable and ductile
- **proven reserve:** a reserve supply of a valuable mineral substance that can be exploited at a future time
- **sedimentary rock:** rock formed from the accumulation of fragments weathered from preexisting rocks or by the precipitation of dissolved materials in water
- **star sapphire/ruby:** a gem that has a starlike effect when viewed in reflected light because of the mineral's fibrous structure
- **strategic resource:** an Earth resource, such as manganese or oil, which would be essential to a nation's defense in wartime

DEFINITION OF NONMETALLIC RESOURCES

The definition of a nonmetal is problematic. To the chemist, a nonmetal is any element not having the character of a metal, including solid elements, such as carbon and sulfur, and gaseous elements, such as nitrogen and oxygen; this is the definition found in most dictionaries. To the economic geologist, however, a nonmetal is any solid material extracted from the earth that is neither a metal nor a source of energy. A nonmetal is valued because of the nonmetallic chemical elements that it contains, or because of some highly desirable physical or chemical characteristic.

Economic geologists consider the following to be major nonmetallic Earth resources: fertilizers, raw materials for the chemical industry, abrasives, gemstones, and building materials. A few minerals, such as asbestos and mica, are used for distinctive or unusual physical properties. As can readily be seen, with the exception of the gemstones, nonmetallic Earth resources lack the glamour of such metals as gold and platinum or such energy sources as oil and uranium. Nevertheless, nonmetallic Earth resources play an essential role in the world economy.

Except for gemstones, nonmetallic Earth resources have certain common characteristics. First, they tend to be more abundant in the earth's crust than metals and are therefore lower in price. Nevertheless, most of them are needed in much larger quantities, so the total value of the substances produced is considerable. This is particularly true of the building materials needed for the construction industry. Second, nonmetallic Earth materials tend to be taken from local sources. Most of them are needed in such large quantities that transportation costs would be excessive if they were brought long distances. As a result, regional variations occur in the types of rock used for building stone and crushed rock. Third, problems of supply are not generally associated with nonmetallic Earth materials. Most of them are fairly abundant at the earth's surface, and few major industrial nations are without deposits of each. None of these resources is classified as a strategic material by the United States. Finally, nonmetallic Earth materials tend to require very little processing before being sent to market. In fact, most of them are used in the raw state.

FERTILIZERS

The first category of major nonmetallic Earth resources is fertilizers. These substances are absolutely essential to a nation's agriculture and food supply. The three most important elements for plant growth are nitrogen, potassium, and phosphorus. For years, nitrogen was obtained from nitrogen-rich Earth materials—either from the famous guano deposits in Peru, which were built up by accumulated bird droppings on coastal islands, or from the nitrate deposits in Chile, which cover the floor of a desert. In 1900, however, German chemist Fritz Haber discovered a way of manufacturing nitrates synthetically, using nitrogen extracted directly from the atmosphere.

Today, the synthetic nitrate industry provides 99.8 percent of the world's nitrogen needs.

The potassium required for fertilizers was originally obtained from wood ashes (hence the common name "potash"), and still is in many developing nations. In 1857, however, potassium-bearing salt beds were discovered in Germany. These had formed as a result of the evaporation of lakes in an arid climate, and they were the world's major source of potassium until 1915, when Germany placed a wartime embargo on their shipment. This embargo forced other countries to explore for replacement deposits, and similar salt beds were eventually found in the former Soviet Union, Canada, and the western United States.

The third element required for fertilizers, phosphorus, was originally obtained from guano or from bones. These sources have been replaced by natural phosphate rock deposits, which are widely distributed around the world. Most of these deposits occur in marine sedimentary rocks, and it is believed that deposition of the phosphate resulted when cool, phosphorus-bearing waters upwelled from the sea floor and were carried into shallow environments, where the phosphorus was precipitated. The largest U.S. phosphate deposits are found in Florida and North Carolina.

RAW MATERIALS FOR CHEMICAL INDUSTRY

The second major category of nonmetallic Earth resources consists of raw materials for the chemical industry. In terms of total production, the most important of these is salt, for which the mineral name is halite. In addition to its use as a dietary ingredient, salt is the raw material from which a number of important chemicals are made, including chlorine gas, hydrochloric acid, and lye. In colder climates, salt is also used for snow and ice control on roads. Salt is produced by the evaporation of seawater and by the mining of underground salt deposits. Although underground salt commonly occurs as deeply buried layers, nature has an interesting way of bringing the salt to the surface. Since salt is lighter in weight than the overlying rocks and is capable of plastic flow, it rises through the surrounding rocks as a salt plug with a circular cross-section, known as a salt dome. Salt domes are particularly common along the Louisiana and Texas Gulf Coast and are even more important as petroleum traps than as sources of salt.

Another important raw material for the chemical industry is sulfur, a soft yellow substance that burns with a blue flame. The major industrial use for sulfur is in the production of sulfuric acid. Large quantities of sulfuric acid are used in converting phosphate rock to fertilizer. Sulfur also has important uses in the manufacture of insecticides. Most of the sulfur used in the United States comes from the salt domes of the Louisana and Texas coast, where it is found in the upper part of the dome. Superheated steam is pumped down through drill holes to melt the sulfur, then the liquid sulfur is brought to the surface by the pressure of compressed air. In recent years, the removal of sulfuric acid from smokestack emissions has sharply reduced the demand for sulfur extracted from salt domes. Additional raw materials for the chemical industry include several that are obtained from the beds of dry desert lakes: sodium carbonate and sodium sulfate, which are used in the manufacture of glass, soaps, dyes, and paper, and borax, which is used in making detergents, certain types of glass, and explosives. Another product is sodium bicarbonate, the familiar baking soda.

ABRASIVES

The third major category of nonmetallic Earth resources is abrasives, which are materials used for grinding, cleaning, polishing, and removing solid material from other substances. Most abrasives are very hard, but those used for cleaning porcelain sinks and silverware need to be fairly soft, so as not to scratch. Abrasives can be either natural or human-made. The natural abrasives are rock and mineral substances that have been extracted from the earth and are then either used in the raw state, such as a block of pumice, or pulverized and bonded into sandpapers, wheels, saws, drill bits, and the like. Artificially made abrasives, however, are gradually coming to dominate the market.

The most common abrasive is diamond, which has a hardness of 10 on a scale of 1 to 10 and is the hardest known natural substance. Most natural diamonds are unsuitable for use as gems, however, so about 80 percent of them are used as abrasives. In 1955, General Electric developed a process to make industrial diamonds synthetically, and by 1986 two-thirds of the world's industrial diamonds were produced synthetically. Other important natural abrasives include the following: corundum, the second hardest natural substance, with a hardness of 9; emery, a gray-to-black mixture of corundum and the iron mineral known as magnetite; and garnet, a reddish-brown mineral with a hardness of approximately 7, which is commonly used in sandpaper. Ninety-five percent of the world's garnet comes from the Adirondack Mountains in New York State.

UNUSUAL PHYSICAL PROPERTIES

Among minerals used for their unusual physical properties are barite and barium sulfate, which is unusually dense for a nonmetallic mineral. It has some medical uses because it absorbs X rays, but its principal use is in the petroleum industry, drilling mud. Powdered barite is mixed with water and poured down oil wells as a coolant and lubricant. Its density helps keep oil and gas from escaping to the surface. Mica, which splits into thin, transparent sheets, is used in electrical applications because of its resistance to breaking and melting. Asbestos is obtained from flame-resistant mineral fibers that can be woven into fireproof cloth or mixed with other substances to make fireproof roofing shingles and floor tiles. Concerns about the health hazards of asbestos use arose in the 1970's. Fine particles of asbestos dust can lodge in the lungs, causing asbestosis and lung cancer. As a result, the U.S. consumption of asbestos has declined markedly since the 1970's.

GEMSTONES

The fourth major category of nonmetallic Earth resources is gemstones. These are used primarily for adornment and decoration. Unlike the other nonmetals, gems are generally not abundant, have a moderate to high value, come in only small quantities, are rarely of local origin, and are often in only short supply. Desirable properties in a gem are color, brilliance, transparency, hardness, and rarity. Gems are categorized into two principal groups. Precious gems are diamond, ruby, sapphire, emerald, and pearl. All of these can be produced synthetically, except for high-quality gem diamonds. Ruby and sapphire are varieties of corundum and may exhibit "stars." Emerald is a variety of beryl with a hardness of 8. Pearls have a hardness of 3 and are technically not true minerals, even though they consist of calcium carbonate, because they are produced by a living organism. Semiprecious gems include some one hundred different substances. Most are minerals, except for amber (hardened resin from a pine tree), jet (a dense variety of coal), and black coral (a substance

produced by a living organism). As attractive and valuable as gems are, the abrasives diamond and corundum are vastly more valuable to technology than they are as gemstones.

BUILDING MATERIALS

The fifth major category of nonmetallic Earth resources is building materials for the construction industry. They include building stones obtained from quarries, such as granite, sandstone, limestone, marble, and slate. There is also a high demand for crushed rock, which is used in highway roadbeds and for concrete aggregate. Sand and gravel also are used in making concrete. In some parts of the country, construction materials are becoming scarce not because of rarity but because of political opposition to the extraction of resources from residential areas. In addition, many useful products are prepared from Earth materials, such as cement, which is made from a mixture of limestone and clay; plaster, which comes from the mineral gypsum; brick and ceramics, which

Asbestos is obtained from flame-resistant flexible mineral fibers and was used to create fire-retardant materials until it was discovered that its inhalation can provoke various cancers. (U.S. Geological Survey)

use clay as their raw material; and glass, which is made from very pure sand or sandstone rock.

ASSURING SUPPLIES OF NONMETALLIC RESOURCES

An important way in which nonmetallic Earth resources are studied is to analyze proven reserves. Proven reserves are supplies of a mineral substance that remain in the ground and are available for future removal. When experts compare the present rates of production of various nonmetallic Earth resources with the same resources' proven reserves, they can predict which resources may be in short supply someday. In the case of the phosphate rock used in making fertilizers, for example, analysts have found that the United States' phosphate reserves will begin to decline by about the year 2010.

Economic geologists also study ways to assure that adequate supplies of nonmetallic Earth resources will be available for future needs. In the case of phosphate, for example, the need for such studies is critical. Such a program of exploration for new phosphate supplies, or for any other nonmetallic resource, must begin with a full understanding of the ways in which the mineral resource originates. Only when geologists understand the conditions under which valuable concentrations of mineral substances form can they successfully search for them.

In the case of phosphate, careful scientific study has shown that the cold waters found in the deep ocean contain thirty times as much dissolved phosphorus as do warm shallow waters. This observation suggests that when cold, deep waters are brought to the surface and warmed (by upwelling from the ocean floor, flowing into coastal zones across shallow submarine banks) the warming effect makes the phosphorus less soluble, and it precipitates. As a result, scientists are exploring for phosphates along present or former continental margins, where such upwelling might have taken place.

Another way in which adequate future supplies of nonmetallic Earth resources can be assured is by creating nonmetals synthetically in the laboratory. A good example of this process was the successful synthesis of industrial diamonds by General Electric in 1955. Before then, the United States had no industrial diamond production or reserves and was totally dependent on supplies purchased in the world market, then controlled by the De Beers group in South Africa. General Electric was able to create

diamonds synthetically by subjecting the mineral graphite—which, like natural diamond, is composed of pure carbon—to incredibly high temperatures and pressures in a special sealed vessel, using molten nickel as a catalyst. By 1986, two-thirds of the world's industrial diamonds were being produced by this process, and the De Beers monopoly on industrial diamond production was broken.

CONTINUED IMPORTANCE

The United States is fortunate in having a plentiful food supply, and most Americans rarely think about the importance of fertilizers. They are essential, however, for successful agricultural operations. Plant growth requires ample mineral matter, partly decomposed organic matter (humus), water, air, and sunlight. Of these, mineral matter is crucial, because it provides the nitrates and phosphates essential for healthy plant growth. These substances are quickly used or washed away, and they must be replenished regularly so that the soil does not become worn out and infertile. Worn-out soils are encountered frequently in developing nations, where farmers are often too poor to buy fertilizers.

Industrial chemicals such as salt, sulfur, and borax appear on grocery shelves in their pure state, but they are also present as ingredients in products where one would never suspect their existence. In addition, they are frequently needed to manufacture everyday products, such as drinking glasses or writing paper. Salt is a good example. In addition to its use as table salt, it is an ingredient in almost every prepared food item on the grocery shelf. Abrasives, too, are common on grocery shelves, although the word "abrasive" may not be written on the package. They are in toothpaste, silver polish, bathroom cleanser, pumice stones, sandpaper, and emery boards.

Nonmetals are used in many common construction materials. The beautiful white buildings in Washington, D.C., are made of pure white Vermont marble. Granite, however, is preferred for tombstones, because it resists weathering more successfully. It is used also for curbstones in northern cities, because it holds up best under the repeated impacts from snowplow blades. The days of buildings faced with cut stone, however, are on the wane; production and transportation costs are simply too high. Today's private dwelling is more likely to be built of cinder blocks manufactured at a plant outside the city, and downtown office towers are sheathed with walls of glass and prefabricated concrete, all of which ultimately depend on nonmetallic mineral resources.

Donald W. Lovejoy

FURTHER READING

Burger, Ana, and Slavko V. Šolar. "Mining Waste of Non-metal Pits and Querries in Slovenia." In *Waste Management, Environmental Geotechnology and Global Sustainable Development*. Ljubljana, Slovenia: Geological Survey of Slovenia International Conference, 2007. Discusses mining processes and the waste material produced. Evaluates types of mines and materials mined in Slovenia to determine methods for reducing mining waste.

CAFTA DR and U.S. Country EIA. *EIA Technical Review Guideline: Non-metal and Metal Mining*. Washington, D.C.: U.S. Environmental Protection Agency, 2011. Discusses alternative mining methods, mineral processing, facilities, and environmental impact, as well as assessment methods and mitigation strategies.

Constantopoulos, James T. *Earth Resources Laboratory Investigations*. Upper Saddle River, N.J.: Prentice Hall, 1997. An excellent guidebook for any laboratory work involving minerals, deposits, and natural resources. Discusses the protocol and procedures used in working with Earth resources in the laboratory setting.

Craig, J. R., D. J. Vaughan, and B. J. Skinner. *Earth Resources and the Environment*. 4th ed. Upper Saddle River, N.J.: Prentice Hall, 2010. A well-illustrated text with numerous black-and-white photographs, color plates, tables, charts, maps, and line drawings. Covers the major categories of nonmetallic Earth resources: fertilizer minerals, chemical minerals, abrasives, gemstones, and building materials. Includes suggestions for further reading and a list of the principal nonmetallic minerals. Suitable for college-level readers or the interested layperson.

Evans, Anthony M. *An Introduction to Economic Geology and its Environmental Impact*. Oxford: Blackwell Scientific Publications, 1997. Intended for undergraduate students. Emphasis is on types of deposits, their environments of formation, and their economic value, along with the impact those deposits have on their environments. Well illustrated, with an extensive bibliography.

Fisher, P. J. *The Science of Gems*. New York: Charles Scribner's Sons, 1966. Includes excellent color photographs of the well-known gems. Topics covered include the history of gems, their origins, their characteristics, gem cutting, and gem identification. A glossary of terms and a useful appendix give detailed information relating to each type of gem. Written for general audiences.

Gray, Theodore. *The Elements: A Visual Exploration of Every Known Atom in the Universe*. New York: Black Dog & Leventhal Publishers, 2009. An easily accessible overview of the periodic table. Written for the general public. Provides a useful introduction of each element for high school students and the layperson. Includes excellent images and index.

Jensen, M. L., and A. M. Bateman. *Economic Mineral Deposits*. 3d ed. New York: John Wiley & Sons, 1981. Provides detailed information on the different metallic and nonmetallic mineral deposits and their modes of formation. Covers the history of mineral use and the exploration and development of mineral properties. Provides cross-sections of individual deposits. Suitable for college-level readers.

Lennox, Jethro, ed. *The Times Comprehensive Atlas of the World*. 13th ed. New York: Times Books, 2011. Includes a large map of world minerals printed in eight colors, showing the world distribution of diamonds, chemical and fertilizer minerals, asbestos, clay, magnesite, and talc. Indicates the relative importance of each mineral deposit. Suitable for high school students.

Pernetta, John, ed. *The Rand McNally Atlas of the Oceans*. Skokie, Ill.: Rand McNally, 1994. A well-written and beautifully illustrated atlas. Discusses nonmetallic resources that can be obtained from the sea, such as phosphorite, salt, sulfur, diamond, shell sands, sand, and gravel. Includes color photographs, maps, and line drawings. Suitable for high school students.

Tennissen, A. C. *The Nature of Earth Materials*. 2d ed. Englewood Cliffs, N.J.: Prentice-Hall, 1983. Contains detailed descriptions of 110 common minerals, with a black-and-white photograph of each. Contains helpful sections on the distribution of mineral deposits and the utilization of various Earth materials. Suitable for the layperson.

U.S. Environmental Protection Agency. *Profile of the Non-metal, Non-fuel Mining Industry*. Washington,

D.C.: U.S. Environmental Protection Agency, 1995. Provides a broad scope of topics related to mining nonmetals. Discusses economic trends, nonmetal processing, and waste release regulations. More recent data are available at the U.S. EPA Web site, from the Toxic Release Inventory to compliance and enforcement data.

Youngquist, Walter Lewellyn. *GeoDestinies: The Inevitable Control of the Earth Resources Over Nations and Individuals.* Portland, Ore.: National Book Company, 1997. Examines the earth's natural, mineral, and power resources, with an emphasis on social and environmental problems that accompany the recovery of such resources. A good book for the layperson. Illustrations and maps.

See also: Aluminum Deposits; Building Stone; Cement; Coal; Diamonds; Dolomite; Fertilizers; Gold and Silver; Industrial Metals; Iron Deposits; Manganese Nodules; Pegmatites; Platinum Group Metals; Salt Domes and Salt Diapirs; Sedimentary Mineral Deposits; Uranium Deposits.

IRON DEPOSITS

Iron, one of the most abundant metals in the earth's crust, is invaluable in modern industry and architecture because it is the primary component in steel. Locating iron deposits requires on-site testing and interpretation of satellite and other data. Deposits occur in many forms and are extracted all over the world in pit or tunnel mines.

PRINCIPAL TERMS

- **alloy:** a metal composed of two or more elements
- **banded iron formations:** the type of iron deposit most commonly mined; composed of iron ores in alternating layers with chert
- **ferric:** describes iron compounds with an oxidation number of +3
- **ferrous:** describes iron compounds with an oxidation number of +2
- **goethite:** a yellowish-brown or reddish-brown iron oxide-hydroxide
- **hematite:** an iron oxide that is usually gray or black, but can also be red; contains 70 percent iron
- **limonite:** an iron oxide-hydroxide of varying composition; includes hematite and goethite
- **magnetite:** an iron oxide that is usually black; contains 72 percent iron
- **siderite:** an iron carbonate that contains 48 percent iron
- **taconite:** a low-grade ore that contains 20 to 30 percent iron

What Is Iron Ore?

The planet Mars derives its red appearance from the iron oxides in its crust. Iron is one of the most abundant metals in the earth's crust, too, and it is the magnetism of the iron in the earth's core that creates the North Pole and the South Pole. Iron, along with cobalt and nickel, is one of three elements that is naturally magnetic, and of the three, iron is the most magnetic.

Iron ores are compounds of iron and other elements, and they can occur in a variety of forms. Ferric iron is iron with an oxidation number of +3, while ferrous iron has an oxidation number of +2. A positive oxidation number results when an iron atom loses electrons from joining with another element, such as oxygen or sulfur. For instance, hematite has an oxidation number of +3. Iron has more than two oxidation states, but +2 and +3 are the most common.

A number of frequently occurring iron compounds exist. Hematite, also called bloodstone, is reddish in color, while magnetite is usually gray or black. Both of these are iron oxides, which means they contain both iron and oxygen. Siderite is an iron carbonate, made of iron and carbon. Iron also can join with sulfur to form a sulfide such as pyrite. These ores are differentiated by the amount of iron they contain. Hematite, for example, contains 70 percent iron, while magnetite contains 72 percent iron.

Iron-rich soil is easily identified by its reddish color. However, though common, it would be impractical to extract that usable iron from soil because of the cost and difficulty of the process. Iron (or any other metal) that is found in a form that can be extracted profitably is called a reserve. A resource, though, is simply the presence of iron (or any other metal), whether or not it is usable. In time, iron deposits that are more accessible and more highly concentrated are exploited. Mining poorer or less accessible deposits requires more energy and larger mining operations, but mining and processing technologies are constantly being improved.

Taconite, which is mined in the United States, contains only 20 to 30 percent iron. It used to be considered a resource because its concentration of iron was too low to make it worth mining, but new technology has allowed it to be processed more efficiently; now, large taconite deposits are considered reserves rather than resources.

The Formation of Iron Deposits

In the past, iron ores called bog iron and ironstones were mined. Bog iron formed (and still forms) in lakes and swamps when chemical reactions precipitate iron from solution, producing thin deposits. Ironstones, which are usually pellets of goethite and hematite, formed in shallow ocean environments during periods of tectonic change and warm climate. Neither of these sources is commonly mined.

Banded iron formations (BIFs), found throughout the world, compose the type of deposit most commonly mined today. BIFs are made up of iron ores, such as magnetite, hematite, pyrite, and siderite, in alternating layers with chert (a silicate). BIFs were formed by sedimentary processes—iron precipitating from shallow ocean waters—in the Precambrian period, primarily 1.8 to 2.5 billion years ago. Because conditions for the formation of BIFs no longer exist, humans can only hypothesize the specifics of their formation.

It is thought that the oxygen-poor conditions of the atmosphere and oceans of the past created ideal conditions for the formation of BIFs because photosynthetic microorganisms flourished. These organisms created oxygen as a waste product, and the oxygen bonded with dissolved iron before the sediment settled on the ocean floors. The population of microorganisms increased and decreased cyclically, causing a corresponding cyclical accumulation of sediment, which could explain the alternating iron-rich and iron-poor layers.

Iron deposits often form at tectonic plate boundaries, where cold water passes through basaltic rocks and is heated by magma. These metal-rich liquids are called hydrothermal solutions. The heated water leaches metals from the basaltic rock and then deposits them in formations called black smokers, which are chimney-like formations reaching heights of up to 60 meters (197 feet). Another hydrothermal process is the formation of ferromanganese nodules, which occur in open ocean conditions where currents prevent dense accumulation of sediment. The formation of ferromanganese nodules occurs slowly. These resources cannot be profitably mined because of the low concentrations of iron and the relative inaccessibility of the deposits.

IDENTIFYING IRON DEPOSITS

For a deposit to be profitably mined, it must contain large amounts of ore (in the case of hematite, tens of millions of tons). It might seem like these giant deposits would be easy to find, but the process requires gathering extensive data and then testing to make sure a deposit will be worth mining.

Iron deposits can be identified in many other ways. Magnetometers can measure both the strength and the direction of magnetic fields. Gravity meters on the ground can locate areas of gravitational attraction created by the presence of heavy elements. Ground crews can also test soil, rivers, and streams for iron particles, which result from the weathering and erosion of iron-containing rock. Taking readings from airplanes or satellites, a process called remote sensing, also can provide geologists with important data.

One method of remote sensing is called hyperspectral imaging. The rocks on the surface of the earth have magnetic, reflective, and radioactive properties that can be measured from the air. Every metal reflects light and emits electromagnetic radiation, and each can be identified by what is called a spectral signature. A spectral signature is similar to a fingerprint in humans: It is a unique identifier. Instead of a pattern in the skin, the spectral signature comprises absorption bands in the visible, infrared, and ultraviolet spectra. In a laboratory, identification is relatively straightforward because conditions are controlled, but the process of identifying metals in situ is complicated by atmospheric attenuation, the presence of vegetation, variations in topography, the quality of the satellite data, and the organic and inorganic elemental make up of rocks and soils.

The use of satellite technology to locate potential iron reserves is relatively new. As late as the 1980s, few mining companies had the money, equipment, or expertise to acquire and interpret satellite imagery, but agencies such as the National Aeronautics and Space Administration were developing hyperspectral imaging that could accurately identify mineral species.

Both ferric and ferrous forms of iron are highly identifiable with hyperspectral imaging because they have broad absorption bands. Hematite and goethite, for instance, have distinct spectral signatures. Limonite encompasses several different types of ferric iron (including hematite and goethite). Identification of surficial limonite is useful for locating iron deposits because the presence of limonite often indicates hydrothermal processes, which are associated with the accumulation of iron.

MINING IRON ORE

All of the above methods can indicate the general area in which iron deposits are present, but to confirm the exact location of the deposits, the land must be drilled for rock samples. Test drilling helps to define the area to be mined by showing the extent and shape of the deposit. A shallower, broader deposit can be exploited more easily with a pit mine, but a deeper, narrower deposit requires tunneling. In the United States, both types of mines are commonly used to obtain iron ore.

Pit mines are generally dug in a bench formation with slightly graded walls to prevent collapse or slides. Because they spread out instead of down, pit mines can become large. The Hull-Rust-Mahoning Mine in Hibbing, Minnesota, one of the largest open pit mines in the world, is more than 5 kilometers (3 miles) long, 3 km (2 mi) wide, and 163 m (535 ft) deep at its most extreme points. It is so large that the town of Hibbing had to be relocated to accommodate the mine. Because of its long mining history, the mine and surrounding area is now a tourist attraction and a national historic landmark.

Shaft or underground mining requires tunneling into the earth and bringing the iron ore to the surface. A shaft is a vertical or near-vertical tunnel, which is excavated adjacent to the deposit. At intervals along the shaft, tunnels called drifts and crosscuts are excavated to allow access for the miners. The drifts and crosscuts are connected by vertical tunnels called raises. This network of passages allows workers to blast into the deposit and remove materials for transport to the surface.

Some iron deposits (soft hematite) are not competent, which means they will not remain in place when the material beneath them is excavated. For this type of ore, a method of extraction called block caving must be used. In block caving, an excavation is made beneath the deposit, and the deposit is allowed to cave in to the level below.

The largest producers of iron ore are Russia, Brazil, China, Australia, India, South Africa, and the United States. Worldwide, high-grade ores have been largely depleted and lower-grade ores are taking their place in the mining and manufacturing cycle. The U.S. Geological Survey reports that primary North American iron reserves will be depleted sometime between 2025 and 2050.

Technology may make underwater mining profitable. Areas of high iron content would be located by surveying with submersibles, instead of with airplanes or satellites, and then drilling for samples. It would not be economical to mine iron in this manner. Because so much iron already has been brought to the surface, recycling is more likely to replace mining as a way to supply human needs. Although not much scrap iron is recycled, more than 67 percent of steel, mostly from the auto industry, is salvaged and reprocessed.

ENVIRONMENTAL ISSUES RELATED TO MINING AND PROCESSING

Pit mines are less costly than underground mines, but they also have a higher environmental impact.

Dust is stirred up on-site during clearing of the land and during blasting, drilling, crushing, and transportation of ore. The machinery used in the iron mining process is also energy intensive and produces emissions. As with all forms of ore mining, buried rock that is exposed to air and water leaches metals into nearby water systems, creating a substance known as acid mine drainage, which can have far-reaching implications when it enters rivers and streams.

A particular iron deposit may be exploited for decades before it is depleted, but when a mine is no longer viable, the mining company must do a number of things to restore the landscape and to prevent pollution and other side effects of the mining process.

A pit mine can be filled in with mullock or overburden (the rock removed to access the ore) after the mine is closed. Mine walls can be reshaped and planted with vegetation. Toxic water and tailings must be contained by dams and neutralized in treatment pools. A possible alternative option is to use bioleaching bacteria to process contaminated water.

In the case of underground mines, mine shafts and adits (horizontal access tunnels) must be capped or filled. If bats are present, bat grates must be used to seal tunnel openings while still allowing bat ingress and egress. Subsidence hazards also must be addressed. Mine exhaustion also has social and economic consequences because towns are built to support the labor and other needs of a mine. When the mine closes, the town's industries and workers are no longer needed by the mining company.

Environmental issues also affect processing and manufacturing. During the manufacture of steel, particulates are kept from the air with dust catchers, electrostatic precipitators, and wet scrubbing systems. Settling basins and clarifiers remove oil and solids from water before it is returned to the local water system. Water used in processing can also be cleaned and recycled within the plant.

J. D. Ho

FURTHER READING

Brandt, Daniel A., and J. C. Warner. *Metallurgy Fundamentals.* Tinley Park, Ill.: Goodheart-Willcox, 1999. Provides background on physical characteristics of various metals, including ferrous metals. Discusses the composition, extraction, and processing of iron ore with detailed descriptions of manufacturing of different types of iron and steel.

Ericsson, Magnus, Anton Löf, and Olle Östensson. "Iron Ore Review 2010/2011." *Nordic Steel and Mining Review* 195 (2011): 14-19. This annual international issue summarizes the state of the global iron-mining industry, including the amount of ore produced by leading iron-producing countries, fluctuation in prices and trade, and corporate players.

Keller, Edward A. *Introduction to Environmental Geology.* 4th ed. Upper Saddle River, N.J.: Pearson Prentice Hall, 2008. Provides an overview of mineral resources, including how deposits form, how humans extract and use them, environmental consequences of mining, recycling, and reclamation of mines. Provides case studies and illustrations of undersea iron deposits, such as black smokers.

Lottermoser, Bernd G. *Mine Wastes: Characterization, Treatment, Environmental Impacts.* 2d ed. New York: Springer, 2007. A thorough and readable survey of various types of mines (ferrous and nonferrous) and the environmental issues associated with their construction and reclamation. Defines key terms related to ore processing and mining.

Rencz, Andrew N., ed. *Remote Sensing for the Earth Sciences.* New York: John Wiley & Sons, 1999. Gives a history of the development of remote sensing, particularly with regard to satellite imaging, and describes the science behind spectral imaging and other processes.

Robb, Laurence. *Introduction to Ore-Forming Processes.* Malden, Mass.: Blackwell, 2005. Discusses the process behind the formation of different types of iron deposits, including bog iron, ironstones, banded iron formations, and ferromanganese nodules. Includes illustrative case studies and diagrams elucidating tectonic processes.

See also: Aluminum Deposits; Geologic Settings of Resources; High-Pressure Minerals; Manganese Nodules; Remote Sensing; Uranium Deposits.

K

KIMBERLITES

Kimberlite is a variety of ultramafic rock that is fine- to medium-grained, with a dull gray-green to bluish color. Often referred to as a mica peridotite, kimberlite originates in the upper mantle under high temperature and pressure conditions. It frequently occurs at the earth's surface as old volcanic diatremes and dikes. Kimberlite can be the source rock for diamonds.

PRINCIPAL TERMS

- **blue ground:** the slaty blue or blue-green kimberlite breccia of the South African diamond pipes
- **diamond:** a high-pressure, high-temperature mineral consisting of the element carbon; it is the hardest naturally occurring substance and is valued for its brilliant luster
- **diatreme:** a volcanic vent or pipe formed as the explosive energy of gas-charged magmas breaks through crustal rocks
- **dike:** a tabular body of igneous rock that intrudes vertically through the structure of the existing rock layers above
- **magma:** a semiliquid, semisolid rock material that exists at high temperatures and pressures and is mobile and capable of intrusion and extrusion; igneous rocks are formed from magma as it cools
- **mantle:** the layer of the earth's interior that lies between the crust and the core; it is believed to consist of ultramafic material and is the source of magma
- **peridotite:** any of a group of plutonic rocks that essentially consist of olivine and other mafic minerals, such as pyroxenes and amphiboles
- **ultramafic:** a term used to describe certain igneous rocks and most meteorites that contain less than 45 percent silica; they contain virtually no quartz or feldspar and are mainly of ferromagnesian silicates, metallic oxides and sulfides, and native metals
- **xenocrysts:** minerals found as either crystals or fragments in some volcanic rocks; they are foreign to the body of the rock in which they occur
- **xenoliths:** various rock fragments that are foreign to the igneous body in which they are present

KIMBERLITE COMPOSITION AND EMPLACEMENT

The rock known as kimberlite is a variety of mica-bearing peridotite and is characterized by the minerals olivine, phlogopite (mica), and the accessory minerals pyroxene, apatite, perovskite, and various opaque oxides such as chromite and ilmenite. Chemically, kimberlite is recognized by its extraordinarily low silicon dioxide content (25-30 percent), high magnesia content (30-35 percent), high titanium dioxide content (3-4 percent), and the presence of up to 10 percent carbon dioxide. It is a dark, heavy rock that often exhibits numerous crystals of olivine within a serpentinized groundmass. Upon weathering, kimberlite is commonly altered to a mixture of chlorite, talc, and various carbonates and is known as blue ground by diamond miners. Occasionally, kimberlites contain large quantities of diamonds, which makes them important economically.

"Kimberlite" was first used in 1887 by Carvill Lewis to describe the diamond-bearing rock found at Kimberly, South Africa. There, it primarily occurs as a breccia found in several deeply eroded volcanic pipes and also in an occasional dike. Unfortunately, the kimberlite is so thoroughly brecciated and chemically altered that it does not lend itself to detailed petrographic study. Instead, the kimberlite of Kimberly, like kimberlite elsewhere, is more often noted for the varied assortment of exotic xenoliths (inclusions of other rocks) and megacrysts (large crystals) it contains. The explosive nature of a kimberlite pipe leads to the removal of country rock (rock surrounding an igneous intrusion) as the magma passes through the crust and thus can provide scientists with samples of material that originated at great depths. Among the many xenoliths found in the kimberlite breccia pipes are ultramafic rocks such as garnet-peridotites and eclogites, along with a variety of high-grade metamorphic rocks. These specimens provide scientists with an excellent vertical profile of the rock strata at various locations, serve to construct a model of the earth's crust at various depths, and provide information on the chemical variations in magma.

The emplacement of kimberlite is believed to be due to the eruption of a gas-rich magma that has intruded rapidly up a network of deep-seated fractures and probably breached the surface as a volcanic eruption. The magma's rate of upward mobility must have been rapid, as evidenced by the occurrence of high-density xenoliths such as eclogite and peridotite within the kimberlite pipes. The final breakthrough of the kimberlite magma may have taken place at a depth of 2-3 kilometers, where contact with groundwater contributed to its propulsion and explosive nature. Brecciation rapidly followed, along with vent enlargement by hydrologic fracturing of the country rock. Fragments of deep-seated rock, along with other country rocks, were then incorporated within the kimberlite magma.

Kimberlite Mineralogy and Texture

As a rock, kimberlite is very complex. Not only does it contain its own principal mineral phases, but it also has multicrystalline fragments or single crystals derived from the various fragmented xenoliths that it collected along the way. These fragments represent upper mantle and deep crustal origins and are further complicated by the intermixing with the mineralogy of a highly volatile fluid. As a result, no two kimberlite pipes have the same mineralogy. The continued alteration of the high-temperature phases after crystallization can produce a third mineralogy that can affect the interpretation of a particular kimberlite's occurrence.

The characteristic texture of kimberlite is inequigranular because of the presence of xenoliths and megacrysts within an otherwise fine-grained matrix. In relation to kimberlite, the term "megacryst" refers to both large xenocrysts and phenocrysts, with no genetic distinctions. Among the more common

megacrysts present are olivine (often altered to serpentine), picro-ilmenite, mica (commonly phlogopite), pyroxene, and garnet. These megacrysts are usually contained in a finer-grained matrix of carbonate and serpentine-group minerals that crystalized at considerably lower temperatures. Among the more common matrix minerals are phlogopite, perovskite, apatite, calcite, and a very characteristic spinel. Found within the textures of these matrix minerals are examples of both rapid and protracted cooling, with the latter evidenced by zoning. Zoning indicates that the matrix liquid cooled after emplacement and that there was sufficient time for crystals present to react with the remaining liquid, which is common to the megacrysts as well. In addition, the megacrysts exhibit an unusual, generally rounded shape that is believed to be a result of their rapid transport to the surface during the eruptive phase.

Kimberlite Occurrence

The geological occurrence of kimberlite, clearly volcanic in origin, takes the form of diatremes, dikes, and sills of relatively small size. In shape, kimberlite diatremes usually have a rounded or oval appearance but can occur in a variety of forms. Quite often, diatremes occur in clusters or as individuals scattered along an elongated zone. They rarely attain a surface area greater than a kilometer but in profile will resemble an inverted cone that descends to a great

Kimberlite. (Geological Survey of Canada)

depth. When it occurs as a dike, kimberlite is quite small—often not more than a few meters wide. It may occur as a simple ring dike or in swarms of parallel dikes. Kimberlite's occurrence as a sill is quite rare and will have a wide variability in thickness.

As compared with other types of igneous rocks, the occurrence of kimberlite is considered to be quite rare. Based on factors such as specific matrix color, density, mineral content, and xenolith size, shape, and number, connections between a certain kimberlite and a specific occurrence can be made. Aside from their scientific value, most kimberlites are economically worthless, except when they contain diamonds.

DIAMOND-BEARING KIMBERLITE

Of the many minerals that constitute kimberlite, diamond is the most noteworthy because of its great economic value. Not all kimberlite contains diamonds, and when diamonds are present, they are not always of gem quality. Even in the most diamondiferous kimberlites, the crystals and cleavage fragments are rare and highly dispersed. Diamond mining from kimberlite requires the removal and processing of enormous amounts of rock to produce relatively few diamonds. This method is expensive and can be dangerous. A much better rate of return can be gained from the placer mining of riverbeds and shorelines where nature has weathered out the diamonds and concentrated them in more readily accessible areas. The discovery of a large diamond in a riverbed in 1866 led to the eventual search for the source rock that produced diamonds. Up to that time, diamonds were recovered only from alluvial deposits and not from the host rock. In 1872, as miners were removing a diamond-bearing gravel, they uncovered a hard bluish-green rock that also contained diamonds. Further mining revealed the now familiar circular structure of the diatreme that continued downward to an undetermined depth. In 1887, the first petrographic description was made of the rock now known as kimberlite, and it was then recognized to be similar to other volcanic breccias from around the world.

Diamond-bearing kimberlite locations can be found around the world, the most famous being in South Africa. The diamond pipes of South Africa have been the leading producer of diamonds for well over a hundred years and are still unrivaled in terms of world production. Other locations of kimberlites that produce diamonds include the pipes of Siberia, western Australia, and Wyoming. For many years, Arkansas had the only commercial diamond mine in the United States, but the site is no longer active and is now preserved as a state park where visitors can hunt for diamonds.

KIMBERLITE ORIGIN HYPOTHESES

The chemistry and mineralogy characteristic of kimberlite indicate a very complex set of conditions that existed during their emplacement and subsequent crystallization. This fact, combined with kimberlite's limited occurrence and the altered nature of its matrix, makes kimberlite a puzzle to scientists. Three hypotheses have been proposed to explain its origin. A zone-refining hypothesis describes a liquid generated at great depth (600 kilometers) that is dynamically unstable and, as a result, rises toward the surface. As it reaches specific lower pressures, its composition is altered through partial crystallization and fractionation. A second hypothesis envisions a residual process: Partial melting of a garnet-peridotite parent, at depths of 80-100 kilometers, produces fractional crystallization of a picritic basalt (olivine-rich, or more than 50 percent olivine) at high pressure, which could lead to the formation of eclogite and a residual liquid of kimberlite composition. The third hypothesis describes kimberlite as either the residual end product of a long fractionation process or the product of a limited amount of partial melting. Each of the three hypotheses has merit but falls short of providing a definitive answer, partly because the near-surface environment where kimberlites are found is one of complex chemical reactions, which makes interpretation difficult. Adding to the problem is that no kimberlite eruption has ever been observed, so the exact processes are still somewhat speculative.

Although kimberlites tantalize scientists with their complexity, the xenoliths and megacrysts that arise with the kimberlites provide substantial data on the relationship between the lower crust and the upper mantle. Kimberlites show the earth's upper mantle to be very complex in terms of petrography, and they define both large- and small-scale areas of chemical and textural heterogeneities. Pressure and temperature conditions have been accurately established for depths down to 200 kilometers through studies based on the specimens brought up during kimberlite eruptions.

KIMBERLITE SPECIMEN ANALYSIS

Analysis of a rock such as kimberlite begins with the collection of a fresh specimen (as unaltered by weathering as possible) and continues with the preparation of a series of thin sections and microprobe sections. In this process, a slice of rock is cut with a diamond saw and glued to a glass plate. A second cut is then made to reduce the specimen's thickness to nearly 0.03 millimeter. A final polishing will achieve this thickness, which will permit light to pass through the specimen and provide the opportunity for microscopic examination. A similar procedure will produce a microprobe-thin section, but it requires extra polishing to ensure a uniform surface.

Once the thin sections have been prepared, a geologist uses a petrographic microscope, which employs polarized light, to make proper identification of the mineral phases present, along with their specific optical parameters. Opaque minerals require reflective light study. Microscopic examination is usually the first step in classifying a rock, and it may include an actual point count of the minerals present to determine their individual ratios. Afterward, an electron microprobe or electron microscope is used to give specific chemical compositions for individual mineral phases. With these devices, the scientist can analyze mineral grains as small as a few microns with a high degree of accuracy. Mineral grains or crystals that are smaller require an analyzing electron microscope to reveal their composition and fine detail. If the individual minerals are large enough and can be extracted, then X-ray diffraction can be used for a definitive identification of a particular mineral phase.

It is also important to understand the bulk chemistry of a rock specimen. Here, the scientist can select several different methods, including neutron activation analysis (NAA) and atomic absorption spectrometry (AAS). Both techniques provide excellent sensitivity and precision for a wide range of elements. A third method, X-ray fluorescence (XRF), is also commonly used and provides an accurate and quick means for analysis.

In NAA, a specimen is activated by thermal neutrons generated in a nuclear reactor. In this process, radioisotopes formed by neutron bombardment decay with characteristic half-lives and are measured by gamma-ray spectrometry. In AAS, a specimen is first dissolved in solution and then identified element by element. XRF, which provides sensitivity and precision for quantitative determination of a wide range of elements (best for those above atomic number 12), is the principal method used by most geologists to gather element amounts in both rocks and minerals.

After all the bulk chemical analyses have been gathered and run through various computer programs to determine mineral compositions, the data are compared to the microscopy results for evaluation and classification of the rock. This evaluation is compared to similar data from other locations. A final check and comparison to other rock types and strata in the field collection area complete the analysis.

SCIENTIFIC AND INDUSTRIAL APPLICATIONS

Although some kimberlites have economic value in their diamond content, most do not. No dramatic breakthroughs from the study of kimberlites will make diamonds more abundant or cheaper in price. The result of these studies is a better understanding of the processes and conditions under which diamonds and semiprecious minerals such as garnet are formed. This knowledge has led to the development of synthetic gems that can be used for industrial applications.

The study of kimberlites also has a direct bearing on scientists' understanding of volcanic activity. The prediction of eruptions is an aim of geology, as it affects the hundreds of thousands of people who live near potentially dangerous volcanoes. Kimberlite magma, by the nature of its chemical composition, is a very explosive material and can produce violent eruptions. Even so, it is true that most kimberlite magmas do not reach the surface, and when they do, they affect only a small area. Kimberlite magmas are more effective at depth, where rock is being fractured by the rising magma and by the hydraulic pressures exerted by the various gases moving through the magma. By examining the results of the movement of a kimberlite magma, scientists can gain a better understanding of volcanic behavior.

Paul P. Sipiera

FURTHER READING

Basaltic Volcanism Study Project. *Basaltic Volcanism on the Terrestrial Planets.* Elmsford, N.Y.: Pergamon Press, 1981. Represents the efforts of a team of scientists to provide the most up-to-date review of the subject of basalt and its relationship to planetary structure. Included are good references to kimberlites. Best suited for undergraduate and graduate students.

Blackburn, William H., and William H. Dennen. *Principles of Mineralogy*. 2d ed. Dubuque, Iowa: Wm. C. Brown, 1993. A basic textbook on the subject of mineralogy. Provides an excellent review of the principles of mineralogy and crystallography, along with the various techniques used in their study. Offers the reader both a broad overview and specific references to mineral families and their relationship to rock groups. Suitable for undergraduate and graduate students.

Boyd, F. R., and H. O. A. Meyer, eds. *Kimberlites, Diatremes, and Diamonds: Their Geology, Petrology, and Geochemistry*. Washington, D.C.: American Geophysical Union, 1979.

_____. *The Mantle Sample: Inclusions in Kimberlites and Other Volcanics*. Washington, D.C.: American Geophysical Union, 1979. Examines the many aspects of kimberlites and their formation conditions. Specialized articles cover a wide range of related geochemical, petrological, and mineralogical topics. Volume 2 also provides an in-depth study of the minerals and foreign rocks brought up by the kimberlite pipes. An excellent reference set suitable for undergraduate and graduate students.

Dawson, J. Barry. *Kimberlites and Their Xenoliths*. New York: Springer-Verlag, 1980. A very technical work that examines the geochemistry and mineralogy of kimberlites, along with their relationship to other rock types. Complete and detailed in its approach. Bibliography provides a wealth of journal article references. An excellent review source on kimberlites for the undergraduate and graduate student.

Decker, Robert, and Barbara Decker. *Volcanoes*. 4th ed. San Francisco: W. H. Freeman, 2005. Suitable for the general reader, this book provides useful background for the understanding of igneous rock bodies.

Le Roex, Anton P., David R. Bell, and Peter Davis. "Petrogenesis of Group I Kimberlites from Kimberley, South Africa: Evidence from Bulk-Rock Geochemistry." *Journal of Petrology* 44 (2003): 2261-2286. Discusses the geochemistry of kimberlite pipes found in South Africa. Major and trace elements have been analyzed to determine the parental rock and formation process of these kimberlite formations. Builds off rock cycle theory and petrogenesis, geochemical analysis techniques, and rare-earth elements.

Mitchell, Roger H. *Kimberlites, Orangeites, and Related Rocks*. New York: Plenum Press, 1995. Provides a good introduction to the study of kimberlites and related rocks, including an extensive bibliography that will lead the reader to additional information.

_____. *Kimberlites, Orangeites, Lamproites, Melilitites, and Minettes: A Petrographic Atlas*. Thunder Bay, Ontario: Almaz, 1997. This book covers the worldwide distribution of kimberlites and other related rock types. Color illustrations and bibliography.

Nixon, Peter H., ed. *Mantle Xenoliths*. New York: John Wiley & Sons, 1987. Covers a wide range of topics related to kimberlites. Considerable attention is paid to regional kimberlite occurrences and the foreign rocks that are brought up by the kimberlite diatremes. Several of the articles are general enough to suit a beginning reader, but the work is best suited to the undergraduate and graduate student.

Sage, R. P., and T. Gareau. *A Compilation of References for Kimberlite, Diamond and Related Topics: Ontario Geological Survey, Open File Report 6067*. Ontario: Queen's Printer for Ontario, 2001. Discusses various topics related to kimberlites. Begins with general mineralogy of kimberlites followed by numerous references to other minerals contained in kimberlite formations. Later sections discuss petrology, melting, crystallization, and research techniques.

Sutherland, Lin. *The Volcanic Earth: Volcanoes and Plate Tectonics, Past, Present, and Future*. Sydney, Australia: University of New South Wales Press, 1995. Provides an easily understood overview of volcanic and tectonic processes, including the role of igneous rocks. Includes color maps, illustrations, and a bibliography.

Tappert, Ralf, and Michelle C. Tappert. *Diamonds in Nature: A Guide to Rough Diamonds*. New York: Springer, 2011. The text discusses the unique properties of diamonds and the processes involved in their formation. The background information and characteristics of kimberlitic rock are discussed. Covers differences in diamond characteristics and the associated kimberlite in which it is found. The text includes color images, references, and indexing.

Wyllie, P. J., ed. *Ultramafic and Related Rocks*. Malabar, Fla.: Kieger Publishing Company, 1979. A collection of specialized articles that deal with the various types of rock that are derived from mantle material. Several of the articles focus specifically on kimberlites, especially as an overall review of the subject. Best suited as a reference work for undergraduate and graduate students.

See also: Andesitic Rocks; Anorthosites; Basaltic Rocks; Batholiths; Carbonatites; Feldspars; Granitic Rocks; Igneous Rock Bodies; Komatiites; Minerals: Physical Properties; Orbicular Rocks; Plutonic Rocks; Pyroclastic Rocks; Rocks: Physical Properties; Xenoliths.

KOMATIITES

Komatiites are volcanic rocks with abundant olivine and pyroxene and little or no feldspar. They also contain large magnesium oxide concentrations (greater than 18 percent). They are most abundant in the lower portion of exceedingly old piles of volcanic rocks. Economic deposits of nickel, copper, platinum group minerals, antimony, and gold have been found in some komatiites.

PRINCIPAL TERMS

- **basalt:** a dark rock containing olivine, pyroxene, and feldspar, in which the minerals often are very small
- **crust:** the veneer of rocks on the surface of the earth
- **feldspar:** calcium, potassium, and sodium aluminum silicate minerals
- **igneous rock:** a rock solidified from molten rock material
- **metamorphism:** the process by which a rock is buried in high temperature and high pressure, transforming the original minerals into new minerals in the solid state
- **olivine:** a magnesium and iron silicate mineral
- **pyroxene:** a calcium, magnesium, and iron silicate mineral
- **silicate mineral:** a naturally occurring element or compound composed of silicon and oxygen with other positive ions to maintain charge balance
- **skeletal crystals:** elongated mineral grains that may resemble chains, plates, or feathers
- **upper mantle:** the region of the earth immediately below the crust, believed to be composed largely of periodotite (olivine and pyroxene rock), which is thought to melt to form basaltic liquids

KOMATIITE COMPOSITION

Komatiites are unusual volcanic rocks. They contain mostly large grains of olivine (magnesium and iron silicate mineral) and pyroxene (calcium, iron, and magnesium silicate mineral) scattered among fine mineral grains that at one time were mostly glass. The glass was unstable and in time slowly converted into individual minerals. The larger minerals often form elongated grains composed of olivine, pyroxene, or chromite (a magnesium-chromium oxide mineral at the top of some lava flows) called skeletal grains. No other rocks form these skeletal grains. The skeletal grains are believed to have formed by quick cooling at the top of the lava flow. The glass must have also formed by quick cooling of the lava, often in contact with water. Because most igneous rocks contain feldspar (calcium, sodium, and potassium aluminum silicate minerals), the lack of feldspar in komatiites also makes them unusual. They also contain magnesium concentrations (magnesium oxide greater than 18 percent) higher than most other igneous rocks. Most important, komatiites are extremely silica-poor rocks (ultramafic rocks) that melt at temperatures hotter than any present-day lava, and currently do not erupt.

KOMATIITE LAVA FLOWS

Each individual lava flow of a komatiite may be a meter to tens of meters thick. A given lava flow can often be traced for a long distance without much variation in thickness. Some of the lava flows contain rounded or bulbous portions called pillows. Pillows can be formed only by extrusion of the lava into water. As lava breaks out of the solid front of a flow, it is quickly cooled into a rounded pillow. Since this process continually takes place, numerous pillows form as the lava advances. The more magnesium-rich komatiites occur as lava flows in the lower portions of these volcanic piles. They gradually become less magnesium-rich in the younger or upper portions of these volcanic piles.

The mineralogy and observed mineral shapes can vary vertically through a given lava flow. The most spectacular and beautiful komatiite lava flows are the ones in which the upper portions contain the skeletal olivine and pyroxene grains. Skeletal grains have an extremely distinctive texture and may resemble chains, plates, or feathers. Some grains grow as large as 3 centimeters. The thickness of these lava flows varies from about 1 to 20 meters. In one type of vertical variation, there are abundant fine minerals formed from the original glass at the top of these flows, along with irregular fractures caused by quick cooling at the top of the lava. Underlying this quickly cooled zone is a layer of large skeletal grains of olivine and pyroxene that increase in grain size downward. These skeletal crystals probably form very rapidly in the lava by growth from the top downward.

The lower portion of these komatiite flows often contains abundant olivine grains that are much more rounded and less elongated than those of the upper zone. The more rounded grains are believed to have formed by the slow crystallization of the olivine from the liquid lava. As the olivine formed from the liquid, the grains slowly sank and piled up on the bottom of the flow until the lava solidified. The base of the lava also may contain very fine-grained, skeletal grains of olivine with irregular cracks, much like the top of the flow. Presumably, the base of the flow formed by quick cooling, as did the top portion.

Observing several cross-sections shows how komatiite flows can vary. There is, for example, a considerable difference in the amount of the flow that contains the skeletal minerals and the flow that contains the more rounded minerals. Most komatiite flows do not contain any skeletal minerals. Instead, they consist mostly of the more rounded mineral grains and have irregular fractures extending throughout the lava flow. These differences could be caused by different cooling rates or by the lava's viscosity (how easily it flows). Other flows are composed mostly of rounded pillows.

Komatiite lava flows are often interbedded with rocks composed of angular fragments that came from all parts of the lavas. Some of these rocks show features that might be found along the edge of a body of water, such as ripples formed by wave action. These rocks have formed by the action of waves breaking up some of the solidified lavas and reworking them after they were deposited. (Rocks that have been reworked by water are sedimentary rocks.)

KOMATIITE OCCURRENCE

Komatiites were discovered in the late 1960's in exceedingly old Archean rocks (more than 2.5 billion years old) in Zimbabwe and South Africa. Only a few komatiites occur in younger rocks. The youngest known komatiite, only about 70 million years old, and the only one less than 500 million years old, is located on Gorgona Island off the coast of Colombia. Komatiites typically occur in the lower or older portions of vast piles of volcanic rocks containing many layers of different lava flows. Komatiites gradually become less abundant farther up the volcanic pile and in younger volcanic rocks. Other dark, feldspar-rich, volcanic rocks called basalt (calcium-rich feldspar and pyroxene rock) are interlayered with komatiites. Basalts gradually become more abundant in the younger volcanic flows, along with more light-colored and more silica-rich rocks such as andesite. Small amounts of sedimentary rocks may be interlayered with the volcanic rocks. Sedimentary rocks are derived by water reworking the volcanic rocks. Komatiites eventually disappear in the upper portions of these volcanic piles.

Many komatiites are located in rather inaccessible regions. One area that is accessible is the Vermilion district of northeastern Minnesota. The age of the

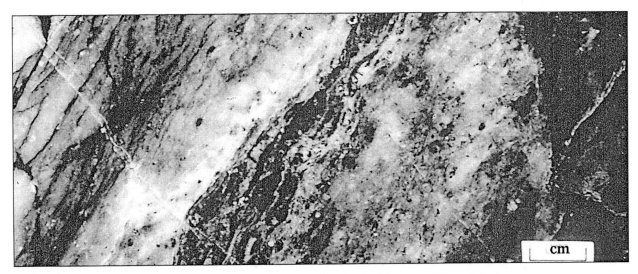

A komatiite flow adjacent to a large rhyolite lapilli fragment. (Geological Survey of Canada)

district is 2.7 billion years old. Here, there are no true komatiites with magnesium oxide concentrations greater than 18 percent. There are, however, very magnesium-rich basalts that contain the skeletal crystals found in true komatiites. All the rocks have been buried deep enough that the original minerals were changed or metamorphosed to new minerals in response to the high temperature and pressure. The temperature of metamorphism was still low enough that the original igneous relations may still be observed.

A second example of a komatiite sequence is located at Brett's Cove in Newfoundland. The komatiites there are much younger than most komatiites (formed during the Ordovician, about 450 million years ago). Those komatiites formed within layers of rocks called ophiolites, which are believed to be sections of ruptured and tilted oceanic crust and part of the upper mantle. The lower part of the ophiolites contains an olivine and/or pyroxene mineralogy thought to compose much of the upper mantle of the earth. Overlying the upper-mantle rocks are rocks of basaltic composition containing coarse crystals of olivine, pyroxene, and feldspar. Above these rocks are numerous basaltic lavas that were extruded at the surface and that built up large piles of lavas on the ocean floor. Some komatiite lavas are interbedded in the lower portion of these mainly basaltic lavas. Those komatiite flows have a lower zone rich in pyroxene and an upper pillow lava that contains skeletal crystals of pyroxene.

STUDY OF FIELD AND CHEMICAL CHARACTERISTICS

The characteristics of komatiites that are exposed at the surface are described carefully during a field study. These characteristics can suggest how komatiites form at the surface. For example, features such as abundant pillows indicate that the komatiite lava was extruded into water. A small amount of reworking of the lavas by moving water suggests that there was little time between eruptions. A sign of a rapid eruption is the spread of lava over a great distance and at a gentle flow rate.

In addition to the field characteristics of komatiites, geologists study their chemical characteristics in order to understand how they form and evolve. Komatiites are igneous rocks, so they form by the melting of another rock. Experiments using furnaces suggest that much of the rock called peridotite

(olivine-pyroxene rock) must melt to form high-magnesium magmas. The magma then may evolve or change in composition by processes such as the crystallization of minerals from the magma, by dissolving some of the solid rock through which it moves, or by mixing with magma of a different composition. The komatiite may even change composition after the magma solidifies because of water vapor or carbon dioxide-rich solutions moving through the solid. It is difficult to assess the relative importance of these processes to modify the composition of a given komatiite.

ANALYZING ELEMENTAL CONCENTRATIONS

One way to study the mineral crystallization of a magma is to plot the elemental concentrations of several analyzed rocks from the same general area against another element, such as magnesium. If the concentration of all elements systematically increases or decreases with increasing magnesium concentrations, then the lavas were probably related by fractional crystallization, because olivine becomes poorer in magnesium and richer in iron as the magma crystallizes. The formation and settling of olivine from lava appears to explain the concentration of most elements. For example, magnesium, chromium, nickel, and cobalt all concentrate in lava-related olivine; thus, a plot of chromium, nickel, and cobalt shows smooth and systematic decreases when compared to magnesium. Elements such as calcium, titanium, aluminum, silicon, iron, and scandium are rejected from lava-related olivine; they gradually increase with decreasing magnesium. Some elements, such as sodium, potassium, barium, rubidium, and strontium, should likewise systematically increase with decreasing magnesium, as they are also rejected from lava-related olivine. Instead, these elements are greatly scattered when they are plotted relative to magnesium. Those elements are notorious for being moved by carbon dioxide or water-vapor-rich fluids. Thus, it is assumed that the scatter of these elements is a result of the movement of these fluids. The fractionation of these elements because of olivine crystallization, therefore, is obscured.

Some of the variations in elemental concentrations in komatiite lavas cannot be explained by crystallization or alteration processes. For example, two rare-earth elements like lutetium and lanthanum should not differ in ratio during olivine crystallization, as those elements are chemically very similar and olivine should not fractionate these elements. Nevertheless, some processes do produce unexplained variations in rare-earth contents.

ECONOMIC VALUE

Komatiites contain several types of economic deposits. They may contain important deposits of nickel sulfides, along with large concentrations of platinum group elements (such as platinum and palladium) and copper. The nickel sulfide was probably formed from a sulfide liquid that separated as an immiscible liquid from the komatiite liquid. This is similar to the way oil and water separate when they are mixed together. Nickel sulfide ores have been found in western Australia, Canada, and Zimbabwe. The most important deposits are found in the komatiite lava flows in the lower portion of a lava pile. The nickel sulfide ore is concentrated in a portion of the base of a lava flow where it is thicker than other portions of the flow. The immiscible and dense nickel-sulfide liquid may have settled in a thick portion of the flow that was not stirred as much as other portions of the flow. The platinum group metals have a stronger affinity for the sulfide liquid than for the komatiite liquid; consequently, they concentrate in the sulfide liquid.

Gold, antimony, and a few other elements are concentrated in some komatiite flows. Running water may alter and rework some komatiite lavas and form sedimentary rocks. Examples of these deposits occur northeast of Johannesburg in South Africa. There, the lower portion of the rock pile is mostly layers of successive komatiite or basaltic lava flows. This portion is overlaid by sedimentary rocks composed of mudrocks changed, at high temperature and pressure, into metamorphic rocks. Some quartz-carbonate rocks within the sedimentary rocks likely formed by the alteration of komatiites, probably because carbon dioxide reacted with other elements to form carbonates, leaving residual silica to form quartz. The quartz-carbonate rocks contain high concentrations of antimony in the mineral stibnite and small particles of gold. Solutions moving through the komatiites probably altered the komatiites, leaving the high concentrations of these elements.

Robert L. Cullers

FURTHER READING

Arndt, N. T., D. Frances, and A. J. Hynes. "The Field Characteristics and Petrology of Archean and Proterozoic Komatiites." *Canadian Mineralogist* 17 (1985): 147-163. An advanced article summarizing the way komatiite lavas occur. A number of diagrammatic representations of these lavas make it easier to visualize what they look like. A college student with a petrology course could read the discussion; someone with a course in introductory geology could read it if he or she were willing to look up terminology in a geologic dictionary.

Arndt, Nicholas, C. Michael Lesher, and Steven J. Barnes. *Komatiite.* New York: Cambridge University Press, 2008. An update on the original *Komatiites* book published in 1982. It is organized into two sections: background information, and interpretations of komatiite properties, geodynamic setting, and composition. References and indexing are well developed.

Arndt, N. T., and E. G. Nisbet. *Komatiites.* Winchester, Mass.: Allen & Unwin, 1982. An advanced book reviewing many aspects of komatiites. Much of the book could be read by students of a petrology course. Someone taking an introductory geology course could read Chapters 1 and 2 on the history and definition of komatiites. A mineral collector can find references to the location of komatiites. The skeletal crystals developed by some komatiites are especially beautiful; includes black-and-white photographs of skeletal crystals.

Best, M. G. *Igneous and Metamorphic Petrology.* Malden, Mass: Blackwell Science, 2003. Offers a short section on komatiites. Several photographs of the skeletal olivine and pyroxene crystals and a diagrammatic cross-section of a komatiite lava flow are included. A person taking an introductory geology course could read the text with the help of a geology dictionary.

Decker, Robert, and Barbara Decker. *Volcanoes.* 4th ed. San Francisco: W. H. Freeman, 2005. Suitable for the general reader, the text provides useful background for the understanding of igneous rock bodies.

Eriksson, P. G., et al. *The Precambrian Earth: Tempos and Events.* Edited by K. C. Condie. Amsterdam: Elsevier, 2004. Focuses on the processes involved in the creation of the earth's crust and development of the various rock types. Discussion of komatiites is spread throughout the text.

Faure, Gunter. *Origin of Igneous Rocks: The Isotopic Evidence.* New York: Springer-Verlag, 2010. Discusses chemical properties of igneous rocks and isotopes within rock formations. Specific locations of igneous rock formations and the origins of these rocks are provided. There are diagrams, drawings, and an overview of isotope geochemistry in the first chapter to make this accessible to undergraduate students as well as professionals.

Grotzinger, John, et al. *Understanding Earth.* 5th ed. New York: W. H. Freeman, 2006. This comprehensive physical geology text covers the formation and development of the earth. Written for high school students and general readers. Includes an index and a glossary of terms.

Hyndman, D. W. *Petrology of Igneous and Metamorphic Rocks.* 2d ed. New York: McGraw-Hill, 1985. Contains a short section on komatiites, including a section on the economic importance of komatiites. A black-and-white photo of the skeletal crystals is provided. A person taking an introductory geology course could read this section with the help of a geologic dictionary.

Mitchell, Roger H. *Kimberlites, Orangeites, and Related Rocks.* New York: Plenum Press, 1995. Mitchell provides a good introduction to the study of kimberlites and related rocks, including an extensive bibliography that will lead the reader to additional information.

Winter, John D. *An Introduction to Igneous and Metamorphic Petrology.* Upper Saddle River, N.J.: Prentice Hall, 2001. Provides a comprehensive overview of igneous and metamorphic rock formation. A basic understanding of algebra and spreadsheets is required. Discusses volcanism, metamorphism, thermodynamics, and trace elements; focuses on the theories and chemistry involved in petrology. Written for students taking a university-level course on igneous petrology, metamorphic petrology, or a combined course.

See also: Andesitic Rocks; Anorthosites; Basaltic Rocks; Batholiths; Carbonatites; Diamonds; Gem Minerals; Granitic Rocks; Igneous Rock Bodies; Kimberlites; Orbicular Rocks; Plutonic Rocks; Pyroclastic Rocks; Rocks: Physical Properties; Xenoliths.

L

LANDFILLS

Humankind has found it convenient throughout history to dump unwanted wastes into nearby ravines, swamps, and pits. With increased emphasis on sanitation, the open dumps have been replaced with landfills in which wastes are placed into excavations and covered with soil. Landfilling is the most common method of disposing of garbage and other unwanted material generated by cities and industry.

PRINCIPAL TERMS

- **aquifer:** a porous, water-bearing zone beneath the surface of the earth that can be pumped for drinking water
- **clay:** a term with three meanings—a particle size (less than 2 microns), a mineral type (including kaolin and illite), and a fine-grained soil that is like putty when damp
- **geomembrane:** a synthetic sheet (plastic) with very low permeability used as a liner in landfills to prevent leakage from the excavation
- **groundwater:** water found below the land surface
- **leachate:** water that has seeped down through the landfill refuse and has become polluted
- **permeability:** the ability of a soil or rock to allow water to flow through it; sands and other materials with large pores have high permeabilities, whereas clays have very low permeabilities
- **pollution:** a condition of air, soil, or water in which substances therein make it hazardous for human use
- **saturated zone:** that zone beneath the land surface where all the pores in the soil or rock are filled with water rather than air
- **vector:** a term used in waste disposal when referring to rats, flies, mosquitoes, and other disease-carrying insects and animals that infest dumps
- **water table:** the upper surface of the saturated zone; above the water table, the pores in the soil and rock containing both air and water

Hazardous versus Nonhazardous Wastes

All human activities produce unwanted by-products called wastes. For normal household living, these unwanted wastes are garbage and trash. Stores, factories, gas stations, and all other businesses also create large amounts of waste materials. Waste materials are classified as hazardous and nonhazardous: Hazardous wastes contain chemicals and other constituents that, if inhaled, eaten, or absorbed by humans and other life-forms, are detrimental to health; nonhazardous wastes may contain small amounts of toxic or hazardous ingredients. Yet they present no threat to the welfare of society if disposed of correctly.

Because of the severe environmental restrictions and legal liabilities associated with burying hazardous materials in the ground, recycling and incineration have become the principal disposal options for hazardous wastes. Sites at which toxic wastes are buried are tightly controlled and carefully monitored, to ensure that no leakage occurs that could pollute the groundwaters.

Land Disposal

One of the primary methods for getting rid of wastes is land disposal. Land disposal may take the form of placing the unwanted material directly on the land surface, especially in low, swampy areas, ravines, and old gravel, or other mined-out pits. These types of land disposal are known as dumps, or open dumps, because the unwanted wastes are haphazardly dumped at the site, and the waste piles often are not covered until some other use is made of the surface. The term "landfill," although sometimes used to describe dumps, refers to the land disposal of wastes in which the disposal site is designed and operated to a specific plan. Modern landfills are designed to receive a certain amount of refuse over a specified period of time, such as twenty years. At the end of that

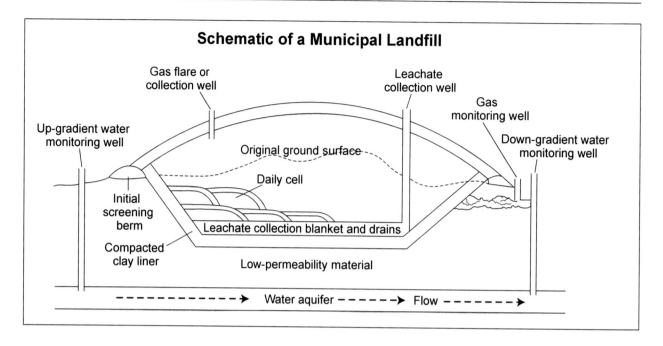

Schematic of a Municipal Landfill

period, the site is reclaimed and converted for a different land use.

In ancient times, when the earth's population was small, waste products were discarded on the ground surface or thrown into a stream at the campsite. The amount was small enough at that time so that there was no significant pollution or adverse effects on the environment. As the population grew and more people began to live in towns, trash and garbage often were thrown in ravines and low, swampy areas as a way of filling in the land. This practice of open dumping has existed in the United States and all other nations up to the present.

The earth's population has soared to more than 7 billion people, with more than 300 million of them living in the United States. The population concentration has shifted from rural to urban areas. To supply the demands of the public, industries have developed and produced a wide variety of materials and chemicals, such as plastics, that end up as waste products after use. More than 10 billion metric tons of solid waste is generated each year in the United States; some 200 million metric tons consists of municipality-generated rubbish and garbage, which amounts to 730 kilograms of nonindustrial waste per person per year.

As the amount of municipal wastes increased, more and more space was needed for open dumps.

Many serious problems arose as the use of dumps conflicted with land use, sanitation, and aesthetics in the surrounding communities. The pollution of rivers and groundwater was directly traced to uncontrolled dumping of toxic wastes in dumps and pits. The federal government, therefore, enacted a series of laws over the years to protect the environment and especially the surface and groundwaters from pollution. The result of the legislation was to define and separate wastes that were hazardous to humans from the municipal wastes in land disposal. Special restrictions were placed on the disposal of hazardous wastes in the earth. Although originally no special restrictions were placed on the land disposal of municipal wastes, the protection of rivers and groundwater from pollution required states to establish a permit process that set limits on where dumps and landfills could be located. Thus, the disposal of municipal and industrial wastes has become expensive.

LANDFILL FACILITIES AND FEATURES

There are thousands of active and inactive nonhazardous waste landfills of various types in the United States. More than 3,000 are classified as municipal, which means that most of the waste placed in the landfill is from households or is general community refuse. Modern landfills cover many acres. Some are more than 100 acres in size.

Although the actual size and shape of each landfill depend upon the site geology and the amount of waste, landfill facilities generally are excavated into the ground to form a pit. The refuse is placed in the pit and gradually builds up throughout the life of the landfill to a predetermined vertical height above the original ground level. Once the design height is reached, a soil cover is added, and the ground surface is reclaimed to form a moundlike hill. The depth of excavation to form the pit will depend upon the thickness of low-permeability clay soils over a buried water-bearing deposit, called an aquifer, which supplies drinking water to households and towns. Enough clay soil must be left to prevent the seepage of any leachate that may escape from the landfill and pollute the aquifer. (Leachate is rainwater that has seeped through the landfill refuse and has become polluted.) Monitoring wells are placed around the landfill into the aquifer to ensure that no pollution occurs.

After the excavation has been dug to its intended depth, a protective liner is placed over the bottom and sides of the pit. The liner may be of clay compacted by a construction roller, or it may be a plastic sheet called a geomembrane. Waste material that is brought to the landfill daily is placed in specified layers called cells. At the close of each day, the exposed waste is covered with soil to prevent odors, blowing debris, and infestation by vectors. "Vector" is a term used to collectively refer to any disease-carrying insects or animals, such as rats, flies, and birds, that would infest the waste material. The daily cells are stacked one on top of another until the final height is reached. Leachate and gas collection pipes are installed throughout the landfill to collect and dispose of dangerous gases and liquids. These perforated pipes are covered with a layer of gravel through which the leachate can easily flow from the landfill waste to the collection network.

SITE EVALUATION

Before a landfill can be constructed and operated at a proposed location, the site must be evaluated to ensure that it satisfies all local, state, and federal regulations relating to protecting the environment and the health and welfare of the citizens. The landfill operator must collect and assess much information before applying to a state pollution control board or similar governing commission for a permit. The information must verify that the landfill will present no danger to the public or to the environment. A large number of factors must be evaluated and those evaluations presented to the permit board.

By far the most important factor for determining whether a site is satisfactory for a landfill is the geology. The local geology must contain thick and continuous clay soils or their rock equivalent, shale. These deposits do not allow water or contaminated water from landfills (leachate) to flow rapidly through them; their low permeability will prevent any leachate that may escape from the pit from flowing into surface streams or underlying aquifers, which supply water to the surrounding communities. Generally, additional lines of defense are installed, such as impermeable plastic barriers and clay liners. It is very important to protect these aquifers; once they become polluted, it is extremely difficult (and costly) to clean them up for public use in a short period of time. Compared to water in surface streams, groundwater in porous aquifers moves quite slowly. Also, because the aquifer is hidden beneath the surface, it is difficult to trace and clean up the polluted water flow. To evaluate the subsurface geology of the site and the surrounding region, the company proposing the landfill will drill holes around the area and take soil samples so that a geologist can identify the different soils and rocks and construct a cross-section illustrating the thicknesses, types, and relationships of the different materials.

Besides the geology, other major factors that must be assessed are climate and weather, flooding, ecology, historical landmarks, nearness to airports, traffic, and land-use and zoning restrictions. The climate and weather describe the rainfall and winds to be expected. All landfills must be above the 100-year flood height or have suitable flood protection. Special emphasis is placed on the ecology: A landfill cannot be built in protected wetlands or compromise the habitat of endangered species of plants and animals. Also, landfills are not allowed to destroy historical landmarks and archaeological sites. Municipal landfills attract birds; therefore, landfills must be more than a mile away from airports. Information on the road system and traffic volumes must be gathered to assess the impact of waste trucks, which add both more traffic and greater loads to the roads. Landfills are not the most desirable form of land use and, therefore, such developments must prove compatible with the surrounding land use. Many communities have zoning restrictions. Thus, evaluating a proposed site for a landfill is a complex and time-consuming job.

LANDFILL DESIGN

A company must supply to the state permit board data and plans outlining the overall design, construction, operation, and reclamation of the land when the landfill is closed—in addition to information and assessment on the suitability of the proposed site. The design information must include drawings showing the depth and size of the pit area. The plans must give details on how the leachate is prevented from escaping from the pit, and where monitoring wells will be drilled around the site to detect any leakage that may occur. Landfills that contain garbage and other organics will generate methane gas that must be collected and either flared into the atmosphere or piped away for fuel. Some landfills use the gas at the site for commercial and light-industrial energy.

The final aspect in a landfill design is the closing of the operation and the reclaiming of the site for a different land use. A final soil cover is placed over all the waste to isolate it from the public. Five to ten feet in thickness, this layer of soil controls vectors, prevents odors, and beautifies the landfill's surface. Periodic checks are made by the owner after the landfill has been reclaimed to sample the monitoring wells and to repair any erosion features.

PUBLIC POLICY

The federal government, the states, and communities have had to reevaluate their policies of uncontrolled dumping of municipal and industrial wastes because of serious pollution to the water supplies. With the enactment of laws to protect the environment and prevent pollution, land disposal of wastes is now controlled and monitored. Hazardous wastes are a severe threat to the health and welfare of communities. As a result, with few exceptions, land disposal of these wastes is no longer the primary disposal option.

Land disposal remains the primary method of disposing of nonhazardous wastes. Liquid nonhazardous wastes usually are disposed of in lagoons. Solid wastes, such as garbage and trash, are placed in landfills. For cities and other urban areas, large acreage of open land is scarce, and citizens generally object to having a landfill constructed nearby. Thus, cities and adjacent communities have been forced to cooperate on large regional landfills. It is therefore necessary for companies wanting to site new landfills to hunt for undeveloped land of a hundred acres or more where the geology can prevent pollution of the public water supplies. Even though municipal landfills, often called sanitary landfills, are considered nonhazardous, they will contain from 5 to 10 percent toxic material. If the landfill is small, such as in rural communities, the amount of hazardous leachate that may escape from the pit site is small and thus will pose no appreciable detrimental effects on the environment and public health. The large landfills that serve cities, however, do pose a threat to the public welfare and health if major amounts of leachate escape from the site. It is extremely difficult and costly to purify groundwater used for public water supplies if an aquifer becomes polluted. When households and communities lose their groundwater supply to pollution, the hardships to the citizens are both severe and costly. Some communities have had to transport water from other areas and impose tight controls on how the water is used.

Waste disposal is a very important part of the lifestyle of modern society; with the quest for more and more conveniences and services, the unwanted by-products of civilization continue to increase and must be disposed of in some way. In addition, the natural environment has deteriorated to a point of serious concern. Therefore, landfilling must be done with the utmost care and planning in order to guard against pollution.

N. B. Aughenbaugh

FURTHER READING

Bagchi, Amalendu. *Design, Construction, and Monitoring of Landfills.* 2d ed. New York: John Wiley & Sons, 1994. Discusses the elements of design and the procedures involved in the construction and upkeep of sanitary landfills. Includes twenty-five pages of bibliographical references and an index.

_____. *Design of Landfills and Integrated Solid Waste Management.* 3d ed. Hoboken, N.J.: John Wiley & Sons, 2004. Expands on the traditional landfill design to discuss integrated solid-waste management. Discusses collection, disposal, and recycling of solid waste, as well as leachate and microbiology of landfills. Debates the health and safety of composting.

California Integrated Waste Management Board, Permitting and Enforcement Division. *Active Landfills*. Publication Number 251-96-001. Sacramento: Author, 1998. A state examination of the protocol and maintenance of active California landfills. Includes a review of sanitation policies and safety measures. A revised edition from 2000 is available online through CalRecycle.ca.gov.

Cook, James. "Not in Anybody's Back Yard." *Forbes* 142 (November 28, 1988): 172-177. One of several articles that *Forbes* magazine has printed on waste disposal and pollution of the environment.

Foreman, T. L., and N. L. Ziemba. "Cleanup on a Large Scale." *Civil Engineering* (August, 1987): 46-48. Covers articles published in technical journals on waste disposal and pollution. Written in an understandable way.

LaGrega, Michael D., Philip L. Buckingham, and Jeffrey C. Evans. *Hazardous Waste Management*. 2d ed. Long Grove, Ill.: Waveland Press, 2001. Covers the fundamentals of hazardous waste, from the technical definition and chemical and physical properties, to federal regulations. Discusses current preventive practices, as well as disposal methods and risk assessment. Technically written and suited for advanced undergraduate and graduate students with some background in engineering.

Lee, G. Fred, and Anne Jones-Lee. *Overview of Subtitle D Landfill Design, Operation, Closure and Postclosure Care Relative to Providing Public Health and Environmental Protections for as Long as the Wastes in the Landfill Will Be a Threat*. El Macero, Calif.: G. Fred Lee & Associates, 2004. Discusses Subtitle D and EPA regulations of solid waste landfills. Easily accessible and loaded with information both historical and current. Includes many figures.

Mulamoottil, George, Edward A. McBean, Frank Rovers, et al., eds. *Constructed Wetland for the Treatment of Landfill Leachates*. Boca Raton, Fla.: Lewis Publishers, 1999. Examines the use of the wetlands for sanitary landfills. Provides an extensive look at the construction and waste treatment procedures involved.

Pichtel, John. *Waste Management Practices: Municipal, Hazardous, and Industrial*. Boca Raton, Fla.: CRC Press, 2005. Discusses various types of waste, the history of waste management, laws and regulations, as well as information on the sampling, chemical properties, collection, transport, and processing of municipal solid wastes. Discusses many of the technical components of recycling and composting. Includes an extensive and detailed chapter on landfills. Covers hazardous waste in multiple chapters. Includes references, suggested readings, questions, and a sizeable index.

Qian, Xuede, Robert M. Koerner, and Donald H Gray. *Geotechnical Aspects of Landfill Design and Construction*. Upper Saddle River, N.J.; Prentice Hall, 2001. Discusses the engineering and geological aspects of landfill design. Written for the engineer; highly technical and includes design equations and protocols. Discusses many issues in landfill construction, such as erosion, stability, landfill liners, and vertical expansion.

U.S. Environmental Protection Agency. *RCRA Subtitle D Study*. Washington, D.C.: Government Printing Office, 1986, 1988. This study on nonhazardous wastes is one of many reports by the U.S. Environmental Protection Agency on all aspects of waste disposal and pollution control. The agency is an excellent source of both general information and specific data. Interested readers can write to the U.S. Environmental Protection Agency, Office of Solid Waste (WH-562), 401 M Street S.W., Washington, D.C. 20460.

U.S. Geological Survey. *Water Fact Sheet: Toxic Waste—Groundwater Contamination Program*. Washington, D.C.: Government Printing Office, 1988. An informational brochure, or fact sheet, for the general public on groundwater supplies and pollution. Interested readers should write to the Hydrologic Information Unit, U.S. Geological Survey, 419 National Center, Reston, Virginia 22092.

See also: Earth Resources; Earth Science and the Environment; Hazardous Wastes; Mining Processes; Mining Wastes; Nuclear Waste Disposal; Oil Chemistry; Soil Chemistry; Soil Profiles; Strategic Resources.

LAND MANAGEMENT

Land management—the control of land use and the preservation of land—is essential to the proper maintenance of this limited natural resource.

PRINCIPAL TERMS

- **land:** in the legal sense, any part of the earth's surface that may be owned as goods and everything annexed to that part, such as water, forests, and buildings
- **land use:** the direct application of a tract of land
- **multiple use:** the simultaneous use of land for more than one purpose or activity
- **subsurface features:** land features or characteristics that are not visible or apparent, such as minerals, oil and gas, and structural features lying beneath the land surface
- **taxation:** a land-management tool that usually reflects the perceived best use of land
- **topography:** the collective physical features of a region or area, such as hills, valleys, streams, cliffs, and plains
- **zoning:** a land-management tool used to limit and define the conditions and extent of land use

TYPES OF LAND USE

Land is not reproducible; it is present in only a fixed amount, a fact often repeated in the adage "They're not making any more real estate." In addition, the location of land—important to its use and value—cannot be changed. Aspects of land include topography, which controls many uses; soil, which is vital to agricultural and other applications; subsurface structure and composition, which can prove to be either beneficial or problematic; and the availability of minerals, oil and gas, and other natural resources.

Land management is the science that has developed in response to the need for control of land use and preservation. Land-management programs attempt to organize, plan, and manage land-use activities. Well-developed programs are concerned with land and water, and address the issues of water-use rights and rights related to the surface, subsurface, and above-surface. Land-management activities can take place at local, state, and federal levels. They can be geared to single-purpose or multipurpose land uses and can be oriented toward rural or urban settings.

Land uses can be either reversible or irreversible. Reversible uses, such as agricultural activities, grazing of livestock, forestry, recreation, and mining, can allow for regression or reversal toward former or alternative uses. These uses are frequently applied in multiple-use programs of land management and are generally compatible. The purpose of programs such as these is to maximize use while allowing for the greatest good for the most people. Irreversible land uses result in permanent changes in the character of the land, such that it cannot revert to a former condition or use. The filling-in of swamps or other water bodies, the building of cities, and the development of nonreclaimable surface mines result in irreversible and permanent changes. Activities such as these generally result in single-use situations and preclude the development of alternative-use plans. If the activity terminates (a mine is exhausted, for example), the land may be put to other uses but will not revert to its previous state.

LAND-USE ISSUES

Major issues that require the application of land-management policies include new growth, declining growth, reclamation, resource exploitation and utilization, preservation of natural or cultural resources, plans for maintenance of stable populations, or environmental, economic, or social concerns. Each land-use issue and attendant land-management policy may require specialized knowledge and specific approaches. Science and technology are vital in developing land-use plans; science provides the knowledge, while technology provides the means to implement that knowledge. Successful application also requires appreciation of the political and economic issues that affect a population's response to land-use policies.

All land-use issues can present or generate one or more uncertainties with respect to future applications of the land; these issues can represent a problem or an opportunity, be subject to the effects of supply and demand, and be dealt with systematically or conceptually. The degree of uncertainty in any land-use issue is a direct result of the availability of data regarding the use and prior experience with the issue. One of the focal points of well-developed land-management policies is to reduce or remove the uncertainties related to land use. Opportunities in land use are those activities that benefit a large segment of the population, either directly or indirectly. Examples range from the establishment of a national park to the development of a new airport. Problems in land use might include the subsurface disposal of radioactive waste or the threat to wildlife habitat from construction of a reservoir system. Supply dictates how much land should be subjected to a specific use in response to the perceived demand.

ECOLOGICAL CONSIDERATIONS

Ecological diversity is an important aspect of land-use planning and also land-management policies. The ultimate goal of all land-management programs should be to put land to its multiple best uses. In the process, the ecological "carrying capacity" of a regional environment should not be exceeded. Natural resources should be maintained in a state of availability, and development should be encouraged only in areas best suited for it; development should be discouraged in areas of significant resource value. Development should also be discouraged in areas of natural or human-made hazards.

Land-management programs and policies are largely a result of the location of the land to be managed and the anticipated impact the programs might have on a given population. Economics will frequently dictate the preferred use of land, sometimes at the expense of wildlife, aesthetic beauty, and other ecological factors. Land-use issues can be addressed at different levels. Factors that help to determine the level at which any particular issue might be addressed include the number of people and localities that might be affected by the issue, the magnitude of the potential cumulative effects that may result from the issue, and the threshold at which an issue becomes significant. The availability (or lack) of water and the effect of water pollution are examples of factors that can have a cumulative effect, while air pollution is an example of an issue that has reached a threshold, elevating it from a local to an international concern.

REGULATORY LAND MANAGEMENT

Land management is an ongoing activity, and policies may require change and/or modification with time. Land has an intrinsic (cash or exchange) value

and an extrinsic (inherent or judgmental) value, both of which must be considered when dealing with or formulating a management plan. Effective planning and subsequent management, public or private, require that land-use controls be regulated and supported by sufficient authority.

Lands are generally managed and their use controlled by taxation, police and regulatory powers, and strategic considerations. Taxation serves as a management tool because taxes levied are generally a reflection of the perceived best use. Changes in tax status frequently result in changes in land use, as in the case of agricultural land that is converted to urban use as a direct result of an increase in taxes and the subsequent cessation of agricultural activities. Even if the tax rate does not change, rising land prices may make it attractive to sell farm land. Police and regulatory powers dictate what can and cannot be done on a specific piece of land. Subdivision regulations, environmental laws, and zoning ordinances are the most common form of regulatory land management. Master plans also control land use and are required by most local governments. They assist cities and counties in coordinating the regional implementation of statutes and/or regulations and include a statement of goals, an outline of societal needs, and a list of specific objectives. They are generally collective plans backed by extensive information and by many independent studies. Master plans also outline mechanisms by which the objectives are to be reached. Strategically located lands can affect the use of adjoining parcels. The presence of industrial areas could, for example, preclude adjoining residential development, while the existence of parks and golf courses could discourage adjacent industrial development. Airports, ski areas, forests, and rangelands can also have strategic value if situated properly.

A drive through cities and suburban developments shows that some areas have been assigned to industrial, commercial, or residential uses. These use areas are the result of zoning, taxation, and other management tools that attempt to encourage certain types of development in relation to the carrying capacity or suitability of the land and its annexed improvements. The availability of deep-water ports and rail transportation is, for example, more important to commercial and industrial development than to residential land use. By the same token, certain soils and other natural factors might favor residential development.

Land reclamation, soil-erosion prevention measures, and imposed land-use limitations are all part of land management. The state of Georgia, for example, requires that all mine sites provide reclamation of as many acres as were actually mined during a given year, although the reclaimed acreage need not be that which was mined. Soil-erosion prevention programs are incorporated as part of nearly every development or activity plan that will result in disturbance or modification of a soil profile, including plans for wilderness or forest roads, residential subdivisions, construction along waterways or coastlines, and agricultural activities.

MANAGEMENT OF FEDERAL LANDS

Federal and state land-management programs are widely recognized forms of land-use planning. These programs have a direct impact on the use of public land, such as parks, forests, seashores, and inland waterways. The U.S. federal government has owned lands that were not otherwise owned by local government or private owners since the nation was founded. Approximately one-fifth of the public domain (lands owned by the federal government) was eventually granted to individual states. These lands were set aside for schools, hospitals, mental institutions, and transportation or were swamps and flooded lands—all part of an overall land-management plan.

The management of federal lands is largely custodial. It is carried out under the provisions of numerous statutes and regulations, including the Multiple Use-Sustained Yield Act of 1960, which legalized the multiple use of federal lands; the Wilderness Act of 1964, which set aside wilderness areas; the Classification and Multiple Use Act of 1964, which allowed for the classification of land for determining the best use and determining the lands that should be retained or discarded; and the National Environmental Policy Act of 1969, which required the filing of an Environmental Impact Statement (EIS) for major actions that would significantly affect the quality of the human environment.

Grazing is the oldest use of federal lands, but oil and gas activities generate the largest revenues. Mining of nonfuel minerals is governed by the Mining Laws of 1866 and 1872, while the Mineral Leasing Act of 1920 provides for the competitive and noncompetitive leasing of land containing oil and gas, oil shale, coal, phosphate, sodium, potash, and sulfur. The United States Forest Service administers all federal forest lands, while the Bureau of Land Management (BLM) administers all other lands.

Issues that must be addressed in all federal land-management programs include fraud and trespass (relating to illegal harvesting of timber or other valuable materials), resource depletion, reserved rights on lands that have been discarded, multiple use of lands, equity for future generations, the ability to maintain lands and retain their value, and the ideal of private land ownership. Policy issues that are closely related include how such land should be acquired by the federal government; how much land should be discarded; to whom the lands should be granted and what rights (such as access to minerals) should be retained and for how long; what the terms of land disposal should be; how much should be spent to maintain lands that are retained; to what use should retained lands be put; who should share in the benefits that accrue from lands retained; and who should develop and execute the land-management plan. Policy issues change with time, as do approaches to land management.

FOREST MANAGEMENT

Forests are an important target of land-management activities because they occupy approximately one-third of the total land area of the United States. Of that area, nearly two-thirds is occupied by commercial forests, while the remainder is reserved from harvest. Forests are used as watershed areas, renewable consumable resources, recreational areas, and wilderness preserves. Policy issues that affect the management of forests include questions regarding how much forest to maintain, how much to restore, how much to withdraw from use, and how they should be harvested. Several criteria must be met to establish practical forest policy: the physical and biological feasibility of an action, economic efficiency and equity, social acceptability, and operational practicality. Not all uses of forests are compatible in a potential multiuse scenario; some uses will necessarily exclude others, which must be considered in a forest-management program. For example, interactive effects must be considered in the harvesting of timber, which affects the watershed, soil, regenerative growth, and wildlife. Policy areas directly related to the maintenance and management of forest resources include taxation, often a large cost of forest ownership; housing programs, which affect the demand for forest products; foreign trade with attendant import duties, quotas, or tariffs that affect the merchantability of forest products; transportation, which affects the marketability of forest products; direct aids to forest development programs, such as research, education, and production subsidies; and the administration of public forests.

COASTAL LAND MANAGEMENT

Coastal land management is as complex as the management of inland areas, if not more so. In Florida, for example, the value of shore properties frequently dictates the reclaiming of lost lands or the creation of new lands for urban use. Swamps and intertidal areas along coastlines may be filled in at the expense of what is frequently a fragile environment. At issue is whether development can take place in such a manner that people can live in an area without destroying the natural features and beauty that attracted them in the first place. One approach has been to set aside land as parks or conservation areas. This approach is increasingly popular in the formulation of land-management policy. Aesthetic concerns aside, coastal land management may discourage development in hazardous areas, such as those prone to storm surges.

On a stroll along the beach, one may observe a person fishing from a jetty or surfcasting from the base of a seawall. Sailboats cruise the inner harbor, protected from the sea by the distant breakwater. All these physical structures—the jetty, seawall, and breakwater—are part of the coastal land-management program. The jetty attempts to prevent beach erosion by the longshore current that runs nearly parallel to the shoreline, while the seawall aids in the maintenance of a stable coastline that might otherwise erode under the constant battering of winter storms. The breakwater helps to maintain quiet waters in the shallow inner harbor area, which would otherwise be subjected to high and frequently damaging waves.

MINING AND AGRICULTURE

Mining activities generally require extensive land-use planning and must be carried out under well-defined land-management policies. These policies control mining activities from the earliest stages of exploration, through the actual mining and production of mineral materials, and finally through reclamation.

Most management policies are in place to help preserve the character of the land to the greatest possible degree. The routing and design of access roads on federal lands must generally be approved by either the Forest Service or the Bureau of Land Management. Some restrictions also apply on private lands, which require special permits.

Mineral-exploration activities frequently are limited in size on federal land so that they do not interfere with natural wildlife habitats or other approved uses of the land. Once a valuable deposit has been identified, mining permits must be applied for, EIS's may be required, and reclamation procedures must be outlined prior to the extraction of the deposit. Once mining has been completed, the land must be reclaimed in accordance with an approved plan. All these activities take place under the land-management plan.

Crop rotation and strip farming are land-management mechanisms employed in agriculture. Different crops require different kinds and levels of nutrients for proper development. Crop rotation, or changing the type of crop grown on a particular tract of land with each growing season, allows for the greatest yield of nutrients from the soil. Strip farming regenerates nutrients and also serves as a soil-erosion prevention measure. Early farming techniques put all lands under cultivation and as a result all were subject to wind erosion. Strip farming leaves alternating strips vegetated or cultivated, helping to prevent erosion.

Kyle L. Kayler

FURTHER READING

Burby, Raymond J., ed. *Cooperating with Nature: Confronting Natural Hazards with Land-Use Planning for Sustainable Communities.* Washington D.C.: Joseph Henry Press, 1998. Covers the role of natural disasters and hazards in the development of new communities. Intended for the reader with some background in the field but also useful to the layperson with an interest in land-management policies. Illustrations, bibliography, and index.

Clawson, Marion. *The Federal Lands Revisited.* Washington, D.C.: Resources for the Future, 1983. Discusses federal land policies and management. Explores all aspects of land management and introduces many new concepts as solutions to the current problems in the field. Discusses major policy issues and present usages, including wildlife, grazing, minerals extraction, oil and gas production, watershed protection, and recreation. Includes numerous data tables, figures, and an index. Written for the nonspecialist.

_____. *Forests for Whom and for What?* Baltimore: Johns Hopkins University Press, 1975. A well-developed discussion of forest management from both a public and a private viewpoint. Addresses timber production, recreational usage, wildlife protection, and watershed management, as well as other economic, social, and environmental concerns. Discusses public forest policy in detail, and the impact of land conversion, restoration, and clearcutting. Includes several data tables, an index, and a bibliography. Geared toward an intellectual, nonspecialist audience.

_____. *Man, Land, and the Forest Environment.* Seattle: University of Washington Press, 1977. Gathers three essays originally delivered as public lectures in 1976. Discusses land-use planning and control, and the private and federal ownership of forested land. Presents suggestions for future land management and discussions of past "mismanagement." Includes a short bibliography, graphs, and data tables. Targeted toward an intellectual, non-specialist audience.

The Conservation Foundation. *State of the Environment: An Assessment at Mid-Decade.* Washington, D.C.: Author, 1984. An insightful report on the status of the environment in 1984. Deals with underlying trends in conditions and policy. Addresses environmental contaminants, natural resources, future problems, and the assessment of environmental risks. Includes an extensive bibliography and an index. Geared toward a diverse audience with interests in politics, statistics, and the environment.

Dasmann, Raymond F. *No Further Retreat.* New York: Macmillan, 1971. Discusses the development of Florida from the late 1960's to 1971. Concentrates on coastal and inland waterway land-management efforts. Discusses Everglades protection, wildlife preservation, and the development of the Florida Keys. Offers reasonable approaches for controlling land use and instituting proper planning. Includes several photographs and maps and is well indexed. Suitable for any interested layperson. Uses no technical language.

Davis, Kenneth P. *Land Use.* New York: McGraw-Hill, 1976. Discusses concepts of land, land ownership, land use and classification, and land-use controls. Examines the planning process with respect to land management and valuation, as well as the attendant decision-making processes. Presents several case histories. Appropriate for many levels of reader interest, from the layperson to the technical specialist.

Economic Commission for Europe. *Land Administration Guidelines: With Special Reference to Countries in Transition.* New York: United Nations, 1996. This United Nations publication describes the land-use and land-tenure policies that have been enacted in Central and Eastern European countries during their transition from communist to democratic governments.

Fabos, Julius Gy. *Land-Use Planning.* New York: Chapman and Hall, 1985. Examines land-use planning from many perspectives. Planning issues are discussed in detail, as are the roles of science and technology in land-use planning. Addresses the evolution of land-use planning from the standpoint of interaction between disciplines, public versus private planning, and types of planning. Discusses regional and local considerations, as well as future prospects in land-use planning. Contains a good bibliography. Appropriate for all levels of interested readership.

Healy, Robert G. *Competition for Land in the American South.* Washington, D.C.: The Conservation Foundation, 1985. Discusses land use and development in the southern United States. Addresses the issues of competition for land, the economic uses of agriculture, wood protection, animal agriculture, and human settlement. Analyzes each land use with respect to future demands. Discusses the impact on soil, water, wildlife, and aesthetics. Includes a reference list and an index. Targeted toward a nonspecialist audience.

Karl, Herman A., et al., eds. *Restoring Lands; Coordinating Science, Politics, and Action.* New York: Springer, 2012. Discusses social, economic, policy, and environmental challenges in land-use management and planning. Addresses these issues from local and global perspectives. Provides a number of tools to aid students and professionals in analyzing policies and developing planning strategies. Highlights the importance of scientists working with the public to make changes.

Knight, Richard L., and Peter B. Landres, eds. *Stewardship Across Boundaries*. Washington D.C.: Island Press, 1998. Analysis of the conservation of natural resources and management of public lands in the United States. Maps and illustrations.

Loomis, John B. *Integrated Public Lands Management*. 2d ed. New York: Columbia University Press, 2002. Discusses federal agency land-management plans, including topics such as policy analysis, economics, natural resource management, and conservation.

Newson, Malcolm. *Land, Water and Development: Sustainable and Adaptive Management of Rivers*. 3d ed. London: Routledge, 2008. Covers land-water interactions. Discusses recent research, study tools and methods, and technical issues, such as soil erosion and damming. Suited for undergraduate students and professionals. Covers concepts in managing land and water resources in the developed world.

Paddock, Joe, Nancy Paddock, and Carol Bly. *Soil and Survival*. San Francisco: Sierra Club Books, 1986. Analyzes what the authors view as a lack of human commitment to the land. Examines the threats to American agriculture, the drive for greater land efficiencies at the expense of natural beauty, and the technical loss of land through erosion, chemical usage, and development. Stresses attitudes about ethics, land stewardship, and environmental concerns. Well indexed and contains footnotes. Geared toward an environmentalist audience.

Steel, Brent S., ed. *Public Lands Management in the West: Citizens, Interest Groups, and Values*. Westport, Conn.: Praeger, 1997. Explores the volatile conflict between environmentalists and "wise use" groups in the western United States. Suitable for readers without any background in environmental studies. Index and bibliography.

The World Bank. *Sustainable Land Management: Challenges, Opportunities, and Trade-offs*. Washington, D.C.: International Bank for Reconstruction and Development, 2006. Discusses causes of land degradation and sustainable land-management policies.

See also: Clays and Clay Minerals; Desertification; Expansive Soils; Land-Use Planning; Land-Use Planning in Coastal Zones; Landslides and Slope Stability; Soil Chemistry; Soil Erosion; Soil Formation; Soil Profiles; Soil Types.

LANDSLIDES AND SLOPE STABILITY

Landslides occur under specific geological conditions that are usually detectable. Site assessments done by qualified geologists are important to land-use planning and engineering design; much of the tragedy and expense of landslides is preventable.

PRINCIPAL TERMS

- **angle of repose:** the maximum angle of steepness that a pile of loose materials such as sand or rock can assume and remain stable; the angle varies with the size, shape, moisture, and angularity of the material
- **avalanche:** any large mass of snow, ice, rock, soil, or mixture of these materials that falls, slides, or flows rapidly downslope
- **cohesion:** the strength of a rock or soil imparted by the degree to which the particles or crystals of the material are bound to one another
- **creep:** the slow and more or less continuous downslope movement of Earth material
- **earthflow:** a term applied to both the process and the landform characterized by fluid downslope movement of soil and rock over a discrete plane of failure; the landform has a hummocky surface and usually terminates in discrete lobes
- **hummocky:** a topography characterized by a slope composed of many irregular mounds (hummocks) that are produced during sliding or flowage movements of earth and rock
- **landslide:** a general term that applies to any downslope movement of materials; includes avalanches, earthflows, mudflows, rockfalls, and slumps
- **mass wasting:** collective term for all forms of downslope movement propelled mostly by gravity, including avalanche, creep, earthflow, landslide, mudflow, and slump
- **mudflow:** both the process and the landform characterized by very fluid movement of fine-grained material with a high water content
- **slump:** a term Lathat applies to the rotational slippage of material and the mass of material actually moved

FALLS, SLIDES, AND FLOWS

Slope failure, or mass wasting, is the gravity-induced downward and outward movement of Earth materials. Landslides involve the failure of Earth materials under shear stress and/or flowage. When slope failures are rapid, they become serious hazards. Areas of the United States that are particularly susceptible to landslides include the West Coast, the Rocky Mountains of Colorado and Wyoming, the Mississippi Valley bluffs, the Appalachian Mountains, and the shorelines and bluffs around the Great Lakes. The downslope movement of soil and rock is a natural result of conditions on the planet's surface. The constant stress of gravity and the gradual weakening of Earth materials through long-term chemical and physical weathering processes ensure that, through geologic time, downslope movement is inevitable.

Slope failures involve the soil, the underlying bedrock, or both. Several types of movements (falling, sliding, or flowing) can take place during

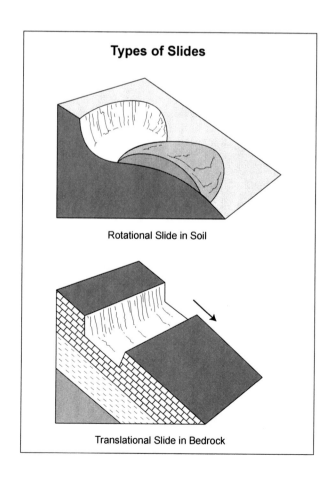

Types of Slides

Rotational Slide in Soil

Translational Slide in Bedrock

the failures. Simple rockfalls, or topples, may occur when rock overhangs a vertical road cut or cliff face. Other failures are massive and include flows and slides. Slides involve failure along a discrete plane. The failure planes in soils are usually curved, as in the illustration of a rotational slide or slump. The failure planes in bedrock can be curved or straight. Failures often follow planes of weakness, such as thin clay seams, joints, or alignment of fabric in the rock. Slides may be slow or rapid but usually involve coherent blocks of dry material. Flows, in contrast, behave more like a fluid and move downslope much like running water. Earthflows, mudflows, sand flows, debris flows, and avalanches occur when soils or other unconsolidated materials move rapidly downslope in a fluidlike manner. The movement destroys the vegetative cover and leaves a scar of hummocky deposits where the flow occurred. Although flows usually involve wet materials, rare exceptions, such as the destructive flows in Kansu, China, occur in certain types of dry materials that are finely grained and loosely consolidated.

Slides and flows are terms applied to failures that produce rapid movement. Rock slides are those slides that involve mostly fresh bedrock; debris slides include those movements that are mostly rock particles larger than sand grains but with significant amounts of finer materials; mudslides involve even finer material and water, but the failure plane is straight. Earthflows involve mostly the soil overburden and move over a slope or into a valley rather than failing along a rock bedding plane; mudflows involve more water than earthflows and have a downslope movement much like flowing water.

CREEP

"Creep" is a term given to very slow movement of rock debris and soils. Creep in itself does not usually pose a life-threatening danger. When creep occurs beneath human-made structures, however, it leads to economic damage that requires repair or reconstruction in a new location. Examples include the gradual cracking and destruction of buildings, misalignment and breaking of power lines and fences, the filling of drains along highways, and the movement of soil into streams and reservoirs. Sometimes creep precedes a more rapid failure, and therefore new evidence of creep requires careful monitoring and an evaluation of the conditions that produce it.

Solifluction is a special type of creep that occurs in permafrost regions, in cold climates where the soil is frozen most of the year. In summer, the ice in the upper layer of soil melts, and the soil becomes waterlogged and susceptible to downslope movement. Solifluction is an important consideration for the design of structures in cold climates: For example, the Alaska pipeline, used for transporting petroleum, could not be buried but had to be set above ground on supports that were anchored at a depth below the seasonal depth of thaw to escape solifluction movement. Houses and other buildings in such areas must be set on supports and insulated in order to keep heat from the structures from melting the underlying soils.

ROLE OF WATER

Water is an important agent in promoting instability in slopes. Where soils are saturated, water in large pores will flow naturally in a downward direction. The resulting pressure of the water pushing against the soil grains is called pore pressure. As pore pressures increase, the grains are forced apart and cohesion decreases. The saturated soils become easier to move downslope. Water flowing along a bedding plane or joint can also exert pressure on either side of the joint and decrease cohesion, causing the block above the discontinuity to move downslope.

Loose volcanic materials often can absorb so much water that they flow quickly down even gentle slopes. A mudflow buried the Roman city of Herculaneum at the base of Mount Vesuvius in 79 A.D. (although the city had already been hit by pyroclastic flows); and mudflows generated by the Mount St. Helens eruption of May 18, 1980, destroyed many properties. An especially destructive mudflow on the volcano Nevado Ruiz in Colombia in 1986 struck an unsuspecting town at night and killed 22,000 people.

ROLE OF HUMAN DEVELOPMENT

Designers may unwittingly assemble, in a human-made structure, conditions that produce slope failures. Dry materials such as mine wastes have sometimes been stacked into piles that are steeper than their angle of repose after saturation. Much later, a rainstorm or Earth tremor can send the piles into motion, destroying all structures around them, as happened at Aberfan, Wales, in 1964. When the supporting toe is removed from the base of a slope, as during such excavations that occur for a highway or building foundation, this action produces many landslides, evidence of which can be found on most highways constructed through hilly terrain. An unstable slope can be set into motion by loading the slope from above, which occurs when a structure such as a building, a storage tank, or a highway is built on materials that cannot remain stable under the load. Catastrophic failures have occurred in which mudflows were produced when dams built from mine tailings burst as a result of slope failure. The mine wastes used for these dams were susceptible to swelling, absorption of water, and weakening over time.

Human development alters the natural drainage of an area and increases runoff. Occasionally, water from storm drains, roof gutters, septic tanks, or leaking water mains reaches a sensitive slope and generates movement. This instability is particularly likely to occur where intensive housing development takes place in several levels on a long slope.

REGIONAL STUDIES

The stability of slopes is evaluated over large regions from aerial photographs, satellite photographs, and images made by remote sensing techniques. Investigators look for telltale signs, such as hummocky topography and old scars left by slides, that may not be evident when viewed from the ground.

Regional studies involve an evaluation of the history of past landslides within the region. History often reveals particular geological formations that have an association with landslides. For example, the shales of the Pierre formation are well known by engineering geologists in the area of Denver, Colorado, as materials in which many slope failures occur. A geologic map that shows where this formation is exposed at the ground surface reveals potentially dangerous areas. Ignoring evidence of past landslides invites disaster.

The regional study also defines loose surficial materials that are likely to fail. Soils rich in clay minerals that swell and expand when wet are notorious for slumping and flowing. Usually, movement occurs in these soils in the spring, when the soil is very saturated from soil thaw and snowmelt. Other soils fail simply because they have low cohesion and large amounts of open space (pore space) between the tiny soil grains. Collapse of these soils requires no wetting; strong vibrations can trigger the movement. The most tragic example of this type took place in 1920, when thick deposits of fine loess (a type of soil deposited by the wind) settled rapidly during an earthquake in Kansu, China, and the resulting flows toppled and buried the many homes built upon them; more than 100,000 people died.

Regional studies also look at the earthquake history of an area, because a tremor, even a fairly mild one, can provide the coup de grâce to a slope that has been resting for decades in a state of marginal instability. Huge blocks of the shoreline slid beneath the ocean at Valdez, Alaska, during the 1964 earthquake when the rotational slumping of materials occurred below sea level. Landslides triggered by that same earthquake destroyed much of Anchorage, Alaska. Much of the failure was due to the sudden liquefaction of buried clay layers.

Finally, the regional study includes a history of weather events. When the right geological conditions exist, periods of intense rainfall can trigger the movement of unstable slopes. In the southern and central Appalachian Mountains, periods of increased slide frequency often coincide with severe local summer cloudbursts and thunderstorms. Intense rainfall events associated with hurricanes that have moved inland also trigger landslides over larger areas. Studies in the Canadian Rockies reveal a definite link between rainstorms and rockfalls, and landslides are particularly abundant during the rainy season along the West Coast of North America.

ON-SITE INVESTIGATION

Once knowledge is collected on the region, more specific questions are considered about the local site itself. The investigator first looks at the steepness of the slopes and the earth materials present. In the case of loose materials and soils, the angle of repose is very important. Dry sand poured carefully onto a table to form an unsupported conical pile cannot achieve a cone with sides steeper than approximately 30 degrees, because the cohesion between loose dry sand grains is not strong enough to allow the material to support a steeper face. The 30-degree angle is the maximum angle of repose for dry sand. The angle of repose changes with water content, mineral content, compaction, grain shape, and sorting. Soils that contain clay may be tough and cohesive when dry and have natural repose angles greater than 30 degrees. When wet, their angle of repose may be only 10 degrees. This is particularly true if the soils contain clay minerals such as montmorillonite that absorb large amounts of water. Those soils, sometimes called "quick clays," can fail instantaneously and flow downslope almost as rapidly as pure water.

The orientation of discontinuities in rocks is as important in determining the stability of a slope as is the type of rock involved. Bedding planes that dip downslope serve as directions of weakness along which failure may occur. Other planar weaknesses may develop along joints and faults and parallel fabrics produced by the alignment of platy and rod-shaped minerals that are oriented downslope.

The investigator will check to see if natural processes are removing the supporting material at the bases of slopes. Landslides are particularly common along stream banks, reservoir shorelines, and large lake and seacoasts. The removal of supporting material by currents and waves at the base of a slope produces countless small slides each year. Particularly good examples are found in the soft glacial sediments along the shores of the Great Lakes of the United States and Canada.

Finally, the investigator will look for evidence of actual creep at the site. Damage to structures already on the site—such as curved tree trunks (where tilting is compensated for by the tree's tendency to resume vertical growth), the offset of fences and power lines, or the presence of hummocky topography on slopes—can demonstrate the presence of recent motion.

Costs and Remediation of Stability Problems

An annual economic loss of between $1 billion and $1.5 billion is a reasonable estimate for costs of landslides within the United States. Expenses include the loss of real estate around large lakes, rivers, and oceans; loss of productivity in agricultural and forest lands; depreciated real estate in areas of slide development; public aid for victims of large landslides; and the contribution of sediment to streams, which decreases water quality, injures aquatic life, and results in the loss of reservoir storage space. Small-scale damage from soil creep is not dramatic but very widespread all the same. In the United States, approximately twenty-five lives are lost each year from landslides. Elsewhere, in densely populated areas, single landslide events cause death tolls in the thousands.

The remediation of slope stability problems involves contributions from both geologists in the investigation of the site and civil engineers in the design of the project. Geologists employed by state geological surveys and the U.S. Geological Survey provide a tremendous service in several ways—constructing geological and slope stability maps based on knowledge of the soils and rock formations, use of remote-sensing methods such as satellite and high-altitude photography, and field study of suspect areas. These maps are made readily available to engineers, contractors, developers, and homeowners; they show color-coded areas of active and potentially active landslides. Such maps have been produced for many areas with a high population density. Residents in the United States may contact their local state's geological survey, which distributes these maps.

Edward B. Nuhfer

Further Reading

Casale, Riccardo, and Claudio Margottini, eds. *Floods and Landslides: Integrated Risk Assessment.* New York: Springer, 1999. Examination based on case studies of the associated risks of floods and landslides in regions, primarily European, prone to such natural disasters. Illustrations and maps.

Close, Upton, and Elsie McCormick. "Where the Mountains Walked." *National Geographic* 41 (May, 1922): 445-464. A graphic account of the most devastating landslide in history, in Kansu, China, 1920.

Cornforth, Derek H. *Landslides in Practice.* Hoboken, N.J.: John Wiley & Sons, 2005. Explains how and why landslides occur. Describes the resulting changes to a landscape and prevention methods. Compares various soil types and slope stability. Examines a number of case studies. Provides the tools and techniques needed to evaluate landslide potential and determine preventative strategies.

Costa, J. E., and V. R. Baker. *Surficial Geology: Building with the Earth.* New York: John Wiley & Sons, 1981. Written for undergraduates in environmental geology and engineering geology courses. Demonstrates the use of simple numerical problems to illustrate concepts quantitatively. Designed for students with a background in algebra, trigonometry, introductory geology, or Earth science. Well illustrated, with a good bibliography.

Costa, J. E., and G. F. Wieczorek, eds. *Debris Flows/ Avalanches: Process, Recognition, and Mitigation.* Reviews in Engineering Geology 7. Boulder, Colo.: Geological Society of America, 1987. A number of case studies from various parts of the United States, Canada, and Japan. The content of the text is intended for professionals. Offers many photographs and illustrations for students and laymen.

Cummans, J. *Mudflows Resulting from the May 18, 1980, Eruption of Mount St. Helens.* U.S. Geological Survey Circular 850-B. Washington, D.C.: Government Printing Office, 1981. Illustrates the devastation caused by mudslides associated with volcanism.

Hays, W. W. *Facing Geologic and Hydrologic Hazards.* U.S. Geological Survey Professional Paper 1240-B. Washington, D.C.: Government Printing Office, 1981. A well-written and beautifully illustrated booklet. Appropriate for readers from grade school through professionals.

Hoek, E., and J. W. Bray. *Rock Slope Engineering.* 3d ed. Brookfield, Vt.: IMM/North American Publications Center, 1981. An engineering reference often used by professionals. Involves a solid quantitative approach; however, the descriptive sections are graphic and well written. Largely appropriate for secondary students, junior and senior undergraduates in geology and civil engineering.

Kalvoda, Jan, and Charles L. Rosenfeld, eds. *Geomorphological Hazards in High Mountain Areas.* Boston: Kluwer Academic, 1998. Examines the processes used in determining which mountain areas are at risk of landslides. Provides a clear understanding of the hazards involved and what to look for in order to predict them.

Keefer, D. K. "Landslides Caused by Earthquakes." *Geological Society of America Bulletin* 95 (April, 1984): 406-421. A good review of the relationship between earthquakes and major landslides.

Keller, E. A. *Environmental Geology.* 9th ed. Upper Saddle River, N.J.: Prentice Hall, 2010. Chapter 7, on landslides and related phenomena, is ideal for beginners. Well illustrated; written in a simple, descriptive manner. References follow each chapter.

Kennedy, Nathaniel T. "California's Trial by Mud and Water." *National Geographic* 136 (October, 1969): 552-573. A graphic account of the interaction between landslides, earthquakes, and heavy seasonal rainfall.

Kiersch, G. A. "Vaiont Reservoir Disaster." *Civil Engineering* 34 (1964): 32-39. Enthralling account of one of the world's most tragic landslides. Excerpts of Kiersch's original article have been reprinted in many engineering geology texts.

McDowell, Bart. "Avalanche!" *National Geographic* 121 (June, 1962): 855-880. A graphic account of avalanches in Peru.

Norris, Joanne E., et al., eds. *Slope Stability and Erosion Control: Ecotechnological Solutions.* New York: Springer, 2010. Discusses erosion processes, mass wasting, and landslides as they occur on varying slopes. Evaluates vegetation as a solution to slope instability. Cites many studies but is not overly technical. Well suited for engineers, landscape architects, ecologists, foresters, and undergraduate students.

Radbruch-Hall, Dorothy H. *Landslide Overview Map of the Conterminous United States.* U.S. Geological Survey Professional Paper 1183. Washington, D.C.: Government Printing Office, 1981. A map, with accompanying text, that illustrates the major landslide areas within the United States.

Rahn, P. H. *Engineering Geology: An Environmental Approach.* 2d ed. New York: Elsevier, 1996. Appropriate for the serious undergraduate or graduate student interested in a solid quantitative approach. Well illustrated and well referenced.

Sassa, Kyoji, and Paolo Canuti, eds. *Landslides: Disaster Risk Reduction.* New York: Springer, 2009. Discusses a wide variety of topics related to landslides. A section discusses projects in the Machu Picchu area. Covers preventative methods and risk reduction.

Schuster, Robert L., ed. *Landslide Dams: Processes, Risk, and Mitigation.* New York: American Society of Civil Engineers, 1986. Includes case studies of slides at Thistle Creek, Utah, and control of the new Spirit Lake, produced by landslide at Mount St. Helens.

Schuster, Robert L., and Keith Turne, eds. *Landslide: Investigation and Mitigation.* Washington, D.C.: National Academy Press, 1996. National Research Council and Transportation Research Board special report. Offers a thorough analysis of landslide hazards and slope stability. Describes the processes involved in pinpointing unstable areas. Illustrations, bibliographical references, and index.

See also: Desertification; Expansive Soils; Land Management; Land-Use Planning; Land-Use Planning in Coastal Zones; Soil Chemistry; Soil Erosion; Soil Formation; Soil Profiles; Soil Types.

LAND-USE PLANNING

Land-use planning is part of the broader comprehensive planning process that deals with the types and locations of existing and future land uses, as well as their impacts on the environment.

PRINCIPAL TERMS

- **derivative maps:** maps that are prepared or derived by combining information from several other maps
- **geographic information system:** a series of data collected and stored in an organized manner in a computer system
- **grid:** a pattern of horizontal and vertical lines forming squares of uniform size
- **landscape:** the combination of natural and human features that characterize an area of the earth's surface
- **remote sensing:** any number of techniques, such as aerial photography or satellite imagery, that can collect information by gathering energy reflected or emitted from a distant source
- **scale:** the relationship between a distance on a map or diagram and the same distance on the earth
- **zoning ordinance:** a legal method by which governments regulate private land by defining zones where specific activities are permitted

LAND USE: OPPORTUNITIES AND LIMITATIONS

Land-use planning is a process that attempts to ensure the organized and wise use of land areas. Two things are certain: First, land has great value in modern society; second, there is only a limited amount of it. Land is indeed a valuable resource, deserving of careful management. It is also true that the physical environment influences the location and types of human settlements, transportation routes, and economic endeavors. The hills, ridges, valleys, and depressions of a landscape create potential opportunities and limitations for human use. It is important to realize that preexisting land uses likely will affect an area's future possibilities. Land-use planning is part of the master-planning process: It deals with the types and the distribution of existing and future land uses; their relationship to other planning areas, such as transportation network; and the interactions between land use and the environment.

This last point deserves elaboration. While it is certain that existing physical and cultural aspects of a landscape affect land use, humans have an enormous capacity to alter their surroundings. The behaviors and ambitions of society greatly influence emerging land uses. Additionally, each land use affects not only the users but the nonusers as well. An industrial park, for example, may result in increased traffic, longer travel times for commuters through the area, and the eventual construction of a multilane highway requiring expenditure of public funds. Furthermore, the effects of land use tend to be cumulative: Even small changes can, over time, combine to produce large and long-lasting impacts.

Almost without exception, everyone is affected by growth, development, and changing land-use patterns. Plans may determine how far residents must drive to shop, where a new park or school is located, or where houses may be built. Land-use plans may be as simple as efforts to protect citizens from hazardous locations. Such plans might call for setback zones along eroding sea cliffs that prohibit construction, thus preventing structures from being destroyed as the cliff retreats. Conversely, plans may be very complex attempts to guide and direct the types, rates, and locations of change in an area. Such plans have the potential to influence the area's economic and cultural characteristics, its environmental quality, and the way its residents will live in the near and distant future.

GOAL DEFINITION AND DATA COLLECTION

In a report issued by the U.S. Department of the Interior, *Earth-Science Information in Land-Use Planning*, William Spangle and others outline five separate phases of the land-use planning process: (1) the identification of problems and definition of goals and objectives; (2) data collection and interpretation; (3) plan formulation; (4) review and adoption of plans; and (5) plan implementation. At each phase, feedback occurs so that modifications can be made along the way. Even implemented plans are subject to review and redefinition as information accumulates.

Once the goals and objectives have been defined, the problems of the acquisition and interpretation of the plan's basic data must be addressed. For example, Earth science information, in the form of an Environmental Impact Statement (EIS), is needed throughout the planning process. At least a basic understanding of the ecology, climate, hydrology, geology, and soils of the area is essential. A considerable amount of data may already be available and may need to be consolidated from existing sources, such as published reports. Most often, however, much of the necessary data must be collected specifically for the proposed plan. All data, whether already available or newly developed, must be evaluated and analyzed. Not all information is of equal quality or compatible with the needs of the plan. Some data may be of poor quality or outdated and must be eliminated. Although it may be of high quality, some data are not useful. For example, in an assessment of an area's capability to support structures of various kinds, a list and discussion of the fossils found in the bedrock are not useful, whereas an account of the engineering properties is important. Other types of data may be incompatible with the needs of the plan because they were collected for different purposes, by different groups, or with different systems. EIS's also must offer sufficient detail and feature the appropriate map scale.

Plan Formulation, Review, and Implementation

Once high-quality and appropriate data are accumulated, they can be used to produce maps that show the capability of the area to support each potential use. These land-capability maps are analyzed together with projections of future growth and with economic, social, and political factors to evaluate alternative land-use patterns. Maps are prime aids for land-use planning because they present the location, size, shape, and distribution patterns of landscape features. Specialized or derivative maps can be prepared by combining several environmental factors—such as geology, soil types, and slope—to illustrate the specific problems or best possible use of an area. Furthermore, maps can be used to guide further development by outlining areas reserved for particular land uses.

A suitable plan is formed based upon the most desirable and feasible courses of action for both immediate and future decisions. The plan must then be reviewed and adopted by the commissioning agency or governmental body. At this phase, the technical personnel responsible for the development of the plan must be available to answer questions, respond to criticism, and make any further changes.

The final and perhaps most critical phase is that of plan implementation. Plans that are adopted but not implemented serve little purpose. There must be ways in which the plan influences the formal and informal processes of decision making. Zoning ordinances, construction regulations, and building codes based on the plan should be in place and enforced. Responsibilities and guidelines for the preparation and evaluation of required reports, impact assessments, and proposals must be established. This requires a staff that not only reviews proposals but also requests additional information or modification of project proposals.

Data Collection Methods

Data are collected for land-use planning purposes in many ways. Some information can be obtained through mailed surveys or surveys conducted door-to-door. Often, technical personnel such as geologists or hydrologists will conduct field investigations in the area. Observations and measurements are made and recorded, and samples are collected for laboratory analysis. Water samples might be analyzed to determine the quality of the water resources. Rock and soil samples may be tested to yield information about engineering properties such as strength. Certain types of information, such as geology, soil, and climate data, may already exist in published form.

Remote-sensing methods are among the most useful resources for collecting data that pertain to land-use studies. Remote sensing is a method of imaging the earth's surface with instruments operated from distant points, such as airplanes or satellites. Many different types of remote-sensing instruments, such as aerial photography equipment and radar, are applicable to land-use purposes. Basically, remote-sensing systems collect and record reflected or emitted energy from the land surface. Aerial photography, for example, collects visible light reflected from the earth's surface to produce images on film. Line-scanning systems can record a wide range of energy types. For example, infrared devices can be used to detect small differences in temperature, which reflect differences of soil, water, or vegetation on the earth's surface. Another technique, side-looking airborne radar (SLAR), uses pulses of microwave energy to locate surface objects by recording the time necessary for the energy transmitted to the object to return to the radar antenna.

Remote-sensing techniques are applicable to many data-collection efforts. Inventories of existing land uses, crop patterns, or vegetation types can be accomplished relatively quickly. The location and extent of environmental hazards, such as floods or forest fires, can be delineated, and landscape changes—crop rotations, shoreline shifts, and forest clearings—can be documented over a period of time.

DATA ANALYSIS

Because they can store and manipulate large data sets, computers are ideally suited for comparing and combining many types of data into a final integrated or interpretive map. The application of a computer system for land-use planning is accomplished by changing EIS's and other data into a form that can be entered into the machine. The data can be entered manually

or by more sophisticated optical scanning methods. One straightforward method uses a grid system to enter data. Basically, maps portraying different types of information of the relevant areas can be subdivided into a grid formed by equally spaced east-west and north-south lines. The size of the grid squares depends on the purpose of the study and the nature of the data. The intersection points of the grid lines or the centers of the grid squares can be used as data entry locations. An illustration of this operation is the overlaying of a grid on a map of soil types and the entry of the soil type at each line intersection or the dominant soil type in each square. This kind of data is called raster data. Linear features like roads and streams are often easier to represent as a series of points; data of this sort are called vector data. Programs for analyzing spatial data can manipulate and combine both forms of data.

When all available types of data are in the computer, they can be analyzed, combined, and applied to indicate the areas best or least suited for a particular use. The information can also be used to make predictions. For example, a planner may need

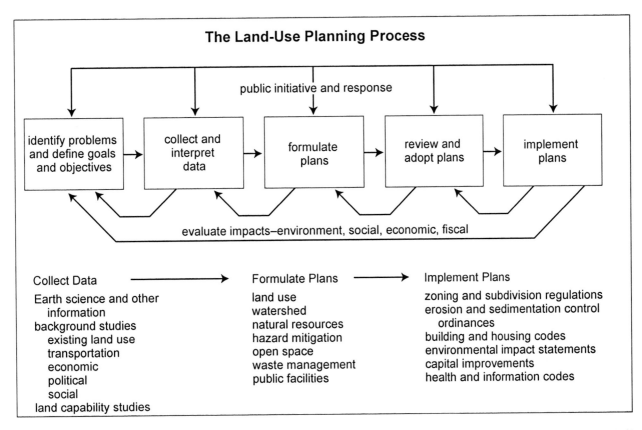

The Land-Use Planning Process

public initiative and response

identify problems and define goals and objectives → collect and interpret data → formulate plans → review and adopt plans → implement plans

evaluate impacts–environment, social, economic, fiscal

Collect Data ⟶ Formulate Plans ⟶ Implement Plans

Collect Data

Earth science and other
    information
background studies
    existing land use
    transportation
    economic
    political
    social
land capability studies

Formulate Plans

land use
watershed
natural resources
hazard mitigation
open space
waste management
public facilities

Implement Plans

zoning and subdivision regulations
erosion and sedimentation control
    ordinances
building and housing codes
environmental impact statements
capital improvements
health and information codes

to suggest the location of future solid-waste disposal sites. Although such a task involves knowledge of waste sources, waste volumes, and transportation distances, the performance capability of various sites is a primary concern. The planner can select those data sets considered to be important to a site's capability to contain waste effectively, and the computer will evaluate the area in terms of those factors. Bedrock geology, soil type and depth, distance to groundwater, distance from surface water bodies, and other factors may be selected. Each factor is assigned a weighted value, a measure of its relative importance to the specific purpose. Depth to the water table, for example, might be considered more important than the distance to surface water bodies. The computer incorporates the relative ranking of the variables into the data analysis to produce a composite map of favorable and unfavorable areas for solid-waste disposal. Such comprehensive systems, known as geographic information systems (GIS's), are being developed widely for many purposes. A GIS consists of computer storage files of data, a program to analyze the data, and another program to map the stored data and produce several forms of output. Many public agencies have converted data to digital form to make it easily available for use with GISs.

EFFECTIVE LAND-USE PLANNING

Land-use planning attempts to ensure the organized and reasonable development of areas as the possibility of multiple uses mounts and the value of the land increases. Planning must be based on a thorough understanding of the environment, and of economic, political, and cultural factors, to produce a prediction of future land-use needs and ways to satisfy those needs.

An effective land-use plan consists of four basic parts. First, it includes a discussion of land-use issues and a statement of goals and objectives. This section provides the background and establishes what the plan seeks to accomplish. Second, there is a discussion of the methods of data collection, evaluation, and analysis. This will include not only a description of the collection techniques, methodology, and information sources but also a consideration of the limitations of the data used in the study. Third, a land classification map is produced. The classification used to complete the map will depend on the goals of the plan and on the needs of each area. This map will serve as the basis for many kinds of decisions,

from the location of new facilities to regulatory policies and tax structure. Finally, a report is included that provides a framework for the plan and discusses sensitive issues. Any environmentally significant sites that may be adversely affected by change are considered, and appropriate policies are recommended.

POLITICAL CONSIDERATIONS

Land-use decisions are made by individuals, groups, industries, and governmental bodies. Individuals want the freedom to choose where they live and what to do with their property, but most do not want an incompatible land use located next door. Land-use plans based on high-quality data and sound interpretations of those data, with abundant public input during the formulation, adoption, and implementation of the plan, can help to avoid costly and time-consuming confrontations. There will seldom be unanimous agreement, but planning can facilitate a decision-making process that will have far-ranging effects.

Land-use planning or management is a controversial subject. Landowners, developers, environmentalists, and government agencies have very different views on the best use of land, and on who should make the decision. These are difficult issues of individual versus group rights and benefits. It is clear that as land resources dwindle and the pressures of land use increase, some form of planning or management is necessary. Similarly, it seems clear that while decisions among multiple uses will often be difficult, the best decisions are those based on high-quality data.

Ronald D. Stieglitz

FURTHER READING

Berke, Philip R., and David R. Godschalk. *Urban Land Use Planning.* 5th ed. University of Illinois Press, 2006. Discusses the concepts and values examined and applied during land-use planning. Covers research, information systems, and types of data used in planning. Provides information on common challenges in executing land-use planning.

Burby, Raymond J., ed. *Cooperating with Nature: Confronting Natural Hazards with Land-Use Planning for Sustainable Communities.* Washington, D.C.: Joseph Henry Press, 1998. Covers the role of natural disasters and hazards in the development of new communities. Intended for the reader with some background in the field but useful for laypersons with an interest in land-management policies. Illustrations, bibliography, and index.

Carr, Margaret, and Paul Zwick. *Smart Land-Use Analysis: The LUCIS Model.* Redlands, Calif.: ESRI Press, 2007. Discusses the use of ArcGIS software to analyze the land suitability. Provides a step-by-step guide to using ArcGIS software. Explains the LUCIS model with case studies to provide examples of potential conflicts of land use.

Davidson, Donald A. *Soils and Land Use Planning.* New York: Longman, 1980. Written for middle-level students. Demonstrates how soil information can be used in land-use planning.

Dluhy, Milan J., and Kan Chen, eds. *Interdisciplinary Planning: A Perspective for the Future.* New Brunswick, N.J.: Center for Urban Policy Research, 1986. Covers land-use planning, along with other types of planning. Illustrates the complex nature of the planning process.

Economic Commission for Europe. *Land Administration Guidelines: With Special Reference to Countries in Transition.* New York: United Nations, 1996. This United Nations publication describes the land-use and land-tenure policies that have been enacted in Central and Eastern European countries during their transition from communist to democratic governments.

Marsh, William M. *Landscape Planning: Environmental Applications.* 5th ed. New York: John Wiley & Sons, 2010. A useful introduction to landscape planning. Discusses a wide range of topics and provides numerous case histories.

McHarg, I. L. *Design with Nature.* Hoboken, N.J.: John Wiley & Sons, 1992. An older but still useful text. Provides the basis for environmentally based planning technologies.

Nolon, John R., and Patricia E. Salkin. *Land Use in a Nutshell.* St. Paul, Minn.: Thomson/West, 2006. Describes the origins of common property laws and property rights, constitutional regulations and amendments. Explains the planning process and historical, cultural, judicial, and environmental issues involved in land-use plans, as well as specialized topics. Provides an Internet research guide as an appendix.

Rajapakse, Ruwan. *Geotechnical Engineering Calculations and Rules of Thumb.* Burlington, Mass.: Butterworth-Heinemann, 2008. Written for the professional geological engineer. Provides explanations of calculations and general practices used by engineers in building shallow and pile foundations, retaining walls, and analyzing geological features and compositions.

Randolph, John. *Environmental Land Use Planning and Management.* Washington, D.C.: Island Press, 2004. Describes basic principles and strategies of land-use planning and management. Discusses various land features, types, and environmental issues, such as soils, wetlands, forests, groundwater, biodiversity, and runoff pollution. Provides case studies and specific examples.

Rhind, D., and R. Hudson. *Land Use.* London: Methuen, 1980. A well-illustrated treatment of land-use issues and the planning process. Discusses data needs and planning models.

Seabrooke, W., and Charles William Noel Miles. *Recreational Land Management.* 2d ed. New York: E. and F. N. Spon, 1993. Examines the management of land intended for recreational use in urban environments. Includes illustrations, bibliographies, and index.

So, Frank S., ed. *The Practice of Local Government Planning.* 3d ed. Washington, D.C.: International City/County Management Association, 2000. A thorough treatment of all aspects of planning at the level of cities and counties. Discussion is clear but advanced.

So, Frank S., Irving Hand, and Bruce McDowell. *The Practice of State and Regional Planning.* Chicago: Planners Press, 1986. Designed as a companion volume to the work above. Thorough treatment of the practices of planning at higher government levels.

Spangle, William, and Associates; F. Beach Leighton and Associates; and Baxter, McDonald and Company. *Earth-Science Information in Land-Use Planning: Guidelines for Earth Scientists and Planners.* Geological Survey Circular 721. Arlington, Va.: U.S. Department of the Interior, Geological Survey, 1976. An excellent discussion of the sources, accuracy, and applications of Earth science information in the land-use planning process.

See also: Desertification; Expansive Soils; Hazardous Wastes; Landfills; Land Management; Landslides and Slope Stability; Land-Use Planning in Coastal Zones; Mining Wastes; Nuclear Waste Disposal; Soil Chemistry; Soil Erosion; Soil Formation; Soil Profiles; Soil Types; Strategic Resources.

LAND-USE PLANNING IN COASTAL ZONES

In order to designate appropriate coastal land use, policymakers must take into consideration geological processes, recognizing that coasts are zones of rapidly evolving landforms, sedimentary deposits, and environments. Much of the American coastal zone has undergone rapid development without the benefit of landscape evaluation. Future development, as well as corrective measures, must be grounded in knowledge of geology and the interactions of climate and oceanography.

PRINCIPAL TERMS

- **coastal wetlands:** shallow, wet, or flooded lowlands that extend seaward from the freshwater-saltwater interface and may consist of marshes, bays, lagoons, tidal flats, or mangrove swamps
- **coastal zone:** coastal waters and lands that exert a measurable influence on the uses of the sea and its ecology
- **estuarine zone:** an area near the coastline that consists of estuaries and coastal saltwater wetlands
- **estuary:** a zone along a coastline, generally a submerged valley, where freshwater system(s) and river(s) meet and mix with an ocean
- **groundwater:** water that sinks into the soil, where it may be stored in slowly flowing underground reservoirs
- **land-use planning:** a process for determining the best use of each parcel of land in an area
- **mass wasting:** the downslope movement of Earth materials under the direct influence of gravity
- **saltwater intrusion:** aquifer contamination by salty waters that have migrated from deeper aquifers or from the sea
- **water table:** the level below the earth's surface at which the ground becomes saturated with water

EVALUATION OF COASTAL ZONES

The dependency of the United States on its coastal zone cannot be exaggerated; national leaders learned this lesson in 1969 when the Stratton Report, "Our Nation and the Sea," revealed that coastal areas contained more than 50 percent of the nation's population (a percentage that has likely grown), seven of the nation's largest cities, 60 percent of the petroleum refineries, 40 percent of industry, and two out of three nuclear or coal-fired electrical generating plants. Not surprisingly, the Stratton Report became the primary impetus for the passage of the National Coastal Zone Management Act (NCZMA) of 1972 and 1980. NCZMA provides federal aid to coastal states and territories for the development and implementation of voluntary, comprehensive programs for the management and protection of coastlines. In 1982, Congress passed the Coastal Barrier Resource Act, which declared millions acres of beachfront on various barrier islands ineligible for federal infrastructure funding. By denying funding for roads and other construction projects, the act discourages development on sensitive coastlines.

Geologic information and analysis, the first step of all land-use planning, is complicated in the coastal zone by the fact that the major value of the area is water; indeed, water is the determining factor of the suitability of the microenvironments under consideration. Landscape evaluation must also recognize that the zone is not static but the locus of rapid geologic change. Such change involves a complex sediment dispersal system that responds rapidly to human modifications. The coastal zone also includes fundamental legal boundaries between private and public ownership, and the boundaries themselves are high-energy geological boundaries. The presence of barrier islands demands a highly specialized set of land-use practices. Natural hazards in the coastal zone include hurricane attacks on the Atlantic and Gulf coasts, landslides on the cliffed coasts of the Pacific, and tsunamis in the Pacific Northwest region and the coastal zones of Alaska, Hawaii, and Puerto Rico. Superimposed upon the dynamic processes of the coastal zone is a global sea-level rise, initiated some 18,000 years ago with the termination of the Late Wisconsin glaciation. Sea-level rise is currently accelerating, possibly in response to such human-induced atmospheric changes as the greenhouse effect.

A major resource of the area is the estuarine zone, encompassing less than 10 percent of the total ocean area but containing 90 percent of all sea life. Estuaries trap the nutrients that rivers wash down from the land and use them to produce an extraordinary quantity of biomass that sustains a variety of marine life. It is estimated that 60 to 80 percent of all edible seafood is dependent on estuaries for survival.

Estuarine zones and coastal wetlands also serve as natural flood control devices by absorbing the energy of damaging storm waves and storing floodwaters. If not overloaded, these wetlands have the ability to remove large quantities of pollutants from coastal waters.

BEACH EROSION AND SHORELINE RETREAT

The great financial potential of beachfront properties makes the consequences of beach erosion—as well as efforts to halt shoreline retreat—of vital concern in coastal zone land-use issues. Although some beach erosion is rightly attributed to sea-level rise, much is the direct result of human activities. The damming of major rivers has essentially cut off the sediment supply to the nation's beaches. Loss of dredged sediment through maintenance of shipping channels results in rapid erosion on the downdrift beaches to which wave action would normally deliver sands. Both jetties and groin fields act as dams to sediment transport and create erosion on adjacent beaches. Dune destruction removes the reservoirs for sand storage and the natural barriers to storm surges. The emplacement of sand on the beach from offshore or inland sources (beach renourishment) and pumping sediment across navigation channels (inlet bypassing) represent more acceptable "soft engineering" alternatives to the armoring of the shoreline.

On the Atlantic and Gulf coasts, hurricanes account for many of the sediment distribution patterns. These tropical cyclones modify barrier islands, removing beach sand offshore and over the dunes. Overwash, in which large quantities of sediment are moved inland, is the primary means by which barrier islands retreat before the rising sea. The inevitability of hurricanes mandates land use that is compatible with predicted flooding patterns, as well as building codes based on hurricane-force winds and tidal storm surges, and the maintenance of up-to-date evacuation plans.

Retreat of the shoreline along the West Coast is more frequently related to seacliff erosion. Erosion along cliffed coasts is caused by a combination of wave scouring at the base and landsliding higher on the bluffs. In winter, the large tides and waves force the rain-weakened cliffs to retreat. The degree of failure depends on wave energy, the hardness of the cliff rock, and internal fractures and faults. Weathering processes further weaken rocks and aid erosion. The human activities that accelerate cliff retreat include septic tank leaching, landscape irrigation, alteration of drainage patterns, and introduction of nonnative vegetation.

GULF AND ATLANTIC COASTAL DEVELOPMENT

The low, sandy Gulf and Atlantic coasts, characterized by estuaries, marshes, and inlets, are separated from the open ocean by barrier islands. Major features of the lower coastal plain include abundant forests, a diversity of wildlife habitats, and a unique potential for water-based recreation. The area is largely unsuited for urbanization because of the high water table, but remnant ancient barrier island chains, or "terraces," represent elevated sites suitable for building. The largely sandy soils are poorly suited for widespread agriculture, though some fertile alluvial, or river valley, soils are present. The high water table makes aquifer pollution an ever-present threat; the abundant forests, which support a major pulp and paper production region, are fire-prone.

Because the broad coastal plains of the depositional coasts were formed by barrier island migration during the sea-level fluctuations of the Pleistocene "ice ages," the resulting sediments are largely barrier island sands or marsh muds and peats. Rapid, unplanned growth in the coastal zone has resulted in widespread use of septic tanks in these unsuitable, sandy soils, allowing waste water to percolate too rapidly and releasing improperly treated effluent. The degradation of coastal waters by septic tanks can be avoided through the tertiary treatment of waste water, which may in turn be used for wetlands recharge or irrigation. A sanitary landfill properly sited in mud or peat has low permeability, which limits the flow of leachate and contaminated landfill runoff. The most heavily developed resort areas with the greatest need for landfill space also have the highest land prices, making suitable land acquisition difficult.

MARSHLAND DEVELOPMENT

The vast marshes of the coastal zone grade from fresh to brackish to salt water. The river-swamp system, interrelated with the marsh-estuary system through flows of water, sediments, and nutrients, plays a major role in determining the impact of inland development on the estuary. In their natural state, swamps buffer the coast from the many impacts of land-use activities; if drained, filled, or altered, this contribution is lost. An increase in stormwater runoff related to urbanization reduces their ability to hold and absorb floodwaters. This problem could be avoided by placing limits on development and requiring that nonhighway "paving" be of permeable material such as crushed limestone instead of asphalt. The river swamp's ability to filter and absorb pollutants also improves the quality and productivity of downstream coastal waters; if overloaded with pollutants from industrial effluents, agricultural practices, or improperly sited septic tanks and landfills, the closing of shellfish beds is inevitable.

The "low marsh," inundated daily by the tides, is well recognized as an area so valuable and productive that conservation is mandatory. Although protected by state laws, estuarine marshland continues to experience some destruction as a result of marina construction and the disposal of dredged sediments from navigation channels. Marshland is also lost to the boat-wake erosion that undercuts the banks, causing slumping of the rooted clays.

At a slightly higher elevation is the "high marsh," sandy marshland occasionally wetted by storms or extreme tides. This marsh, unprotected by law, is known to be an important wildlife habitat but has not been widely studied. The role of the high marsh and its relationship to the low marsh is unknown. The widespread exploitation of this area for residential and commercial development will make it impossible for the low marsh to reestablish itself landward in response to sea-level rise. The resultant loss in a real extent of the salt marsh will cause a great decline in estuarine-dependent marine life and a corresponding rise in pollution.

BARRIER ISLAND DEVELOPMENT

The barrier islands that front low-lying coastal plains consist of such dynamic geologic features that they are clearly unsuited for widespread development. These islands are composed of unconsolidated sediments that continuously seek to establish equilibrium with the waves, winds, currents, and tides that shape them. Early settlements were wisely constructed on the landward-facing, or back-barrier, portion of the islands. The beachfront was considered too dangerous because of hurricanes.

Ideally, high-density barrier island development should be confined to the back-barrier part of the island, protected from storms and hurricanes.

Low-density development should thin out in both directions, and open island should be preserved along the high-energy beachfront. This would ensure the scenic and recreational value of the island and protect the sand supply and natural maintenance of the system. If stabilization of inlets for shipping is necessary, the sand built up on the updrift side should be bypassed downdrift to reenter the system. Many islands should be preserved free of development for sand storage, as well as for educational, recreational, and aesthetic benefits.

The barrier island uplands, or high ground, consist of maritime forest on ancient oceanfront dune ridges; these dunes were left behind as the island built seaward, a process that was caused by a fortuitous combination of sediment supply and fluctuating sea level. This forested region is appropriate for environmentally sensitive development as long as care is taken to conserve ample wildlife preserves. Freshwater sloughs and ponds are located in the low swale areas between dune ridges, where rainfall floats on salt water within the porous material of the island subsurface. These wetlands represent important wildlife habitat and, on developed islands, serve as natural treatment plants for storm runoff and natural storage areas for floodwaters. They may also be recharged with tertiary treated waste water, if not overloaded. Unfortunately, there has been a great loss of freshwater wetlands on the mainland and on the islands to developers who have drained and filled them to produce construction sites.

BEACHFRONT DEVELOPMENT

Beachfront construction should first be considered in the context of available data on historical change to the shoreline. If the area under consideration has a suitable history of stability or accretion, development should be limited to a setback line that is landward of the heavily vegetated and stable dunes. Construction should be compatible with historical hurricane storm-surge data, and, at elevations below recorded storm surges, dwellings should be elevated on a foundation of pilings.

Access to the beach should be via dune walkovers to avoid damage to the plants that stabilize the fragile dune environment. The dunes that are closest to the ocean and tied to the beach for their windblown sand supply are constantly changing landforms. Storm waves scour away the front of the dunes, and fair-weather waves return the lost sand to the beach to rebuild them. These ephemeral features, vital as buffers to storm attack, should be protected not only from construction but also from any sort of human activity.

The updrift and downdrift inlet beachfronts are best utilized as nature preserves or public beaches, free of construction, because of the natural instability of both zones. The updrift inlet is constantly re-establishing its channel in response to tide-deposited sand mounds, causing alternating and far-flung advances and retreats of the north-end shoreline. The downdrift end represents the geologically younger and most recently constructed part of the island. Because the area is barely above sea level, hurricane storm surges can readily cut through an elongated downdrift spit, creating a new inlet and freeing the sand to migrate downdrift.

PACIFIC COASTAL DEVELOPMENT

The geologic instability of the rocky Pacific coast demands land-use planning strategies that will protect its rugged beauty and minimize the threats of natural hazards. Development in the available coastal zone regions of cliff tops, dune fields, and beachfront is potentially hazardous.

On the beachfront, a wide beach between cliff and shoreline should not be mistaken for a permanent feature. Developed beachfront areas at the bases of cliffs are subject to periodic floodings by large waves combined with high tides; after much of the beach sand is removed by the storm, the wave energy is expended against the buildings. Houses in this zone collapse when their foundations are undermined or their pilings smashed through after being uplifted by waves. If beachfront development is allowed, it should be based on a comparison of the site elevation and expected tidal ranges, storm surges, and storm-wave heights. A secure piling foundation below any potential wave scour should elevate the structure above any maximum inundation.

Although the presence of dune fields is restricted on young mountain range coasts, existing dunes have not escaped urbanization, and the resultant problems are identical to those of the East Coast. Where homes and condominiums have been built on conventional foundations, periodic dune erosion has undermined and threatened the structures, forcing emplacement of sea walls constructed of boulders. The dynamic nature of sand dunes makes the total prohibition of construction there the only appropriate course of action for wise land-use planning.

The zone consisting of bluff or cliff tops is the West Coast environment that faces the greatest potential for development. Construction should not proceed here without a large enough setback behind the cliff edge, so that the structure can endure for at least one hundred years based on long-term erosion rates. The increased runoff associated with urbanization, if not collected and diverted away from the seacliff, results in serious slope failure. Homes, patios, swimming pools, and other construction also decrease the stability of the seacliff by increasing the driving forces, forces that tend to make Earth materials slide. Although seacliff erosion is a natural process that cannot be controlled completely, regardless of financial investment, it can be minimized by sound land-use practices.

Martha M. Griffin

FURTHER READING

Burby, Raymond J., ed. *Cooperating with Nature: Confronting Natural Hazards with Land-Use Planning for Sustainable Communities.* Washington D.C.: Joseph Henry Press, 1998. Discusses the role of natural disasters and hazards in the development of new communities. Designed for the reader with some background in the field; also useful for the layperson with an interest in land-management policies. Illustrations, bibliography, and index.

Carr, Margaret, and Paul Zwick. *Smart Land-Use Analysis: The LUCIS Model.* Redlands, Calif.: ESRI Press, 2007. Discusses the use of ArcGIS software to analyze the land suitability. Provides a step-by-step guide for using ArcGIS software. Explains the LUCIS model with case studies to provide examples of potential conflicts of land use.

Dolan, Robert. "Barrier Islands: Natural and Controlled." In *Coastal Geomorphology,* edited by Donald R. Coates. Binghamton: State University of New York, 1972. Compares the stabilized islands (Cape Hatteras National Seashore) and the natural islands (Core Banks) of the Outer Banks of North Carolina. Systems' responses to coastal processes, particularly major storms, are of great interest for both preservation and management reasons. Suitable for college-level readers.

Griggs, Gary, and Lauret Savoy, eds. *Living with the California Coast.* Durham, N.C.: Duke University

Press, 1985. Part of the Living with the Shore series, sponsored by the National Audubon Society. Addresses the problems resulting from coastal development during a relatively storm-free period without analysis of historical storms or long-term erosion rates. Includes conclusions and recommendations of coastal geologists and other specialists. Suitable for those with an interest and concern for the California coast.

Hoyle, Brian, ed. *Cityports, Coastal Zones, and Regional Change: International Perspectives on Planning and Management.* New York: Wiley, 1996. Covers the planning and management of coastal zones, with particular emphasis on harbors and commercial ports. Illustrations and maps.

Kaufman, Wallace, and Orrin H. Pilkey, Jr. *The Beaches Are Moving: The Drowning of America's Shorelines.* Durham, N.C.: Duke University Press, 1983. An introduction to environmental coastal geology, nearshore processes, and the effects of human modifications to the shorelines of the United States. Covers the basic issues applied to specific shorelines. Suitable for those interested in the shorelines of the United States.

Keller, Edward A. *Environmental Geology.* 9th ed. Upper Saddle River, N.J.: Prentice Hall, 2010. Introduces physical principles basic to applied geology and reviews major natural processes and geologic hazards, including how society deals with them. Does not require previous exposure to geological sciences. Suitable for college-level readers.

Knight, Richard L., and Peter B. Landres, eds. *Stewardship Across Boundaries.* Washington D.C.: Island Press, 1998. Analyzes the conservation of natural resources and management of public lands in the United States, including coastal regions. Maps and illustrations.

Kraus, Nicholas C. *Shoreline Changed and Storm-Induced Beach Erosion Modeling.* Springfield, Va.: National Technical Information Service, 1990. Details the U.S. Army Engineer Waterways Experiment Station's work in developing mathematical models to study coastal changes, beach erosion, and storm surges, and their effects on coastal development.

McHarg, Ian L. *Design with Nature.* Hoboken, N.J.: John Wiley & Sons, 1992. An explanation of the ecological land-use planning method, proving that necessary human structures can be accommodated within the existing natural order. Suitable for high-school-level readers.

Neumann, A. Conrad. "Scenery for Sale." In *Coastal Development and Areas of Environmental Concern.* Raleigh: North Carolina State University Press, 1975. A statement on the science, scenery, and selling of a barrier island system like the Outer Banks of North Carolina; cleverly illustrated by the author. Free of technical terminology; designed for all readers.

Newson, Malcolm. *Land, Water and Development: Sustainable and Adaptive Management of Rivers.* 3d ed. London: Routledge, 2008. Presents land-water interactions. Discusses recent research, study tools and methods, and technical issues, such as soil erosion and damming. Covers concepts in managing land and water resources in the developed world. Suitable for undergraduates and professionals.

Nolon, John R., and Patricia E. Salkin. *Land Use in a Nutshell.* St. Paul, Minn.: Thomson/West, 2006. Describes the origins of common property laws and rights, and constitutional regulations and amendments. Explains the planning process and historical, cultural, judicial, and environmental issues involved in land-use plans, as well as such specialized topics as community building, zoning, and flexible use and redevelopment. Provides an online research guide as an appendix.

Randolph, John. *Environmental Land Use Planning and Management.* Washington, D.C.: Island Press, 2004. Describes basic principles and strategies of land-use planning and management. Discusses various land features, types, and environmental issues, such as soils, wetlands, forests, groundwater, biodiversity, and runoff pollution. Provides case studies and specific examples.

See also: Desertification; Expansive Soils; Landfills; Land Management; Landslides and Slope Stability; Land-Use Planning; Remote Sensing; Soil Chemistry; Soil Erosion; Soil Formation; Soil Profiles; Soil Types.

LIMESTONE

Limestone, the third most common sedimentary rock, is composed mostly of calcium carbonate, typically of organic origin. Limestone is usually fossiliferous and thus contains abundant evidence of organic evolution; it is also important as a construction material, groundwater aquifer, and oil reservoir.

PRINCIPAL TERMS

- **calcite:** the main constituent of limestone, a carbonate mineral consisting of calcium carbonate
- **carbonates:** a large group of minerals consisting of a carbonate anion (three oxygen atoms bonded to one carbon atom, with a residual charge of 2) and a variety of cations, including calcium, magnesium, and iron
- **cementation:** the joining of sediment grains, which results from mineral crystals forming in void spaces between the sediment
- **deposition:** the settling and accumulation of sediment grains after transport
- **depositional environment:** the environmental setting in which a rock forms; for example, a beach, coral reef, or lake
- **diagenesis:** the physical and chemical changes that occur to sedimentary grains after their accumulation
- **grains:** the individual particles that make up a rock or sediment deposit
- **lithification:** compaction and cementation of sediment grains to form a sedimentary rock
- **texture:** the size, shape, and arrangement of grains in a rock
- **weathering:** the disintegration and decomposition of rock at the earth's surface as the result of the exertion of mechanical and chemical forces

LIMESTONE IDENTIFICATION AND IMPORTANCE

Limestones are a diverse group of sedimentary rocks, all of which share a common trait: They contain 50 percent or more calcium carbonate, either as the mineral calcite or as aragonite. Both are composed of calcium carbonate; however, they have different atomic arrangements. Other carbonate minerals may also be present; siderite (iron carbonate) and dolomite (calcium-magnesium carbonate) are especially common. Although carbonate minerals can form other rocks, limestone is easily the most common and important carbonate rock. Many geologists use the terms "carbonate rock" and "limestone" almost interchangeably, because most carbonate rocks are limestones.

Limestone may be of chemical or biochemical (organic) origin, and can form in a wide variety of depositional environments. A limestone's texture and grain content are often useful clues for determining how and where it formed; however, diagenesis can easily obscure or destroy this evidence. Texture and grain content remain the basis for naming numerous varieties of limestone. These include dolomitic limestone, fossiliferous limestone, and crystalline limestone. Other common varieties include chalk, a very soft, fine-grained limestone; travertine, a type of crystalline limestone that forms in caves; and calcareous tufa, which forms by precipitation of calcium carbonate at springs.

Most limestones contain fossils, and many are highly fossiliferous. Limestones are perhaps our best record of ancient life and its evolutionary sequence. They are important sources for building and crushed stone and often contain large supplies of groundwater, oil, and natural gas. Weathering of limestone helps to develop distinctive landscapes as well.

LIMESTONE FORMATION

Limestones form in one of three ways: chemical precipitation of crystalline grains, biochemical precipitation and accumulation of skeletal and nonskeletal grains, or accumulation of fragments of preexisting limestone rock. Chemical precipitation occurs when the concentration of dissolved calcium carbonate in water becomes so high that the calcium carbonate begins to come out of solution and form a solid, crystalline deposit. The concentration of calcium carbonate in the water may change for a number of reasons. For example, evaporation, increase in water temperature, influx of calcium or carbon dioxide, and decreasing acidity can all cause precipitation. Crystalline limestone forms in the ocean, in alkaline lakes, and in caves, and also as a precipitate in arid climate soils—a variety known as caliche.

Certain marine organisms are responsible for the formation of many kinds of limestone. Their calcareous (calcium carbonate) skeletons accumulate after death, forming carbonate sediment. Many limestones are nothing more than thousands of skeletal grains joined to form a rock. The organisms that contribute their skeletons to carbonate sediments are a diverse group and include both plants and animals. Among these are algae, clams, snails, corals, starfish, sea urchins, and sponges. Some marine animals also produce nonskeletal carbonate sediments. An animal's solid wastes, or fecal pellets, may accumulate to form limestones if they contain abundant skeletal fragments or compacted lime mud. Limestones composed of skeletal grains, or of nonskeletal grains produced by living organisms, are called "organic limestones."

Recycling of preexisting limestones is a third source for carbonate grains. Weathering and erosion produce limestone fragments, or clasts, that may later be incorporated into new limestone deposits. Limestones consisting of clasts are clastic, or detrital, limestones; they are probably the least common of the three types of limestone.

LITHIFICATION

The processes that turn loose sediment grains into sedimentary rock are known as "lithification." These may include either compaction, cementation, or both. The grains (crystals) in a chemically formed limestone are usually joined together into an interlocking, solid matrix when they precipitate; thus, they do not undergo further lithification. The grains in organic and clastic limestones, however, are usually loose, or unconsolidated, when they first accumulate and so must be lithified to form rock.

Limestones, unlike most other sedimentary rocks, are believed to undergo lithification during shallow burial rather than when deep below the earth's surface. Some may be lithified within a meter or two of the surface, or even at the surface. Therefore, lithification in most limestones consists of cementation without significant compaction of grains. In most cases, the cement is calcium carbonate. If the spaces, or pores, between the grains become cement-filled without much compaction, cement can be as much as 50 percent or more of the volume of a limestone. The cement forms by precipitation, much like the formation of a crystalline limestone.

FORMATION AND PRESERVATION FACTORS

A number of factors control the formation and preservation of carbonate sediments. These include water temperature and pressure, the amount of agitation, concentrations of dissolved carbon dioxide, noncarbonate sedimentation, and light penetration. Cold, deep water with high levels of dissolved carbon dioxide tends to discourage the formation and accumulation of carbonate sediment. Warm, clear, well-lit, shallow water tends to promote formation and accumulation.

Certain periods of geologic history also favored limestone formation. Generally, the greatest volumes of ancient limestones formed when the global sea level was higher than today, so that seas covered large areas of the continents, and when global temperatures were also higher than at present. This combination of factors was ideal for producing thick, extensive deposits of carbonate rocks. Such limestones are exposed throughout the world today and provide a glimpse into Earth's distant past. Their abundant marine fossils are especially useful to paleontologists and biologists, as they allow them to piece together the sequence of biological evolution for a variety of plants and animals.

Modern carbonate sediments accumulate in ocean waters ranging in depth from less than 1 meter to more than 5,000 meters and at nearly all latitudes. However, most ancient limestones now exposed at the earth's surface formed in low-latitude, tropical, shallow marine environments; for example, in reefs or lagoons.

One of the world's largest modern accumulations of carbonate sediment and rock is located in the Great Barrier Reef off the northeast coast of Australia. This reef tract, the largest in the world, contains thick sequences of carbonate sediment deposited during the last few thousand years draped over even older carbonate rocks formed by coral reef organisms in the more distant past. As long as these reefs continue to thrive, carbonate sediment production will also continue. As a result, this mass of limestone will grow even thicker, and the older rock will continue to subside, sinking deeper into the subsurface.

DIAGENESIS

No matter where they form or what their origin, carbonate sediments are all subject to diagenesis. Diagenesis consists of those processes that alter the composition or texture of sediments after their formation and burial and before their eventual reexposure at the earth's surface. Therefore, lithification is a part of diagenesis, and weathering is not.

One of the great mysteries of geology concerns the origin of dolomite, the calcium-magnesium carbonate mineral, and dolostone, the dolomitic equivalent of limestone. Many geochemists believe dolomite and dolostone owe their origin to the diagenesis of limestone. Dolostones are relatively common in the ancient rock record, yet the formation of dolomite by direct crystallization is rare. This creates a dilemma: Where did all this ancient dolomite come from? Many geochemists believe that the answer lies in alteration (diagenesis) of relatively pure limestone to form dolostone, which contains at least 15 percent dolomite. This process can involve the mixing of fresh and marine water, the enrichment of magnesium by evaporation, and the circulation of warm fluids after diagenesis and burial.

WEATHERING OF LIMESTONE

Most rocks contain fractures known as joints. When rainwater enters joints, chemical weathering occurs, and the joints quickly widen. In the subsurface, horizontal and vertical joints widen, as downward-flowing surface water and laterally flowing groundwater dissolve away limestone, creating increasingly large void spaces, or caves, in the rock. The largest and most extensive cave systems in the world, such as Mammoth Cave in Kentucky, form in limestones. The largest caves usually form where multiple joints intersect in the subsurface.

Where subterranean cavities collapse just below the surface, they form sinkholes. Sinkholes may be exposed at the surface or covered by a layer of soil. Some sinkholes grow and subside only very slowly, while others may collapse in one rapid, catastrophic event. They range in size from a few meters wide and deep to sinkholes large enough to swallow several large buildings should they collapse. The resulting irregular, pockmarked landscape is called "karst topography." Karst topography is easily recognized by the presence of sinkholes, disappearing streams (which flow into sinkholes), caves, and springs.

CARBONATE PETROGRAPHY

Geologists study limestones for a variety of reasons and at a variety of scales. Most early studies of limestones focused on their fossils. Many limestones contain abundant, well-preserved fossils; some are famous for the exceptional quality of the specimens they contain. Visible, or macroscopic, fossils provide evidence for the sequence of evolution of many invertebrate organisms. Fossils are clues to a limestone's depositional environment as well; however, more detailed information concerning depositional environments can often be gathered by examining limestones at either smaller or larger scales.

Carbonate petrography involves the study of limestones for the purpose of description and classification. This usually involves using a microscope to determine a limestone's grain content—that is, the types of carbonate and noncarbonate grains present and their mineral composition (mineralogy). Carbonate petrology deals with the origin, occurrence, structure, and history of limestones. This involves petrographic studies of limestone as well as field studies of one or more outcrops. Carbonate stratigraphy applies the concepts of petrology at even larger scales and attempts to determine the physical and age relationships between rock bodies that may be separated by great distances.

Carbonate petrography is commonly performed by observing a thin slice of rock through a light microscope. A small block of the rock is cemented onto a microscope slide, then ground down and polished until the slice is about 30 microns (0.03 millimeter) thick. The slice is then thin enough for light to pass easily through it. Microscopic examination of a thin section can reveal a limestone's mineralogy and its microfossil or other grain content. Other observable traits include cement types, the presence or absence of lime mud, the purity of the limestone, and the types and degree of diagenesis. Use of special stains along with microscopy can reveal even more details of mineralogy. Stains allow easy identification of particular minerals. For example, Alzarin Red S colors calcite red and dolomite purple, so that their percentages can be determined.

STUDY OF DEPOSITIONAL ENVIRONMENTS

The field study of limestone exposures, or outcrops, also provides useful information. Along with the macroscopic fossil studies mentioned previously, geologists can study sedimentary structures to learn about a limestone's depositional environment. Sedimentary structures are mechanically or chemically produced features that record environmental conditions during or after deposition and before lithification. For example, ripple marks indicate water movements by either currents or waves, and their shape and spacing suggests the depth of water and velocity of water movement. The careful study of sedimentary structures provides methodical observers detailed information.

Outcrops also contain evidence of the lateral and vertical sequence of environments responsible for limestone deposition. By studying the lateral changes in a series of limestone outcrops, it is possible to interpret the distribution of environments, or paleogeography, in an area. For example, a researcher might determine in what direction water depths increased, or where a coastline might have been located. The vertical sequence of limestones at an outcrop indicates the paleogeography through time. By interpreting changes in sedimentary structures and other characteristics, a vertical sequence of limestones may indicate that, during deposition, a lagoon existed initially but gave way first to a coral reef and finally to an open ocean environment.

Many researchers conduct even larger-scale studies of limestone sequences. Using advanced technology developed for locating and studying petroleum reservoirs, geophysicists can produce cross-sections showing limestone distribution in the subsurface. This research technique, sequence stratigraphy, allows geologists to see very large-scale features located thousands of meters below the surface and so provides even better insights into regional paleogeography.

STUDY BY GEOCHEMISTS, MINERALOGISTS, AND ENGINEERS

Geochemists and mineralogists study limestones to determine their mineral composition. Simple techniques might involve dissolving a sample of limestone in stages, using a series of different acids. At each stage, the scientist weighs the remaining solid material. From this, approximate percentages of the limestone's mineral components can be determined. More advanced techniques can involve the use of X rays and high-energy particle beams to determine a mineral's atomic structure and precise composition. Such analysis might, for example, allow a chemist to suggest new industrial uses for a particular limestone deposit.

Engineers also study limestones, usually to determine their suitability as a construction or foundation material. Numerous tests are available; engineering tests generally involve determination of physical and chemical properties such as composition, strength, durability, porosity, permeability, solubility, and density. Results from such tests help to predict the behavior of limestones under certain conditions. For example, testing may indicate how a particular limestone would perform as a building foundation in an area with a humid climate and highly acidic soils.

Clay D. Harris

FURTHER READING

Boggs, Sam, Jr. *Petrology of Sedimentary Rocks.* New York: Cambridge University Press, 2009. Begins with a chapter explaining the classification of sedimentary rocks. Remaining chapters provide information on different types of sedimentary rocks. Describes siliciclastic rocks and discusses limestones, dolomites, and diagenesis.

Brown, G. C., C. J. Hawkesworth, and R. C. L. Wilson, eds. *Understanding the Earth: A New Synthesis.* Cambridge, England: Cambridge University Press, 1992. Chapter 17 summarizes how limestone deposition, mineralogy, and diagenesis have changed through geologic time. Contains numerous small diagrams and photographs that help to illustrate ideas presented in the text. A short bibliography of more advanced sources is included. Suitable for college-level readers.

Chesterman, C. W. *The Audubon Society Field Guide to North American Rocks and Minerals.* New York: Alfred A. Knopf, 1978. Well written and organized, beautifully illustrated with color photographs. Contains systematic descriptions of occurrences, chemical formulas, physical traits, and more. Contains a mineral identification key, a glossary of terms, a short bibliography, and a list of localities from the text. Suitable for high school readers.

Dixon, Dougal. *The Practical Geologist.* New York: Simon & Schuster, 1992. Covers all areas of geology at an introductory level. Related topics are presented in a methodical manner and integrated with other concepts through illustrations and examples. Discusses the tools geologists use to investigate minerals, rocks, and fossils and how they use them. Contains a short glossary of terms. Suitable for high school readers.

Grotzinger, J. P. "New Views of Old Carbonate Sediments." *Geotimes* 38 (September, 1993): 12-15. Discusses how ancient limestones, unlike modern ones that are produced primarily by organic activity, may contain evidence of inorganic carbonate production and a record of the development of carbonate-producing organisms. Suitable for college-level readers.

Mackenzie, F. T., ed. *Sediments, Diagenesis, and Sedimentary Rocks*. Amsterdam: Elsevier, 2005. Volume 7 of the Treatise on Geochemistry series, edited by H. D. Holland and K. K. Turekian. Compiles articles on such subjects as diagenesis, chemical composition of sediments, biogenic material recycling, geochemistry of sediments and cherts, and green clay minerals.

Middleton, Gerard V., ed. *Encyclopedia of Sediments and Sedimentary Rocks*. Dordrecht: Springer, 2003. Cites a vast number of scientists. Subjects range from biogenic sedimentary structures to Milankovitch cycles. Offers an index of subjects. Designed to cover a broad scope and a degree of detail useful to students, faculty, and geology professionals.

Oates, Joseph A. H. *Lime and Limestone: Chemistry and Technology, Production and Uses*. New York: Wiley-VCH, 1998. Examines the geochemical makeup, evolution, and use of limestones. The emphasis on chemistry, geochemistry, and geophysics makes this book useful for the reader with some scientific background. However, the book is filled with useful illustrations, charts, and maps that clarify difficult concepts.

Oldershaw, Cally. *Rocks and Minerals*. New York: DK, 1999. Useful to new students who may be unfamiliar with the rock and mineral types discussed in classes or textbooks.

Prothero, Donald R., and Fred Schwab. *Sedimentary Geology: An Introduction to Sedimentary Rocks and Stratigraphy*. 2d ed. New York: W. H. Freeman, 2003. A thorough treatment of most aspects of sediments and sedimentary rocks. Well illustrated with line drawings and black-and-white photographs, it also contains a comprehensive bibliography. Chapters 11 and 12 focus on carbonate rocks and limestone depositional processes and environments. Suitable for college-level readers.

Scholle, P. A., D. G. Bebout, and C. H. Moore. *Carbonate Depositional Environments*, Memoir 33. Tulsa: American Association of Petroleum Geologists, 1983. Well illustrated with color photographs, figures, and tables. Each chapter covers a specific example of a different depositional environment and contains a general description of the physical traits of the environment and the limestones that form there, as well as a thorough bibliography for that environment. Suitable for college-level readers.

Waltham, Tony. *Great Caves of the World*. Buffalo, N.Y.: Firefly Books, 2008. Discusses twenty-eight cave systems from around the world. Includes examples of maze caves, icefield and underwater caves, and the world's deepest cave. Describes the cave environment, location, and formation process of each cave.

Wilson, J. L. *Carbonate Facies in Geologic History*. New York: Springer-Verlag, 1975. A comprehensive treatment of most aspects of limestone deposition throughout geologic time. Includes examples from a variety of environments and geographic locations. Contains many line drawings and black-and-white photographs, as well as a detailed bibliography. Suitable for college-level readers.

See also: Biogenic Sedimentary Rocks; Carbonates; Chemical Precipitates; Clays and Clay Minerals; Contact Metamorphism; Diagenesis; Metamorphic Rock Classification; Minerals: Structure; Oil and Gas Origins; Regional Metamorphism; Rocks: Physical Properties; Sedimentary Mineral Deposits; Sedimentary Rock Classification; Siliciclastic Rocks.

LITHIUM

Lithium, an alkali metal, is one of only three elements to have been produced during the big bang. Lithium is highly reactive, and it has wide-reaching applications, including ceramics, nuclear power and warfare, optics, mental health pharmacology, greases and lubricants, lightweight metal alloys, and air purification.

PRINCIPAL TERMS

- **alkali metals:** soft, shiny, highly reactive metals that readily lose their one valence electron; group 1 of the periodic table of elements
- **disposable battery:** a type of battery, also called a primary cell battery, in which the electrochemical reaction is irreversible and the battery cannot be recharged
- **electrolysis:** a method of separating elements from ores and other sources by using a direct electric current to catalyze a nonspontaneous chemical reaction
- **flame test:** a chemistry test that detects the presence of some metal ions based on the color of the flame when the material in question is held in it; lithium turns a flame crimson
- **isotopes:** variants of an element, both stable and unstable (radioactive); isotopes have the same number of protons but a different number of neutrons
- **Mohs scale:** a scale that characterizes the hardness of minerals based on the ability of one mineral to scratch another
- **orbitals:** imaginary rings around atoms' nuclei in which electrons can be found; mathematical functions that describe the probable location of electrons relative to an atom's nucleus
- **rechargeable battery:** a type of battery, also called a secondary cell battery, in which the electrochemical reaction is reversible and the battery can be recharged
- **salt:** an electrically neutral ionic compound; lithium salts consist of lithium cations paired with a variety of anions
- **stellar nucleosynthesis:** a nuclear reaction within a star that results in the production of elemental nuclei heavier than hydrogen
- **valence electron:** an electron that is free to combine with the valence electrons of other atoms, resulting in a bond; in alkali metals and many other elements; only the outermost electrons can be valence electrons

OVERVIEW

Lithium, commonly abbreviated Li, is the element that holds the third position on the periodic table of elements, after helium and before beryllium. Although its name is derived from the Greek word *lithos*, meaning "stone," it is a soft metal; the term *lithium* derives from its discovery as a mineral.

Lithium is a member of the alkali-metal group of elements, a group that also includes sodium, potassium, rubidium, cesium, and francium. All of these elements reside in the first column of the periodic table; one other element exists in the same column: hydrogen, a nonmetal.

Lithium is an abundant element with many applications in modern life, particularly in batteries but also in optics, nuclear weapons, ceramics, air purification, and mental health. The following sections will examine lithium's chemical and physical properties and a number of its applications.

According to the big bang theory, several different nuclei were produced immediately following the big bang, including two stable lithium isotopes, Li-6 and Li-7. Hydrogen and its heavier isotope deuterium also were produced at this time, along with several helium isotopes.

Lithium was first discovered not in its elemental form but as a salt. Swedish chemist Johan August Arfwedson found lithium in 1817 while analyzing the ore of petalite, a compound that had been discovered in a Swedish mine in 1800. Arfwedson later found lithium in several other elements. Four years later, an English chemist, William Thomas Brande, was the first to isolate elemental lithium; he accomplished this by electrolysis of lithium oxide. In 1855, German chemist Robert Bunsen used electrolysis to isolate lithium from lithium chloride.

Commercial production of lithium did not begin until 1923, when a German company, Metallgesellschaft, began using electrolysis to isolate the element from a molten mixture of two salts, potassium chloride and lithium chloride. The lithium was used to improve Bahnmetall (railway metal), a material the company produced for use in train bearings.

In 1932, lithium atoms were transmuted to helium in the world's first human-made nuclear reaction, setting the stage for lithium's role in nuclear weapons.

For stockpiling nuclear weapons during the Cold War, the United States became a top producer of lithium. During World War II, scientists found that various lithium compounds could be combined to make a grease that worked well to lubricate high-temperature machinery, such as aircraft engines.

The mid-1990s marked the discovery that lithium could be extracted from brine, a much less expensive solution than mining. By the early twenty-first century, the use of lithium-ion batteries was on the rise.

PHYSICAL PROPERTIES OF LITHIUM

Lithium is a silver-white metal. Under standard temperature and pressure conditions, it is the lightest metal. It also is the least dense element that is not a gas at room temperature (its density is 0.534 gram per cubic centimeter [g/cm^3]) and it floats on water and oil. For comparison, water's density is 1.000 g/cm^3.

Lithium rates a 0.6 on the Mohs scale of mineral hardness, softer than even talc, which defines position 1 on the Mohs scale. Lithium is soft enough to be scratched with a fingernail or even cut with a knife, like the other alkali metals, and it yields a shiny metallic surface when cut. Lithium is highly reactive, and the cut surface will quickly tarnish when exposed to moisture in the air. Because it is so reactive, lithium needs to be stored carefully, often in a viscous hydrocarbon, such as petroleum jelly.

Lithium conducts both heat and electricity well, just like its alkali-metal group mates. Among the alkali metals, lithium has the highest melting point (80.54 degrees Celsius, or 356.97 degrees Fahrenheit), but this is low compared with most other metals. Similarly, its boiling point, 1,342 degrees Celcius (2,448 degrees Fahrenheit), is the highest of the alkali metals. Lithium's specific heat capacity is the highest of all of the solid elements, explaining its usefulness in high-temperature applications such as aircraft engines and train bearings.

ATOMIC AND CHEMICAL PROPERTIES OF LITHIUM

Lithium's atomic number is 3 and its electron configuration is $1s^2 2s^1$, so its first two electrons fill the 1s orbital and its third electron is alone in the 2s orbital. This valence electron readily reacts with other compounds, leaving the cation Li^+, which explains lithium's high reactivity and flammability. All of the alkali metals have just one valence electron, and all are quite reactive, although lithium is actually the least reactive of the group.

In addition to its classification as a group 1 element and an alkali metal, lithium is classified as a period 2 element, which refers to the second row of the periodic table. This period also includes beryllium, boron, carbon, nitrogen, oxygen, fluorine, and neon.

As its final periodic table classification, lithium belongs to the s-block, which includes the alkali metals (most of group 1) and the alkaline earth metals (group 2). In each group within the s-block, most chemical and physical properties move down the column in a trend-like way, so while lithium has the highest boiling point of the alkali metals, sodium has the second highest, followed by potassium, down to francium at the bottom of the column.

Lithium's high reactivity is especially pertinent in the presence of oxygen and water or even moist air; it can ignite or explode. However, the likelihood is greater with the other alkali metals, which are more reactive than lithium. Generally, under normal temperature and pressure conditions, lithium just tarnishes upon contact with moisture; the reaction produces lithium hydroxide, lithium nitride, and lithium carbonate, the combination of which appears as a dark coating.

Held in a flame, lithium ions cause the flame to appear distinctly crimson as the lithium's electrons absorb energy from the heat, jump up to higher orbitals, and then fall back down, releasing the extra energy as light (which appears as a visible color). The other metals will produce different colors in the flame test; sodium, for example, yields a bright orange color while potassium yields a lilac or violet color.

The stable Li-7 isotope makes up about 92.5 percent of naturally occurring lithium, while Li-6 makes up the rest. There are several radioactive isotopes as well, but they are unstable. (The most stable radioactive isotope, Li-8, has a half-life of about 800 milliseconds.) Li-6 plays an important role in nuclear reactors (and nuclear weapons), as it is involved in one of several reactions that can create tritium, a heavy radioactive isotope of hydrogen.

SOURCES OF LITHIUM

Because it is so reactive, lithium does not occur naturally in its elemental form, only as part of compounds such as salts. Lithium compounds are found virtually everywhere: on land, in the water, within living organisms, and even in outer space. Lithium also can be synthesized commercially by electrolysis of various compounds.

Terrestrially, lithium can be found in the earth's crust in an abundance of about 70 parts per million, similar to the crustal abundance of nickel and lead. Crustal lithium is most often found in granites, particularly within the pegmatitic crystalline masses that can intrude upon granites. Worldwide, Chile has one of the largest reserves of lithium, approximately 7.5 million tonnes (metric tons), and it mines more than 12,000 more tonnes each year. Bolivia also has several million tonnes of reserves.

Lithium also can be obtained from clays and brines in the ocean, where it is present in an abundance of about 0.25 part per million; the ratio increases to about 7 parts per million around hydrothermal vents on the ocean floor. Extraction of lithium from brine did not begin until the mid-1990s and has proved to be a cheaper alternative to mining.

Lithium was created during the big bang, and about 10 percent of the lithium in the Milky Way galaxy is probably derived from that initial creation. Lithium is still created by stellar nucleosynthesis, which involves nuclear reactions within stars that produce the nuclei of elements. Star-created lithium is found in some red giants, brown dwarfs, and orange stars, and it is possibly a result of the radioactive decay of unstable beryllium-7.

Lithium also is found in trace amounts in many living organisms, including plants and invertebrates and vertebrates. While its physiological role in living organisms is unclear, studies have suggested that it does play a role of some kind.

USE OF LITHIUM IN BATTERIES

Lithium's most familiar common use is in batteries, especially lithium-ion batteries. Other types of batteries, both rechargeable and disposable, also contain lithium.

Lithium-ion batteries are rechargeable batteries that are prevalent in portable electronics such as mobile phones, and their usage is increasing in other areas, too, such as in electric vehicles. Lithium-ion batteries have several advantages over other batteries. For one, their energy density is efficient, meaning they are able to store large amounts of energy in small amounts of space. They also have a low self-discharge rate; other rechargeable batteries, particularly most models containing nickel, can lose a significant percentage of their charge while not in use because of internal chemical reactions within the battery. Also, lithium-ion batteries do not have the "memory effect" problem of some rechargeable nickel cadmium batteries, in which the total energy capacity of the battery can permanently decrease if the battery is recharged before being fully discharged.

A typical lithium-ion battery has four main components: a positive electrode (made of a metal oxide), a negative electrode (made of carbon, often in the form of graphite), a nonaqueous electrolyte (a lithium salt in an organic solvent), and a separator, which keeps electrons from flowing directly between the electrodes. The positive and negative electrodes can both play the role of either the anode or the cathode, depending on the direction of the current flow through the battery. Commercial production of lithium-ion batteries began in 1991 by the Japanese companies Sony and Asahi Kasei.

Lithium-ion polymer batteries evolved from standard lithium-ion batteries; the main difference involves the electrolyte, which exists in a solid polymer instead of an organic solvent. In the late 2000's, improvements to the design yielded shorter charging times and faster discharge rates, and it is likely that continued improvements will have far-reaching positive implications for electric vehicles, consumer electronics, and other products.

A class of disposable (nonrechargeable) batteries known simply as lithium batteries features an anode made of either a lithium metal or a lithium compound. These batteries have quite long lives (some specialized types can last years) and are used in pacemakers, toys, clocks, cameras, and other devices.

OTHER LITHIUM APPLICATIONS

Aside from batteries, lithium has a wide variety of other applications, such as ceramics, nuclear power and warfare, optics, air purification, greases and lubricants, and even mental health. By the numbers, the use of lithium compounds in ceramics and glass is actually the biggest use worldwide, even beyond batteries.

As an additive, a lithium compound can reduce melting temperatures of ceramics and glass, decreasing energy costs required in the production of those materials. Lithium also makes the finished product more durable. Lithium is especially important in the production of glass cookware that must not expand as it undergoes temperature changes.

In nuclear power and nuclear weapons, the Li-6 isotope acts as a fusion fuel, aiding in the production of the heavy hydrogen isotope tritium. While this use has dangerous implications as a weapon, it also brings the promise of clean electricity generation in the future.

Lithium makes an appearance in optical equipment in the form of artificially created lithium fluoride crystals, which are transparent and have a low refractive index. They can be used in telescope focal lenses and in other specialized optics applications. In the mental health field, lithium compounds have long been used as mood stabilizers in the treatment of bipolar disorder and various forms of depression.

In enclosed places like spacecraft and submarines, lithium is a useful air purification aid because of its ability to remove carbon dioxide. Lithium also is used in some types of oxygen candles, apparatus that generate oxygen in submarines. Lithium is a useful component of lubricating greases used in high-temperature environments, such as in the engine of an airplane. It also is found in aircraft as part of a variety of lightweight but strong alloys, most often with aluminum.

Lithium technology has seen many advances, particularly with regard to lithium-ion batteries and lithium-ion polymer batteries. It remains likely that distinct improvements will be made in the quality of consumer electronics, electric cars, and many other technologies.

Rachel Leah Blumenthal

FURTHER READING

Dahlin, Greger R., and Kalle E. Strøm. *Lithium Batteries: Research, Technology, and Applications.* New York: Nova Science, 2010. Focuses on the development of more effective cathodes in lithium batteries using different compounds. This highly technical text is part of a series on electrical engineering developments; it is most useful for persons in the industry.

Fletcher, Seth. *Bottled Lightning: Superbatteries, Electric Cars, and the New Lithium Economy.* New York: Hill and Wang, 2011. This nonfiction book (not a textbook) explores the many ways in which the rise of lithium batteries will affect the future, particularly in terms of consumer electronics and electric cars.

Lew, Kristi. *The Alkali Metals: Lithium, Sodium, Potassium, Rubidium, Cesium, Francium.* New York: Rosen Central, 2010. This fairly basic text provides an easy-to-understand general overview of the properties and the real-world applications of the elements in the alkali-metal group, including lithium.

Nazri, Gholam-Abbas, and Gianfranco Pistoia. *Lithium Batteries.* New York: Springer, 2009. A comprehensive and highly technical overview of lithium batteries examines medical and industrial applications of lithium batteries and trends in research and development.

Ozawa, Kazunori. *Lithium Ion Rechargeable Batteries.* Hoboken, N.J.: Wiley, 2009. This esoteric text will be most useful to those in the industry. Explores new materials that could be used for the main components of lithium-ion batteries, the cathode, the anode, the separator, and the electrolyte.

Yuan, Xianxia, Hansan Liu, and Jiujun Zhang. *Lithium-Ion Batteries: Advanced Materials and Technologies.* Boca Raton, Fla.: CRC Press, 2012. As part of a series on green chemistry and chemical engineering, this text focuses on the different materials that can be used for each of the main components of lithium-ion batteries; it also examines the production of those materials.

See also: Aluminum Deposits; Crystals; Iron Deposits; Minerals: Physical Properties; Minerals: Structure; Quartz and Silica; Radioactive Materials; Uranium Deposits.

M

MANGANESE NODULES

Manganese nodules are rough spheres of manganese and other minerals deposited by seawater on the seabed. These nodules, about the size of a potato, are being considered as a potential source of manganese, which is essential for steel production. Manganese also is used to make pigments, alkaline batteries, and fertilizers.

PRINCIPAL TERMS

- **abyssal plain:** vast, flat underwater plains
- **crystal:** a solid with a repeating structure
- **crystallization:** the process by which crystals are formed when a substance comes out of solution
- **electronegative:** a measure of how tightly an atom holds its electrons
- **electrowinning:** a process by which metals are pulled from solution or a liquid
- **ion:** an atom with a net charge through either electron addition or electron loss
- **ionic bond:** a bond in which molecules are held together by an electrostatic bond created when the constituent atoms transfer electrons
- **ooze:** seafloor sediments
- **solution:** a mixture in which a solute is dissolved in a solvent
- **tailings:** waste from mining operations

MANGANESE

Manganese nodules are vaguely spheroid rocks made of deposited sediments, primarily the eponymous manganese. These nodules also contain enough other metals so that they are referred to as polymetallic nodules. Nodules form through millions of years of deposition, and the highest concentrations are on the abyssal plains of the oceans, at about 4,000 to 6,000 meters, or 2.5 miles (mi) deep. Manganese nodules have been suggested as potential sources of manganese for industry.

CHEMISTRY OF MANGANESE NODULES

Uses for manganese nodules come from their chemical properties. Manganese is a transition metal, like iron or nickel, and has several common oxidation states, with the most stable being +2. Manganese forms from a variety of minerals, though the majority of the minerals in a nodule consists of manganese hydroxides.

Seawater, an important element in the formation of manganese nodules, is not pure dihydrogen monoxide. Instead, it includes many other dissolved substances, but mostly such minerals as sodium chloride (table salt). Water is a good solvent because it is a polar molecule. Water is a polar molecule not simply because of its bent shape, but also because of its electronegativity. The oxygen in water is more electronegative than is water's hydrogen; thus, the electrons are closer to the oxygen. Oxygen is more electronegative because of various properties of atoms and electron orbitals; the atoms with eight electrons in their outermost orbits are the most stable. Atoms seek to have eight electrons, or to have none. Thus, hydrogen "gives" electrons to the oxygen. The electrons are thus closer to the oxygen than the hydrogen, giving it a net charge.

Limits exist to the amount of any substance that can be dissolved, however. The maximum amount of solute depends on properties such as temperature and pressure but varies for every material combination. Some substances increase in solubility with temperature and others decrease, all due to the binding characteristics of the compounds. The bonds of compounds are either more ionic or more covalent. As a metal, manganese compounds tend to be ionic and to dissolve in water. When the solubility of the water changes, either through cooling or through the addition of more solutes, the dissolved manganese precipitates out. It then crystallizes around a core, such as a fragment of a nodule, a shark's tooth, microfossils, or basalt fragments.

Crystallization is the formation of crystals of a substance as they precipitate from solution. Crystals are ordered arrangements of atoms. They are organized by the bonding properties of the atoms in question.

However, the precise pattern is still difficult to predict (computers and algorithms are leading to progress in this field, however). At any rate, continued deposition of the component of the crystal will lead to crystal growth.

NODULE FORMATION AND LOCATION

Nodules can be found in any body of water suitable for their growth, including some lakes, but most nodules are on the abyssal plains of the oceans. In particular, the northern regions of the Pacific equatorial seabed and the Atlantic have been found to have high concentrations. The Indian Ocean also appears to have concentrations of the nodules north of the equator.

These patterns in Pacific distribution seem to be a result of seafloor composition. North of the equator, the tropical Pacific is mostly siliceous in content, made of the noncarbonate portions of the skeletons of plankton. Because the sea floor here is deeper than the calcite compensation depth (calcite is calcium carbonate, which is what plankton shells are made of), the plankton dissolve. Calcium carbonate dissolves in water naturally, but because sufficient amounts exist in surface-level water, shells made of calcium carbonate are virtually insoluble. This explains why so many marine creatures make their shells of calcium carbonate.

At lower depths, calcium carbonate solubility increases dramatically. As a result the sea floor at lower depths is made not of calcium carbonate but of other materials, such as silicon, which does not dissolve. This effect leads to the creation of siliceous ooze. Calcareous sediments are formed where the seabed is above the calcite compensation level. The calcium compensation level depends upon many factors, but it is at about 42 to 500 meters (m), or 138 to 1,640 feet (ft), in the Pacific, except in the equatorial upwelling zone (where it is at 5,000 m, or 3 mi). Farther from the tropics, the sea floor tends to be made of red pelagic clay and has little biological material.

These differences in seafloor composition mean that the composition of nodules can be different, too. Where calcareous ooze, formed by sedimentation of calcareous shells, predominates, concentrations of nodules are lower. Equally, deposits in red clay regions have higher iron content than other deposits.

The sedimentation rate of siliceous ooze and pelagic clay regions is far faster than the rate of growth of nodules; therefore, the means by which nodules remain on the seabed and avoid getting buried are unknown. It has been suggested that deep-sea fauna might occasionally run into the nodules, pushing them over and thus preventing them from being buried. It also is possible that bottom currents push the nodules to the top or that buried nodules dissolve and are redeposited on the surface nodules.

THE ABYSSAL PLAIN

The abyssal plain is the sea floor beneath most of the ocean, typically at a depth of between 3,000 and 6,000 m (1.8 to 3.7 mi). The plain covers more than one-half the earth's surface and is among the flattest landscapes on the planet.

The plain also is one of the least explored surfaces on Earth. Despite the overall flatness, the plain also contains midocean ridges and oceanic trench subduction zones. These features are caused by plate subduction and crust formation.

In a phenomenon known as marine snow, organic detritus, such as dead organisms, fecal matter, land sediment, and inorganic sediment, continually fall from the upper levels of the ocean, making this detritus the main source of nutrients for the abyssal plain's ecosystem; the detritus also provides much of the sediment for magnesium nodules. Once thought lifeless, the abyssal plain is biologically diverse. Many species are bottom-dwelling species that subsist on the marine snow. Despite recent discoveries, however, much remains to be learned about these organisms.

Experts fear that manganese-nodule mining could negatively affect the seafloor environment. Disruptions to the seabed could increase the toxicity of the water near the seabed. Another possible problem is the formation of sediment plumes. A sediment plume is caused when tailings from mining are dropped back into the water.

DEEP SEA MINING

Deep-sea mining of manganese remains speculative. Modern prospecting methods tend to use remotely operated vehicles to collect samples from prospective mining areas. After a site has been found, a mining ship or platform is placed into position.

Two main mining methods are being considered: the continuous-line bucket system (CLB) and the hydraulic suction system. CLB is the primary method under consideration and is akin to a giant conveyer belt. It operates by using dredging buckets to pull material from the seabed. The material is processed on a platform and the waste materials, called tailings, are returned to the seabed in the dredging buckets. The other method uses hydraulic suction to pull the materials through a large tube. Another tube returns the waste material to the seabed.

The CLB system was put into use by a syndicate of thirty companies in the 1970's using an 8-kilometer (5-mi) cable launched from a former whaling vessel. The system worked but was prone to tangling.

The hydraulic system was tried by an American consortia in the Pacific Ocean. The system used a dredge on the sea floor attached through a tube to a surface object, such as a ship or a platform. This allowed it to reach an area of seabed. The initial tests ran into problems because the cables to the dredge could not be fully waterproofed; later dredges had problems that included muck and hurricanes. However, the system was demonstrated successfully in 1979.

Many current designs work on variants of this system. The fully operational system would have a central mining platform with a fleet of ships to supply and transport ore back to shore. To be of economic interest, however, the abundance of manganese must be 10 kilograms, or 22 pounds (lb) per square meter or more, with an average of 15 kg (33 lb) per square meter in several tenths of a square kilometer.

Sonar is used to map the sea floor for mining, too. Modern sonar systems can survey strips of more than 20 km (12.5 mi) at a time from the surface. This mapping can be supplemented with the use of arrays, which are towed above the seabed. Additionally, free-fall devices can take readings as they descend and return to the surface on their own. They even can take samples and photographs from the sea floor. Cable-operated apparatus also can perform this function.

MANGANESE NODULE PROCESSING

The idea for processing manganese nodules was first proposed in the 1960's by John Mero in his landmark paper on the subject, "Mineral Resources of the Sea." The paper received considerable attention, leading to several efforts to utilize the manganese resources.

Once on shore, several methods exist for refining. Some of the most popular methods include the cuprion process, sulfuric leaching, and smelting. The cuprion process, developed by Kennecott, an American mining company, involves grinding the nodules into slurry. The slurry is reduced by carbon monoxide in the presence of ammonia, in a low-temperature tank. The metals are made soluble and then are electrowinned, or separated by electrolysis. (Electrolysis is the process by which ions are pulled out of solution by an electric field to be deposited on a collector. Because atoms have different electrical properties, different voltages can pull out different materials.) Recovery of manganese from the remaining ferromanganese residue remains problematic for the mining industry.

Sulfuric leaching works by dissolving the nodules in sulfuric acid at high temperature and pressure. Copper, nickel, and cobalt are precipitated with hydrogen sulfide. The resultant materials are then refined. These products are electrowinned, and the ferromanganese residue is smelted after drying.

CURRENT EFFORTS

Harvesting manganese nodules remains expensive, given available technology. Many countries are placing claims on regions of the sea floor, however, to use in the event manganese mining becomes viable.

Gina Hagler

FURTHER READING

Chang, Raymond. *Chemistry*. Boston: McGraw-Hill Higher Education, 2007. A lucidly written comprehensive textbook for readers studying the chemistry of precipitation.

Cronan, D. S. *Handbook of Marine Mineral Deposits*. Boca Raton, Fla.: CRC Press, 2000. A detailed contextual examination of the nature of marine mineral deposits, including manganese deposits.

Halfar, Jochen, and Rodney M. Fujita. *Science* 316, no. 5827 (2007): 987. A detailed analysis of the problems associated with deep-sea mining, including manganese nodule exploitation. Brief but comprehensive.

Koslow, J. A. *The Silent Deep: The Discovery, Ecology, and Conservation of the Deep Sea*. Chicago: University of Chicago Press, 2007. A detailed discussion of the deep-sea environment, its creatures, and threats to those creatures. Accessible and recommended for all who need further information on the subject.

Rona, Peter A. "Resources of the Seafloor." *Science* 299, no. 5607, n.s. (2003): 673-674. Provides details of the methods by which oceanic ores reach the sea. Brief and accessible.

See also: Chemical Precipitates; Crystals; Geologic Settings of Resources; Unconventional Energy Resources; Water and Ice.

METAMICTIZATION

Metamictization breaks down the original crystal structure of certain rare minerals to a glassy state by radioactive decay of uranium and thorium. It is accompanied by marked changes in properties and often water content. Many minerals start to lose their radioactive components upon metamictization. Those that do not may be candidates for synthetic rocks grown from high-level nuclear wastes in order to isolate them until they are safe.

PRINCIPAL TERMS

- **alpha particle:** a helium nucleus emitted during the radioactive decay of uranium, thorium, or other unstable nuclei
- **crystal:** a solid made up of a regular periodic arrangement of atoms; its form and physical properties express the repeat units of the structure
- **glass:** a solid with no regular periodic arrangement of atoms; that is, an amorphous solid
- **granite:** a light-colored igneous rock made up mainly of three minerals (two feldspars, and quartz), with variable amounts of darker minerals
- **isotopes:** atoms of the same element with identical numbers of protons but different numbers of neutrons in their nuclei
- **isotropic:** having properties that are the same in all directions—the opposite of anisotropic, having properties that vary with direction
- **pegmatite:** a very coarse-grained granitic rock, often enriched in rare minerals
- **silicate:** a substance whose structure includes silicon surrounded by four oxygen atoms in the shape of a tetrahedron
- **spontaneous fission:** uninduced splitting of unstable atomic nuclei into two smaller nuclei, an energetic form of radioactive decay
- **X ray:** a photon with much higher energy than light and a much shorter wavelength; its wavelength is about the same as the spacing between atoms in crystal structures

METAMICT MINERAL PROPERTIES

Metamict minerals are an anomaly. They have the form of crystals, but in all other ways they resemble glasses. They fracture like glass, are optically isotropic like glass, and to all appearances are noncrystalline. Yet minerals cannot grow without an ordered crystalline arrangement of their constituent atoms—even metamict ones. The term "metamict" is from the Greek roots *meta* and *miktos*, meaning "after mixed." It aptly describes minerals that must originally have grown as crystals and subsequently have been rendered glassy by some process that has destroyed their original crystallinity. The discovery that all metamict minerals are at least slightly radioactive and that metamict grains have more uranium or thorium than do their nonmetamict equivalents led to the realization that radiation damage resulting from the decay of uranium and thorium causes metamictization. Although controversial at first, the concept of radiation damage is now so easy for scientists to accept that the puzzle is not why some minerals become metamict, but why others with relatively large concentrations of uranium and thorium never do.

The varieties and chemical compositions of minerals that undergo metamictization are quite diverse. Yet all metamict minerals share several common properties: They are all radioactive, with measurable contents of uranium, thorium, or both; they are glassy, brittle, and fracture like glass; they are usually optically isotropic for both visible and infrared light; they are amorphous to X-ray diffraction; and they often have nonmetamict equivalents with the same form and essentially the same composition. Finally, compared to their nonmetamict equivalents, they have lower indices of refraction, are more darkly colored, are softer, are less dense, are more soluble in acids, and contain more water. In those minerals that exhibit a range of metamictization from crystalline to metamict, partially metamict samples have intermediate properties.

For example, X-ray diffraction patterns of zircon show that radiation damage causes it to swell markedly in proportion to accumulated radiation damage up to the point of total metamictization. Beyond that point, the structure is so disordered that it can no longer diffract X rays, even though a continued decrease in density indicates further expansion.

Changes in other properties parallel the decrease in density. Partially metamict zircons that are heated at temperatures well below the melting point recrystallize readily to grains that are aligned with their original form, while those that are completely metamict recrystallize just as readily, but not to the original alignment. Several other metamict minerals recrystallize in the same way, but many produce mixtures of different minerals on heating because either they are outside their fields of stability or their compositions have changed subsequent to metamictization.

METAMICT MINERAL GROUPS

Metamict minerals belong to only a few broad groups of chemical compounds. In each group, some have uranium or thorium as the dominant metal ion present, but most have only small amounts substituting in the crystal structure for other ions, such as zirconium, yttrium, and the lanthanide rare-earth elements (REE), which just happen to have similar sizes and charges. This is known as isomorphous substitution and is common in most mineral groups.

The largest group of metamict minerals are yttrium-, REE-, uranium-, and thorium-bearing complex multiple oxides of niobium, tantalum, and titanium. All the metal-oxygen bonds in these minerals have about equal strength and about equal susceptibility to radiation damage. As it happens, most of these minerals have little resistance to metamictization and are commonly found in the metamict state.

Samarskite, brannerite, and columbite are examples of complex multiple oxide minerals. Samarskite is always totally metamict and compositionally altered with added water. It usually recrystallizes to a mixture of different minerals on heating, making its original structure and chemical formula hard to determine. Brannerite can range from partially to totally metamict. It has been found to be the primary site for uranium in some 1,400-million-year-old granites from the southwestern United States. In those rocks, it was recrystallized as recently as 80 million years ago, but even in that geologically short period of time, it has become metamict again. Columbite is never more than partially metamict. Apparently, the presence of iron in its formula helps prevent metamictization from progressing as far as it does in other niobates. These minerals are generally found in coarse-grained pegmatites associated with granites. They may also be found as accessory minerals within the granites themselves. Even when they account for most of the rock's uranium or thorium, however, they are so rare that it takes special concentrating techniques simply to find a few grains.

Silicates are the largest group of minerals in the earth's crust, but they account for only the second largest number of metamict minerals. They are characterized by having very strongly bonded silicon-oxygen groups in their crystal structures. Each silicon ion is surrounded by four oxygen ions in a tetrahedral (three-sided pyramid) shape. The chemical bonding between silicon and oxygen in the tetrahedral group is considerably stronger than the bonding between oxygen and any other metal ions in the structure, and is more resistant to radiation damage. The only silicates susceptible to metamictization are those in which the tetrahedral groups are not linked to form strong chains, sheets, or networks that make the structure resistant to radiation damage. The most commonly occurring metamict mineral is zircon, a zirconium silicate with isolated tetrahedral groups. Only a small amount of uranium can substitute for zirconium in the structure—up to about 1.5 percent, but usually less than 0.5 percent. Up to about 0.1 percent of thorium can also be present. These small amounts are sufficient to cause zircon to occur in a wide range of degrees of metamictization. Zircon is only one member of a group of silicates that grow with essentially the same crystal structure and subsequently become metamict. Thorite (thorium silicate) and coffinite (uranium silicate) also belong to that group. They each require about the same amount of radiation damage as zircon to become metamict, but because their concentrations of radioactive elements are much higher, metamictization occurs much more quickly.

Phosphates are the smallest group of metamict minerals, and they are usually found only partially metamict. One of particular interest is xenotime (yttrium phosphate), which has the same crystal structure as zircon. It commonly occurs in the same rocks as zircon, with equal or even greater contents of uranium and thorium, but is usually much less radiation damaged. Similarly, monazite (a rare-earth phosphate) takes up about the same amount of uranium and considerably more thorium than zircon but is seldom more than slightly damaged. Apparently, the substitution of phosphorus for silicon in these structures makes the phosphates less susceptible to metamictization than the equivalent silicates.

RADIATION DAMAGE TO MINERAL STRUCTURE

In order to understand metamictization fully, one must also understand how the radioactive decay of uranium and thorium damages crystal structures. In nature, uranium and thorium are both made up of a number of isotopes. Their most common isotopes are uranium-238, uranium-235, and thorium-232. Each of these isotopes decays, through a series of emissions of alpha particles, into an isotope of lead—lead-206, lead-207, and lead-208, respectively. The alpha particle is emitted from the decaying nucleus with great energy, and the emitting nucleus recoils simultaneously in the opposite direction. The energy transmitted to the mineral structure by the alpha particle and the recoiling nucleus is the major cause of radiation damage. The alpha particle has a very short range in the mineral structure (only about 20 wavelengths of light) and imparts most of its energy to it by ionizing the atoms it passes. Near the end of its path, when it has slowed down enough, it can collide with hundreds of atoms. The much larger recoil nucleus has a path that is about one thousandth as long, but it collides with ten times as many atoms. Thus, the greatest amount of radiation damage is caused by the recoil nucleus rather than by the alpha particle itself. Both particles introduce an intense amount of heat in a very small region of the structure, disrupting it but also increasing the rate at which the damage is spontaneously repaired. The accumulation of radiation damage depends on a balance between the damage and self-repair processes. In simple terms, radioactive minerals that remain crystalline have high rates of self-repair, while those that become metamict do not.

Another contribution to radiation damage in uranium-bearing minerals is the spontaneous fission of uranium-238, in which its nucleus splits into two separate nuclei of lighter elements. That process is much more energetic than is alpha emission, but it happens at a much lower rate. It probably produces only about one-tenth to one-fifteenth the damage that alpha emission and alpha recoil cause. The radiation damage done by the decay of uranium and thorium is not always confined to the metamict mineral itself. The decay process also involves the emission of gamma rays that penetrate the surrounding rock, often causing the development of dark halos in the host minerals. Metamict grains are also often surrounded by a thin, rust-colored, iron-rich rim at their contact with other minerals. These rims are enriched in uranium and lead that have leaked out of the damaged grain as well as the iron from the surrounding rock. These phenomena draw dark outlines around radioactive grains and make them easy to spot in most granites and pegmatites.

X-RAY STUDIES

X-ray diffraction results from the reinforcement of X-ray reflections by repeated planes of atoms in a crystal. The diffraction pattern is characteristic of each crystal structure and depends on the spacing of the planes. The strength of diffraction peaks depends on the atomic density in the diffracting planes and on the regularity of the crystal structure. If the structure is damaged, both the intensity and the sharpness of the diffraction peaks decrease. Glasses produce no X-ray diffraction pattern because there is no long-range order or regularity in their structures, and thus no planes of atoms exist on which X rays can diffract; they are said to be "X-ray amorphous."

X-ray diffraction studies of metamict minerals have shown them to resemble glasses in being devoid of a regular crystal structure. In partially metamict samples, the spacing between planes increases and the regularity of the structure decreases with progressive radiation damage. For example, in some 570-million-year-old zircons, 0.25 percent uranium is enough to make them X-ray amorphous—completely metamict. Much older zircons from other localities with higher uranium contents, however, yield X-ray patterns that reveal them to be only slightly damaged. In those areas, a relatively recent geologic event caused the zircons to be completely recrystallized with partial loss of lead, produced by radioactive decay of uranium and thorium, but little or no loss of uranium or thorium themselves.

Once a mineral is X-ray amorphous, X-ray diffraction is not as effective as is X-ray absorption spectroscopy for studying its structure. The way that a material absorbs X rays depends on the spacing of the atoms in it. X-ray absorption spectroscopy has shown that the spacing between adjacent ions is little changed in metamict minerals, but the regularity at greater distances has been lost. It has also shown that among complex multiple oxides, their crystal structures are highly regular and easily distinguished, but their metamict structures are nearly indistinguishable. Thus, the oxides all approach the same glassy state on metamictization.

ELECTRON MICROSCOPY AND INFRARED SPECTROSCOPY

Electron diffraction from single grains of metamict minerals yields results very similar to those of X-ray diffraction techniques. For example, it has been shown that partially metamict zircons are made up of misaligned crystalline domains that are destroyed with progressive metamictization.

High-resolution electron microscopy allows the scientist to examine a mineral's structure directly. While actual atoms cannot be seen, the regularity of the array of atoms making up the structure can. Applied to zircons, this technique has shown that fission particles from the spontaneous fission of uranium-238 leave long tracks of disruption behind them as they pass through the crystal structure. In partially metamict zircons, highly disordered damaged patches (called domains) are interspersed with undamaged domains. As metamictization proceeds, the damaged domains begin to overlap and eventually wipe out all remaining crystalline domains.

Infrared spectroscopy is a powerful tool for studying the bonds between atoms in minerals. The absorption of infrared light by a crystal structure depends on the strength and regularity of bonds within the structure. Infrared study has shown that the silicon-oxygen bonds of the tetrahedral groups in zircon remain intact to a large extent even in metamict samples, while the regularity of virtually every zirconium-oxygen bond has been disturbed. Thus, silicon remains surrounded by four oxygen atoms even in the most metamict zircons when the regularity of the formerly crystalline structure has been destroyed. Infrared spectroscopy is also often used to study the occurrence of water in minerals. Because most metamict minerals are enriched in water to some extent, its role in producing or stabilizing the metamict state is a question of some interest. Infrared studies of zircons in a wide range of metamict states show that while there is no water in their structures, small amounts of hydroxyl are common. The existence of "dry" zircons in all states from crystalline to metamict, however, shows that neither water nor hydroxyl is necessary to the metamictization process. The tendency for metamict zircons to have more hydroxyl than less damaged samples suggests that hydroxyl enters zircons only after metamictization has opened up their structures sufficiently for it to diffuse in.

ROLE IN ECONOMY AND PUBLIC SAFETY

Metamict minerals are important to people for two principal reasons. First, some gemstones, such as zircon, occur in the metamict state, and several others, such as topaz, can be enhanced in appearance by irradiation. Metamict gemstones are often considerably more valuable than the crystalline varieties because the anisotropic optical properties of the crystalline gems make them look "fuzzy" inside and less beautiful than the isotropic metamict stones. Radiation damage also often imparts color to the gemstone, increasing its value. (Artificial means of imparting color to gemstones by irradiation have been developed, but the gem often fades back to its original color with time. Consequently, it is best to know the entire history of a colored gem, or have it examined by a knowledgeable and trustworthy expert, before investing large amounts of money in it.)

Second, and more importantly, the understanding of metamictization may someday be invaluable to the safe disposal of high-level nuclear wastes. Ways must be found to keep high-level wastes from leaking into the environment and poisoning groundwater for a period of at least ten thousand years, after which they will have lost most of their radioactivity. The only way that scientists can determine what could conceivably lock up hazardous, highly radioactive atoms for that length of time is to examine results of long-term experiments on potentially leak-proof containers—solid materials that could be grown from the radioactive wastes that would be impervious to alteration, breakdown, or damage. The "experiments" of that duration are those that have already occurred in nature: radioactive minerals that have been damaged internally for millennia. Geochemists have found that some minerals retain their radioactive elements over millions of years, despite metamictization, while others do not. Synthetic analogues of the minerals that do may be grown from a mixture of the radioactive elements and added compounds to produce artificial rocks that are resistant to leaching and that have the potential to trap the hazardous substances for the required time.

James A. Woodhead

FURTHER READING

Black, Malcolm, and Joseph A. Mandarino. *Fleischer's Glossary of Mineral Species.* 10th ed. Tucson, Ariz.: The Mineralogical Record, 2008. An alphabetical listing of all currently accepted mineral names and formulas, along with the now-accepted assignments of discredited names. Includes separate lists of important mineral groups as well as the chemical word-formulas for several minerals. Not indexed or illustrated and has a minimum of text. Suitable for anyone needing the formula of a particular mineral, but less suitable for identifying a mineral from its composition.

Blake, Alexander J., and Jacqueline M. Cole. *Crystal Structure Analysis: Principles and Practices.* 2d ed. New York: Oxford University Press, 2009. Written for advanced undergraduate and graduate students. Covers diffraction, crystal structures, data collection, and crystal-structure determination and analysis.

Cavey, Christopher. *Gems and Jewels.* London: Studio, 1992. Provides information about gems and jewels and corrects many of the misconceptions concerning their study. Includes many illustrations.

Deer, W. A., R. A. Howie, and J. Zussman. *Rock-Forming Minerals.* 2d ed. 5 vols. London: Pearson Education Limited, 1992. Considered essential for geology students as a resource and laboratory handbook. Organized by mineral group for good accessibility. Appendices include a table of atomic and molecular weights, and other material used for chemical analysis and mineral analysis.

Hurlbut, C. S., Jr., and Robert C. Kammerling. *Gemology.* 2d ed. New York: John Wiley & Sons, 1991. A well-illustrated and complete introduction to gemstones, describing the methods of study of gems. A good introduction to mineralogy for the nonscientist, this volume includes a section with descriptions of minerals and other materials prized as gems. Suitable for high school level readers.

Mitchell, R. S. "Metamict Minerals: A Review, Part 1. Chemical and Physical Characteristics, Occurrence." *The Mineralogical Record* 4 (July/August, 1973): 177. Reviews the process of metamictization without becoming overly technical. Includes photographs.

_____. "Metamict Minerals: A Review, Part 2. Origin of Metamictization, Methods of Analysis, Miscellaneous Topics." *The Mineralogical Record* 4 (September/October, 1973): 214. Covers the process of metamictization in great depth without becoming overly technical. Written for lay mineral and gem collectors as well as Earth scientists. Aims to cover the famous mineral and gem localities of the world with state-of-the-art photography. Several issues have become collector's items. Suitable for anyone interested in gems or minerals.

Smith, David G., ed. *The Cambridge Encyclopedia of Earth Sciences.* New York: Crown Publishers, 1981. Chapter 5 offers a good description of the areas of mineralogy essential to an understanding of metamictization, although there is no discussion of metamict minerals. Suitable for college-level readers not intimidated by somewhat technical language. A well-illustrated and carefully indexed reference volume.

Tilley, Richard J. D. *Crystals and Crystal Structures.* Hoboken, N.J.: John Wiley & Sons, 2006. Discusses the characteristics, structure, and chemical dynamics of crystals. Provides information on specific crystals and specific structure types. Includes appendices, practice problems and exercises, bibliographies, and indexes.

Webster, Robert F. G. A. *Gems: Their Sources, Descriptions, and Identification.* 5th ed. Boston: Butterworth-Heinemann, 1994. A technical college-level text that is an important reference for the gemologist or the senior jeweler. Better for sources and descriptions than for identification. Essential for the gemologist.

Zhang, Ming, et al. "Metamictization of Zircon: Raman Spectroscopic Study." *Journal of Physics: Condensed Matter* 12 (2000): 1915-1925. Examines the crystal structure of zircon. Includes information on the molecular structure as well. A good example of crystal structure research.

See also: Aluminum Deposits; Biopyriboles; Carbonates; Clays and Clay Minerals; Feldspars; Gem Minerals; Hydrothermal Mineralization; Iron Deposits; Minerals: Physical Properties; Minerals: Structure; Mining Processes; Mining Wastes; Non-silicates; Oxides; Radioactive Minerals; Sedimentary Mineral Deposits.

METAMORPHIC ROCK CLASSIFICATION

Metamorphic rocks bear witness to the instability of the earth's surface. They reveal the long history of interaction among the plates that constitute the surface and the deep-seated motions within the plates. Among the metamorphic rocks are found many ores and stones of value to human civilization.

PRINCIPAL TERMS

- **contact metamorphism:** metamorphism characterized by high temperature but relatively low pressure, usually affecting rock in the vicinity of igneous intrusions
- **facies:** a part of a rock, or a group of rocks, that differs from the whole formation in one or more properties, such as composition, age, or fossil content
- **foliation:** a texture or structure in which mineral grains are arranged in parallel planes
- **metamorphic facies:** an assemblage of minerals characteristic of a given range of pressure and temperature; the members of the assemblage depend on the composition of the protolith
- **metamorphic grade:** the degree of metamorphic intensity as indicated by characteristic minerals in a rock or zone
- **metamorphism:** changes in the structure, texture, and mineral content of solid rock as it adjusts to altered conditions of pressure, temperature, and chemical environment
- **pelitic rock:** a rock whose protolith contained abundant clay or similar minerals
- **pressure-temperature regime:** a sequence of metamorphic facies distinguished by the ratio of pressure to temperature, generally characteristic of a given geologic environment
- **protolith:** the original igneous or sedimentary rock later affected by metamorphism
- **regional metamorphism:** metamorphism characterized by strong compression along one direction, usually affecting rocks over an extensive region or belt
- **texture:** the size, shape, and relationship of grains in a rock

METAMORPHISM

A significant part of the earth's surface is made up of rocks quite different from sedimentary or igneous rocks. Many of them have distinctive textures and structures, such as the wavy, colored bands of gneiss or the layered mica flakes of schist. They often contain certain minerals not found or not common in igneous or sedimentary rocks, such as garnet and staurolite. Studies of their overall chemical composition and their relationships to other rocks in the field show that they were once igneous or sedimentary rocks, but, after being subjected to high pressure and temperature, have been altered or recrystallized through a process called metamorphism. Metamorphism involves both mechanical distortion and recrystallization of minerals present in the original rock, the protolith. It can cause changes in the size, orientation, and distribution of grains already present, or it can cause the growth of new and distinctive minerals built mostly from materials provided by the destruction of minerals that have become unstable under the changed conditions. The chemical components in the rock are simply reorganized into minerals that are more stable under higher pressure and temperature.

DESCRIPTIVE CLASSIFICATION OF METAMORPHIC ROCKS

Metamorphic rocks can be classified in a purely descriptive fashion according to their textures and dominant minerals. Because the growing understanding of metamorphic processes can be applied to interpret the origin and history of the rocks, they are also classified according to features related to these processes. The most common classification schemes, in addition to the purely descriptive schemes, categorize the rocks by general metamorphic processes (metamorphic environments), the original rocks (protoliths), metamorphic intensities (grades), the general pressure and temperature conditions (facies), and the ratios of pressure to temperature (pressure-temperature regimes).

The oldest classification is purely descriptive, based on rock texture (especially foliation) and mineral content. Foliation is an arrangement of mineral grains in parallel planes. The most common foliated rocks are slate, schist, and gneiss. In slate, microscopic flakes of mica or chlorite are aligned so that the rock breaks into thin slabs following the easy cleavage of the flakes. Schist contains abundant, easily visible flakes of mica, chlorite, or talc arranged in parallel; it breaks easily along the flakes and has a highly reflective surface. Gneiss contains little mica, but its minerals (commonly quartz, feldspar, and amphibole) are separated into different-colored, parallel bands, which are often contorted or wavy. The foliated rocks can be described further by naming any significant minerals present, such as "garnet schist." Nonfoliated metamorphic rocks lack parallel structure and are usually named after their dominant minerals. Common types are quartzite (mostly quartz), marble (mostly calcite), amphibolite (with dominant amphibole), serpentinite (mostly serpentine), and hornfels (a mixture of quartz, feldspar, garnet, mica, and other minerals). Quartzite breaks through its quartz grains, whose fracture surfaces give the break a glassy sheen. Marble breaks mostly along the cleavage of its calcite crystals, so that each flat cleavage surface has its own glint. Hornfels often exhibits a smooth fracture with a satiny luster reminiscent of horn.

CLASSIFICATION BY PROCESSES OR ENVIRONMENTS

The second classification of metamorphic rocks is based on the general processes that formed them or the corresponding environments in which they are found. The recognized categories are usually named regional, contact, cataclastic, burial, and hydrothermal metamorphism.

Regional metamorphism is characterized by compression along one direction that is stronger than the pressure resulting from burial. The compression causes foliation, typified by the foliated rocks slate, schist, and gneiss. These rocks are found in extensive regions, often in long, relatively narrow belts parallel to folded mountain ranges. According to the theory of plate tectonics, folded mountain ranges like the Appalachians and the Alps began as thick beds of sediments deposited in deep troughs offshore from continents. The sediments were later caught up between colliding continents, strongly compressed, and finally buckled up into long, parallel folds. The more deeply the original sediment is buried, the more intense is the metamorphism. Clay in the sediment recrystallizes to mica, oriented with the flat cleavage facing the direction of compression. Thus, pelitic (clay-bearing) sediments become slates and schists

with foliation parallel to the folds of the mountains. At higher temperatures, the mica recrystallizes into feldspar, and the feldspar and quartz migrate into light-colored bands between bands of darker minerals so that the rock becomes gneiss. At yet higher temperatures, some of the minerals melt (a process called anatexis), and the rock, called a migmatite, becomes more like the igneous rock granite.

Contact metamorphic rocks are commonly found near igneous intrusions. Heat from the intrusive magma causes the surrounding rock, called country rock, to recrystallize. Though the rock is under pressure because of burial, there is usually no tendency toward foliation because the pressure is equal from all directions. Some water may be driven into the rock through fine cracks, or conversely water may be driven from the rock by the heat. Mobile atoms such as potassium especially can migrate into the rock and combine with its minerals to form new crystals (a process called metasomatism). Usually, however, the chemical content is not greatly altered, and recrystallization chiefly involves atoms from smaller crystals or the cement migrating into larger crystals or forming new minerals. In quartz sandstone, for example, the quartz crystals grow to fill all the pore space in a tight, polygonal network called crystalloblastic texture, and the rock becomes quartzite. Similarly, the tiny crystals in limestone or dolomite grow into space-filling calcite crystals, forming marble or, if other minerals are present, the mixed rock called skarn. Pelitic rocks recrystallize to hornfels, containing a variety of minerals such as quartz, feldspar, garnet, and mica.

Cataclastic (or dynamic) metamorphism occurs along fault zones, where both the rock and individual grains are intensely sheared and smeared out by stress. In deep parts of the fault, the sheared grains recrystallize to a fine-grained, finely foliated rock called mylonite.

Burial metamorphism occurs in very deep sedimentary basins, where the pressure and temperature, along with high water content, are sufficient to form fine grains of zeolite minerals among the sedimentary grains. The process is intermediate between diagenesis, which makes a sediment into a solid rock, and regional metamorphism, in which the texture of the rock is modified.

Hydrothermal metamorphism (which many prefer to call alteration) is caused by hot water infiltrating the rock through cracks and pores. It is most

Metamorphic Rock Classification by Texture and Composition

	non-foliated			foliated			
	non-layered			layered	non-layered		
Texture	fine to coarse grained	fine to coarse grained	fine grained	coarse grained	coarse grained	fine grained	very fine grained
Composition	calcite					chlorite	
			mica		mica		
		quartz					
			feldspar				
			amphibole				
			pyroxene				
Rock Name	marble	quartzite	hornfels	gneiss	schist	phyllite	slate

common near volcanic and intrusive activity. The water itself, or substances dissolved in the water, may be incorporated into the crystals of certain minerals. One important product is serpentine, formed by the addition of water to olivine and pyroxene, which is significant in sub-seafloor metamorphism.

CLASSIFICATION BY ORIGINAL ROCKS

A third classification is based on the original rocks, or protoliths. This classification is possible because relatively little material is added to or lost from the rock during metamorphism, except for water and carbon dioxide, so the assemblage of minerals present depends on the overall chemical composition of the original rock. The most abundant protoliths are pelitic rocks (from clay-rich sediments, usually with other sedimentary minerals), basaltic igneous rocks, and limestone or other carbonate rocks. Each kind of protolith recrystallizes into a different characteristic assemblage of minerals. The categories can be named, for example, metapelites, metabasalts, and metacarbonates.

CLASSIFICATION BY GRADE

Metamorphic intensity, or grade, is the basis for a fourth classification scheme. As pressure and temperature increase, certain minerals become unstable, and their chemical components reorganize into new, more stable minerals in the surrounding conditions. The presence of certain minerals, called index minerals, indicates the intensity of pressure and temperature. The grades most commonly used are named for index minerals in pelitic rocks; in the late nineteenth century, they were described by George Barrow in zones of metamorphic rocks in central Scotland. These Barrovian grades, in order of increasing intensity, are marked by the first appearances of chlorite, biotite, garnet, staurolite, kyanite, and sillimanite.

Metamorphism is a slow process, however (especially at low temperatures), and conditions sometimes change too rapidly for the mineral assemblage to come to equilibrium. It is not uncommon to find crystals only partially converted into new minerals or to find lower-grade minerals coexisting with those of higher grade. At low grade, some structures of the original rock, such as bedding, may be preserved. Even the outlines of earlier crystals may be seen, filled in with one or more new minerals. High-grade metamorphism usually destroys earlier structures.

All metamorphic rocks available for study are at surface conditions, so the pressures and temperatures that caused them to recrystallize have been relieved. Usually the loss of water and carbon dioxide prevents metamorphic reactions from reversing, but if conditions were relieved slowly enough, and particularly if water was available, the rock may have undergone retrograde metamorphism, reverting to a lower grade and thus adjusting to the less intense pressure and temperature. Retrograde metamorphism is usually not very complete, and some evidence of the most intense conditions almost always remains. For example, the distinct outline of a staurolite crystal might be filled with crystals of quartz, biotite mica, and iron oxides.

CLASSIFICATION BY FACIES

A fifth classification scheme categorizes the rocks according to the intensity of pressure and temperature, or facies, without reference to protoliths. The concept of facies was developed by Pentti Eskola, working in Finland around 1915, who enlarged on the work of Barrow in Scotland and V. M. Goldschmidt in Norway. Eskola realized that each protolith has a characteristic mineral assemblage within a given facies, or range of pressure and temperature. The facies are named for one of the assemblages within a specified range of conditions; for example, the greenschist facies, named for low-grade metamorphosed basalt, refers to equivalent low-grade assemblages from other protoliths as well. Other examples are the amphibolite facies, the range of pressures and temperatures that would give the staurolite and kyanite grades in pelitic rocks; and the granulite facies, which corresponds to extreme conditions bordering on anatexis. Facies is a more precise version of the concept of grade, and is defined by combinations of minerals that define specific chemical reactions rather than single minerals that may form via many different reactions.

CLASSIFICATION BY PRESSURE-TEMPERATURE REGIMES

Finally, the facies themselves, or the metamorphic assemblages in them, can be classified according to ratios of pressure to temperature, called pressure-temperature regimes. The greenschist, amphibolite, and granulite facies include mineral assemblages of increasing metamorphic intensity whose pressure and temperature rise together approximately as they would with increasing depth under most areas of the earth's surface. This sequence is sometimes referred to as the Barrovian pressure-temperature regime because Barrow's metamorphic grades in pelitic rocks fall in these facies. In another regime, called the Abukuma series after an area in Japan, the temperature is much cooler for any given pressure. The blueschist facies, characterized by blue and green sodium-rich amphiboles and pyroxenes, is typical of this series. The converse situation, in which temperature rises much faster than pressure, corresponds to contact metamorphism and is called the hornfels facies.

INVESTIGATING ROCK FEATURES AND HISTORIES

Initial studies of metamorphic rocks are almost always done in the field. The tectonic or structural nature of the region suggests the processes to which the rock has been subjected. For example, an area of folded mountains can be expected to exhibit regional metamorphism; a volcanic area, some contact metamorphism; and a fault zone, some cataclastic metamorphism. Some features are obvious at the scale of an outcrop, such as banding and foliation, or the halo of recrystallized country rock abutting an igneous intrusion. Some textural features—such as foliation, large crystals, or the luster of a fracture surface, as in quartzite or marble—are easily seen in a hand specimen. Similarly, a preliminary estimate of mineral content can be made from a hand specimen.

Many features, however, are best seen in thin section under a petrographic microscope. Usually all but the finest grains can be identified. From the relative abundance of the various minerals and their known chemical compositions the overall chemical composition of the rock can be calculated. The protolith can then be identified by comparing the calculated composition to the known compositional ranges of igneous and sedimentary rocks.

Textures seen under the microscope reveal much about the history of the rock. Foliation, for example, usually indicates regional metamorphism. A space-filling, polygonal texture can show contact or hydrothermal recrystallization, and a crumbled, smeared-out cataclastic texture indicates faulting.

More recent methods of investigation sometimes applied to metamorphic rocks are X-ray diffraction and electron microprobe analysis. The pattern of X

rays scattered from crystals depends on the exact arrangement and spacing of atoms in the crystal structure, which is useful for identifying minerals. X-ray diffraction can be used to identify crystals that are too small or too poorly formed to be identified with a microscope. The microprobe can analyze the chemical composition of crystals even of microscopic size. Determination of exact composition or of variation in composition within growth zones of a single crystal can be especially useful for identifying variations in conditions during crystal formation.

RECOGNIZING METAMORPHIC ROCKS

Most people encounter metamorphic rocks while traveling through mountains and other scenic regions. Recognizing these rocks is easier if one has a general idea of where the various types occur and how they appear in outcrops.

Regional metamorphic rocks are best exposed in two kinds of localities: the continental shields and the eroded cores of mountain ranges. Ancient basement rocks of the continental platform, composed of regionally metamorphosed and igneous rocks, are exposed in shields without a cover of sedimentary rock. The Canadian Shield, extending from northern Minnesota through Ontario and Quebec to New England, is the major shield of North America. Similar shields are exposed in western Australia and on every other continent. The old, eroded mountains of Scotland and Wales contain abundant outcrops in which some of the pioneering studies of metamorphic rocks were conducted. The Appalachians have even larger exposures, extending from Georgia into New England. Somewhat smaller outcrops occur in many parts of the Rocky Mountains and the Coast Ranges. The foliated rocks schist, slate, and gneiss make up the bulk of these exposures. Rock cleavage parallel to the foliation is an important clue to recognizing outcrops of these rocks; slopes parallel to foliation tend to be fairly smooth and straight, while slopes eroded across the foliation are ragged and steplike. Bare slopes of schist can reflect light strongly from the many parallel flakes of mica. Slate is usually dull and dark-colored but characteristically splits into ragged slabs. The colorful, contorted bands of gneiss are easily recognized.

Contact metamorphic rocks are much less widespread and are generally confined to areas of active or extinct volcanism. The Cascades and the Sierra

Nevada show many examples, but some of the best exposures are found near ancient intrusive rocks in the Appalachians and New England. Contact metamorphic rocks are more of a challenge to recognize because they generally lack foliation. The best clue is physical contact with a body of igneous rock. They often form a shell or halo around an igneous body, most intensely recrystallized at the contact and extending outward a few centimeters to a few kilometers (depending mostly on the size of the igneous body), until they eventually merge into the surrounding unaltered country rock. The halo, called an aureole, is generally similar to the country rock, but because it is recrystallized it is usually harder, more compact, and more resistant to erosion. Broken surfaces can be distinctive. Depending on the nature of the country rock, one might look for the glassy sheen of fractured quartzite, the glinting cleavage planes of marble, or the smooth, hornlike fracture of hornfels.

Hydrothermally altered rocks from sub-seafloor metamorphism, when finally exposed on land, are found among regional metamorphic rocks and appear much like them except for the greenish-gray colors of chlorite, serpentine, and talc. Terrestrial hydrothermal alteration is most easily recognized in areas of recent volcanic activity. Good examples are exposed in the southwestern United States, such as the so-called porphyry copper deposits. Many such areas contain valuable deposits, and so may have been mined or prospected. The outcrops are often much fractured and veined near the intrusion. Where alteration is most intense, the rock may appear bleached; farther from the intrusion, it may have a greenish hue because of low-grade alteration. Quartz and sulfide minerals such as pyrite are common in the veins, but weathering often leaves a rusty-looking, resistant cap over the deposit called an iron hat or gossan.

Rocks formed by cataclastic and burial metamorphism require specialized equipment for their recognition. Zones many miles wide containing mylonite, the product of cataclastic metamorphism, are exposed along the Moine fault in northwestern Scotland (where mylonite was first studied) and along the Brevard fault, extending along the Appalachians from Georgia into North Carolina. Examples of burial metamorphic rocks are found under the Salton Sea area of California and the Rotorua area of New Zealand.

James A. Burbank, Jr.

FURTHER READING

Bates, Robert L. *Geology of the Industrial Rocks and Minerals.* Mineola, N.Y.: Dover, 1969. Offers technical but practical descriptions of the geological occurrence and production of metamorphic rocks and minerals, listing their principal uses. Representative rather than comprehensive. Includes topical bibliography and good index.

Best, Myron G. *Igneous and Metamorphic Petrology.* 2d ed. Malden, Mass.: Blackwell Science Ltd., 2003. A popular university text for undergraduate majors in geology. A well-illustrated and fairly detailed treatment of the origin, distribution, and characteristics of igneous and metamorphic rocks. Chapter 4 treats granite plutons and batholiths, and Chapter 12 treats a broad spectrum of metamorphic topics.

Blatt, Harvey, Robert J. Tracy, and Brent Owens. *Petrology: Igneous, Sedimentary, and Metamorphic.* 3d ed. New York: W. H. Freeman, 2005. Appropriate for readers with some familiarity with minerals and chemistry. Thorough, readable discussion of most aspects of metamorphic rocks. Abundant illustrations and diagrams, good bibliography, and thorough indices.

Compton, Robert. *Geology in the Field.* New York: John Wiley & Sons, 1985. A standard undergraduate text with a chapter devoted to the interpretation of metamorphic rocks and structures in the field. Some knowledge of mineralogy is assumed.

Fettes, Douglas, and Jacqueline Desmons, eds. *Metamorphic Rocks: A Classification and Glossary of Terms.* New York: Cambridge University Press, 2007. Discusses the classification of metamorphic rocks. Feldspars are mentioned throughout the book in reference to the various types of metamorphic rocks and basic classifications.

Lambert, David. *The Field Guide to Geology.* 2d ed. New York: Checkmark Books, 2007. Written for high school students and amateur geologists. Discusses a number of topics in geology and is a useful resource for identifying rocks, minerals, and landscape features. Many diagrams to illustrate geological processes and features.

Oldershaw, Cally. *Rocks and Minerals.* New York: DK, 1999. Useful for new students who may be unfamiliar with the rock and mineral types discussed in classes or textbooks.

Perchuk, L. L., ed. *Metamorphic and Magmatic Petrology.* New York: Cambridge University Press, 1991. Although intended for the advanced reader, several of the essays in this multiauthored volume will serve to familiarize new students with the study of metamorphic rocks. The bibliography will lead the reader to other useful material.

Philpotts, Anthony, and Jay Aque. *Principles of Igneous and Metamorphic Petrology.* 2d ed. New York: Cambridge University Press, 2009. Discusses igneous rock formations of flood basalts and calderas. Covers processes and characteristics of igneous and metamorphic rock.

Pough, Frederick H. *A Field Guide to Rocks and Minerals.* 5th ed. Boston: Houghton Mifflin, 1998. The best of the most widely available field guides; authoritative but easy to read. Offers color plates of representative mineral specimens, and sufficient data to be useful for distinguishing minerals. Brief description of rocks, including metamorphic rocks. Elementary crystallography and chemistry are presented in the introduction.

Strahler, Arthur N. *Physical Geology.* New York: Harper & Row, 1981. The chapter on metamorphic rocks is a good intermediate-level approach to classification and metamorphic processes. Related chapters on geological environments may interest the reader. Excellent bibliography for the beginning student, with a thorough glossary and index.

Tarbuck, Edward J., Frederick K. Lutgens, and Dennis Tasa. *Earth: An Introduction to Physical Geology.* 10th ed. Upper Saddle River, N.J.: Prentice Hall, 2010. For beginning college or advanced high school readers. Good elementary treatment of metamorphic rocks and related environments. Color pictures throughout are excellent. Bibliography, glossary, and a short index.

Vernon, Ron H. *A Practical Guide to Rock Microstructure.* New York: Cambridge University Press, 2004. Investigates the microscopic structures of sedimentary, igneous, metamorphic, and deformed rocks. Discusses techniques and common difficulties with microstructure interpretations and classifications. Includes color illustrations, references, and indexing.

Winter, John D. *An Introduction to Igneous and Metamorphic Petrology.* Upper Saddle River, N.J.: Prentice Hall, 2001. Provides a comprehensive overview of igneous and metamorphic rock formation. Discusses volcanism, metamorphism, thermodynamics, and trace elements; focuses on the theories and chemistry involved in petrology. Written for students taking a university-level course on igneous petrology, metamorphic petrology, or a combined course.

See also: Blueschists; Contact Metamorphism; Metamorphic Textures; Metasomatism; Pelitic Schists; Regional Metamorphism; Sub-Seafloor Metamorphism; Ultra-High-Pressure Metamorphism.

METAMORPHIC TEXTURES

Metamorphic textures are important criteria in the description, classification, and understanding of the conditions under which metamorphic rocks form. Textures vary widely and develop as a result of the interaction between deformation, recrystallization, new mineral growth, and time.

PRINCIPAL TERMS

- **coherent texture:** an arrangement allowing the minerals or particles in a rock to stick together
- **foliation:** a planar texture in metamorphic rocks
- **mica:** a silicate mineral (one silicon atom surrounded by four oxygen atoms) that splits readily into thin, flexible sheets
- **strain:** deformation resulting from stress
- **stress:** force per unit of area
- **tectonics:** the study of the processes and products of large-scale movement and deformation within the earth

METAMORPHISM TYPES

In the classification and description of sedimentary, igneous, or metamorphic rocks, texture is a very important factor. Texture has a relatively straightforward definition—it is the size, shape, and arrangement of particles or minerals in a rock. How texture is related to the rock fabric and rock structure may be confusing. Rock structure may refer only to features produced by movement that occurs after the rock is formed. Rock fabric sometimes refers to both the structure and the texture of crystalline (most igneous and metamorphic) rocks. There is so much disagreement that the three terms are used interchangeably. Although there is no agreement on the categories of metamorphism, some are more generally accepted than others. Regional, contact, and cataclastic are three principal types. Shock, burial, sub-seafloor, and hydrothermal metamorphism are other varieties of metamorphism.

Metamorphic textures develop in response to several factors, the most influential being elevated temperatures and pressures and chemically active fluids. In each type of metamorphism, the relative influence of each of these three factors varies greatly. Generally, the effect of temperature on metamorphism is easy to understand. Converting dough to bread by baking provides a common analogy. The roles of pressure and fluids, however, are more difficult to explain. Two types of pressure, confining and directed, play a role in metamorphism. Confining pressure is the force exerted on a rock in all directions by the overlying material. Directed pressure, as the name implies, is a force exerted more strongly in some directions than others by some external process, usually tectonic in origin. Of these two types, directed pressure produces more apparent textural changes in metamorphic rocks. The primary role of chemically active fluids is to facilitate the metamorphic reactions and textural changes that occur.

REGIONAL METAMORPHISM

Of all the types of metamorphism, the most pervasive is regional metamorphism. Regional metamorphism occurs in the roots of mountain belts as they are forming. Thousands of square kilometers can be involved. Temperature and directed pressure are the critical agents in producing the new mineral assemblages and textures distinctive of this type of metamorphism. The most diagnostic feature in regionally metamorphosed rocks is a planar fabric, which forms as platy and elongate minerals develop a preferred orientation. This planar fabric, or foliation, generally forms only in the presence of the directed pressure, and in a perpendicular orientation to it. Directed pressure may also impart a linear texture to some rocks. This lineation may develop because elongate minerals or groups of minerals are recrystallized, deformed into a preferred direction, or both. Lineations may also develop where two planar fabrics intersect. Of all the different rock types, shales and other fine-grained rocks are the most sensitive to increasing conditions of metamorphism, and they show marked textural changes as metamorphic conditions increase. Not all regionally metamorphosed rocks develop foliations or lineations; many rock types show very little textural change from the lowest to highest conditions of metamorphism.

As metamorphism begins, shale is converted to slate, which diagnostically splits along smooth surfaces called cleavage. Most cleavage is pervasive or penetrative fabric; directed pressure influences every portion of the rock, with most of the micaceous minerals undergoing recrystallization. A variety of penetrative fabrics characterize regionally meta-morphosed rocks formed at different metamorphic conditions. In so-called fracture cleavage, the rock contains distinct planar fractures or cracks that are separated by discrete, relatively undeformed seg-ments of rock. A large amount of fracture cleavage is probably the result of solution activity removing por-tions of the rock.

At the lowest grades of regional metamorphism, and as the penetrative fabric develops in slate, very fine-grained micaceous minerals start to grow in a preferred orientation, thus producing planes of weaknesses along which the slate readily splits. At slightly higher metamorphic conditions, the mica-ceous minerals continue to form. The rock typically takes on a pronounced sheen, although the indi-vidual mineral grains remain mostly too small to see with the unaided eye. This texture is referred to as a phyllitic texture, and the rock is called a phyllite. As metamorphic conditions continue to increase at ever greater depths, the micaceous minerals eventu-ally grow large enough to be seen with the naked eye, and the preferred orientation of these platy minerals becomes obvious. This texture is referred to as schis-tosity, and the rock itself is called a schist. As the con-ditions of metamorphism approach very high levels, the micaceous minerals start to break down to form other minerals that are not platy, and the rock begins to lose its property to split along foliation surfaces. When rocks are metamorphosed, very large crystals of minerals such as andalusite, garnet, and cordi-erite may develop in an otherwise fine-grained rock. These large crystals constitute a porphyroblastic texture and are themselves called porphyroblasts. A new type of foliation develops, however, because light and dark minerals tend to separate into alternating bands. This banded texture is characteristic of rocks called gneisses, and the texture is referred to as a gneissosity. At even higher conditions, rocks begin to melt in a process called anatexis. Anatexis occurs at conditions that are considered to be at the interface between metamorphism and igneous processes.

CONTACT METAMORPHISM

Although not as widespread as regional metamor-phism, contact metamorphism is of interest because many contact metamorphic mineral deposits are economically important. Contact metamorphism occurs in areas of relatively shallow depths where a hot, molten igneous body comes in contact with the cooler rock that it has invaded. The metamorphism takes place outside the margins of the igneous mass, with temperature as the key agent of metamorphism and recrystallization as the dominant process modi-fying the original rock. The area metamorphosed around the intrusion is called the aureole. The size of this aureole is a function of the size, composition, and temperature of the igneous body, as well as the composition of the host rock and depth of the intru-sion. The intensity of the metamorphism is greatest at the contact between the two rock bodies and de-creases away from the source of heat. Aureoles may be as thin as a meter or, on rare occasions, as wide as 2 to 3 kilometers; usually, however, they are tens of meters wide.

A number of rock types and textures are character-istic of contact metamorphism. Most rocks produced by contact metamorphism are fine-grained because the processes by which they were formed are rela-tively short-lived. Typically, the aureole is made of a hard, massive, and fine-grained rock called hornfels. Characteristically, hornfels is more fine-grained than metamorphosed rock. The uniform grain size gives the rock a granular appearance, particularly when magnified; this texture is termed granoblastic or hornfelsic. Because hornfels undergoes no deforma-tion, textures in the original rock may be preserved, even when recrystallization is complete. If the au-reole contains silicon-rich carbonate rocks (impure limestones), much coarser textures are likely to form at or near the contact of the igneous body. These coarser-grained rocks are termed tactites or skarns. In some cases the intrusion is emplaced forcefully and pushes the surrounding rock out of the way. Under these conditions, contact metamorphism may produce foliation or lineation.

CATACLASTIC METAMORPHISM

The third type of metamorphism is cataclastic or dynamic metamorphism, which occurs in fault or deep-shear zones. Directed pressure is the key agent, with temperature and confining pressure playing vari-able roles. The textures in cataclastic rocks are divided into four groups: incoherent, nonfoliated, mylonitic, and foliated-recrystallized. Incoherent texture de-velops at very shallow depths and low confining pres-sures. Based on the degree of cataclasis (from least to most), fault breccia and fault gouge are the rock types formed. Nonfoliated (or mortar) texture forms coherent rocks typified by the presence of porphyroclastic minerals (large crystals that have survived cataclasis) surrounded by finely ground material. Rocks with this texture lack obvious foliations. Also included in this group are cataclastic rocks with glassy textures. Unusually intense cataclasis produces enough frictional heating to melt portions of the rock partially. Mylonitic texture occurs in coherent rocks that show a distinct foliation. The foliation typically forms because alternating layers show differing intensities of grinding. Textural subdivisions and rock names are based on the degree of grinding, such as protomylonite (least amount of grinding), orthomylonite, and ultramylonite (most amount of grinding). The final group is the foliated-recrystallized textures. Cataclastic rocks that illustrate considerable recrystallization are also typically foliated. The degree of grinding determines the rock name. Mylonite gneiss shows the least cataclasis, whereas the blastomylonite is more crushed.

SHOCK METAMORPHISM

Shock metamorphism is the rarest type of metamorphism. It is characteristically associated with meteorite craters and astroblemes. Astroblemes are circular topographic features that are inferred to represent the impact sites of ancient meteorites or comets. Some geologists consider shock metamorphism to be a type of dynamic metamorphism because directed pressure plays the essential role in both, whereas temperature may vary from low to extremely high.

A number of textural features are associated with shock metamorphism. Brecciation (breaking into angular fragments), fracturing, and warping of crystals are common. On a microscopic scale, the presence of two or more sets of deformation lamellae in quartz crystals is considered conclusive evidence of a meteorite impact. (Deformation lamellae are closely spaced microscopic parallel layers that are partially or totally changed to glass or are sets of closely spaced dislocations within mineral grains.) Shock metamorphism is also detected when minerals are partially or completely turned to glass (presumably with and without melting) and also when shatter cones are present. Shatter cones are cone-shaped rock bodies or fractures typified by striations that radiate from the apex.

STRUCTURAL ANALYSIS

Although much of the work of describing metamorphic textures comes from the field of metamorphic petrology, the means of studying how these textures, fabrics, or structures are generated is more within the area of structural geology called structural analysis. Structural analysis relies heavily on the fields of metallurgy, mechanics, and rheology to describe and interpret the process by which rocks are deformed. Structural analysis considers metamorphic textures and larger features, such as faults and folds, from three different perspectives: descriptive analysis, kinematic analysis, and dynamic analysis. Synthesis of the information obtained from these three analyses typically leads to comprehensive models or hypotheses that explain problems concerning metamorphic textures.

Descriptive analysis is the foundation of all studies of metamorphic textures. In descriptive analysis, geologists consider the physical and geometric aspects of deformed rocks, which include the recognition, description, and measurement of the orientations of textural elements. Typically, descriptive analysis involves geologic mapping, in which geologists measure the orientations of metamorphic structures with an instrument that is a combination compass-clinometer. (A clinometer is an instrument or scale used to measure the angle of an inclined line or surface in the vertical plane.) These structural measurements are plotted in a variety of ways to determine if the data show any statistical significance. Field studies may also include analysis of aerial photographs or satellite imagery to determine if textures or structures that are evident on the microscopic scale, or those of a rock outcrop, are related to much larger patterns.

In the laboratory, descriptive analysis of metamorphic textures can involve several techniques. One of the most important involves the petrographic microscope, in which magnified images of textures result from the transmission of polarized light through thin sections of properly oriented rock specimens. The thin section is a paper-thin slice of rock that is produced by gluing a small cut and polished block of the rock specimen to a glass slide. This block, commonly a little less than 2.7 millimeters by 4.5 millimeters, is then cut and ground to the proper thickness.

Kinematic analysis is the study of the displacements or deformational movements that produce features such as metamorphic textures. Several types of movement are recognized, including distortion and dilation. The study of distortion (change of shape) and dilation (change of volume) in a rock is called strain analysis. Strain analysis involves the quantitative evaluation of how original sizes and shapes of geological features are changed. Although strain commonly expresses itself as movement along preexisting surfaces, distortion also creates new surfaces such as cleavages and foliations in metamorphic rocks. Often deformation reorients the crystal structures of deformed minerals. The alignment of mineral structures can be studied with the universal stage, a device that allows slides under the microscope to be tilted to any desired orientation.

Dynamic analysis attempts to express observed strains in terms of probable patterns of stress. Stress is the force per unit of area acting upon a body, typically measured in kilograms per square meter. One important area of study in dynamic analysis involves experimental deformation of rocks under various levels of temperature, confining pressure, and time. Such experiments are typically conducted on short cylindrical specimens on a triaxial testing machine, which attempts to simulate natural conditions with the variables controlled. By jacketing specimens in impermeable coverings, pressures can be created hydraulically to exceed 10 kilobars, which are comparable to those near the base of the earth's crust, or outermost layer. Dynamic analysis also involves scale-model experiments, using clays and other soft substances to attempt to replicate naturally occurring structural features.

APPLICATIONS IN BUILDING AND CONSTRUCTION

Although a number of metamorphic minerals and rocks are used by society, in most cases the metamorphic textures of these materials play no role in their utility. One notable exception is slate. Because metamorphic processes impart a strong cleavage to this very fine-grained rock, it readily splits apart along these thin, smooth surfaces. The combination of grain size, cleavage, and strength makes slate useful in a number of products, including roofing shingles, flagstones, electrical panels, mantels, blackboards, grave vaults, and billiard tables. Most other metamorphic rocks, however, have foliations or other textural features that make the rocks structurally weak, and limit their usefulness.

In other cases, however, metamorphic rocks are useful because they lack pronounced foliations or do not readily break along planes of weakness. Many widely used marbles, particularly those quarried in Italy, Georgia, and Vermont, are products of regional metamorphism. They are prized in part because they are generally massive (lacking texturally induced planes of weakness) and can be cut into very large blocks. Several quartzites and gneisses also tend to be massive enough to be used as dimension stone or as aggregate because of high internal strength and relative chemical inertness. Some gneisses are prized as decorative stones because of their complex patterns.

Aside from direct usage, problems can develop in mountainous or hilly regions when construction occurs where weak, intensely foliated metamorphic rocks exist. When road or railway cuts are excavated through foliations inclined toward these cuts, rock slides are possible. In some cases, the roads can be relocated, or, if not, slide-prone slopes may be modified or removed. In other places, slide-prone exposures or mine walls and ceilings can be pinned and anchored.

Construction of buildings in metamorphic areas also requires careful evaluation. Where weak foliations run parallel to slopes, it must be determined whether the additional weight of structures and water for lawns and from runoff are likely to induce rock slides or other forms of slope instability. If so, land-use plans and zoning restrictions need to be adopted to indicate that these areas are potentially hazardous.

Ronald D. Tyler

FURTHER READING

Barker, A. J. *Introduction to Metamorphic Textures and Microstructures.* 2d ed. London: Routledge, 2004. An undergraduate text covering recognition and interpretation of metamorphic textures as well as microstructures.

Best, Myron G. *Igneous and Metamorphic Petrology.* 2d ed. Malden, Mass.: Blackwell Science Ltd., 2003. Advanced college-level text that should not be beyond the general reader researching metamorphic textures. Discusses mineralogy, chemistry, and the structure of metamorphic rocks. Discusses both field relationships and global-scale tectonic associations.

Blatt, Harvey, Robert J. Tracy, and Brent Owens. *Petrology: Igneous, Sedimentary, and Metamorphic.* 3d ed. New York: W. H. Freeman, 2005. Upper-level college text deals with the descriptions, origins, and distribution of igneous, sedimentary, and metamorphic rocks. Includes an index and bibliographies for each chapter.

Chernicoff, Stanley. *Geology: An Introduction to Physical Geology.* 4th ed. Upper Saddle River, N.J.: Prentice Hall, 2006. A good overview of the scientific understanding of the geology of the earth and surface processes. Includes a Web address that provides regular updates on geological events around the globe.

Davis, George H. *Structural Geology of Rocks and Regions.* 2d ed. New York: John Wiley & Sons, 1996. Upper-level text provides an excellent treatment of the deformational features in metamorphic rocks. Covers descriptive, kinematic, and dynamic analyses. Provides a well-illustrated treatment of cleavage, foliation, and lineation. Good bibliography and author and subject indexes.

Dietrich, Richard V., and B. J. Skinner. *Rocks and Rock Minerals.* New York: John Wiley & Sons, 1979. Provides a relatively brief but excellent treatment of regional, contact, and cataclastic metamorphic rocks, with good coverage of metamorphic textures. Very well illustrated. Includes a subject index and a modest bibliography.

Dolgoff, Anatole. *Physical Geology.* Boston: Houghton Mifflin, 1999. Although this is an introductory text for college students, it is written in a style appropriate for the interested layperson. Extremely well illustrated, with a glossary and index.

Ernst, W. G. *Earth Materials.* Englewood Cliffs, N.J.: Prentice-Hall, 1969. Chapter 7 in this compact, introductory-level book provides a succinct treatment of the topic of metamorphism. Discusses metamorphic structures, cataclastic rocks, contact metamorphism, and regional metamorphism. Includes a subject index and a short bibliography.

Fettes, Douglas, and Jacqueline Desmons, eds. *Metamorphic Rocks: A Classification and Glossary of Terms.* New York: Cambridge University Press, 2007. Discusses the classification of metamorphic rocks. Feldspars are mentioned throughout the book in reference to the various types of metamorphic rocks and basic classifications.

Hyndman, Donald W. *Petrology of Igneous and Metamorphic Rocks.* 2d ed. New York: McGraw-Hill, 1985. A college-level text intended for the undergraduate geology student. Covers the traditional themes (categories, processes, and conditions) necessary to understand metamorphic rocks. Provides information on the locations and descriptions of metamorphic rock associations. Includes an index and an exhaustive bibliography.

Newton, Robert C. "The Three Partners of Metamorphic Petrology." *American Mineralogist* 96 (2011): 457-469. Discusses field observations, experimental petrology, and theoretical analysis of the earth's crust. Highlights the relevance of collaboration to petrology.

Spry, A. *Metamorphic Textures.* Oxford, England: Pergamon Press, 1969. Advanced college-level text that treats metamorphism in terms of textural changes and largely disregards chemical interactions. Several chapters deal with very basic principles. Provides excellent illustrations and photographs. Includes an extensive bibliography and author and subject indexes.

Suppe, John. *Principles of Structural Geology.* Englewood Cliffs, N.J.: Prentice-Hall, 1985. A well-illustrated college text intended for the geology major. Chapters 10 and 11 provide clear discussions of "fabrics" and "impact structures," respectively, that are very readable and do not require an extensive geology background to understand. Includes a bibliography and subject index.

Vernon, Ron H. *A Practical Guide to Rock Microstructure.* New York: Cambridge University Press, 2004. Investigates the microscopic structures of sedimentary, igneous, metamorphic, and deformed rocks. Discusses techniques and common difficulties with microstructure interpretations and classifications. Includes color illustrations, references, and indexing.

Williams, Howel, F. J. Turner, and C. M. Gilbert. *Petrology: An Introduction to the Study of Rocks in Thin Sections.* 2d ed. San Francisco: W. H. Freeman, 1982. Provides descriptions of textures and mineral associations in igneous, sedimentary, and metamorphic rocks beneath the petrographic microscope. Also provides excellent, comprehensible descriptions of the major types of metamorphism and associated textures.

See also: Blueschists; Contact Metamorphism; Hydrothermal Mineralization; Metamorphic Rock Classification; Metasomatism; Pelitic Schists; Regional Metamorphism; Sub-Seafloor Metamorphism; Ultra-High-Pressure Metamorphism.

METASOMATISM

Metasomatism is produced by the circulation of aqueous solutions through rock undergoing metamorphic recrystallization. The solutions cause chemical losses and additions by dissolving and precipitating minerals along their flow paths. Metasomatic processes have produced the world's major ore bodies of tin, tungsten, copper, and molybdenum, as well as smaller deposits of many other metals.

PRINCIPAL TERMS

- **aqueous solution, hydrothermal fluid, and intergranular fluid:** synonymous terms for fluid mixtures that are hot and have a high solvent capacity, permitting them to dissolve and transport chemical constituents; they become saturated upon cooling and may precipitate metasomatic minerals
- **contact metasomatism:** metasomatism in proximity to a large body of intrusive igneous rock, or pluton
- **density:** the ratio of rock mass to total rock volume; usually measured in grams per cubic centimeter
- **diffusion:** process whereby atoms move individually through a material
- **meteoric water:** water that originally came from the atmosphere, perhaps in the form of rain or snow, as contrasted with water that has escaped from magma
- **permeability:** the capacity to transmit fluid through pore spaces or along fractures; high-fracture permeability is generally requisite for metasomatism, as porosity is greatly reduced by metamorphic recrystallization
- **porosity:** the ratio of pore volume to total rock volume; usually reported as a percentage
- **prograde metamorphism:** recrystallization of solid rock masses induced by rising temperature; differs from metasomatism in that bulk rock composition is unchanged except for expelled fluids
- **recrystallization:** a solid-state chemical reaction that eliminates unstable minerals in a rock and forms new stable minerals; the major process contributing to rock metamorphism
- **regional metasomatism:** large-scale metasomatism related to regional metamorphism

PROCESSES OF CHEMICAL CHANGE

Metasomatism is an inclusive term for processes that cause a change in the overall chemical composition of a rock during metamorphism. Such processes may be described as positive or negative depending upon whether a net gain or a loss is produced in the affected rock body. Where chemical changes are slight, the minerals in the original rock remain unchanged or register only very subtle changes. In contrast, intense metasomatism may result in total destruction of the original mineral assemblage and its replacement by new, and different, minerals. In these extreme cases, metasomatism is usually difficult to detect, particularly if large rock volumes are affected. Metasomatic effects are most obvious when the original rock texture and mineral assemblage is partially destroyed. In such cases, the resulting rock will exhibit an unusually large variety of minerals as well as microscopic evidence of incomplete chemical reactions.

In normal metamorphic reactions, chemical migration occurs on the scale of a single mineral grain (a few millimeters at most) during recrystallization. In contrast, metasomatism involves chemical transport on the scale of a few centimeters or more. In areas where metasomatism has been intense, it can often be demonstrated that the scale of chemical transport ranged from about 100 meters to as much as several kilometers. It is the movement of chemical components through rocks on this larger scale that distinguishes metasomatism from metamorphism.

Dehydration (water-releasing reaction) and decarbonation (carbon dioxide-releasing reaction) are the most common types of chemical reactions during metamorphism at the higher grades. These reactions certainly produce significant changes in rock composition and involve large-distance chemical transport, but because they typify normal prograde metamorphism, they do not constitute metasomatism. Conversely, if water and/or carbon dioxide were reintroduced into a rock that had previously experienced prograde metamorphism, this would constitute a fairly common type of metasomatism.

DIFFUSION AND INFILTRATION METASOMATISM

The reshuffling of the chemical components into new mineral assemblages during metamorphism, particularly when gaseous phases are lost, leads to

285

major changes in rock volume (usually volume reductions) and a corresponding change in rock porosity and density. Metasomatic replacement of a metamorphic rock will induce additional changes in volume, porosity, and density. Although metasomatism is defined as a process of chemical change, physical changes in rock properties also occur, and these are an essential aspect of metasomatism. For convenience, introductory textbooks ignore volume changes in metamorphic and metasomatic reactions. This assumption may hold true in specific instances, but it has no general validity, particularly when volatile constituents are involved in the reactions. It has been demonstrated on theoretical grounds that chemical transport on the scale of a few centimeters or more requires the presence of an intergranular pore fluid that can effectively dissolve existing minerals and deposit others while the rock as a whole remains solid.

The relative mobility of this fluid phase leads to two theoretical types of metasomatism: diffusion metasomatism and infiltration metasomatism. In diffusion metasomatism, the chemical components move through a stationary aqueous pore fluid permeating the rock by the process of diffusion. The effects are limited to a distance of a few centimeters from the surface of contact between rocks of sharply contrasting composition. Because this process acts over small distances, it cannot produce large-scale metasomatic effects. Infiltration metasomatism involves a mobile aqueous fluid that circulates through pores and fractures of the enclosing rock and carries in it dissolved chemical components. This fluid actively dissolves some existing minerals and deposits new ones along its flow path. The scale of infiltration metasomatism is thus determined by the circulation pattern of the fluid, and rock compositions over distances of several kilometers can be easily altered. For convenience, the term "metasomatism" will be used in place of "infiltration metasomatism" and the effects of diffusion metasomatism will be ignored.

CONDITIONS FAVORABLE TO METASOMATISM

The degree of chemical change that accompanies recrystallization is closely related to the fluid/rock ratio prevailing during the process. Since regional metamorphism is a deep-seated process, a low fluid/rock ratio generally prevails, and the process is "rock-dominated" in the sense that minerals dominate the composition of the fluid circulating through the rocks. A significant degree of metasomatism under such conditions is unlikely unless the fluid is very corrosive. Small quantities of fluorine or chlorine can produce corrosive aqueous fluids, but these elements are rarely important in regional metamorphism. Contact metamorphism, however, occurs close to the earth's surface, where large volumes of groundwater circulate in response to gravity. Groundwater will mix with any water given off by a high-level crystallizing pluton; it follows that contact metamorphism takes place under conditions of relatively high fluid/rock ratio (although the amount of fluid is always much less than the amount of solid rock). Such a process will be "fluid-dominated" in the sense that the circulating fluid will govern the compositions of the minerals formed by recrystallization. These conditions are highly favorable for metasomatism.

Intense, pervasive metasomatism will develop under the following conditions: when energy is available to provide temperature and pressure gradients to sustain fluid movement; when a generally high fluid/rock ratio prevails; when the aqueous fluid has the solvent capacity to dissolve minerals in its flow path; and when the enclosed rocks possess, or develop, sufficient permeability to permit fluid circulation. These conditions must be sustained for a sufficient time period, or recur with sufficient frequency, to produce metasomatism.

GREISENIZATION AND FENITIZATION

Such conditions are commonly attained in rocks adjacent to large bodies of intrusive igneous rock, or plutons, particularly those that expel large quantities of water. As an example of contact metamorphism, consider the effects produced in the fractured roof zone of a peraluminous, or S-type granite pluton (meaning "derived from the melting of sedimentary parent material"). Hot water vapor, concentrated below the roof by crystallization, is often enriched in corrosive fluorine accompanied by boron, lithium, arsenic, silicon, tungsten, and tin. This reactive fluid migrates up fractures, enlarging pathways by dissolving minerals and seeping into the adjacent rock. The result is a network of quartz-muscovite-topaz-fluorite replacement veins that may contain exploitable quantities of tin, tungsten, and base metal ore (copper, zinc, lead, and so on). Fluorine-dominated metasomatism is known as greisenization. Greisen effects may extend 5 to 10 meters into the wall rocks from vein margins. Within this zone, the original rock textures, minerals, and chemical composition will be profoundly modified by metasomatism. In many cases, the entire roof of the granite pluton is destroyed and replaced by greisen minerals. Some of the world's major tin and tungsten deposits—such as those in Nigeria, Portugal, southwest England, Brazil, Malaysia, and Thailand—were formed by just such a process.

Spectacular sodium and potassium metasomatism develops adjacent to intrusions of ijolite and carbonatite plutons. Ijolite is an igneous rock extremely rich in sodium and potassium, and carbonatites are intrusions consisting of carbonate minerals rather than silicates. This intense alkali metasomatism is called fenitization and is developed on a regional scale in the vicinity of Lake Victoria in East Africa. In this region, ijolite-carbonatite complexes are plentiful, and aureoles, or ring-shaped zones, of fenite extend outward 1 to 3 kilometers from the individual plutons. The width of the fenite zone around a source pluton is largely determined by the fracture intensity in the surrounding rocks; where they are highly fractured, large-scale and even regional fenitization is present. Massive, unfractured rocks resist fenitization, and the resulting aureoles are narrow.

Metasomatism around ijolite intrusions is dominated by the outflow of sodium dissolved in an aqueous fluid expelled by the magma and an apparent back-flow migration of silicon to the source intrusion. A typical result would be an inner zone, 20 to 30 meters wide, of coarse-grained "syenite fenite" composed of aegirine, sodium feldspar, and sometimes nepheline. Beyond the outer limit of this zone, there would be a major aureole of shattered host rock veined by aegirine, sodium amphibole, albite, and orthoclase, which might extend well beyond 1 kilometer from the parent intrusion. Fenitization around carbonatite intrusions is dominated by potassium metasomatism, which converts the country rocks into a coarse-grained metasomatic rock composed almost exclusively of potassium feldspar called orthoclase. The resulting fenite aureole is typically less than 300 meters wide and is roughly proportional to the diameter of the parent carbonatite. The chemical composition of the country rocks appears to exert little influence on the progression of fenitization. The rocks adjacent to ijolite intrusions are driven toward a bulk chemical composition approaching that of ijolite, while those adjacent to carbonatite intrusions are driven toward orthoclase regardless of their initial composition. Not all fenites are associated with intrusions, although perhaps the source intrusions are deeper underground.

METASOMATIC MINERAL ZONES

Intensely metasomatized rocks usually exhibit mineral zoning, which is more or less symmetrical around the passageways that controlled fluid migration. These passages may be networks of vein-filled fissures, major fault zones, shattered roof zones of igneous plutons, or the fractured country rocks adjacent to such plutons. Emphasis is placed on fracture permeability as a control of fluid migration, as most metamorphosed rocks have negligible porosity. The metasomatic mineral zones are often dominated by a single, coarse-grained mineral species, which is obviously "exotic" with respect to the original mineral assemblage. Inner zones often cut sharply across outer zones, and the resulting pattern reflects a systematic increase in metasomatic intensity as the controlling structure is approached. By means of foot traverses across the metasomatized terrain, geologists carefully map the mineral zoning pattern and its controlling fractures. Such maps provide insight into the fracture history of the area, show the distribution and volume of the metasomatic products, and indicate the relative susceptibility of the various rock types present to the metasomatic process. If ore bodies are present, geologic maps provide essential information for exploring the subsurface extent of the ores through drilling. They also provide a basis for systematically sampling the metasomatic zones as well as the country rock beyond the limit of metasomatism; the unaltered country rock is called the protolith.

The sample collection provides material for study in the laboratory after field studies are complete. A paper-thin slice is cut from the center of each rock sample and is mounted on a glass slide for microscopic examination. From such examinations, the scientist can identify both fine- and coarse-grained minerals and can determine the order of metasomatic replacement of protolith minerals.

STUDY AND SAMPLING OF PROTOLITHS

The objectives in a study of metasomatism are to determine the chemical changes that have taken place in the altered rocks and to reconstruct the history of fluid-rock interaction. To this end, it is essential to determine the chemical compositions of the various protoliths affected by metasomatism so that additions and losses in the metasomatized rocks may be calculated. The ideal protolith is a rock unit that is both chemically uniform over distances comparable to the scale of zoning and highly susceptible to metasomatism. Considerable attention must, therefore, be devoted to the study and sampling of protoliths during the field stage of a project. Ideally, the samples should be collected just beyond the outer limit of metasomatism in order to avoid metasomatic contamination and to minimize the effect of a lack of protolith uniformity. Unfortunately, it is only possible to approximate the position of this outer limit in the field, because the decreasing metasomatic effects merge imperceptibly with the properties of the unaltered protolith. Sampling problems are further compounded in study areas where rock exposures are poor or where exposed rocks are deeply weathered. Rock weathering promotes chemical changes that are at times indistinguishable from metasomatism, so weathered samples cannot be accepted for chemical study. The sample requirements for metasomatic research are more stringent than for any other type of geological study, and meeting them always taxes the ingenuity of a geologist.

The samples, having been cut in half for the microscope slides, are then prepared for chemical analysis and for density measurement. One half is stored "as is" in a reference collection for future use. The other half is cleaned of all traces of surface weathering and plant material. The samples are then oven-dried, and bulk densities are obtained by weighing and coating them with molten paraffin (to seal pore spaces), followed by immersion in water to determine their displacement volumes. Next, the samples are crushed and ground to fine powder. The average mineral grain density is determined by weighing a small amount of rock powder and measuring its displacement volume in water. The porosity of each sample is determined, and the powdered samples (20-30 grams each) are then sent to a laboratory specializing in quantitative chemical analysis. Because it is desirable to study as many as sixty-five different chemical elements, the laboratory will use several modern instrumental techniques as well as the traditional "wet method" to determine their concentrations.

Modern studies use the mass balance approach, which relates the physical, volumetric, and chemical properties between the altered rocks and their protoliths. As an example, consider a particular element in a given volume of protolith. The mass of this element is the product of volume × bulk density × element concentration. If the element is totally insoluble in the circulating fluid, then its mass must remain constant during the metasomatism of the enclosing rock. It follows that for this immobile element, the product of volume × bulk density × element concentration in the altered rock must equal that of the parent protolith. This provides an objective test for determining the exact elements that were mobile and immobile during the metasomatic event. From this point, it is a simple matter to calculate the gains and losses of each element for the altered rocks.

SKARN DEPOSITS

Metasomatic deposits of a wide range of metals and industrial minerals are commonly found at or near the contact between igneous plutons and preexisting sedimentary rocks. Ore deposits concentrated by contact metasomatic processes are collectively known as skarn deposits. Skarn deposits are major sources of tungsten, tin, copper, and molybdenum. Important quantities of iron, zinc, cobalt, gold, silver, lead, bismuth, beryllium, and boron are also mined from skarn deposits. Additionally, such deposits are a source of the industrial minerals fluorite, graphite, magnetite, asbestos, and talc. For the most part, skarn deposits are found in relatively young mobile belts that are not yet deeply eroded. The most productive skarns are generally those in which granitic magma has invaded sedimentary sequences dominated by layers of carbonate rocks.

The physical and chemical principles of metasomatism have been deduced by research geologists over many decades from thousands of individual field and laboratory studies. The knowledge derived from the studies is put to practical use by economic geologists who explore remote areas in search of new skarn deposits or who exploit known skarn ores at producing mines. These economic geologists test, on a daily working basis, the theoretical hypotheses and generalizations formulated by research geologists regarding metasomatism.

Gary R. Lowell

FURTHER READING

Augustithis, S. S. *Atlas of Metamorphic-Metasomatic Textures and Processes*. Amsterdam: Elsevier, 1990. A thoughtful treatment of metamorphic rocks, mineralogy, and metasomatism. Excellent illustrations add to understanding concepts and processes. Appropriate for the college student or layperson with an interest in Earth sciences.

Blatt, Harvey, Robert J. Tracy, and Brent Owens. *Petrology: Igneous, Sedimentary, and Metamorphic*. 3d ed. New York: W. H. Freeman, 2005. An upper-level college text that deals with the descriptions, origins, and distribution of igneous, sedimentary, and metamorphic rocks. Includes an index and bibliographies for each chapter.

Burnham, C. W. "Contact Metamorphism of Magnesian Limestones at Crestmore, California." *Geological Society of America Bulletin* 70 (1959): 879-920. A classic account of metasomatism at one of the best-known mineral-collecting localities in the world. Written for professional geologists but comprehensible for those with some knowledge of geology.

Carlson, R. W., ed. *The Mantle and Core*. Amsterdam: Elsevier Science, 2005. Suited for researchers and graduates. Discusses mantle composition, tectonics, mantle volatiles, and volcanic processes. Plate subduction, mantle convection and exchange between the mantle and core.

Chernicoff, Stanley. *Geology: An Introduction to Physical Geology*. 4th ed. Upper Saddle River, N.J.: Prentice Hall, 2006. A good overview of the scientific understanding of geology and surface processes. Includes a Web address that provides regular updates on geological events around the globe.

Coltorti, M., and M. Gregoire. *Metasomatism in Oceanic and Continental Lithospheric Mantle*, Special Publication no. 293. London: Geological Society of London, 2008. A collection of papers discussing major and trace elements, fluid/rock ratios, and supra-subduction. Also includes intra-plate settings relating metasomatic signatures to one event.

Einaudi, M. T., L. D. Meinert, and R. J. Newberry. "Skarn Deposits." In *Economic Geology: Seventy-fifth Anniversary Volume*. El Paso, Tex.: Economic Geology Publishing, 1981. A comprehensive review of skarn deposits and skarn theory written for professional geologists. The extensive bibliography is useful to those interested in learning more about specific types of skarn and skarn localities.

Fettes, Douglas, and Jacqueline Desmons, eds. *Metamorphic Rocks: A Classification and Glossary of Terms*. New York: Cambridge University Press, 2007. Discusses the classification of metamorphic rocks. Feldspars are mentioned throughout in reference to the various types of metamorphic rocks and basic classifications.

Fyfe, W. S., N. J. Price, and A. B. Thompson. *Fluids in the Earth's Crust*. New York: Elsevier, 1978. The best overview of the theoretical side of fluid-rock interaction. Argues that subduction and seafloor spreading may be viewed as large-scale metasomatic processes, the end product of which is the earth's crust. Written for advanced students of geology or chemistry. Extensive bibliography.

Grotzinger, John, et al. *Understanding Earth*. 5th ed. New York: W. H. Freeman, 2006. Introduces the high school and college reader without a geology background to the subject of metamorphism and metasomatism. Metasomatism is treated in very general terms. Good index and illustrations.

LeBas, Michael John. *Carbonatite-Nephelinite Volcanism: An African Case History*. New York: John Wiley & Sons, 1977. A detailed account of the geology in the Lake Victoria region of East Africa, containing excellent descriptions of volcanoes and calderas formed by recent alkalic and carbonatitic magmas. Fenites are discussed throughout in great detail. Can be understood by those with a modest background in geology, provided that they first familiarize themselves with the nomenclature of alkalic rocks.

Mason, Roger. *Petrology of the Metamorphic Rocks*. 2d ed. Berlin: Springer, 2011. A very good college-level text dealing with major aspects of metamorphism. Chapter 5 is devoted to contact metamorphism and metasomatism. Excellent glossary and index.

Oldershaw, Cally. *Rocks and Minerals*. New York: DK, 1999. Useful to new students who may be unfamiliar with the rock and mineral types discussed in classes or textbooks.

Sigurdsson, Haraldur, ed. *Encyclopedia of Volcanoes*. San Diego, Calif.: Academic Press, 2000. Contains a complete summary of the scientific knowledge of volcanoes. Also contains eighty-two well-illustrated overview articles, each of which is accompanied by a glossary of key terms. Written in a clear and comprehensive style that makes it generally accessible. Cross-references and index.

Taylor, Roger G. *Geology of Tin Deposits*. New York: Elsevier, 1979. An overview of tin deposits. Chapter 6 deals authoritatively with metasomatism and its application to exploration for tin. Requires some background in geology and chemistry.

See also: Blueschists; Contact Metamorphism; Metamorphic Rock Classification; Metamorphic Textures; Pelitic Schists; Regional Metamorphism; Sub-Seafloor Metamorphism.

MINERALS: PHYSICAL PROPERTIES

Many minerals are readily identified by their physical properties, but identification of other minerals may require instruments designed to examine details of their chemical composition or crystal structure. The characteristic physical properties of some minerals, such as hardness, malleability, and ductility, make them commercially useful.

PRINCIPAL TERMS

- **cleavage:** the tendency for minerals to break in smooth, flat planes along zones of weaker bonds in their crystal structure
- **crystal:** a solid bounded by smooth planar surfaces that are the outward expression of the internal arrangement of atoms; crystal faces on a mineral result from precipitation in a favorable environment
- **density:** in an informal sense, the relative weight of mineral samples of equal size; it is defined as mass per unit volume
- **luminescence:** the emission of light by a mineral
- **luster:** the reflectivity of the mineral surface; there are two major categories of luster: metallic and nonmetallic
- **Mohs hardness scale:** a series of ten minerals arranged in order of increasing hardness, with talc as the softest mineral known (1) and diamond as the hardest (10)
- **tenacity:** the resistance of a mineral to bending, breakage, crushing, or tearing

COLOR, STREAK, AND LUSTER

Minerals have diagnostic physical properties resulting from their chemistry and crystal structure. Physical properties of minerals include color, streak, luster, crystal shape, cleavage, fracture, hardness, and density, or specific gravity. Several minerals have additional diagnostic physical properties, including tenacity, magnetism, luminescence, and radioactivity. Some minerals, notably halite (common table salt), are easily identified by their taste. Other minerals such as calcite effervesce or fizz when they come into contact with hydrochloric acid.

Color is an obvious physical property, but it is one of the least diagnostic for mineral identification. In some minerals, color results from the presence of major elements in the chemical formula; in these minerals, color is a diagnostic property. For example, malachite is always green, azurite is always blue, and rhodochrosite and rhodonite are always pink. In other minerals, color is the result of trace amounts of chemical impurities or defects in the crystal lattice structure. Depending on the impurities, a particular species of mineral can have many different colors. For example, pure quartz is colorless, but quartz may be white (milky quartz), pink (rose quartz), purple (amethyst), yellow (citrine), brown (smoky quartz), green, blue, or black.

Streak is the color of the mineral in powdered form. Streak is more definitive than mineral color, because although a mineral may have several color varieties, the streak will be much more consistent in color. Streak is best viewed after rubbing the mineral across an unglazed porcelain tile. The tile has a hardness of approximately 7, so minerals with a hardness of greater than 7 will not leave a streak, although their powdered color may be studied by crushing a small piece. This property is useful because it effectively reveals the color of very thin pieces of the mineral, which tends to be much more uniform than the color of large pieces.

Luster refers to the reflectivity of the mineral surface. There are two major categories of luster: metallic and nonmetallic. Metallic minerals include metals (such as native copper and gold), as well as many metal sulfides, such as pyrite and galena. Nonmetallic lusters can be described as vitreous or glassy (characteristic of quartz and olivine), resinous (resembling resin or amber, characteristic of sulfur and some samples of sphalerite), adamantine or brilliant (diamond), greasy (appearing as if covered by a thin film of oil, including nepheline and some samples of massive quartz), silky (in minerals with parallel fibers, such as malachite, chrysotile asbestos, or fibrous gypsum), pearly (similar to an iridescent pearl-like shell, such as talc), and earthy or dull (as in clays).

CRYSTAL SHAPE, CLEAVAGE, AND FRACTURE

Crystal shape is the outward expression of the internal three-dimensional arrangement of atoms in the crystal lattice. Crystals are formed in a cooling or evaporating fluid as atoms begin to slow down, move closer, and bond together in a particular geometric lattice. If minerals are unconfined and free to grow, they will form well-shaped, regular crystals. Conversely, if growing minerals are confined by other, surrounding minerals, they may have irregular shapes. Some of the common shapes or growth habits of crystals include acicular (or needlelike, as in natrolite), bladed (elongated and flat like a knife blade, as in kyanite), blocky (equidimensional and cube-like, such as galena and fluorite), and columnar or prismatic (elongated or pencil-like, such as quartz and tourmaline). Other crystal shapes are described as pyramidal, stubby, tabular, barrel-shaped, or capillary.

Cleavage is the tendency for minerals to break in smooth, flat planes along zones of weaker bonds in their crystal structure. Cleavage is one of the most important physical properties in identifying minerals because it is so closely related to the internal crystal structure. Cleavage is best developed in minerals that have particularly weak chemical bonds in a given direction. In other minerals, differences in bond strength are less pronounced, so cleavage is less well developed. Some minerals have no planes of weakness in their crystal structure; they lack cleavage and do not break along planes. Cleavage can occur in one direction (as in the micas, muscovite, and biotite) or in more than one direction. The number and orientation of the cleavage planes are always the same for a particular mineral. For example, orthoclase feldspar has two directions of cleavage at right angles to each other.

Fracture is irregular breakage that is not controlled by planes of weakness in minerals. Conchoidal fracture is a smooth, curved breakage surface, commonly marked by fine concentric lines, resembling the surface of a shell. Conchoidal fracture is common in broken glass and quartz. Fibrous or splintery fracture occurs in asbestos and sometimes in gypsum. Hackly facture is jagged with sharp edges and occurs in native copper. Uneven or irregular fracture produces rough, irregular breakage surfaces.

HARDNESS, DENSITY, AND SPECIFIC GRAVITY

Hardness is the resistance of a mineral to scratching or abrasion. Hardness is a result of crystal structure; the stronger the bonding forces between the atoms, the harder the mineral. The Mohs hardness scale, devised by German mineralogist Friedrich Mohs in 1822, is a series of ten

minerals arranged in order of increasing hardness. The minerals on the Mohs hardness scale are talc (the softest mineral known), gypsum, calcite, fluorite, apatite, potassium feldspar (orthoclase), quartz, topaz, corundum, and diamond (the hardest mineral known). A mineral with higher hardness number can scratch any mineral of equal or lower hardness number. The relative hardness of a mineral is easily tested using a number of common materials, including the fingernail (a little over 2), a copper coin (about 3), a steel nail or pocket knife (a little over 5), a piece of glass (about 5.5), and a steel file (6.5).

Density is defined as mass per unit volume (typically measured in terms of grams per cubic centimeter). In a very informal sense, density refers to the relative weight of samples of equal size. Quartz has a density of 2.6 grams per cubic centimeter, whereas a "heavy" mineral such as galena has a density of 7.4 grams per cubic centimeter (about three times as heavy). Specific gravity (or relative density) is the ratio of the weight of a substance to the weight of an equal volume of water at 4 degrees Celsius.

The terms "density" and "specific gravity" are sometimes used interchangeably, but density requires the use of units of measure (such as grams per cubic centimeter), whereas specific gravity is without unit. Specific gravity is an important aid in mineral identification, particularly when studying valuable minerals or gemstones, which might be damaged by other tests of physical properties. The specific gravity of a mineral depends on the chemical composition (type and weight of atoms) as well as the manner in which the atoms are packed together.

TENACITY

Tenacity is the resistance of a mineral to bending, breakage, crushing, or tearing. A mineral may be brittle (breaks or powders easily), malleable (may be hammered out into thin sheets), ductile (may be drawn out into a thin wire), sectile (may be cut into thin shavings with a knife), flexible (bendable, and stays bent), and elastic (bendable, but returns to its original form). Minerals with ionic bonding, such as halite, tend to be brittle. Malleability, ductility, and sectility are diagnostic of minerals with metallic bonding, such as gold. Chlorite and talc are flexible, and muscovite is elastic.

MAGNETISM AND LUMINESCENCE

Magnetism causes minerals to be attracted to magnets. Magnetite and pyrrhotite are the only common magnetic minerals; they are called ferromagnetic. Lodestone, a type of magnetite, is a natural magnet. When in a strong magnetic field, some minerals become weakly magnetic and are attracted to the magnet; these minerals are called paramagnetic. Examples of paramagnetic minerals include garnet, biotite, and tourmaline. Other minerals are repelled by a magnetic field and are called diamagnetic minerals. Examples of diamagnetic minerals include gypsum, halite, and quartz.

Luminescence is the term for emission of light by a mineral. Minerals that glow or luminesce in ultraviolet light, X rays, or cathode rays are fluorescent minerals. The glow is the result of the mineral changing invisible radiation to visible light, which happens when the radiation is absorbed by the crystal lattice and then reemitted by the mineral at lower energy and longer wavelength. Fluorescence occurs in some specimens of a mineral but not all. Examples of minerals that may fluoresce include fluorite, calcite, diamond, scheelite, willemite, and scapolite. Some minerals will continue to glow or emit light after the radiation source is turned off; these minerals are phosphorescent. Some minerals glow when heated, a property called thermoluminescence, present in some specimens of fluorite, calcite, apatite, scapolite, lepidolite, and feldspar. Other minerals luminesce when crushed, scratched, or rubbed, a property called triboluminescence, present in some specimens of sphalerite, corundum, fluorite, and lepidolite and, less commonly, in feldspar and calcite.

RADIOACTIVITY AND ELECTRICAL CHARACTERISTICS

Radioactive minerals contain unstable elements that change spontaneously to other kinds of elements, releasing subatomic particles and energy. Some elements come in several different forms, differing by the number of neutrons present in the nucleus. These different forms are called isotopes, and one isotope of an element may be unstable (radioactive), whereas another isotope may be stable (not radioactive). Radioactive isotopes include potassium-40, rubidium-87, thorium-232, uranium-235, and uranium-238. Examples of radioactive minerals include uraninite and thorianite.

Some minerals also have interesting electrical characteristics. Quartz is a piezoelectric mineral, meaning that when squeezed, it produces electrical charges. Conversely, if an electrical charge is applied to a quartz crystal, it will change shape and vibrate as internal stresses develop. The oscillation of quartz is the basis for its use in digital quartz watches.

STUDY TECHNIQUES

In many cases, the physical properties of minerals can be studied using relatively common, inexpensive tools. The relative hardness of a mineral may be determined by attempting to scratch one mineral with another, thereby bracketing the unknown mineral's hardness between that of other minerals on the Mohs hardness scale. Streak may be determined by rubbing the mineral across an unglazed porcelain tile to observe the color (and sometimes odor) of the streak, if any. Color and luster are determined simply by observing the mineral.

The angles between adjacent crystal faces may be measured using a goniometer. There are several types of goniometers, but the simplest is a protractor with a pivoting bar, which is held against a large crystal so that the angles between faces can be measured. There are also reflecting goniometers, which operate by measuring the angles between light beams reflected from crystal faces.

Density can be determined by measuring the mass of a mineral and determining its volume (perhaps by measuring the amount of water it displaces in a graduated cylinder), then dividing these two measurements. Specific gravity (SG) is density relative to water. It is usually determined by first weighing the mineral in air and then weighing it while it is immersed in water. When immersed in water, it weighs less because it is buoyed up by a force equivalent to the weight of the water displaced. A Jolly balance, which works by stretching a spiral spring, can measure specific gravity. For tiny mineral specimens weighing less than 25 milligrams, a torsion balance (or Berman balance) is useful for accurate determinations. Heavy liquids, such as bromoform and methylene iodide, are also used to determine the specific gravity of small mineral grains. The mineral grain is placed into the heavy liquid and then acetone is added to the liquid until the mineral grain neither floats nor sinks (that is, until the specific gravity of the mineral and the liquid are the same). Then, the specific gravity of the liquid is determined using a special balance called a Westphal balance, which uses a calibrated float to measure the density of the liquid.

An ultraviolet light source (with both long and short wavelengths) is used to determine whether minerals are fluorescent or phosphorescent. A portable ultraviolet light can be used to prospect for fluorescent minerals. Thermoluminescence can be triggered by heating a mineral to 50 to 100 degrees Celsius. Radioactivity is measured using a Geiger counter or scintillometer.

COMMERCIAL APPLICATIONS

The physical properties of minerals affect their usefulness for commercial applications. Minerals with great hardness—such as diamonds, corundum (sapphire and ruby), garnets, and quartz—are useful as abrasives and in cutting and drilling equipment. Other minerals are useful because of their softness, such as calcite, which is used in cleansers because it will not scratch the surface being cleaned. Also, calcite in the form of marble is commonly used for sculpture because it is relatively soft and easy to carve. Talc (hardness 1) is used in talcum powder because of its softness.

Most metals, such as copper, are ductile, which makes them useful for the manufacture of wire. Copper is one of the best electrical conductors. Copper is also a good conductor of heat and is often used in cookware. Gold is the most malleable and ductile mineral. Because of its malleability, gold can be hammered into sheets so thin that 300,000 of them would be required to make a stack 1 inch high. Because of its ductility, 1 gram of gold (about the weight of a raisin) can be drawn into a wire more than a mile and a half long. Gold is the best conductor of heat and electricity known, but it is generally too expensive to use as a conductor; however, it is widely used in computers because of its corrosion resistance and because it takes only a small amount of gold to wire a computer connection.

Other minerals are valuable because they do not conduct heat or electricity. They are used as electrical insulators or for products subjected to high temperatures. For example, kyanite, andalusite, and sillimanite are used in the manufacture of spark plugs and other high-temperature porcelains. Muscovite is also useful because of its electrical and heat-insulating properties; sheets of muscovite are often used as an insulating material in electrical devices.

Cleavage is the property responsible for the use of graphite as a dry lubricant and in pencils. Graphite has perfect cleavage in one direction, and is slippery because microscopic sheets of graphite slide easily over one another. A "lead" pencil is actually a mixture of graphite and clay; it writes by leaving tiny cleavage flakes of graphite on the paper.

The color of the streak (or crushed powder) of many minerals makes them valuable as pigments. Hematite has a red streak and is used in paints and cosmetics. Silver is essential as a raw material for photographic films and papers because in the form of silver halide, it is light-sensitive and turns black. After developing and fixing, metallic silver remains on the film to form the negative. No other element is as useful for photography, and at one time there was real concern about the supply of photographic silver. Digital photography has greatly relieved fears of silver shortages.

The uranium-bearing minerals (uraninite, carnotite, torbernite, and autunite) are used as sources of uranium, which is important because its nucleus is susceptible to fission (splitting or radioactive disintegration), producing tremendous amounts of energy. This energy is used in nuclear power plants for generating electricity. Pitchblende, a variety of uraninite, is a source of radium, which is used as a source of radioactivity in industry and medicine. The high specific gravity of barite makes it a useful additive to drilling muds to prevent oil well gushers or blowouts. It is also opaque to X rays and is used in medicine for "barium milkshakes" before patients are X-rayed, so that the digestive tract will show up clearly.

Pamela J. W. Gore

FURTHER READING

Blackburn, W. H., and W. H. Dennen. *Principles of Mineralogy.* 2d ed. Dubuque, Iowa: Wm. C. Brown, 1993. Divided into three parts. The first part is theoretical and includes crystallography and crystal chemistry, along with a section on the mineralogy of major types of rocks. The second part is practical and includes chapters on physical properties of minerals, crystal geometry, optical properties, and methods of analysis. The third part contains systematic mineral descriptions. Designed for an introductory college course in mineralogy, but should be useful for amateurs as well.

Bloss, F. D. *An Introduction to the Methods of Optical Crystallography*. New York: Holt, Rinehart and Winston, 1961. For the advanced student of mineralogy who is interested in the ways in which the crystal structure of a mineral changes the characteristics of a beam of light passing through it, as studied with the petrographic microscope. May be of interest to persons with a background in physics or geology.

Cepeda, Joseph C. *Introduction to Minerals and Rocks*. New York: Macmillan, 1994. Provides a good introduction to the structure of minerals for students just beginning their studies in Earth science. Includes illustrations and maps.

Chesterman, C. W., and K. E. Lowe. *The Audubon Society Field Guide to North American Rocks and Minerals*. New York: Alfred A. Knopf, 1988. Contains 702 color photographs of minerals, grouped by color, as well as nearly one hundred color photographs of rocks. Descriptive information follows all mineral photographs, with minerals grouped by chemistry. Lists distinctive features and physical properties for each of the minerals, and provides information on collecting localities. Includes a glossary. Suitable for the layperson.

Deer, William A., R. A. Howie, and J. Zussman. *An Introduction to Rock-Forming Minerals*. 2d ed. London: Pearson Education Limited, 1992. A standard reference on mineralogy for advanced college students and above. Each chapter contains detailed descriptions of chemistry and crystal structure, usually with chemical analyses. Discussions of chemical variations in minerals are extensive.

Desautels, P. E. *The Mineral Kingdom*. New York: Madison Square Press, 1972. An oversize, lavishly illustrated coffee-table book with useful text supplementing the color photographs. Covers how minerals are formed, found, and used, and includes legends about minerals and gems as well as scientific data. Provides a broad introduction to the field of mineralogy. Makes fascinating reading for the professional geologist.

Frye, Keith. *Mineral Science: An Introductory Survey*. New York: Macmillan, 1993. Intended for the college-level reader, the text provides an easily understood overview of mineralogy, petrology, and geochemistry, including descriptions of specific minerals. Illustrations, bibliography, and index.

_____. *Modern Mineralogy*. Englewood Cliffs, N.J.: Prentice-Hall, 1974. Addresses minerals from a chemical standpoint and includes chapters on crystal chemistry, structure, symmetry, physical properties, radiant energy and crystalline matter, the phase rule, and mineral genesis. Designed as an advanced college textbook for a student with some familiarity with mineralogy. Includes short descriptions of minerals in a table in the appendix.

Hammond, Christopher. *The Basics of Crystallography and Diffraction*. 2d ed. New York: Oxford University Press, 2001. Covers crystal form, atomic structure, physical properties of minerals, and X-ray methods. Illustrations help to clarify some of the more mathematically complex concepts. Includes bibliography and index.

Haussühl, Siegfried. *Physical Properties of Crystals: An Introduction*. Weinheim: Wiley-VCH, 2007. Begins with foundational information, and discusses tensors. Contains some mathematical equations. Best suited for advanced students, professionals, and academics.

Kerr, P. F. *Optical Mineralogy*. New York: McGraw-Hill, 1977. Designed to instruct advanced mineralogy students in the study and identification of minerals using a petrographic microscope. The first part concerns the basic principles of optical mineralogy, and the second part details the optical properties of a long list of minerals.

Klein, C., and C. S. Hurlbut, Jr. *Manual of Mineralogy*. 23d ed. New York: John Wiley & Sons, 2008. One of a series of revisions of the original mineralogy textbook written by James D. Dana in 1848. Discusses crystallography and crystal chemistry, with shorter chapters on the physical and optical properties of minerals. Provides a classification and detailed, systematic description of various types of minerals, with sections on gem minerals and mineral associations. Considered to be the premier mineralogy textbook for college-level geology students; many parts will be useful for amateurs.

Lima-de-Faria, José. *Structural Minerology: An Introduction*. Dordrecht: Kluwer, 1994. Provides a good college-level introduction to the basic concepts of crystal structure and the classification of minerals. Illustrations, extensive bibliography, index, and a table of minerals on a folded leaf.

Pellant, Chris. *Smithsonian Handbooks: Rocks and Minerals*. New York: Dorling Kindersley, 2002. An excellent resource for identifying minerals and rocks. Contains colorful images and diagrams, a glossary and index.

Prinz, Martin, George Harlow, and Joseph Peters. *Simon & Schuster's Guide to Rocks and Minerals.* New York: Simon & Schuster, 1978. Fully illustrated with color photographs of 276 minerals. Provides background information on physical properties, environment of formation, occurrences, and uses of each mineral. A sixty-page introduction to minerals provides sophisticated technical coverage that will be of interest to both mineralogy students and amateurs. Includes a glossary.

Pough, F. H. *A Field Guide to Rocks and Minerals.* 5th ed. Boston: Houghton Mifflin, 1998. This well-written and well-illustrated book is suitable for readers of nearly any age and background. One of the most readable and accessible sources, it provides a fairly complete coverage of the minerals. Designed for amateurs.

Zussman, J., ed. *Physical Methods in Determinative Mineralogy.* 2d ed. New York: Academic Press, 1978. A reference book that describes technical methods used in the study of rocks and minerals, including transmitted and reflected light microscopy, electron microscopy, X-ray fluorescence spectroscopy, X-ray diffraction, electron microprobe microanalysis, and atomic absorption spectroscopy. Written for the professional geologist or advanced student.

See also: Biopyriboles; Carbonates; Clays and Clay Minerals; Feldspars; Gem Minerals; Hydrothermal Mineralization; Metamictization; Minerals: Structure; Non-silicates; Nuclear Waste Disposal; Oxides; Radioactive Minerals.

MINERALS: STRUCTURE

The discovery of the internal structures of minerals by the use of X-ray diffraction was pivotal in the history of mineralogy and crystallography. X-ray analysis revealed that the physical properties and chemical behavior of minerals are directly related to the highly organized arrangements of their atoms, and this knowledge has had important scientific and industrial applications.

PRINCIPAL TERMS

- **cleavage:** the capacity of crystals to split readily in certain directions
- **crystal:** externally, a solid material of regular form bounded by flat surfaces called faces; internally, a substance whose orderly structure results from a periodic three-dimensional arrangement of atoms
- **ion:** an electrically charged atom or group of atoms
- **ionic bond:** the strong electrical forces holding together positively and negatively charged atoms
- **mineral:** a naturally formed inorganic substance with characteristic physical properties, a definite chemical composition, and, in most cases, a regular crystal structure
- **X ray:** radiation that can be interpreted in terms of either very short electromagnetic waves or highly energetic photons (light particles)

THE FATHER OF CRYSTALLOGRAPHY

During the seventeenth and eighteenth centuries, natural philosophers used two basic ideas to explain the external structures of minerals: particles, in the form of spheres, ellipsoids, or various polyhedra; and an innate attractive force, the emanating "glue" needed to hold particles together. These attempts to rationalize mineral structures still left a basic problem unanswered. Namely, how does one explain the heterogeneous physical and chemical properties of minerals with homogeneous particles? This problem was not answered satisfactorily until the twentieth century, but the modern understanding grew out of the work of scientists in the eighteenth and nineteenth centuries. The most important of these scientists was René-Just Haüy, often called the "father of crystallography."

Haüy, a priest who worked at the Museum of Natural History in Paris, helped to make crystallography a science. Before Haüy, the study of crystals had been a part of biology, geology, or chemistry; after Haüy, the science of crystals was an independent discipline. His speculations on the nature of the crystalline state began when he accidentally dropped a calcite specimen, which shattered into fragments. He noticed that the fragments split along straight planes that met at constant angles. No matter what the shape of the original piece of calcite, he found that broken fragments were rhombohedra (slanted cubes). He reasoned that a rhombohedron, similar to the ones he obtained by cleaving the crystal, must be in the fundamental shape of the crystal. For Haüy, then, the cleavage planes existed in the crystal like the mortar joints in a brick wall. When he discovered similar types of cleavage in a variety of substances, he proposed that all crystal forms could be constructed from submicroscopic building blocks. He showed that there were several basic building blocks, which he called primitive forms or "integral molecules," and that they represented the last term in the mechanical division of a crystal. With these uniform polyhedra, he could rationalize the many mineral forms observed in nature. Haüy's building block was not, however, the same as what later crystallographers came to call the unit cell, the smallest group of atoms in a mineral that can be repeated in three directions to form a crystal. The unit cell is not a physically separable entity, such as a molecule; it simply describes the repeat pattern of the structure. In contrast, for Haüy, the crystal was a periodic arrangement of equal molecular polyhedra, each of which might have an independent existence.

GEOMETRICAL AND SYMMETRICAL ANALYSES

In the nineteenth century, Haüy's ideas had many perceptive critics. For example, Eilhardt Mitscherlich, a German chemist, discovered in 1819 that different mineral substances could have the same crystal form, whereas Haüy insisted that each substance had a specific crystal structure. Some crystallographers shunned the concrete study of crystals (leaving it to mineralogists), and they defined their science as the study of ordered space. This mathematical analysis bore fruit, for crystallographers were able to show that, despite the great variety of possible mineral structures, all forms could be classified into six crystal systems on the basis of certain geometrical features, usually axes. The cube is the basis of one of these systems, the isometric, in which three identical axes intersect at right angles. Symmetry was another factor in describing these crystal systems. For example, a cube has fourfold symmetry around an axis passing at right angles through the center of any of its faces. As some crystallographers were establishing the symmetry relationships in crystal systems, others were working on a way to describe the position of crystal faces. In 1839, William H. Miller, a professor of mineralogy at the University of Cambridge, found a way of describing how faces were oriented about a crystal, similar to the way a navigator uses latitude and longitude to tell where his ship is on the earth. Using numbers derived from axial proportions, Miller was able to characterize the position of any crystal face.

Friedrich Mohs, a German mineralogist best remembered for his scale of the hardness of minerals, was famous in his lifetime for his system of mineral classification, in which he divided minerals into genera and species, similar to the way biologists organized living things. His system was based on geometrical relationships that he derived from natural mineral forms. He wanted to transform crystallography into a purely geometrical science, and he showed that crystal analysis involved establishing certain symmetrical groups of points by the rotation of axes. When these point groups were enclosed by plane surfaces, crystal forms were generated. The crystallographer's task, then, was to analyze the symmetry operations characterizing the various classes of a crystal system.

BRAVAIS LATTICES

Beginning in 1848, Auguste Bravais, a French physicist, took the same sort of mathematical approach in a series of papers dealing with the kinds of geometric figures formed by the regular grouping of points in space, called lattices. Bravais applied the results of his geometric analysis to crystals, with the points interpreted either as the centers of gravity of the chemical molecules or as the poles of interatomic electrical forces. With this approach, he demonstrated that there is a maximum of fourteen kinds of lattices, which differ in symmetry and geometry, such that the environment around any one point is the same as that around every other point. These fourteen Bravais lattices are distributed among the six crystal systems. For example, the three isometric

Bravais lattices are the simple (with points at the vertices of a cube), body-centered (with points at the corners, along with a point at the center of a cube), and face-centered (with corner points and points at the centers of the faces of the cube). With the work of Bravais, the external symmetry of a mineral became firmly grounded on the idea of the space lattice, but just how actual atoms or molecules were arranged within unit cells remained a matter of speculation.

In the latter part of the nineteenth century, various European scientists independently advanced crystallography beyond point groups and Bravais lattices by recognizing additional forms of symmetry. In the late 1880's, the Russian mineralogist Evgraf Fedorov introduced the glide plane, in which a reflection in a mirror plane is combined with a translation without rotation along an axis. Bricks in a wall are a familiar example (although they have additional symmetry as well). Using various symmetry elements, Fedorov derived the 230 space groups, which represent all possible distributions that atoms can assume in minerals.

PAULING'S RULES

Shortly after the work of Fedorov, William Barlow, an English chemist, began to consider the problem of crystal symmetry from a more concrete point of view. He visualized crystals not in terms of points but in terms of closely packed spherical atoms with characteristic diameters. In considering atoms to be specifically sized spheres, he found that there are certain geometric arrangements for packing them efficiently. One can appreciate his insight by thinking about arranging coins in two dimensions. For example, six quarters will fit around a central quarter, but only five quarters will fit around a dime. Barlow showed that similar constraints hold for the three-dimensional packing of spherical atoms of different diameters.

As scientists determined more and more mineral structures, they became convinced that minerals are basically composed of spherical atoms or ions, each of characteristic size, packed closely together. Silicate minerals were of central concern to William Lawrence Bragg in England and Linus Pauling in the United States. The basic unit in these minerals is the tetrahedral arrangement of four oxygen atoms around a central silicon atom. Each tetrahedral unit has four negative charges, so one would expect that electric repulsion would force these tetrahedral building blocks to fly apart. In actual silicate minerals, however, these units are linked in chains, rings, or sheets and in ways that bring about charge neutralization and stability. These tetrahedra may also be held together by such positively charged metal ions as aluminum, magnesium, and iron. These constraints lead to a fascinating series of structures. Pauling devised an enlightening and useful way of thinking both about these silicate structures and about complex inorganic substances in general. In the late 1920's, he proposed a set of principles (now known as Pauling's Rules) that govern the structures of ionic crystals—that is, crystals in which ionic bonding predominates. The silicate minerals provide striking examples of his principles. One of his rules deals with how a positive ion's electrical influence is spread among neighboring negative ions; another rule states that highly charged positive ions tend to be as far apart as possible in a structure. Pauling's Rules allowed him to explain why certain silicate minerals exist in nature and why others do not.

TEMPERATURE AND PRESSURE STUDIES

In the 1960's the structure determination of minerals became an important activity in some large geology departments. By this time, through computerized X-ray crystallography, it was possible to determine, quickly and elegantly, the exact atomic positions of highly complex minerals. In the 1970's and 1980's, scientific interest shifted to the study of minerals at elevated temperatures and pressures. These studies often showed that temperature and pressure changes cause complex internal structural modifications in the mineral, including shifts in distances between certain ions and their orientation to others. New minerals continue to be discovered and their structures determined. Structural chemistry has played an important role in deepening understanding both of these new minerals and of old minerals under stressful conditions. This knowledge of mineral structure has benefited not only mineralogists, crystallographers, and structural chemists but also inorganic chemists, solid-state physicists, and many Earth scientists.

STUDY OF EXTERNAL STRUCTURE

The first methods for examining the external structure of minerals were quite primitive. In the seventeenth century, Nicolaus Steno cut sections from crystals and traced their outlines on paper. A century later, Arnould Carangeot invented the contact goniometer. This device, which enabled crystallographers to make systematic measurements of interfacial crystal angles, was basically a flat, pivoted metal arm with a pointer that could move over a semicircular protractor. William Hyde Wollaston invented a more precise instrument, the reflecting goniometer, in 1809. This device used a narrow beam of light reflected from a mirror and directed against a crystal to make very accurate measurements of the angles between crystal faces. The reflecting goniometer ushered in a period of quantitative mineralogy that led to the multiplication of vast amounts of information about the external structure of minerals.

The discovery of the polarization of light in the nineteenth century led to another method of mineral investigation. Ordinary light consists of electromagnetic waves oscillating in all directions at right angles to the direction of travel, but a suitable material can split such light into two rays, each vibrating in a single direction (this light is then said to be plane polarized). Various inventors perfected the polarizing microscope, a versatile instrument using plane-polarized light to identify minerals and to study their fine structure. Even the darkest minerals could be made transparent if sliced thinly enough. These transparent slices produced complex but characteristic colors because of absorption and interference when polarized light passed through them.

STUDY OF INTERNAL STRUCTUREINT

In the twentieth century, X-ray diffraction provided scientists with a tool vastly more powerful than anything previously available for the investigation of internal mineral structures. Before the development of X-ray methods in 1912, the internal structure of a mineral could be deduced only by reasoning from its physical and chemical properties. After X-ray analysis, the determination of the detailed internal structures of minerals moved from speculation to precise measurement. The phenomenon of diffraction had been known since the seventeenth century. It can be readily observed when a distant street light is viewed through the regularly spaced threads of a nylon umbrella, causing colored fringes around the light source to be seen. In a similar way, Max von Laue reasoned that the closely spaced sheets of atoms in a crystal should diffract X rays, with closely spaced sheets diffracting X rays at larger angles than more widely spaced ones. William Lawrence Bragg then showed how this technique could be used to provide detailed information about the atomic structure of minerals. The play of colors on a compact disk is a familiar example of the diffraction of visible light. If a laser pointer is aimed at the disk, multiple spots will appear on nearby surfaces because the regular spacing of tiny pits on the disk causes diffracted light to be reinforced in certain directions. X-ray diffraction is exactly the same phenomena applied to the arrays of atoms in a crystal.

The powder method of X-ray diffraction consists of grinding a mineral specimen into a powder that is then formed into a rod by gluing it to a thin glass fiber. As X rays impinge on it, this rod is rotated in the center of a cylindrical photographic film. The diffraction pattern on the film can then be interpreted in terms of the arrangement of atoms in the mineral's unit cell.

Although X rays have been the most important type of radiation used in determining mineral structures, other types of radiation—in particular infrared (with wavelengths greater than those of visible light)—have also been effective. Infrared radiation causes vibrational changes in the ions or molecules of a particular mineral structure, which permits scientists to map its very detailed atomic arrangement. The technique of neutron diffraction makes use of relatively slow neutrons from reactors to determine the locations of the light elements in mineral structures (the efficiency of light elements in scattering neutrons is generally quite high).

In recent decades, scientists have continued to develop sophisticated techniques for exploring the structure of minerals. Each of these methods has its strengths and limitations. For example, the electron microprobe employs a high-energy beam of electrons to study the microstructure of minerals. This technique can be used to study very small amounts of minerals as well as minerals in situ, but the strong interaction between the electron beam and the crystalline material produces anomalous intensities, and thus electron-microprobe studies are seldom used for a complete structure determination. Many new techniques have helped scientists to perform structural studies of minerals in special states—for example, at high pressures or temperatures near the melting point—but the most substantial advancements in determining mineral structures continue to involve X-ray analysis.

SCIENTIFIC AND ECONOMIC APPLICATIONS

A central theme of modern mineralogy has been the dependence of a mineral's external form and basic properties on its internal structure. Because the arrangement of atoms in a mineral provides a deeper understanding of its mechanical, thermal, optical, electrical, and chemical properties, scientists have determined the atomic arrangements of many hundreds of minerals by using the X-ray diffraction technique. This great amount of structural information has proved to be extremely valuable to mineralogists, geologists, physicists, and chemists. Through this information, mineralogists have gained an understanding of the forces that hold minerals together, and have even used crystal-structure data to verify and correct the formulas of some minerals. Geologists have been able to use the knowledge of mineral structures at high temperatures and pressures to gain a better understanding of the eruption of volcanoes and other geologic processes. Physicists have used this structural information to deepen their knowledge of the solid state. Through crystal-structure data, chemists have been able to expand their understanding of the chemical bond, the structures of molecules, and the chemical behavior of a variety of substances.

Because minerals often have economic importance, many people besides scientists have been interested in their structures. Rocks, bricks, concrete, plaster, ceramics, and many other materials contain minerals. In fact, almost all solids except glass and organic materials are crystalline. That is why knowledge of the structure and behavior of crystals is important in nearly all industrial, technical, and scientific enterprises. This knowledge has, in turn, enabled scientists to synthesize crystalline compounds to fill special needs, such as high-temperature ceramics, electrical insulators, semiconductors, and many other materials.

Robert J. Paradowski

FURTHER READING

Blake, Alexander, J., and Jacqueline M. Cole. *Crystal Structure Analysis: Principles and Practices.* 2d ed. New York: Oxford University Press, 2009. Written for advanced undergraduate and graduate students. Covers diffraction, crystal structures, data collection, and crystal-structure determination and analysis.

Bragg, William Lawrence. *Atomic Structure of Minerals.* Ithaca, N.Y.: Cornell University Press, 1937. Primarily a discussion of mineralogy based on the vast amount of new data generated by the successful application of X-ray diffraction analysis to crystalline minerals. Text is highly readable, but the reader's knowledge of elementary physics and chemistry is assumed. Useful to mineralogists, physicists, chemists, and any scientists interested in the physical and chemical properties of minerals.

Bragg, William Lawrence, G. F. Claringbull, and W. H. Taylor. *The Crystalline State.* The Crystal Structures of Minerals 4. Ithaca, N.Y.: Cornell University Press, 1965. A comprehensive compilation of crystal-structure information on minerals. Analyses of structures are authoritative. Accessible to anyone with a basic knowledge of minerals, as crystallographic notation is kept to a minimum and the actual structures take center stage.

Cepeda, Joseph C. *Introduction to Minerals and Rocks.* New York: Macmillan, 1994. Provides a good introduction to the structure of minerals for students just beginning their studies in Earth science. Includes illustrations and maps.

Deer, William A., R. A. Howie, and J. Zussman. *An Introduction to Rock-Forming Minerals.* 2d ed. London: Pearson Education Limited, 1992. Standard references on mineralogy for advanced college students and above. Each chapter contains detailed descriptions of chemistry and crystal structure, usually with chemical analyses. Discussions of chemical variations in minerals are extensive.

Evans, Robert Crispin. *An Introduction to Crystal Chemistry.* 2d ed. New York: Cambridge University Press, 1964. Analyzes crystal structures in terms of their correlation with physical and chemical properties. Discusses only those structures that are capable of illustrating basic principles that govern the behavior of these crystals. Assumes some knowledge of elementary chemistry and physics on the part of the reader.

Ferraris, Giovanni, Emil Makovicky, and Stefano Merlino. *Crystallography of Modular Materials.* New York: Oxford University Press, 2004. Contains advanced discussions of crystal structure. Includes discussion of OD structures, polytypes, and modularity. Contains a long list of references.

Hammond, Christopher. *The Basics of Crystallography and Diffraction.* 2d ed. New York: Oxford University Press, 2001. Covers crystal form, atomic structure, physical properties of minerals, and X-ray methods. Illustrations help clarify some of the more mathematically complex concepts. Includes bibliography and index.

Haussühl, Siegfried. *Physical Properties of Crystals: An Introduction.* Weinheim: Wiley-VCH, 2007. Begins with foundational information and discusses tensors. Text contains some mathematical equations and is best suited for advanced students, professionals, and academics.

Lima-de-Faria, José. *Structural Mineralogy: An Introduction.* Dordrecht: Kluwer, 1994. Provides a good college-level introduction to the basic concepts of crystal structure and the classification of minerals. Illustrations, extensive bibliography, index, and a table of minerals on a folded leaf.

Lipson, Henry S. *Crystals and X-Rays.* New York: Springer-Verlag, 1970. Written for high school students and college undergraduates. Stresses the observational and experimental by showing, for example, how the X-ray diffraction technique was used to determine the structures of some simple minerals.

Pauling, Linus. *The Nature of the Chemical Bond and the Structure of Molecules and Crystals: An Introduction to Modern Structural Chemistry.* 3d ed. Ithaca, N.Y.: Cornell University Press, 1960. The beginner will encounter difficulties, but readers with a good knowledge of chemistry will find this book informative and inspiring.

Sinkankas, John. *Mineralogy for Amateurs.* New York: Van Nostrand Reinhold, 1966. Intended primarily for the amateur mineralogist. Has become popular with nonprofessionals and includes a good chapter on the geometry of crystals, in which the basic ideas of mineral structure are cogently explained. Well illustrated with photographs and drawings.

Smith, David G., ed. *The Cambridge Encyclopedia of Earth Sciences.* New York: Crown, 1981. Written by authorities from England and the United States. Primarily a reference work, the text is both readable and informative. Some knowledge of elementary physics and chemistry is needed for a full understanding of most sections. Profusely illustrated with helpful diagrams and photographs.

Tilley, Richard J. D. *Crystals and Crystal Structures.* Hoboken, N.J.: John Wiley & Sons, 2006. Discusses the characteristics, structure, and chemical dynamics of crystals. Provides information on specific crystals and specific structure types. Includes appendices, practice problems and exercises, bibliographies, and indexes.

See also: Biopyriboles; Carbonates; Clays and Clay Minerals; Diamonds; Earth Resources; Feldspars; Gem Minerals; Gold and Silver; High-Pressure Minerals; Hydrothermal Mineralization; Industrial Metals; Industrial Nonmetals; Metamictization; Minerals: Physical Properties; Non-silicates; Oxides; Radioactive Minerals.

MINING PROCESSES

Mining techniques are methods whereby ore is extracted from the earth. Surface mining techniques are employed when ore is found close to the earth's surface. Underground methods are employed whenever valuable solid minerals occur at depths too far beneath the land surface to be recovered at a profit using other techniques.

PRINCIPAL TERMS

- **dragline:** a large excavating machine that casts a rope-hung bucket, collects the excavated material by dragging the bucket toward itself, elevates the bucket, and dumps the material on a spoil bank, or pile
- **grade:** the classification of an ore according to either material content or value
- **level:** all connected horizontal mine openings at a given elevation; generally, levels are 30 to 60 meters apart and designated by their vertical distance below the top of the shaft
- **ore:** any rock or material that can be mined at a profit
- **overburden:** the material overlying the ore in a surface mine
- **panel:** an area of underground coal excavation for production rather than development; the coal mine equivalent of a stope
- **pillar:** ore, coal, rock, or waste left in place underground to support the wall or roof of a mined opening
- **raise:** a vertical or steeply inclined excavation of narrow dimensions that connects subsurface levels; unlike a winze, it is bored upward rather than sunk
- **scraper:** a digging, hauling, and grading machine that has a cutting edge, a carrying bowl, a movable front wall, and a dumping or ejecting mechanism
- **shaft or winze:** a vertical or steeply inclined excavation of narrow dimensions; shafts are sunk from the surface, and winzes are sunk from one subsurface level to another
- **stope:** an underground excavation to remove ore, as opposed to development; the outlines of a stope are determined either by the limits of the ore body or by raises and levels
- **vein:** a well-defined, tabular mineralized zone that may or may not contain ore bodies

SURFACE MINING METHODS

There are eleven varieties of surface-mining methods: open-pit, mountaintop removal, conventional contour, box-cut contour, longwall strip, multiseam, block-cut, area, block-area, multiseam scraper, and terrace mining. Open-pit mining involves removing overburden and ore in a series of benches from near surface to pit floor. Expansion can occur outward and downward from the initial dig-in point. Many metallic deposits and industrial minerals are mined by this method; one example is the Bingham Canyon mine in Utah. The chief disadvantage of this method is the near impossibility of reclaiming large pits, such as at Bingham, where nearly all the extracted material has been processed or transported away.

Mountaintop removal mining was popular among mining companies, if not among environmentalists, in eastern Kentucky. It involves removing an entire mountaintop, dumping the overburden in valleys on the first cut, and leaving a fairly flat floor upon which to reclaim overburden from succeeding cuts. A post-mining surface of gently rolling terrain will replace the rugged premine surface. The disadvantages of this method include limited valley fill material, increased capital costs for additional equipment, and a need for extensive mine planning, so as to ensure maximum productivity at minimum cost. It is, however, more effective than is contour mining.

In conventional contour mining, waste material is dumped downslope from the active pit, often resulting in toxic material resting on native ground and causing erosion, poor vegetative reestablishment, and potential acid-mine drainage. Mining continues into the hillside until the volume of waste rock makes further mining uneconomical. This type of mining results in a long strip-like bench running around the hillside, and has been discouraged by stricter environmental laws in Kentucky and West Virginia. It is generally favored by smaller operators because expenses are lower and premine and reclamation planning efforts are reduced, but the method disturbs much more acreage than other steep-slope methods. A method similar to conventional contour mining, but with somewhat improved environmental impacts, is box-cut contour mining, in which a boxlike initial pit is dug to receive waste from later mining. Spoil segregation and terrace regrading help to inhibit toxic down slope material and aid revegetation. The low wall of the box-cut helps to support most of the spoil.

Longwall strip mining has been attempted in some parts of West Virginia. This innovative method involves removing a small pit by conventional stripping methods, followed by the setup of a continuous miner (or excavator) and a conveyor. Although production costs are higher, land may be more easily restored. This mining method can also help to prevent acid-mine drainage when geology is favorable.

Multiseam mining is done in many locations throughout the world. Frequently, a bottom and top coal seam are separated by about 2 to 6 meters of rock or more. The distance means that this "interburden" material also must be removed, and often drilled and blasted as well. Yet, even though the overall ratio of waste rock to coal may be higher, it may be beneficial and economical to extract both seams at once. Permitting and assessment costs are lower as well. Block-cut mining is also suitable in areas of moderate to steep topography. A single box- or block-cut is excavated and mining progresses outward from the initial cut in two directions. Environmental disturbances are minimized as the first cut is regraded and revegetated during mining of subsequent cuts.

Large areas of flat or gently rolling terrain are ideal for area mining, in which draglines or large stripping shovels remove overburden. Succeeding pits are dug normally or parallel to the strata-bound ore body. This method is the most common type of mining attempted in the major lignite and subbituminous coal fields of western North Dakota, Montana, and northeastern Wyoming.

Block-area mining is similar in many respects to both area and block-cut methods. It is designed to recover seams as thin as 0.3 meter. Capital requirements are less than in some methods, and overburden removal can be sequenced with reclamation regrading.

Multiseam scraper mining involves uncovering large blocks of coal (up to 122 meters wide by 305 meters long) by scraper. After the coal is removed, scrapers again remove the parting between succeeding seams until all the coal is mined. This method is ideal for the coal and thin parting sequences found in the upper Midwest.

Terrace mining involves overburden and ore removal in a series of benches, or terraces. These benches may follow natural features, as in stream or lake terraces. This type of mining has been effective in the recovery of diamonds, sand, and gravel.

COSTS OF SURFACE MINING

Surface mining comes with a price and is, in many cases, far from easy. Surface mines are frequently located far from populated areas. In arctic regions, operations may be inhibited for much of the year by unfavorable weather. Some surface mines experience difficult mining conditions year-round.

Surface mining has left ugly scars on the landscape in many developing countries and in some parts of the eastern United States, which has prompted tougher environmental legislation. Since the 1950's in the United States, toxic materials have been specially handled, and surface and groundwater problems are being seriously addressed.

UNDERGROUND MINING METHODS

When the valuable mineral to be extracted occurs at a depth too great to allow profitable surface mining, underground mining methods are used. The depth at which profitability decreases depends on the value of the mineral deposit to be mined. Thus, some metallic minerals have been surface (open-pit) mined to depths that equal or exceed the deepest underground coal mines. A number of physical factors influence the selection of underground mining methods, including size, shape, continuity, and depth of the mineral deposit; range and pattern of ore quality; strength, hardness, and structural characteristics of the deposit and the surrounding and overlying rock; groundwater conditions; subsurface temperatures (some mines are so hot they must be air-conditioned); local topography and climate; and environmental protection considerations. If solution mining (flooding the pit) is being considered, the chemical composition of the ore mineral will be an important consideration as well. Additionally, technologic and economic factors must be considered, such as availability of workers and worker safety; availability of water, power, and transportation; and, most important, a market for the mineral.

Once a decision has been made to develop an underground mining operation, a method is chosen from one of the following four categories: a method involving predominantly self-supporting openings, a method predominantly dependent on artificial supports, a caving method, or a solution mining method. In solution mining, men do not work underground to mine the valuable mineral. Instead, wells are drilled and liquid flows or is pumped down the wells and through the deposit to dissolve the mineral. Later, the mineral is recovered by processing the pregnant solution (liquid containing the dissolved mineral) after it has been pumped to the surface. Copper and uranium have both been recovered this way.

SELF-SUPPORTING METHODS

Methods involving predominantly self-supporting openings include open stoping, sublevel stoping, shrinkage stoping, and room-and-pillar mining. Prior to production of the mineral in marketable quantities, the mine must be developed. Development work begins with gaining access to the mineral deposit by sinking a shaft, driving a horizontal tunnel-like excavation (adit) into a hillside, or driving a sloping tunnel-like excavation (decline) from the surface. Once accessed, portions of the sought-after deposit are blocked out by bounding them with a three-dimensional network of horizontal tunnel-like excavations (levels) and vertical shaft-like excavations (raises or winzes). Each of these blocked-out areas defines a stope to be mined by the stoping method selected for the particular mine.

Open stoping is any mining method in which a stope is created by the removal of a valuable mineral without the use of timber, or other artificial supports, as the predominant means of supporting the overburden. Open stopes include both isolated single openings, from which pockets of ore have been extracted, and pillared open stopes, in which a deposit of considerable lateral extent has been mined with ore pillars left for support. In sublevel stoping, large blocks of ore between the levels and raises bounding the stope are partitioned into a series of smaller slices or blocks by driving sublevels. The ore is drilled and blasted by miners in the sublevels in a sequence that causes the sublevels to retreat *en echelon* (in step formation), with each successively lower sublevel being mined slightly ahead of the one immediately above it. This process allows all of the broken ore to fall directly to the bottom of the stope, where it passes through funnel-like openings (called draw points) to the haulage level below for transport out of the mine.

In shrinkage stoping, the ore in the stope is mined from the bottom upward, with the broken ore allowed to accumulate above the draw points to provide a floor from which blasting operations can be conducted. Periodically, the broken ore is drawn down from below (shrunk) to maintain the proper headroom for blasting. Room-and-pillar mining is a form of pillared open stoping employed where the mineral deposit is relatively thin, flat-lying, and of great lateral extent. It is employed most commonly in the mining of coal but is also used to mine other nonmetals, such as rock salt, potash, trona, and limestone.

ARTIFICIALLY SUPPORTED METHODS

Stull stoping, cut-and-fill stoping, square-set stoping, longwall mining, and top slicing are among the methods predominantly dependent on artificial support. Stull stoping is a method that can be considered transitionary between those methods involving predominantly self-supporting openings and those predominantly involving artificially supported openings. It is used chiefly in narrow, steeply sloping (dipping) vein deposits. Timbers (called stulls) are placed for support between the lower side wall (called the footwall) and the upper side wall (called the hanging wall) of the vein.

The cut-and-fill stoping method, like shrinkage stoping, has a working-level floor of broken rock; however, in this case, the broken rock is waste rather than ore. Instead of draw points, heavily timbered ore chutes are constructed through the waste rock to pass the broken ore to the haulage level for transport from the mine. The solid ore is fragmented by blasting so that it falls on top of the waste rock floor, where it is collected and dumped down the ore chutes. Waste rock, often from development work elsewhere in the mine, is placed in the stope to build up the floor and maintain the desired headroom for blasting.

Square-set stoping is a method employing extensive artificial support. "Square sets" (interlocking cube-shaped wooden frames of approximately 2.5 to 3 meters on a side) are constructed to form a network that resembles monkey bars on a playground to provide support for the stope. As mining progresses, the square sets are filled immediately with waste rock—except for those sets kept open for ventilation, ore passes, or passageways for miners (called manways). In the longwall method, a massive system of props or hydraulic supports is used to support the roof over a relatively long, continuous exposure of solid mineral. This method allows for virtually complete extraction, and the overlying roof behind the line of support collapses into the mined void as the mining process advances. A method similar to longwall is top slicing, in which ore is extracted in horizontal timbered slices, starting at the top of the ore deposit and working downward. A timbered mat is placed in the first cut, and the overburden is caved. As subsequent cuts advance, caving is induced by blasting out props (timbers) behind the face, while working room is continuously maintained under the mat. Top slicing differs from longwall mining in that several levels are developed *en echelon* rather than progressing on a single level.

CAVING METHODS

Two caving methods are commonly used in underground mining: sublevel caving and block caving. In sublevel caving, work begins in the uppermost sublevel of the stope and progresses downward. As the ore is blasted and collapses onto the sublevel floor for removal, the wall rock of the stope immediately caves behind the ore. The broken ore is removed through the sublevels. In block caving operations, the ore body is induced to cave downward because of gravity for the entire height of the stope (usually more than 30 meters). The process is initiated by undercutting the block of ore and allowing it to collapse. The collapsed ore is removed through draw points at the bottom of the stope.

ADAPTATION TO OTHER USES

The relevance of mining methods to society is not limited simply to the production of useful and valuable minerals. The ideas and concepts developed for underground mining have been applied and adapted to other uses as well. Techniques used in underground mining are used in the construction industry, especially on projects involving underground openings for transportation, electric power generation, fuel storage, and military purposes. The construction of highway, railroad, and subway tunnels can be cited as examples of transportation applications. Large hydroelectric dams and similar projects utilize rock tunnels and chambers to channel the falling water

used to turn the turbines that generate the electricity. Underground caverns created through underground mining technology are being utilized for petroleum and liquefied natural gas storage, and mine-like underground excavations in rock are also being investigated for the deep underground storage or disposal of nuclear wastes. Military applications include underground chambers for housing intercontinental ballistic missiles, and nuclear weapon-proof command facilities. One of the earliest military applications of mining was tunneling under enemy fortifications, either to collapse walls ("sapping") or to place explosives. On June, 7 1917, the British set off 455 tons of explosives under German lines at Messines, Belgium, killing nearly 10,000 German troops, one of the largest non-nuclear explosions ever.

Paul S. Maywood and Dermot M. Winters

FURTHER READING

CAFTA DR and U.S. Country EIA. *EIA Technical Review Guideline: Non-metal and Metal Mining.* Washington, D.C.: U.S. Environmental Protection Agency, 2011. Discusses alternative mining methods, mineral processing, facilities, and environmental impact, as well as assessment methods and mitigation strategies.

Chironis, N. P., ed. *Coal Age Operating Handbook of Coal Surface Mining and Reclamation.* New York: McGraw-Hill, 1978. A nontechnical introduction to surface coal mining techniques. Well illustrated, and the photographic reproductions are of high quality. Includes many site-specific case studies.

Crickmer, D. F., and D. A. Zegeer, eds. *Elements of Practical Coal Mining.* 2d ed. New York: Society of Mining Engineers of the American Institute of Mining, Metallurgical, and Petroleum Engineers, 1981. Intended to be a beginners' book and to give the reader an overview of the fundamentals of coal mining. Appropriate for use in vocational schools, high schools, community colleges, and libraries, and by persons in the coal mining industry.

Culver, David C., and William B. White, eds. *Encyclopedia of Caves.* Academic Press, 2004. Covers a range of cave-related topics, such as saltpeter mining, cave types, cave ecology, archaeology, and hydrogeology.

Freese, Barbara. *Coal: A Human History.* Cambridge, Mass.: Penguin, 2004. Provides a history of coal usage throughout the world. Includes information on early usage, the Industrial Revolution, mining in the past and present, and current environmental concerns.

Gayer, Rodney A., and Jierai Peesek, eds. *European Coal Geology and Technology*. London: Geological Society, 1997. Examines the technologies used in European coal mines and the coal mining industry. Illustrations, maps, index, and bibliographical references.

Ghose, Ajoy K., ed. *Mining on a Small and Medium Scale: A Global Perspective*. London: Intermediate Technology, 1997. Collates the proceedings of the Global Conference on Small and Medium Scale Mining held in Calcutta, India, in 1996. Focuses on the social and environmental aspects of mining in developing countries.

Institution of Mining and Metallurgy. *Surface Mining and Quarrying*. Brookfield, Vt.: IMM/North American Publications Center, 1983. Contains thirty-eight papers on a variety of topics related to surface mining. Includes illustrations and some technical discussions. Appropriate for college-level students.

Peters, W. C. *Exploration and Mining Geology*. 2d ed. New York: John Wiley & Sons, 1987. Provides a short, but thorough, summary of surface and underground mining methods. The remainder of the book provides an overview of the geologist's work in mineral discovery and mineral production.

Rygle, Kathy J., and Stephen F. Pedersen. *The Treasure Hunter's Gem and Mineral Guides to the U.S.A.* 5th ed. Woodstock, Vt.: GemStone Press, 2008. Provides information on locating and collecting gems, precious stones, and minerals from all regions of the United States. Discusses mining techniques, equipment, and safety. Lists locations by state, with information on the minerals found there. Offers useful information throughout on techniques, tips, and trivia.

Stefanko, Robert. *Coal Mining Technology Theory and Practice*. New York: Society of Mining Engineers, 1983. Covers underground coal mining, but also offers sketches illustrating the basics behind different types of surface mining. The language is for the most part nontechnical. Suitable for high school-level readers.

Stout, K. S. *Mining Methods and Equipment*. Chicago: Maclean-Hunter, 1989. Profusely illustrated with sketches, drawings, and photographs. Provides a thorough introduction to the fundamental concepts of mining and their applications in producing ore and coal. Suitable for an entry-level course in mining engineering and for persons with more than a casual interest in mining. Includes a number of high-quality illustrations, which make the technical details less daunting.

Thomas, L. J. *An Introduction to Mining: Exploration, Feasibility, Extraction, Rock Mechanics*. Rev. ed. Sydney: Methuen of Australia, 1978. An introductory text written for students of mining and mining engineering. Covers the entire mining process, from initial exploration to mine operation. Provides lucid descriptions of the principal underground mining methods.

U.S. Department of Labor, Mine Safety and Health Administration. *Coal Mining*. Washington, D.C.: Government Printing Office, 1997. Examines the state of coal mines and mining in the United States. Focuses on the health risks associated with coal mining. Illustrations and maps.

U.S. Environmental Protection Agency. *Profile of the Non-metal, Non-fuel Mining Industry*. Washington, D.C.: U.S. Environmental Protection Agency, 1995. Provides a broad scope of topics related to mining nonmetals. Discusses economic trends, nonmetal processing, and waste release regulations. The U.S. EPA Web site offers more recent data by specific industry sectors, including the Toxic Release Inventory and compliance and enforcement data.

See also: Clays and Clay Minerals; Earth Resources; Earth Science and the Environment; Hazardous Wastes; Landfills; Land-Use Planning; Mining Wastes; Nuclear Waste Disposal; Strategic Resources.

MINING WASTES

Mining has produced vast areas of disturbed land and disrupted ecosystems. The magnitude of extraction activities, combined with inadequate responses by mining companies, has inspired reclamation laws and research.

PRINCIPAL TERMS

- **ecological succession:** the process of plant and animal changes, from simple pioneers such as grasses to stable, mature species such as shrubs or trees
- **ecosystem:** a self-regulating, natural community of plants and animals interacting with one another and their nonliving environment
- **erosion:** the movement of soil and rock by natural agents such as water, wind, and ice, including chemicals carried away in solution
- **landforms:** surface features formed by natural forces or human activity, normally classified as constructional, erosional, or depositional
- **orphan lands:** unreclaimed strip mines created prior to the passage of state or federal reclamation laws
- **reclamation:** all human efforts to improve conditions produced by mining wastes—mainly slope reshaping, revegetation, and erosion control
- **topsoil:** in reclamation, all soil that will support plant growth, but normally the 20 to 30 centimeters of the organically rich top layer

BY-PRODUCTS OF MINING

Over the years, humankind has found more and more uses for the minerals within the earth and has been able to develop deposits of lower and lower economic content. Miners remove 15 meters or more of overburden (soil and rock) to obtain 30 centimeters of coal; some copper mines process ores present in concentrations as low as 0.5 percent. These endeavors produce literally mountains of waste, referred to as spoil. All human exploitation of Earth resources produces waste as a by-product, but none produces more than the mining of coal. Another inevitable by-product is ecosystem disruption—the alteration of plant and animal habitat—and changes in hydrology, which includes all aspects of water. Reclamation is the effort to heal the altered environment through landform modification, revegetation, and erosion control.

Until the mid-twentieth century, a "use it" ethic of land use was dominant. Nature was all too often seen as an enemy to be conquered. Since World War II, however, and with the rise of environmental concerns, the United States has recognized the need to restore these disturbed lands. Nevertheless, coal strip mining has increased with the growing demand for energy. Between the 1930's and early 1970's, electrical demand doubled every decade. When the 1973 energy crisis hit, the huge deposits of western coal were seen as one pathway to energy independence. Additionally, the abrupt halt in orders for new nuclear power plants meant that almost all new electrical power plants since 1973 have been fueled by coal.

GROWING CONCERN FOR THE ENVIRONMENT

Reclamation efforts in the past have been slight due to issues of apathy, cost, and the relatively small areas involved. Current efforts are directly linked to the growth of conservation and environmental movements in response to population growth and the development of available lands. In the United States, population and economic growth following World War II resulted in pollution conditions that shocked the nation. The first large-scale environmental conference was held at Princeton University in June 1955. Entitled "The International Symposium on Man's Role in Changing the Face of the Earth," it set the agenda for rising environmental concerns across the nation. Stewart Udall, then secretary of the interior, made an instrumental contribution in the 1967 publication *Surface Mining and Our Environment: A Special Report to the Nation*, a clear call to confront the growing problem of unreclaimed coal strip mines (described in the text as "orphan" lands).

The pivotal piece of legislation regarding mining is the Mining Law of 1872, which created liberal and inexpensive conditions for establishing mining claims on federal lands in order to promote western metal mining. Opponents of mining regulation have resisted all efforts to amend the law, so mining laws since can be mostly described as nibbling around the edges (so to speak) of the 1872 law. Issues emerge in attempting to regulate resources, such as coal, not covered in the law and mining on land not covered by the 1872 law, and so on. Since coal generates such a large proportion of mining waste, the ability to regulate coal mining has allowed the federal government to control some of mining's worst problems.

The creation of the Environmental Protection Agency in 1970 led the way to what came to be called "the environmental decade." The Surface Mining Control and Reclamation Act of 1977, though covering only coal mining, was landmark legislation. It requires revegetating and restoring the land to its approximate original contour; bans the mining of prime agricultural land in the West and in locations where owners of surface rights object; requires operators to minimize the impact on local watersheds and water quality; establishes a fee on each ton of coal to help reclaim orphan lands; and delegates enforcement responsibility to the states, except where they fail to act or where federal land is involved. Major efforts are underway to correct the land disruptions associated with coal mining, including research that may apply to other mining operations. Although mining has become more expensive, reclamation gives substantial hope for restoring the land.

WATER QUALITY AND HYDRAULIC IMPACTS

Important research on water quality dates from the 1950's and was focused on acid mine-drainage problems in Appalachia, where it has long been recognized as a problem. Acid is produced in any type of earth movement where iron pyrite and other sulfides are removed from a reducing environment to an oxidizing one. Within the earth these sulfides are stable, but when exposed to oxygen and water they readily oxidize, producing sulfuric acid, the main ingredient in acid-mine drainage. Coal strip mines can produce acid when the sulfide-rich layers above and below the coal (often called fire clay) are scattered on the surface or left exposed in the final cut. A major issue of reclamation is the need to identify these acid-producing materials and bury them. Acids speed up the release of dangerous metals such as aluminum and manganese, which can be toxic to vegetation; they also cause deformities in fish.

Hydrologic impacts mainly involve changes in runoff and sediment. A major study was conducted by the U.S. Geological Survey in Beaver Creek basin, Kentucky. This area was selected because mining operations were just beginning there under typical Appalachian conditions. Data collected from 1955 to 1963 revealed large increases in runoff and sediment. By contrast, another study in Indiana demonstrated the tremendous water-holding capacities of the disturbed lands. The surface topography had much to do with the different results because a more level condition existed in Indiana. Another study was conducted at a basin in northeastern Oklahoma while the area was being subjected to both contour and area strip mining. After initial increases in runoff and erosion, huge decreases occurred as much of the drainage became internalized. When reclamation began under a new state law, however, erosion rates comparable to unprotected construction sites were measured (up to 13 percent sediment by weight). Clearly, slopes lacking a protective vegetation cover will produce much more runoff, which in turn will carry enormous sediment loads. Nevertheless, in several orphan lands, water and sediment are trapped internally because of the drainage obstacles and depressions created by the mining operation.

USES OF MINING WASTE LANDFORMS

Early research on mining wastes involved description of the new landforms created by human activity. The landforms produced by strip mining that are still visible in orphan lands and have yet to be reclaimed are spoil banks, final-cut canyons (which are often filled with ponds), headwalls (above the final cuts), and transport roads. Mining waste landforms have some interesting uses. Perhaps most important is recreation, especially fishing in final-cut ponds. Many abandoned quarries are used by scuba divers for training. In some North Dakota mines, natural revegetation has formed refuges for deer and other wildlife. One site in northeastern Oklahoma, given to the local Boy Scouts, now supports a dense population of trees and birds and is a wonderful small wilderness. Other uses include forestry in the East and sanitary landfills near urban areas. Most softwoods, such as pine, prefer more acidic soils, which are common in eastern strip mines. Use in sanitary landfills is aided by the already disturbed condition of the land and by the fact that the clays and shales commonly found in coal strip mines provide relatively impermeable conditions. One oft-neglected potential use of mined lands is education, as mine cuts can provide access to rock outcrops and geologic features not otherwise accessible.

The linkage between mining waste landforms, vegetation/soil conditions, and erosion was recognized early. One study in Great Britain found that even in successfully revegetated strip mines, erosion rates were 50 to 200 times those of undisturbed areas. Revegetation is the key to reclamation; it, in turn, is heavily dependent upon the quality of the soil. Because of the diverse nature of soil and local ecosystems, a specific reclamation plan is never made easily.

EFFECTIVE RECLAMATION PLANNING

Effective reclamation plans require sufficient data on soil pH (a measure of acidity or alkalinity), soil chemistry for fertility and toxicity problems, water retention capability, soil organisms, and useful native plants, especially perennials. The best way to provide the necessary soil conditions is to save and rapidly replace the topsoil and establish a vegetation cover quickly in order to protect the soil from rain and gully erosion. Even thin layers of good topsoil are a major improvement over the deeper subsoils, which tend to be low in organic matter and rich in acid-producing sulfides.

Detailed planning is now an integral part of the reclamation process. When planning is completed before mining begins, some of the impact can be minimized and reclamation can proceed in concert with mining, shortening the interval before replanting. Since erosion is highly destructive to revegetation efforts, erosion control systems are vital. They are designed to direct the flow of water and to dissipate its energy, enhancing soil moisture for vegetation.

Also vital to a good reclamation plan is a thorough understanding of local ecosystems and the potential success or failure of restoration efforts. Revegetation is mainly an exercise in ecological succession. Once an ecosystem is disrupted, it is difficult to reestablish. In some cases, it may be wiser to pursue a different course. Because ecological succession is slow, fifty years or more may be needed to gauge success or failure accurately.

NEED FOR CONSERVATION AND RECYCLING

The need for conservation and recycling is supported by the ecological axiom that, in nature, matter cycles but energy flows; by recycling matter, large energy savings result from the reuse of materials such as steel, aluminum, and glass. Because energy flows, and therefore cannot be recycled, conservation is the best source of additional energy. These savings reduce the pressure to disturb more land for energy-producing coal. A key factor in the environmental impact of mining is the adaptability of natural ecosystems. How much stress can they tolerate, and what kind of conditions can be expected from the consequent ecological succession? Ecosystems are remarkably resilient and adaptable. Still, the pace of human activity is so rapid and the recovery of ecosystems so slow that the human race would be wise to err on the side of conservation.

Questions are currently being raised about the impact of coal burning on potential atmospheric heating. Because coal is nearly pure carbon and because carbon dioxide (produced when carbon is burned) is a major greenhouse gas (one that absorbs and thus traps heat radiating from the earth's surface), the world's increasing dependence on coal for electrical power production might lead to climatic changes that could affect agriculture. Revegetation efforts can help in this area because growing plants remove carbon dioxide from the atmosphere. One suggested replacement for coal, nuclear power, has its own set of mining waste problems, along with a need for long-term storage of radioactive wastes.

Two opposing land-use viewpoints provide a framework for discussion: the historic "use it" ethic and the growing "preserve it" ethic. Many prefer a middle road of scientific conservation or sustainable Earth practices, both of which include reclamation. A political perspective raises the issues of tolerance levels and affordability. The government has a responsibility to protect the health and welfare of its citizens. Where clear dangers exist, the government has a responsibility to act. Many pollution issues, however, are controversial. How much is too much, and at what level does the cost of abatement become so high as to be unaffordable? Humanity cannot use Earth's resources without also causing pollution, and there are limits to what can be done to abate it. Still, public support remains strong for pollution control, cleanup of past pollution problems, and reclamation of disturbed land.

Nathan H. Meleen

FURTHER READING

Abandoned Mines and Mining Waste. Sacramento: California Environmental Protection Agency, 1996. Focuses on the environmental and social aspects of mines that are no longer operating. Accompanied by an easy-to-read fact sheet provided by the U.S. Department of Health and Human Services.

Burger, Ana, and Slavko V. Šolar. "Mining Waste of Non-metal Pits and Querries in Slovenia." In *Waste Management, Environmental Geotechnology and Global Sustainable Development.* Ljubljana, Slovenia: Geological Survey of Slovenia International Conference, 2007. Discusses mining processes and the waste material produced. Evaluates various types of mines and materials mined in Slovenia, and outlines methods of reducing mining waste.

CAFTA DR and U.S. Country EIA. *EIA Technical Review Guideline: Non-metal and Metal Mining.* Washington, D.C.: U.S. Environmental Protection Agency, 2011. Discusses alternative mining methods, mineral processing, facilities, and environmental impact. Reviews assessment methods and mitigation strategies.

Collier, C. R., et al., eds. *Influences of Strip Mining on the Hydrologic Environment of Parts of Beaver Creek Basin, Kentucky, 1955-66.* U.S. Geological Survey Professional Paper 427-C. Washington, D.C.: Government Printing Office, 1970. One of the first basin-wide studies over an extended period of time. Attempts to cover all aspects of hydrologic impacts. A fundamental reference for any serious student of strip mining. Parts 427-A (basin description) and 427-B (covering 1955-59) were published in 1964.

Dalverny, Louis E. *Pyrite Leaching from Coal and Coal Waste.* Pittsburgh, Pa.: U.S. Department of Energy, 1996. Examines coal-mining policies and protocol, as well as the state of disposal procedures used by coal-mine operators. Illustrations and bibliography.

Freese, Barbara. *Coal: A Human History.* Cambridge, Mass.: Penguin, 2004. Provides a history of coal usage throughout the world. Includes information on early usage, the Industrial Revolution, mining in the past and present, and current environmental concerns.

Hanson, David T. *Waste Land: Meditations on a Ravaged Landscape.* New York: Aperture, 1997. This photo essay shows hazardous waste sites and mines, and allows the reader to see the effects of mining on the environment.

Law, Dennis L. *Mined-Land Rehabilitation.* New York: Van Nostrand Reinhold, 1984. A heavily illustrated hardback with an attractive layout and print style. Provides extensive coverage of reclamation issues. Extensively footnoted and contains useful references and tables. Covers such subjects as mining procedures, environmental impacts, surface rehabilitation, revegetation, and planning.

Lottermoser, Bernd. *Mine Wastes.* 2d ed. New York: Springer, 2010. Discusses the environmental impacts of mining. Provides information on specific resources mined and their resulting waste. Includes chapter summaries, references, and indexing.

Miller, G. Tyler, Jr. *Living in the Environment.* 17th ed. Belmont, Calif.: Brooks/Cole, Cengage Learning, 2012. The dominant textbook in environmental science for college students. Suitable for use by any serious high school student or adult. Covers a variety of topics ranging from ecosystems to politics. Contains a useful glossary, extensive bibliography, and index. An excellent primer on ecosystems, land use, and energy. Contains much information about environmental legislation.

Narten, Perry F., et al. *Reclamation of Mined Lands in the Western Coal Region.* U.S. Geological Survey Circular 872. Alexandria, Va.: U.S. Geological Survey, 1983. Offers sixty pages of photos and detailed information about the current status of reclamation research. Well written, informative, and thorough; a must for anyone interested in reclamation, especially revegetation.

Thomas, William L., Jr., ed. *Man's Role in Changing the Face of the Earth.* Chicago: University of Chicago Press, 1956. Proceedings of the international symposium by the same name, with contributions by many well-known scientists of the period. Considers issues for the past, present, and future, concluding with summary remarks by the conference cochairmen. A classic in environmental literature. Topical coverage is extensive. Few illustrations, but a thorough index.

U.S. Department of the Interior. *Surface Mining and Our Environment: A Special Report to the Nation.* Washington, D.C.: Government Printing Office, 1967. A classic for those interested in surface mining. Features vivid photo-essays and striking illustrations. Alerted the nation to the problems of surface mining and appealed for action. Topics include the nature and extent of surface mining, its environmental impact and related problems, past achievements and future goals, existing laws, and recommendations for action.

U.S. Environmental Protection Agency. *Erosion and Sediment Control: Surface Mining in the Eastern U.S.* EPA Technology Transfer Seminar Publication. Washington, D.C.: Government Printing Office, 1976. This two-part paperback manual ("Planning" and "Design") was written for technicians, professionals, and laypersons. Includes a useful glossary but lacks a bibliography. Provides an understanding of the underlying mechanics and rationale of erosion control and basic information on procedures and design. Illustrated with photos and diagrams.

_____. *Profile of the Non-metal, Non-fuel Mining Industry.* Washington, D.C.: U.S. Environmental Protection Agency, 1995. Provides a broad scope of the topics related to mining nonmetals. Discusses economic trends, nonmetal processing, and waste-release regulations. Recent data are available at the U.S. EPA Web site, including the Toxic Release Inventory and compliance and enforcement data.

Vogel, Willis G. *A Guide for Revegetating Coal Minesoils in the Eastern United States.* USDA Forest Service General Technical Report NE-68. Broomall, Pa.: Northeast Forest Experiment Station, 1981. Offers useful information about revegetation under humid conditions. A must for anyone interested in revegetation of mining waste.

See also: Building Stone; Coal; Earth Resources; Earth Science and the Environment; Evaporites; Fertilizers; Hazardous Wastes; Hydrothermal Mineralization; Industrial Metals; Industrial Nonmetals; Iron Deposits; Landfills; Mining Processes; Nuclear Waste Disposal; Oil Shale and Tar Sands; Sedimentary Mineral Deposits; Strategic Resources; Uranium Deposits.

N

NESOSILICATES AND SOROSILICATES

Nesosilicates are a large and diverse group of rock-forming minerals. The group contains a number of minerals of geologic importance, among them olivine, which may be the most abundant mineral of the inner solar system. Nesosilicates have a few special but limited industrial uses and include a number of important gemstones. Sorosilicates are a small group of silicates with related structures.

PRINCIPAL TERMS

- **anion:** an atom that has gained electrons to become a negatively charged ion
- **cation:** an atom that has lost electrons to become a positively charged ion
- **crystal structure:** the regular arrangement of atoms in a crystalline solid
- **igneous rock:** a rock formed by the solidification of molten, or partially molten, rock
- **metamorphic rock:** a rock formed when another rock undergoes changes in mineralogy, chemistry, or structure owing to changes in temperature, pressure, or chemical environment within a planet
- **mineral:** a naturally occurring solid, inorganic compound with a definite composition and an orderly internal arrangement of atoms
- **silicates:** minerals containing both silicon and oxygen, usually in combination with one or more other elements
- **solid solution:** a solid that shows a continuous variation in composition in which two or more elements substitute for each other on the same position in the crystal structure
- **tetrahedron:** a four-sided pyramid made out of equilateral triangles

CHEMICAL STRUCTURE

Nesosilicates are one of the major groups of silicate minerals. Oxygen and silicon are the two most common elements in the rocky outer layers of earthlike inner planets. Thus, it is not surprising that silicates are the most abundant rock-forming minerals on these planets. They are not only very abundant but also very diverse: It has been estimated that almost one-third of the roughly 3,000 known minerals are silicates.

In all but a few of these minerals, each silicon atom is surrounded by a cluster of four oxygen atoms, which are distributed around the silicon in the same way that the corners of a tetrahedron are distributed around its center. Silica tetrahedra can join together by sharing one oxygen at a corner. In this way, two or more silica tetrahedra can link together to form pairs, rings, chains, sheets, and three-dimensional frameworks. This linking provides the basis for the classification of silica structures: Sorosilicates contain pairs of tetrahedra, cyclosilicates contain rings, inosilicates contain chains, phyllosilicates contain sheets, and tectosilicates contain three-dimensional frameworks. In nesosilicates, however, silica tetrahedra do not link to each other. Each silica tetrahedra is isolated from the others as if it were an island, and hence these minerals are often known as nesosilicates, from the Greek word for "island." In the structure of some minerals, isolated silica tetrahedra are mixed with silicate pairs formed when two tetrahedra share an oxygen at a common corner. Minerals with pairs of silica tetrahedra are called sorosilicates.

CHEMICAL BONDING

The electrostatic attraction between negatively charged ions (anions) and positively charged ions (cations) is the basis of the ionic chemical bond. The chemical bonding in nesosilicates is predominantly ionic. Under normal geological conditions, silicon loses four electrons to become a cation with a charge of +4, while oxygen gains two electrons to become an anion with a charge of −2. The ionic bond between oxygen and silicon is usually the strongest bond in nesosilicate structures. For a mineral to be stable it must be electrically neutral; in other words, the total number of negative charges on the anions must be equal to the total number of positive charges on the cations. It would take two oxygen anions to balance the charge on one silicon cation. In the silica tetrahedra, however, the silicon cation is surrounded by four oxygen anions, so in the group as a whole there are four excess negative charges. In the nesosilicate structures this excess negative charge is balanced by the presence of cations outside the silica tetrahedra. It is the bond between the oxygen in the tetrahedra and these other cations that holds the tetrahedra together to form a coherent, three-dimensional structure. The most common cations to play this role are aluminum with a charge of +3, iron with a charge of +3 or +2, calcium with a charge of +2, and magnesium with a charge of +2. Aluminum, iron, and calcium are, respectively, the third, fourth, and fifth most abundant elements in the earth's crust, while magnesium is the seventh most abundant. Conspicuously absent from the common nesosilicate structures are the alkali cations sodium and potassium, which are the sixth and eighth most abundant elements in the earth's crust.

Of all the different kinds of silicates, nesosilicates have the lowest ratio of silicon to oxygen. As a result they often form in environments relatively low in silicon. The atoms in nesosilicates tend to be packed closer together than the atoms in many other silicates, causing them to be somewhat denser. Greater densities are favored at higher pressure; hence a number of the more important minerals that prove stable at high pressures are nesosilicates. Nesosilicates also tend to be harder (resistant to being scratched) than the average silicate. This property helps give many nesosilicates high durability (resistance to wear), which contributes to their use as gemstones. The nesosilicates are a large and diverse group, and a description of more than the most common of them is well beyond the scope of this article.

OLIVINE

The mineral olivine is the most common and widespread of the nesosilicates. Indeed, it is probably the most abundant mineral in the inner solar system. Olivine contains magnesium (Mg) and iron (Fe), in addition to silicon (Si) and oxygen (O). The chemical formula of olivine is generally written as $(Mg,Fe)_2SiO_4$. The parentheses in this formula indicate that olivine is a solid solution; in other words, magnesium and iron can substitute for each other in the olivine structure. Pure magnesium olivines (Mg_2SiO_4) are known as

forsterite, and pure iron olivines (Fe_2SiO_4) are known as fayalite. Most olivines have both magnesium and iron, and hence have compositions that lie between these two extremes. Each iron and magnesium atom is surrounded by six oxygen atoms, while each oxygen atom is bonded to three iron or magnesium atoms and one silicon atom, thereby creating an extended three-dimensional structure. Olivine is usually a green mineral with a glassy luster and a granular shape.

Rocks made up mostly of olivine are known as peridotites. Although peridotites are relatively rare in the earth's crust (that layer that begins at the surface and extends to depths of between 5 and 80 kilometers), this rock makes up most of the earth's uppermost mantle. The olivine-rich layer begins at depths ranging between 5 and 80 kilometers, and extends downward to depths of roughly 400 kilometers. Evidence indicates that the moon and the other inner planets (Mercury, Venus, and Mars) also have similar olivine-rich layers. Olivine can also be an important mineral in basalts and gabbros, which are the most abundant igneous rocks in the crusts of the inner planets. Additionally, it is an abundant mineral in different kinds of meteorites and some kinds of metamorphic rocks.

GARNETS

The garnets are a group of closely related nesosilicate minerals, each of which is a solid solution. The chemistry of common garnets can be fairly well represented by the somewhat idealized general formula $A_3B_2Si_3O_{12}$, where the A stands for either magnesium, iron (with a charge of +2), manganese (Mn), or calcium (Ca), and the B stands for either aluminum (Al), iron (with a charge of +3), or chromium (Cr). In the crystal structures of these minerals, the cations in the A site are surrounded by eight oxygen anions, while the cations in the B site are surrounded by six oxygens. Most garnets can be described as a mixture of two or more of the following molecules: pyrope ($Mg_3Al_2Si_3O_{12}$), almandine ($Fe_3Al_2Si_3O_{12}$), spessartine ($Mn_3Al_2Si_3O_{12}$), grossular ($Ca_3Al_2Si_3O_{12}$), andradite ($Ca_3Fe_2Si_3O_{12}$), and uvarovite ($Ca_3Cr_2Si_3O_{12}$).

The most abundant and widespread garnets are almandine-rich garnets, which form during the metamorphism of some igneous rocks and sediments rich in clay minerals. Grossular-rich and andradite-rich garnets are found in marbles formed through the metamorphism of limestone. The formation of spessartine-rich or uvarovite-rich garnets occurs during the metamorphism of rocks with high concentrations of manganese or chrome. Rocks with these compositions are relatively unusual, and hence these garnets are fairly rare. Pyrope-rich garnets are widespread in the earth's mantle, although they typically do not occur in abundance (more than 5 or 10 percent of the rock). Garnets that have weathered out of other rocks are sometimes found in sands and sandstones. Garnets may also occur in small amounts in some igneous rocks.

ALUMINOSILICATES

Minerals are called polymorphs if they are made up of the same kinds of atoms in the same proportions but in different arrangements. Polymorphs are minerals with the same compositions but different crystal structures. Aluminosilicates are a group of nesosilicate minerals containing three polymorphs: kyanite, sillimanite, and andalusite. Each of these minerals has the chemical formula Al_2SiO_5. The differences in the structures of these minerals are best illustrated by considering the aluminum atoms; in kyanite, all the aluminum atoms are surrounded by six oxygen atoms; in sillimanite, half of the aluminum atoms are surrounded by six oxygens, while the other half are surrounded by four oxygens; in andalusite, half the aluminum atoms are surrounded by four oxygens, and half are surrounded by five oxygens. Kyanite usually forms elongated rectangular crystals with a blue color. Sillimanite typically occurs as white, thin, often fibrous crystals. Andalusite is most commonly found in elongated crystals with a square cross-section and a red to brown color. Aluminosilicates typically form during the metamorphism of clay-rich sediments; such rocks have the relatively high ratios of aluminum to silicon necessary for the formation of these minerals. The identity of the aluminosilicate formed depends upon the temperature and pressure of metamorphism; kyanite forms at relatively high pressures, sillimanite forms at relatively high temperatures, and andalusite forms at low to moderate temperatures and pressures.

TOPAZ, ZIRCON, TITANITE, AND EPIDOTE

Topaz is another aluminum-rich nesosilicate; however, this mineral also contains fluorine (F) and/or the hydroxyl molecule (OH). The chemical formula of topaz is $Al_2SiO_4(F,OH)_2$. As the parentheses indicate, this mineral is a solid solution in which fluorine and hydroxyl can substitute for each other. Topaz is formed during the late stages in the solidification of a granitic liquid.

The mineral zircon contains the relatively rare element zirconium (Zr). Zircon has the chemical formula $ZrSiO_4$; it also generally contains small amounts of uranium and thorium. The decay of these radioactive elements can be used to determine the age of a rock, making zircon particularly important to geologists. It is most commonly found as brown rectangular crystals with a pyramid on either end. It is a widespread mineral in igneous rocks, although it generally occurs in relatively minor amounts.

Titanite ($CaTiSiO_5$), sometimes known as sphene, is one of the most common minerals bearing titanium (Ti). It is a fairly widespread mineral, occurring in many different kinds of igneous and metamorphic rocks, but it is rarely present in abundance.

Epidote ($Ca_2(Al,Fe)Al_2 Si_3O_{12}(OH)$) contains both isolated silica tetrahedra and tetrahedral pairs. Epidote, a fairly common sorosilicate mineral, most typically forms during low-temperature metamorphism in the presence of water. It most often occurs as masses of fine-grained, pistachio-green crystals. A pistachio-green mineral in granite is almost certainly epidote.

ANALYTICAL TECHNIQUES

To characterize a mineral requires both its chemical composition and its crystal structure. There are many different analytical techniques that will give chemical compositions; probably the most popular for silicates is electron microprobe analysis. In this technique, part of the mineral is bombarded by a high-energy beam of electrons, which causes it to give off X rays. Different elements in the mineral give off X rays of different wavelengths, and the intensities of these different X rays depend on the abundance of their elements. By measuring the wavelengths and intensities of the X rays, the composition of the mineral can be obtained. This technique has the advantage of being nondestructive (the mineral is still available and undamaged after the analysis) and applicable to very small spots on a mineral: The typical electron microprobe analysis gives the composition of a volume of mineral only a few tens to hundreds of cubic microns (millionths of a meter) in size.

The crystal structures of silicates are generally obtained using the technique of X-ray diffraction. In this technique an X-ray beam is passed through the mineral. As the beam interacts with the atoms in the mineral, it breaks up into many smaller, diffracted beams traveling in different directions. The intensity and direction of these diffracted beams depend on the positions of atoms in the mineral. By analyzing the diffraction pattern, a scientist can discover the crystal structure of a mineral.

SCIENTIFIC AND ECONOMIC VALUE

Nesosilicates are one of the important building blocks of the planets of the inner solar system, and this reason alone provides an important scientific rationale for studying them. Garnet and aluminosilicates also provide important clues about the temperatures and pressures at which rocks formed. Despite their importance in nature, these minerals have had only a limited technological use. The relatively high hardness of garnet makes it suitable as an abrasive, and it is used in some sandpapers and abrasive-coated cloths. Aluminosilicates are used in the manufacture of a variety of porcelain, which is noted for its high melting point, resistance to shock, and low electrical conductivity. This material is used in spark plugs and brick for high-temperature furnaces and kilns. Zircon is mined in order to obtain zirconium oxide and zirconium metal. Zirconium oxide has one of the highest known melting points and is used in the manufacture of items that have to withstand exceptionally high temperatures. Zirconium metal is used extensively in the construction of nuclear reactors. Titanite is mined as a source of titanium oxide. Titanium oxide has a number of uses but is most familiar as a white pigment in paint.

Probably the most widespread use of nesosilicates is gemstones. Especially fine, transparent crystals of olivine make a beautiful green gem known generally as peridot, although the names chrysolite and evening emerald are sometimes used instead. Relatively transparent garnets also make very beautiful gems. The most common garnets are a deep red, and this is the color usually associated with the stone. Yet gem-quality garnets can also be yellow, yellow-brown, orange-brown, orange-yellow, rose, purple, or green. Garnet is a relatively common mineral and is typically among the least valuable gemstones. The major exception is the green variety of andradite garnet. Known in the gem trade as demantoid, this relatively rare material is one of the more valuable gems. When

properly cut, zircon, a popular gemstone, has a brilliancy (ability to reflect light) and fire (the ability to break white light up into different colors) second only to diamond. Topaz is also widely used as a gem. The most valuable topaz is orange-yellow to orange-brown in color. Unfortunately, all yellow gems are sometimes incorrectly referred to as topaz: When this practice is followed, true topaz is generally known as precious topaz or oriental topaz. Gem-quality topaz may also be colorless, faintly green, pink, red, blue, and brown.

Edward C. Hansen

FURTHER READING

Deer, W. A., R. A. Howie, and J. Zussman. *Rock-Forming Minerals.* 2d ed. 5 vols. London: Pearson Education Limited, 1992. Considered essential for geology students as a resource and laboratory handbook. Organized by mineral group. Appendices include a table of atomic and molecular weights, and other material used for chemical analysis and mineral analysis. Volume 1 is one of the most complete treatments of nesosilicates. An excellent source for detailed information on individual minerals. Most suitable for college-level audiences.

Frye, Keith. *Mineral Science: An Introductory Survey.* New York: Macmillan, 1993. Provides an easily understood overview of mineralogy, petrology, and geochemistry, including descriptions of specific minerals. Illustrations, bibliography, and index.

Gait, R. I. *Exploring Minerals and Crystals.* Toronto: McGraw-Hill Ryerson, 1972. A lower-level introduction to the science of mineralogy. More scientific than the typical rock-hound manual but less detailed than the introductory textbook. Suitable for high school students.

Hammond, Christopher. *The Basics of Crystallography and Diffraction.* 2d ed. New York: Oxford University Press, 2001. Covers crystal form, atomic structure, physical properties of minerals, and X-ray methods. Illustrations help clarify some of the more mathematically complex concepts. Includes bibliography and index.

Hurlbut, Cornelius S., Jr., and Robert C. Kammerling. *Gemology.* 2d ed. New York: John Wiley & Sons, 1991. An introductory textbook for readers interested in gems. Includes an introduction to the chemistry, physics, and mineralogy needed for the study of gems; a discussion of the general technical aspects of gemology; and a description of individual gemstones. Suitable for upper-level high school students.

Klein, Cornelis, and Cornelius S. Hurlbut, Jr. *Manual of Mineralogy*. 23d ed. New York: John Wiley & Sons, 2008. A popular introduction to the scientific study of minerals. Contains a very good sixteen-page section on the nesosilicates. Suitable for upper-level high school students.

Lauf, Robert. *The Collector's Guide to Silicate Crystal Structures*. Atglen, Pa.: Schiffer Publishing, Ltd., 2010. Contains many photos and diagrams. Discusses crystal structures, crystallography, and silicate classifications. Written for the mineral collector. Useful for students of mineralogy.

Lima-de-Faria, José. *Structural Mineralogy: An Introduction*. Dordrecht: Kluwer, 1994. Provides a good college-level introduction to the basic concepts of crystal structure and the classification of minerals. Illustrations, extensive bibliography, index, and a table of minerals on a folded leaf.

Ribbe, P. H., ed. *Orthosilicates: Reviews in Mineralogy*. Vol. 5. Washington, D.C.: Mineralogical Society of America, 1980. Consists of eleven articles, all but one of which specialize in a specific mineral or mineral group. Requires some previous exposure to mineralogy or chemistry. Suitable for college-level students.

Sinkankas, J. *Mineralogy for Amateurs*. New York: Van Nostrand Reinhold, 1964. Contains much more scientific detail than the average book on minerals written for amateurs. Suitable for high school students.

Webster, Robert F. G. A. *Gems: Their Sources, Descriptions, and Identification*. 5th ed. Boston: Butterworth-Heinemann, 1994. A technical college-level text that is an important reference for the gemologist or the senior jeweler. Better for sources and descriptions rather than for identification. Essential for the gemologist.

Wenk, Hans-Rudolf, and Andrei Bulakh. *Minerals: Their Constitution and Origin*. Cambridge, England: Cambridge University Press, 2004. Covers the structure of minerals, physical characteristics, processes, and mineral systematic. Includes multiple chapters devoted to non-silicates, followed by chapters discussing silicates.

See also: Biopyriboles; Carbonates; Clays and Clay Minerals; Feldspars; Gem Minerals; Hydrothermal Mineralization; Metamictization; Minerals: Physical Properties; Minerals: Structure; Non-silicates; Oxides; Radioactive Minerals.

NON-SILICATES

Non-silicate minerals, although not as abundant as the silicates in the part of the earth that is accessible to humankind, are important because they are the major sources of many of the critical elements and compounds upon which civilized society is based.

PRINCIPAL TERMS

- **crystalline:** a property of a chemical compound with an orderly internal atomic arrangement that may or may not have well-developed external faces
- **ion:** an atom that has a positive or negative charge
- **metal:** an element with a metallic luster, high electrical and thermal conductivity, ductility, and malleability
- **mineral:** a naturally occurring, solid chemical compound with a definite composition and an orderly internal atomic arrangement
- **ore:** mineral or minerals present in large enough amounts in a given deposit to be profitably mined for the metal(s)
- **rock-forming mineral:** the common minerals that compose the bulk of the earth's crust (outer layer)
- **semimetal:** elements that have some properties of metals but are distinct because they are not malleable or ductile

CLASSIFICATION OF MINERALS

The earth is divided into several distinct layers (crust, mantle, and core), each with unique physical and chemical properties. Since only the crust is readily accessible to scientists, this discussion is largely restricted to this outermost layer. The crust is dominated by only eight elements (oxygen, silicon, aluminum, iron, magnesium, calcium, sodium, and potassium).

One of the most important elements is silicon. Geologists commonly divide minerals into two broad categories, silicates and non-silicates, partly because of the sheer abundance of silicon and oxygen, but also because of the immense variety of atomic structures in the silicates. Silicates are those minerals that contain silicon as an essential part of the composition; non-silicates lack silicon in their formulas. The silicates make up the vast majority of minerals in the crust, with all the other classes accounting for only 3 percent of the total.

A common classification scheme for minerals divides them into chemical families, or suites. One suite is silicates (usually subdivided into several groups).

The other principal suites (or non-silicates) include the native elements sulfides and sulfosalts, oxides and hydroxides, carbonates, halides, nitrates, borates, phosphates, arsenates, vanadates, sulfates and chromates, and tungstates and molybdates. Generally the ending "ide" denotes an anion element by itself (sulfide, halide, oxide), and "ate" denotes a complex anion made of an element combined with oxygen.

NATIVE ELEMENTS

Of all the elements, about twenty, the native elements, are known to occur in the earth in the free state. These elements can be separated into metals, semimetals, and nonmetals. Within the metals, based on atomic structure, three groups are recognized: the gold group (gold, silver, copper, and lead), the platinum group (platinum, palladium, iridium, and osmium), and the iron group (iron and nickel-iron). Mercury, tantalum, tin, and zinc are metals that have also been identified. Within the semimetals, two groups are commonly recognized: the arsenic group (arsenic, antimony, and bismuth) and the tellurium group (selenium and tellurium). Sulfur and two forms of carbon (graphite and diamond) are the nonmetallic native element minerals. On the microscopic level, other elements have been found as well; for example, native silicon can be formed in tiny amounts by lightning striking quartz sand.

SULFIDES AND SULFOSALTS

The sulfides include a great number of minerals, many of which are important economically as sources of metals. Although sulfur is the dominant anion (negatively charged ion), this group also includes compounds with arsenic (arsenides), tellurium (tellurides), selenium (selenides), antimony, and bismuth as the anion. Commonly included with the sulfides are the sulfosalts. These sulfosalts are generally distinct because they contain the semimetals arsenic and antimony in place of sulfur. Only four sulfides are considered to be rock-forming minerals: pyrite, marcasite, chalcopyrite, and pyrrhotite. The most common sulfosalt is arsenopyrite, which, as the name implies, is much like pyrite but contains arsenic as well as sulfur.

HALIDES, NITRATES, AND BORATES

The halides are minerals that contain one of four anions—fluorine, chlorine, bromine, and iodine. The chlorides are the most abundant halides, with the fluorides second in abundance, but the bromides and iodides are very rare. The only two halides that are considered to be rock-forming minerals are halite and fluorite.

The nitrates include minerals made up of one nitrogen ion surrounded by three oxygen ions. None of the nitrates is considered common, and they are relatively few in number. Most occur only in deposits in very arid regions because they are extremely soluble.

The borates are minerals that have boron strongly bound to either three or four oxygen ions. All the borates are restricted to dry lake deposits in extremely arid regions. None is considered a common rock-forming mineral, but borax is probably the most readily recognized.

PHOSPHATES AND SULFATES

In phosphates, the phosphorus cation (positively charged ion) is surrounded by four oxygens. Although this class includes many minerals, most are extremely rare; only one, apatite, is common. Apatite is also the principal ingredient of bone. Arsenates and vanadates have similar atomic structures but contain arsenic or vanadium in place of phosphorus.

The sulfate minerals are another large class in which a sulfur cation is surrounded by four oxygens. Two main subgroups of sulfates are recognized: the anhydrous sulfates and the hydrous sulfates. Although this class includes many minerals, very few are considered to be common. Examples of anhydrous sulfates are barite and anhydrite. The most common hydrous sulfate is gypsum.

TUNGSTATE AND MOLYBDATE

The tungstate, chromate, and molybdate minerals are very similar to the sulfates, with either chromium, tungsten, or molybdenum cations surrounded by four oxygens in a pattern slightly different from that in the sulfates. None of the tungstates or molybdates is a common rock-forming mineral, and all are relatively rare but significant ore minerals.

STUDY OF PHYSICAL PROPERTIES

A number of physical properties are studied not only because they help identify particular minerals but also because the physical properties dictate whether minerals have any commercial use. Measurements of the angular relationships between faces with the optical goniometer are helpful in describing minerals. The systematic study of the external form of minerals is commonly called morphological crystallography. Other properties that are studied include the tendency of minerals to fracture (break along irregular surfaces) or cleave (break along straight planar surfaces that represent planes of internal weakness), as well as tenacity (resistance or response to attempts to break, bend, or cut). Hardness, another essential property, refers to the resistance of a substance to abrasion. Hardness is commonly evaluated on a scale of relative hardness, called the Mohs hardness scale, which uses a set of common minerals of different hardness for comparison. Other properties that may have industrial applications include magnetism and electrical properties. Minerals containing uranium and thorium exhibit another property, radioactivity, which is measurable with a Geiger counter or scintillation counter.

Barite, an anhydrous sulfate, is a non-silicate of the sulfates group. It is used in drilling fluids and as an additive to rubber, glass, and plastics. (U.S. Geological Survey)

A variety of thermal properties are commonly determined. Differential thermal analysis (DTA) measures the temperatures at which compositional and structural changes take place in minerals. The DTA curve, which is a graphic recording of the thermal changes in a mineral, is a characteristic of many minerals. Density (mass per unit of volume) is another key property. Density is normally determined as specific gravity (the ratio of the density of the substance to that of an equal volume of water at 4 degrees Celsius and 1 atmosphere of pressure). Depending on the amount and size of the mineral sample available, three methods may be used to determine density: Jolly balance, Berman balance, or pycnometer.

The petrographic microscope is an important analytical tool used to study the optical properties of minerals in polarized light transmitted through the specimen. Both crushed samples and thin sections (0.030 millimeter thick) are commonly utilized. Many of the metal-bearing minerals, particularly the native metals and sulfides, do not readily allow polarized light to pass through them, so microscopic analysis is conducted on highly polished samples in reflected light.

STRUCTURAL AND CHEMICAL ANALYSIS

The study of the orderly internal atomic arrangement within minerals is commonly called structural crystallography. The most common method used to study this internal morphology is the X-ray diffraction powder method, in which a small, finely powdered sample is bombarded with X rays. Like the human fingerprint, every mineral has unique diffraction patterns (caused by X rays interacting with atomic planes). Several other important X-ray diffraction methods involve single crystals (oscillations, rotation, Weissenberg, and precession methods) and are primarily used not for identification purposes, but for refining the complex internal geometries of minerals.

In addition to the physical properties of minerals, a variety of chemical methods can be used to identify and study non-silicate minerals. Until the 1960's, quantitative chemical analysis methods such as colorimetric tests, X-ray fluorescence spectrography, and atomic absorption spectrography were commonly used for mineral analysis. Since that time, however, most mineral analyses have been produced by electron microprobe analysis.

ECONOMIC IMPORTANCE OF NATIVE ELEMENTS

A great number of the non-silicates under consideration touch people's lives in a multitude of ways. Of the native elements, one only needs to look over the commodity reports in the daily newspaper to understand the importance of the likes of gold, silver, and platinum to world economies. Most of the gold in the world is owned by national governments, which commonly use it to settle international monetary accounts. Like silver and platinum, gold is also becoming increasingly popular as a form of investment. In addition, gold is used in jewelry, scientific instruments, and electroplating. Silver is used in the photography, electronics, refrigeration, jewelry, and tableware industries. Most copper is used in the production of electrical wire and in a variety of alloys, brass (copper and zinc) and bronze (copper, tin, and zinc) being the most common. Platinum is a very important metal because of its high melting point, hardness, and chemical inertness. It is primarily used by the chemical industry as a catalyst in the production of chemicals but is also used in jewelry and surgical and dental tools. Sulfur has a wide variety of uses in the chemical industry (production of insecticides, fertilizers, fabrics, paper, and soaps). In addition to their use as gems, diamonds are used in a variety of ways as cutting, grinding, and drilling agents.

USES OF OTHER NON-SILICATES

Like the native elements, the sulfides are also the source of a number of metals. The most important ores of silver, copper, lead, zinc, nickel, mercury, arsenic, antimony, molybdenum, and arsenic are sulfides. The government of the United States considers most of these commodities to be critical in time of war and has huge stockpiles of them in reserve throughout the country, since most are mined in other countries.

The two most important halides, halite and fluorite, are widely used. Halite, or ordinary table salt, is a source of sodium and sodium compounds, chlorine, and hydrochloric acid, which are all important in the chemical industry. It is also used as a de-icing compound, in fertilizers, in livestock feeds, and in the processing of hides. Most fluorite is used to make hydrofluoric acid for the chemical industry or as a flux in the production of steel and aluminum.

Of the borates, borax is used to produce glass, insulation, and fabrics. It is also used in medicines, high explosives and propellants, detergents and soaps, and as a preservative.

Of the phosphates, only apatite is a common mineral, but at least three others are important for the elements that they contain. Monazite is the primary source of thorium, which is a radioactive element with considerable potential as a source of nuclear energy. Autunite, a complex phosphate, and carnotite, a complex vanadate, are both important sources of uranium. A fine-grained form of apatite called collophane (a source of phosphorus) is most widely used in fertilizers and detergents.

Of the sulfates, barite, anhydrite, and gypsum are the three most commonly used. The bulk of the barite is used as a drilling mud in the mineral and energy exploration industry. Barite is unusually dense for a nonmetallic mineral and forms a dense pasty fluid that helps keep oil and gas contained in wells. Anhydrite and gypsum occur in similar geological conditions and have similar compositions. Both are used as soil conditioners. Gypsum is mainly used for the manufacture of plaster of Paris, which is used in wallboard.

In the tungstates, wolframite and scheelite are the main ores of tungsten, which is used in lamp filaments, in tungsten carbide for cutting tools, and as a hardening alloy.

Ronald D. Tyler

FURTHER READING

Berry, L. G., B. Mason, and R. V. Dietrich. *Mineralogy: Concepts, Descriptions, Determinations.* 2d ed. San Francisco: W. H. Freeman, 1983. A college-level introduction to the study of minerals that focuses on the traditional themes necessary to understand minerals: how they are formed and what makes each chemically, crystallographically, and physically distinct from others. Descriptions and determinative tables include almost two hundred minerals (more than half of which are non-silicates).

Bowles, J. F. W., et al. *Rock-Forming Minerals: Non-Silicates; Oxides, Hydroxides and Sulfides.* 2d ed. London: Geological Society Publishing House, 2011. Organized by mineral into oxides, hydroxides, and sulfides. Includes physical and chemical characteristics of each mineral followed by experimental work with the mineral. Well indexed.

Cepeda, Joseph C. *Introduction to Minerals and Rocks.* New York: Macmillan, 1994. Provides a good introduction to the structure of minerals for students just beginning their studies in Earth sciences. Includes illustrations and maps.

Chesterman, C. W., and K. E. Lowe. *The Audubon Society Field Guide to North American Rocks and Minerals.* New York: Alfred A. Knopf, 1988. Contains 702 color photographs of minerals grouped by color, as well as nearly a hundred color photographs of rocks. All the mineral photographs are placed at the beginning of the book, and descriptive information follows, with the minerals grouped by chemistry. Distinctive features and physical properties are listed for each of the minerals, and information on collecting localities is also given. A section at the back of the book discusses various types of rocks. A glossary is also included. Suitable for the layperson.

Deer, William A., R. A. Howie, and J. Zussman. *An Introduction to Rock-Forming Minerals.* 2d ed. London: Pearson Education Limited, 1992. A standard reference on mineralogy for advanced college students and above. Each chapter contains detailed descriptions of chemistry and crystal structure, usually with chemical analyses. Discussions of chemical variations in minerals are extensive.

Dietrich, Richard V., and B. J. Skinner. *Rocks and Rock Minerals.* New York: John Wiley & Sons, 1979. This short, readable college-level text provides a relatively brief but excellent treatment of crystallography and the properties of minerals. Although the descriptions of minerals focus on the silicates, the important rock-forming non-silicates are also considered. Very well illustrated and includes a subject index and modest bibliography.

Ernst, W. G. *Earth Materials.* Englewood Cliffs, N.J.: Prentice-Hall, 1969. The first four chapters of this compact introductory text deal with minerals and the principles necessary to understand their physical and chemical properties, as well as with their origins. Chapter 3 specifically deals with a number of the important rock-forming non-silicate minerals. The text includes a subject index and short bibliography.

Frye, Keith. *Mineral Science: An Introductory Survey.* New York: Macmillan, 1993. This basic text, intended for the college-level reader, provides an easily understood overview of mineralogy, petrology, and geochemistry, including descriptions of specific minerals. Illustrations, bibliography, and index.

Hammond, Christopher. *The Basics of Crystallography and Diffraction*. 2d ed. New York: Oxford University Press, 2001. Covers crystal form, atomic structure, physical properties of minerals, and X-ray methods. Illustrations help clarify some of the more mathematically complex concepts. Includes a bibliography and an index.

Hurlbut, C. S., Jr., and Robert C. Kammerling. *Gemology*. 2d ed. New York: John Wiley & Sons, 1991. A well-illustrated introductory textbook for readers with little scientific background. Covers the physical and chemical properties of gems, their origins, and the instruments used to study them. Later chapters treat methods of synthesis, cutting, and polishing, and descriptions of gemstones.

Klein, C., and C. S. Hurlbut, Jr. *Manual of Mineralogy*. 23d ed. New York: John Wiley & Sons, 2008. An excellent second-year college-level text for use as an introduction to the study of minerals. The topics discussed include external and internal crystallography, crystal chemistry, properties of minerals, X-ray crystallography, and optical properties. The book also systematically describes more than one hundred non-silicate minerals.

Lima-de-Faria, José. *Structural Minerology: An Introduction*. Dordrecht: Kluwer, 1994. Provides a good college-level introduction to the basic concepts of crystal structure and the classification of minerals.

Illustrations, extensive bibliography, index, and a table of minerals on a folded leaf.

Nespolo, Massimo, and Giovanni Ferraris. "A Survey of Hybrid Twins in Non-Silicate Minerals." *European Journal of Mineralogy* 21 (2009): 673-690. Discusses the hybrid nature of several non-silicates, as well as hybrid twinning, reticular merohedrism, polyholohedry, and concurrent (quasi)-restored sublattice.

Pough, Frederick H. *A Field Guide to Rocks and Minerals*. 5th ed. Boston: Houghton Mifflin, 1998. One of the most popular and easily accessible books dealing with non-silicate minerals. Intended for the reader with no scientific background, it includes chapters on collecting and testing minerals, descriptions of environments of formation, physical properties, classification schemes, and mineral descriptions.

Tennissen, A. C. *Nature of Earth Materials*. 2d ed. Englewood Cliffs, N.J.: Prentice-Hall, 1983. Written for the nonscience student and treats minerals from the perspective of both the internal relationships (atomic structure, size, and bonding) and external crystallography. It includes an excellent overview of the physical properties of minerals and classification and description of 110 important minerals.

See also: Biopyriboles; Carbonates; Clays and Clay Minerals; Feldspars; Gem Minerals; Hydrothermal Mineralization; Metamictization; Minerals: Physical Properties; Minerals: Structure; Oxides; Radioactive Minerals.